Lecture Notes in Mathematics

1217

Transformation Groups Poznań 1985

Proceedings of a Symposium
held in Poznań, July 5–9, 1985

Edited by S. Jackowski and K. Pawałowski

Springer-Verlag
Berlin Heidelberg New York London Paris Tokyo

Editors

Stefan Jackowski
Instytut Matematyki
Uniwersytet Warszawski
Pałac Kultury i Nauki IXp.
00-901 Warszawa, Poland

Krzysztof Pawałowski
Instytut Matematyki
Uniwersytet im. A. Mickiewicza w Poznaniu
ul. Matejki 48/49
60-769 Poznań, Poland

Mathematics Subject Classification (1980): 57SXX; 57S10; 57S15; 57S17; 57S25; 57R67; 57R80; 20J05

ISBN 3-540-16824-9 Springer-Verlag Berlin Heidelberg New York
ISBN 0-387-16824-9 Springer-Verlag New York Berlin Heidelberg

Printed in Germany

Printing and binding: Druckhaus Beltz, Hemsbach/Bergstr.
2146/3140-543210

Dedicated to the memory of

A. Jankowski

and

W. Pulikowski

P R E F A C E

The Symposium on Transformation Groups supported by the Adam Mickiewicz University in Poznań was held in Poznań, July 5-9, 1985. The symposium was dedicated to the memory of two of our teachers and friends, Andrzej Jankowski and Wojtek Pulikowski on the tenth anniversary of their deaths.

These proceedings contain papers presented at the symposium and also papers by mathematicians who were invited to the meeting but were unable to attend. All papers have been refereed and are in their final forms. We would like to express our gratitude to the authors and the many referees.

The participants and in particular the lecturers contributed to the success of the symposium and we are most grateful to all of them. Special thanks are due to our colleagues Ewa Marchow, Wojtek Gajda, Andrzej Gaszak, and Adam Neugebauer for their help with the organizational work and to Barbara Wilczyńska who handled the administrative and secretarial duties.

The second editor thanks Sonderforschungsbereich 170 in Göttingen for its hospitality which was very helpful in the preparation of the present volume. Finally, we would like to thank Marrie Powell and Christiane Gieseking for their excellent typing.

Stefan Jackowski

Krzysztof Pawałowski

Poznań/Warszawa, 20.06.1986

ANDRZEJ JANKOWSKI (1938-1975) WOJCIECH PULIKOWSKI (1947-1975)

Andrzej graduated in 1960 from the Nicolaus Copernicus University in Toruń. Topology was his passion and his interests were very broad. Andrzej worked on algebraic and differential topology, his main papers being concerned with operations in generalized cohomology theories and with formal groups. His was not an easy task. Andrzej worked essentially alone. Polish topologists were at that time continuing the tradition of their pre-war school. Andrzej's friend and Ph.D. student wrote*): "He wanted to understand the deepest and most difficult theorems found by his contemporaries. At that beautiful time of great discoveries Andrzej faced the difficult obstacle of being alone. He put a lot of effort into overcoming this difficulty, and also conveying his knowledge to others." Andrzej began to lecture on algebraic topology and to organize seminars as soon as he joined the University of Warsaw in 1962. For nine years, from 1967, he was the *spiritus movens* of the Summer School on Algebraic Topology held annually in Gdańsk. He moved to Gdańsk in 1971. From 1969 until his death he led a seminar on transformation groups. Wojtek Pulikowski was one of the participants.

Wojtek graduated in 1969 from Poznań and moved to Warszawa and then to Gdańsk. In 1973 Wojtek obtained his Ph.D. for the work on equivariant bordism theories indexed by representations and returned to Poznań. He invested great effort into organizing seminars, summer schools and meetings on various topics in algebraic and differential topology. At the same time he continued teaching his students and before long directed their research towards transformation groups. Wojtek was a born teacher, able to convey not only his knowledge but also his passion, enthusiasm, and interest in the subject. He wrote a number of papers on equivariant homology theories, but he spent most of his time in teaching - which he did with joy and love. His friends and students all owe him a great deal.

Besides mathematics, both Andrzej and Wojtek had another passion - mountains. And in the mountains both of them met their death in August 1975, Andrzej climbing the Tirach Mir peak in the Hindu Kush mountains and Wojtek in an accident in the Beskidy mountains in Poland.

---

*) R.Rubinsztein: "Andrzej Jankowski (1938-1975)", Wiadomości Matematyczne, vol. XXIII (1980), pp. 85-91.

TABLE OF CONTENTS

CHRONOLOGICAL LIST OF TALKS

A.Liulevicius (Chicago):	Duality of symmetric powers of cycles
S.Illman (Helsinki):	Product formula for equivariant Whitehead torsion
P.Löffler (Göttingen):	Realization of exotic linking numbers of fixed point sets in representation forms
W.Marzantowicz (Gdańsk):	The S^1-equivariant topology and periodic solutions of ordinary differential equations
A.Szczepański (Gdańsk):	Euclidean space forms with the first Betti number equal to zero
W.Browder (Princeton):	Actions on projective varieties
V.Puppe (Konstanz):	Bounds on the torus rank
Z.Marciniak (Warszawa):	Idempotents in group rings and cyclic homology
E.Laitinen (Helsinki):	Unstable homotopy theory of homotopy representations
J.Kania-Bartoszyńska (Warszawa):	Classification of involutions on 2-handlebodies
R.Vogt (Osnabrück):	Coherence theory and group actions
K.H.Dovermann (West Lafayette):	Symmetries of complex projective spaces
M.Lewkowicz (Wrocław):	Nonabelian Lie group actions and positive scalar curvature
A.Assadi (Charlottesville):	Homotopy actions and G-modules
E.K.Pedersen (Odense):	The bounded and thin h-cobordism theorems
M.Sadowski (Gdańsk):	Injective S^1-actions on manifolds covered by $\mathbb{R}^n$
J.Tornehave (Aarhus):	Units in Burnside rings and the Kummer theory pairing
S.Weintraub (Baton Rouge):	Group actions and certain algebraic varieties
K.Pawałowski (Poznań):	Smooth group actions on disks and Euclidean spaces
R.Oliver (Aarhus):	A transfer map for compact Lie group actions
S.Jackowski (Warszawa):	A fixed point theorem for p-group actions
J.Ewing (Bloomington):	Symmetries of surfaces and homology

CURRENT ADDRESSES OF AUTHORS AND PARTICIPANTS

Christopher Allday
Department of Mathematics
University of Hawaii at Manoa
Honolulu, HI 96822, USA

Paweł Andrzejewski
Instytut Matematyki
Uniwersytet Szczeciński
ul.Wielkopolska 15
70-451 Szczecin, Poland

Amir H. Assadi
Department of Mathematics
University of Wisconsin
Madison, WI 53706, USA

Grzegorz Banaszak
Instytut Matematyki
Uniwersytet Szczeciński
ul. Wielkopolska 15
70-451 Szczecin, Poland

Agnieszka Bojanowska
Instytut Matematyki
Uniwersytet Warszawski
PKiN, IX p.
00-901 Warszawa, Poland

William Browder
Department of Mathematics
Princeton University
Princeton, NJ 08544, USA

Gunnar Carlsson
Department of Mathematics
Princeton University
Princeton, NJ 08544, USA

Tammo tom Dieck
Mathematisches Institut
Universität Göttingen
Bunsenstraße 3-5
3400 Göttingen, West Germany

Ryszard Doman
Instytut Matematyki
Uniwersytet im. A.Mieckiewicza
ul. Matejki 48/49
60-769 Poznań, Poland

Karl Heinz Dovermann
Department of Mathematics
University of Hawaii at Manoa
Honolulu, HI 96822, USA

John Ewing
Department of Mathematics
Indiana University
Bloomington, IN 47405, USA

Wojciech Gajda
Instytut Matematyki
Uniwersytet im. A.Mickiewicza
ul. Matejki 48/49
60-769 Poznań, Poland

Andrzej Gaszak
Instytut Matematyki
Uniwersytet im. A.Mickiewicza
ul. Matejki 48/49
60-769 Poznań, Poland

Jean-Pierre Haeberly
Department of Mathematics
University of Washington
Seattle, WA 98195, USA

Akio Hattori
Department of Mathematics
Faculty of Science
University of Tokyo
Hongo, Tokyo, 113 Japan

Sören Illman
Department of Mathematics
University of Helsinki
Hallituskatu 15
00100 Helsinki 10, Finland

Stefan Jackowski
Instytut Matematyki
Uniwersytet Warszawski
PKiN, IX p.
00-901 Warszawa, Poland

Tadeusz Januszkiewicz
Instytut Matematyki
Uniwersytet Wrocławski
Pl. Grunwaldzki 2/4
50-384 Wrocław, Poland

Jan Jaworowski
Department of Mathematics
Indiana University
Bloomington, IN 47405, USA

Joanna Kania-Bartoszyńska
Department of Mathematics
University of California
Berkeley, CA 94720, USA

Gabriel Katz
Department of Mathematics
Ben Gurion University
Beer-Sheva 84105, Israel

Katsuo Kawakubo
Department of Mathematics
Osaka University
Toyonaka, Osaka, 560 Japan

Tadeusz Koźniewski
Instytut Matematyki
Uniwersytet Warszawski
PKiN, IX p.
00-901 Warszawa, Poland

Piotr Krasoń
Instytut Matematyki
Uniwersytet Szczeciński
ul. Wielkopolska 15
70-451 Szczecin, Poland

Erkki Laitinen
Department of Mathematics
University of Helsinki
Hallituskatu 15
00100 Helsinki 10, Finland

Marek Lewkowicz
Instytut Matematyki
Uniwersytet Wrocławski
Pl. Grunwaldzki 2/4
50-384 Wrocław, Poland

Arunas Liulevicius
Department of Mathematics
University of Chicago
Chicago, IL 60637, USA

Peter Löffler
Mathematisches Institut
Universität Göttingen
Bunsenstr. 3-5
3400 Göttingen, West Germany

Ewa Marchow
Instytut Matematyki
Uniwersytet im. A.Mickiewicza
ul. Matejki 48/49
60-769 Poznań, Poland

Zbigniew Marciniak
Instytut Matematyki
Uniwersytet Warszawski
PKiN, IX p.
00-901 Warszawa, Poland

Wacław Marzantowicz
Instytut Matematyki
Uniwersytet Gdański
ul. Wita Stwosza 57
80-952 Gdańsk, Poland

Mikiya Masuda
Department of Mathematics
Osaka City University
Osaka 558, Japan

Takao Matumoto
Department of Matheamtics
Faculty of Science
Hiroshima University
Hiroshima 730, Japan

J. Peter May
Department of Mathematics
University of Chicago
Chicago, IL 60637, USA

Janusz Migda
Instytut Matematyki
Uniwersytet im. A.Mickiewicza
ul. Matejki 48/49
60-769 Poznań, Poland

Adam Neugebauer
Instytut Matematyki
Uniwersytet im. A.Mickiewicza
ul. Matejki 48/49
60-769 Poznań, Poland

Krzysztof Nowiński
Instytut Matematyki
Uniwersytet Warszawski
PKiN, IX p.
00-901 Warszawa, Poland

Robert Oliver
Matematisk Institut
Aarhus Universitet
Ny Munkegade
8000 Aarhus C, Denmark

Murad Özaydin
Department of Mathematics
University of Wisconsin
Madison, WI 53706, USA

Krzysztof Pawałowski
Instytut Matematyki
Uniwersytet im. A.Mickiewicza
ul. Matejki 48/49
60-769 Poznań, Poland

Erik Kjaer Pedersen
Matematisk Institut
Odense Universitet
Campusvej 55
5230 Odense M, Denmark

Jerzy Popko
Instytut Matematyki
Uniwersytet Gdański
ul. Wita Stwosza 57
80-952 Gdańsk, Poland

Józef Przytycki
Instytut Matematyki
Uniwersytet Warszawski
PKiN, IX p.
00-901 Warszawa, Poland

Volker Puppe
Fakultät für Mathematik
Universität Konstanz
Postfach 5560
7750 Konstanz, West Germany

Andrew Ranicki
Department of Mathematics
Edinburgh University
King's Buildings, Mayfield Rd.
Edinburgh EH9 3JZ, Scotland, UK

Martin Raussen
Institut for Elektroniske Systemer
Aalborg Universitetscenter
Strandvejen 19
9000 Aalborg, Denmark

Melvin Rothenberg
Department of Mathematics
University of Chicago
Chicago, IL 60637, USA

Sławomir Rybicki
Instytut Matematyki
Politechnika Gdańska
ul. Majakowskiego 11/12
80-952 Gdańsk, Poland

Michał Sadowski
Instytut Matematyki
Uniwersytet Gdański
ul. Wita Stwosza 57
80-952 Gdańsk, Poland

Jan Samsonowicz
Instytut Matematyki
Politechnika Warszawska
Pl. Jedności Robotniczej 1
00-661 Warszawa, Poland

Roland Schwänzl
Fachbereich Mathematik
Universität Osnabrück
Albrechtstraße 28
4500 Osnabrück, West Germany

Masahiro Shiota
Department of Mathematics
Faculty of General Education
Nagoya University
Nagoya 464, Japan

Jolanta Słomińska
Instytut Matematyki
Uniwersytet im. M.Kopernika
ul. Chopina 12
87-100 Toruń, Poland

Andrzej Szczepański
Instytut Matematyki
Politechnika Gdańska
ul. Majakowskiego 11/12
80-952 Gdańsk, Poland

Jørgen Tornehave
Matematisk Institut
Aarhus Universitet
Ny Munkegade
8000 Aarhus C, Denmark

Paweł Traczyk
Instytut Matematyki
Uniwersytet Warszawski
PKiN, IX p.
00-901 Warszawa, Poland

Rainer Vogt
Fachbereich Mathematik
Universität Osnabrück
Albrechtstraße 28
4500 Osnabrück, West Germany

Steven H. Weintraub
Department of Mathematics
Louisiana State University
Baton Rouge, LA 70803, USA

Bounds on the torus rank

C. Allday and V. Puppe

For a topological space X let $rk_o(X) := \max\{\dim T$, where T is a torus which can act on X almost freely (i.e. with only finite isotropy subgroups)$\}$ be the torus rank of X . Stephen Halperin has raised the following question (s.[11]):

(H$\mathbb{Q}$) Is it true that $\dim_{\mathbb{Q}} H^*(X;\mathbb{Q}) \geq 2^{rk_o(X)}$ for any simply connected reasonable space X ?

In this context "reasonable" (s. [11]) is a technical condition which assures that one can apply the A. Borel version of P.A. Smith theory (s.[4],[5],[12]) and Sullivan's theory of minimal models (s.[13],[10], [14]). In particular any connected finite CW-complex is certainly "reasonable", but X being connected, paracompact, finitistic (s.[5] p. 133) and of the rational homotopy type of a CW-complex would also suffice.

In the first section we give some lower bounds for $\dim_{\mathbb{Q}} H^*(X;\mathbb{Q})$ if X allows an almost free action of an n-dimensional torus $G = T^n$. These results are obtained using only the additive structure in $H^*(X;Q)$ (and a version of the localization theorem (s. [2])) and hold for rather general spaces X , e.g. simply connectedness is not needed; but the bounds we get are far below the desired $2^{rk_o(X)}$.

The second section gives bounds on the torus rank in terms of the cohomology of X , where a very special structure of the cohomology ring $H^*(X;\mathbb{Q})$, i.e. X being a rational cohomology Kähler space, is used.

The third section is concerned with relations between properties of the minimal model M(X) of X (in particular the rational homotopy Lie algebra $L_*(X)$), $rk_o(X)$ and $\dim_{\mathbb{Q}} H^*(X;\mathbb{Q})$. Halperin observed (s.[11], 1.5) that the results of [1], in particular the inequality $rk_o(X) \leq -\chi_\pi(X)$, where $\chi_\pi(X)$ is the rational homotopy Euler characteristic (s.[1], Theorem 1), implies an affirmative answer to his question if X is a homogenous space G/K, $K \subset G$ compact, connected Lie groups. Among other things we describe another class of spaces for which $\dim_{\mathbb{Q}} H^*(X;\mathbb{Q}) \geq 2^{rk_o(X)}$ holds, but the bound on the torus rank given by the rational homotopy Euler characteristic is not sharp in many cases (compare also [11], 4.4) and does not suffice to answer (H$\mathbb{Q}$). Indeed, for this class the knowledge of the additive structure of $\pi_*(X) \otimes \mathbb{Q}$ is not enough; it is essential to use the Lie algebra structure of $L_*(X) \cong \pi_*(\Omega X) \otimes \mathbb{Q}$.

1. Let X be a connected, paracompact, finitistic space which has the rational homotopy type of a CW-complex and on which a torus $G = T^n$ of dimension n acts almost freely. If M(X) is the minimal model of X over the field $\mathbb{C}$ of complex numbers and $R := H^*(BG;\mathbb{C}) \cong \mathbb{C}[t_1,\dots,t_n]$, then the R-cochain algebra $C_G(X) := R \tilde{\otimes}_{\mathbb{C}} M(X)$ (where the twisting of the boundary, indicated by "~", reflects the G-action) is a model for the Borel construction X_G.

For any $\alpha = (\alpha_1,\dots,\alpha_n) \in \mathbb{C}^n$ we denote by $\mathbb{C}^\alpha$ the field $\mathbb{C}$ together with the R-algebra structure given by the evaluation map ε^α: $R = \mathbb{C}[t_1,\dots,t_n] \to \mathbb{C}$, $t_i \longrightarrow \alpha_i$ for $i = 1,\dots,n$. The cochain algebra $C_G(X)^\alpha$ (over $\mathbb{C}$) is defined to be the tensor product $C_G(X)^\alpha :=$ $\mathbb{C}^\alpha \otimes_R C_G(X)$. Theorem (4.1) of [2] implies that $H^*(C_G(X)^\alpha) = 0$ for all $\alpha \neq 0$ (since the G-action is assumed to be almost free). It follows from a theorem of E.H. Brown (s. [7], (9.1), compare [2], (2.3)) that there exists a twisted boundary on $D_G(X) := R \tilde{\otimes} H^*(X;\mathbb{C})$ which makes $R \tilde{\otimes} H^*(X;\mathbb{C})$ homotopy equivalent to $R \tilde{\otimes} M(X)$ as R-cochain complexes. We therefore get (for an almost free action) that $H(D_G(X)^\alpha) = 0$ for all $\alpha \neq 0$ and we shall use this information to obtain the following proposition:

<u>(1.1) Proposition</u>: Under the above hypothesis one has

a) $\dim_{\mathbb{Q}} H^*(X;\mathbb{Q}) \geq 2n$ for all $n = 1,2,\dots$

b) $\dim_{\mathbb{Q}} H^*(X;\mathbb{Q}) \geq 2(n+1)$ for all $n \geq 3$

<u>Proof</u>: We can of course assume that $\dim_{\mathbb{Q}} H^*(X;\mathbb{Q})$ is finite, and the fact that the action has no fixed point implies that the Euler characteristic $\chi(X)$ is zero. Let $x_1,\dots,x_k$ be a homogenous $\mathbb{Q}$-basis of $H^{ev}(X;\mathbb{Q})$ with $|x_1| \geq \dots > |x_k| = 0$ and $y_1,\dots,y_k$ a homogenous $\mathbb{Q}$-basis of $H^{odd}(X;\mathbb{Q})$ with $|y_1| \geq \dots \geq |y_k| > 0$ ($|\ |$ denotes degree). Since $|t_i| = 2$ for $i = 1,\dots,n$ the twisted boundary $\tilde{d}$ on $R \tilde{\otimes} H^*(X;\mathbb{C})$ is given by two $k \times k$-matrices $P = (p_{ij})$ and $Q = (q_{ij})$, where the entries p_{ij}, q_{ij} are homogenous polynomials in the variables $t_1,\dots,t_n$ of degree $\neq 0$, i.e.

$$\begin{aligned} \tilde{d}y_1 &= p_{11}x_1 + \dots + p_{1k}x_k & \qquad \tilde{d}x_1 &= q_{11}y_1 + \dots + q_{1k}y_k \\ &\vdots & &\vdots \\ \tilde{d}y_k &= p_{k1}x_1 + \dots + p_{kk}x_k & \qquad \tilde{d}x_k &= q_{k1}y_1 + \dots + q_{kk}y_k . \end{aligned}$$

If $p_{ij} \not\equiv 0$ (resp. $q_{ij} \not\equiv 0$) then $|y_i| > |x_j|$ and $|p_{ij}| = |y_i|-|x_j|+1$ (resp. $|x_i| > |y_j|$ and $|q_{ij}| = |x_i|-|y_j|+1$), in particular $\tilde{d}x_k \equiv 0$ (i.e. $q_{kj} \equiv 0$ for all $j = 1,\ldots,k$).

The equation $\tilde{d}\circ\tilde{d} = 0$ is equivalent to $PQ = QP = 0$ and the vanishing of $H(\mathcal{D}_G(X)^\alpha)$ for any $\alpha \in \mathbb{C}^n\setminus\{0\}$ then means that $\operatorname{rk} P(\alpha) + \operatorname{rk} Q(\alpha) = k$ for all $\alpha \in \mathbb{C}^n\setminus\{0\}$, where $\operatorname{rk} P(\alpha)$ denotes the rank of the $k\times k$-matrix over $\mathbb{C}$ obtained from P by evaluating the polynomials p_{ij} at the point $\alpha \in \mathbb{C}^n$ (similar for $\operatorname{rk} Q(x)$). The semi-continuity of $\operatorname{rk} P(\alpha)$ and $\operatorname{rk} Q(\alpha)$ (as a function of α)(together with $\operatorname{rk} P(\alpha) + \operatorname{rk} Q(\alpha) = k$ for $\alpha \neq 0$) then implies that $\operatorname{rk} P(\alpha)$ and $\operatorname{rk} Q(\alpha)$ have to be constant on $\mathbb{C}^n\setminus\{0\}$.

To prove part a) on only needs to observe that the variety $V(p_{1k},\ldots,p_{kk})$ can only consist of the point $0 \in \mathbb{C}^n$. If the polynomials p_{ik}, $i = 1,\ldots,k$ would have a common zero $\alpha \in \mathbb{C}^n\setminus\{0\}$ then "at the point α" the cycle x_k could not be a boundary and hence $H(\mathcal{D}_G(X)^\alpha)$ would not vanish. Since the p_{ik}, $i = 1,\ldots,k$ are k polynomials in n variable one gets $k \geq n$ (otherwise $V(p_{1k},\ldots,p_{kk}) \cap \mathbb{C}^n\setminus\{0\} \neq \emptyset$).

To get the slight improvement b) a considerably more involved argument is necessary:
We assume $k = n$ and will show that this implies $n \leq 2$.

Case 1: Let $|y_1| > |x_i|$ for all i, i.e. the top dimensional classes have odd degree. Again $V(p_{1n},\ldots,p_{nn}) = 0$ and it now follows that $p_{1n},\ldots,p_{nn}$ is a regular sequence in $R = \mathbb{C}[t_1,\ldots,t_n]$. Therefore the condition $QP = 0$ implies that all the q_{ij}'s are contained in the ideal $\langle p_{1n},\ldots,p_{nn}\rangle \subset R$ generated by p_{in}, $i = 1,\ldots,n$. (From $\sum_{j=1}^{n} q_{ij}p_{jn}=0$ it follows that the equivalence class of $q_{ij}\, p_{jn}$ in $R/(p_{1n},\ldots,\hat{p}_{jn},\ldots,p_{nn})$ is zero. The regularity of the sequence $p_{1n},\ldots,p_{nn}$ then implies that the class of q_{ij} is already zero in $R/(p_{1n},\ldots,\hat{p}_{jn},\ldots,p_{nn})$. Hence $q_{ij} \in (p_{1n},\ldots,\hat{p}_{jn},\ldots,p_{nn}) \subset (p_{1n},\ldots,p_{nn})$ for all $i,j = 1,\ldots,n$.) Since $|q_{ij}| = |x_i|-|y_j|+1 < |y_1|+1 = |p_{1n}|$ one actually has $q_{ij} \in (p_{2n},\ldots,p_{nn})$ for all $i,j = 1,\ldots,n$. (This is where we use the assumption that the top classes have odd degree and - as one sees from the above inequality - the weaker assumption "$|y_1|+|y_k|>|x_1|$" would suffice.) Choose $\alpha \in V(p_{2n},\ldots,p_{nn}) \cap (\mathbb{C}^n\setminus\{0\})$. Then $q_{ij}(\alpha) = 0$ for $i,j = 1,\ldots,n$ and hence $\operatorname{rk} Q(\alpha) = 0$. Since $\operatorname{rk} Q$ is constant on $\mathbb{C}^n\setminus\{0\}$ we get $Q \equiv 0$ and $\operatorname{rk} P$ must therefore be maximal $(= n)$ on $\mathbb{C}^n\setminus\{0\}$. Since $\det P$ is a polynomial in the variables $t_1,\ldots,t_n$ this can only happen if $n = 1$.

Case 2: Let $|x_1| > |y_i|$ for all i, i.e. the top classes have even degree. We have $q_{nj} \equiv 0$ for $j = 1,\dots,n$; $V(p_{1n},\dots,p_{nn}) = 0$, i.e. $p_{1n},\dots,p_{nn}$ is a regular sequence (as before), and in addition $p_{i1} \equiv 0$ for $i = 1,\dots,n$ (for degree reasons); $V(q_{11},\dots,q_{1n}) = 0$, i.e. $q_{11},\dots,q_{1n}$ is a regular sequence, since otherwise x_1 would give a non-zero element in $H(\mathcal{D}_G(X)^\alpha)$ for any $\alpha \in V(q_{11},\dots,q_{1n}) \cap (\mathbb{C}^n \setminus \{0\})$. Analogous to case 1 we get from $QP = PQ = 0$ that $q_{ij} \in (p_{1n},\dots,p_{nn})$ and $p_{ij} \in (q_{11},\dots,q_{1n})$ for all $i,j = 1,\dots,n$. In particular $(p_{1n},\dots,p_{nn}) = (q_{11},\dots,q_{1n})$. Since $|p_{1n}| > |p_{ij}|$ for all (i,j) with $j < n$ it follows that $p_{ij} \in (p_{2n},\dots,p_{nn})$ if $(i,j) \neq (1,n)$. For $n > 1$ choose $\alpha \in V(p_{2n},\dots,p_{nn}) \cap (\mathbb{C}^n \setminus \{0\})$, then $\operatorname{rk} P(\alpha) = 1$ and therefore $\operatorname{rk} Q(\alpha) = n-1$. This implies $\operatorname{rk} Q = n-1$ on $\mathbb{C}^n \setminus \{0\}$. Since $q_{nj} \equiv 0$ for all $j = 1,\dots,n$ the $(n-1) \times (n-1)$ minors $Q_1,\dots,Q_n$ of the matrix

$$\begin{matrix} q_{11} & \cdots & q_{1n} \\ \vdots & & \vdots \\ q_{n-1\,1} & \cdots & q_{n-1\,n} \end{matrix}$$

have to form a regular sequence (Q_j is obtained by skipping the j-th column)

The expension formula for the determinant (with respect to the first row) of the matrix

$$\begin{matrix} q_{11} & \cdots & q_{1n} \\ q_{11} & \cdots & q_{1n} \\ q_{21} & \cdots & q_{2n} \\ \vdots & & \vdots \\ q_{n-1\,1} & \cdots & q_{n-1\,n} \end{matrix}$$

gives $q_{11}\, Q_1 - q_{12}\, Q_2 + \dots + (-1)^{n+1}\, q_{1n}\, Q_n = 0$.

As above one gets $q_{ij} \in (Q_1,\dots,Q_n)$ for all $i,j = 1,\dots,n$. This is only possible if $(n-1) = 1$ (otherwise $|q_{11}| < |Q_j|$ for all j, which gives a contradiction). This finishes the proof of (1.1).

<u>(1.2) Corollary</u>: If X is a paracompact, finitistic space which has the rational homotopy type of a CW-complex, then

a) $\dim_{\mathbb{Q}} H^*(X;\mathbb{Q}) \geq 2\, \operatorname{rk}_o(X)$

b) $\dim_{\mathbb{Q}} H^*(X;\mathbb{Q}) \geq 2(\operatorname{rk}_o(X)+1)$, if $\operatorname{rk}_o(X) \geq 3$.

<u>(1.3) Remark</u>: G. Carlsson has asked the analogous question to $(H\mathbb{Q})$ concerning spaces X on which an elementary abelian 2-group $G = (\mathbb{Z}/_2)^n$ acts freely (such that X becomes a finite G-CW complex). Results of Carlsson [8] and - using different methods - of W. Browder [6] in this direction imply in particular, that if G acts trivially in cohomology with $\mathbb{Z}/2$ coefficients s. [8] (resp. $\mathbb{Z}_{(2)} = \mathbb{Z}$ localized at 2 s. [6]) then the cohomology of X (with the corresponding coefficients) is non-zero in at least n+1 different dimensions.

The methods used to prove proposition (1.1) above can be applied in a similar fashion to free $(\mathbb{Z}/_2)^n$-actions (compare [2], 2.). One then obtains for a finite, free $(\mathbb{Z}/_2)^n$-CW complex X with trivial action on $H^*(X;\mathbb{Z}/_2)$:

a) $\dim_{\mathbb{Z}/_2} H^*(X;\mathbb{Z}/_2) \geq n+1$ for all n

b) $\dim_{\mathbb{Z}/_2} H^*(X;\mathbb{Z}/_2) \geq n+2$ for $n \geq 2$

Combining our approach with Carlsson's result leads to

c) $\dim_{\mathbb{Z}/_2} H^*(X;\mathbb{Z}/_2) \geq 2n$ for all n.

Browder's methods also work for $G = (\mathbb{Z}/_p)^n$, p an odd prime and he proves results analogous to the case p = 2 also for odd primes. The above approach would also work for p an odd prime (compare [2], 3.) but there are some technical complication arising from the more complicated structure of $H^*(B(\mathbb{Z}/_p)^n; \mathbb{Z}/_p)$ in case p is odd.

2. Let X be a connected paracompact finitistic space which is a rational Poincaré duality space of formal dimension 2m. In this section H^* denotes sheaf (or Alexander-Spanier or Cech) cohomology and H^*_Δ denotes singular cohomology. X will be called an agreement space if the natural transformation $H^*(X;\mathbb{Q}) \to H^*_\Delta(X;\mathbb{Q})$ is an isomorphism (e.g. X reasonable, as above).

<u>(2.1) Definition</u>: (compare [4]) The space X is said to be a rational cohomology Kähler space (CKS) if

(i) there exists $\omega \in H^2(X;\mathbb{Q})$ such that ω^m is non-zero in $H^{2m}(X;\mathbb{Q}) \cong \mathbb{Q}$; and

(ii) the cup-product with ω^j: $H^{m-j}(X;\mathbb{Q}) \to H^{m+j}(X;\mathbb{Q})$ is an isomorphism for $0 \leq j \leq m$.

<u>(2.2) Theorem</u>: If X is a CKS, then $rk_o(X) \leq \alpha_1(X)$:= the maximal number of algebraically independent elements in $H^1(X;\mathbb{Q})$. In particular, $\dim_{\mathbb{Q}} H^*(X;\mathbb{Q}) \geq 2^{rk_o(X)}$. Furthermore if X is an agreement space and if a torus G acts almost-freely on X, then $H^*(X;\mathbb{Q})$ and $H^*(X/G;\mathbb{Q}) \otimes H^*(G;\mathbb{Q})$ are isomorphic as graded $\mathbb{Q}$-algebras.

<u>Proof</u>: Suppose $G = T^n$ acts almost-freely on X. Let $EG \to BG$ be a universal principal G-bundle, and let $X_G = (X \times EG)/G$ be the Borel construction. Let $E_r^{p,q}$ be the rational cohomology Leray-Serre spectral sequence of $X_G \to BG$; and let s be the rank of the linear map d_2: $E_2^{o,1} = H^1(X;\mathbb{Q}) \to E_2^{2,o} \cong H^2(BG;\mathbb{Q})$. Choose $y_1,\ldots,y_s \in H^1(X;\mathbb{Q})$ such

that $d_2(y_i) = a_i$, $1 \le i \le s$, are linearly independent. Then, for $1 \le i_1 < \ldots < i_{j+1} \le s$, $d_2(y_{i_1} \ldots y_{i_{j+1}}) = \sum_{k=1}^{j+1} \pm a_{i_k} \otimes y_{i_1} \ldots \hat{y}_{i_k} \ldots y_{i_{j+1}}$.

Hence it follows by induction that $y_1 \ldots y_s \neq 0$. I.e. $s \le \alpha_1(X)$. Now let K be the subtorus such that the ideal $(a_1, \ldots, a_s) = \ker[H^*(BG;\mathbb{Q}) \to H^*(BK;\mathbb{Q})]$. In particular, $\dim K = n-s$. In the Leray-Serre spectral sequence of $X_K \to BK$, then $d_2\colon E_2^{0,1} \to E_2^{2,0}$ is zero. Thus, by Blanchard ([3]), the spectral sequence collapses; and so $X^K \neq \emptyset$. So K is trivial, and $n = s \le \alpha_1(X)$.

If X is an agreement space, then it follows from the fibre bundle $X \to X_G \to BG$ that X_G is an agreement space also. Now, above, $d_2\colon H^1(X;\mathbb{Q}) \to H^2(BG;\mathbb{Q})$ is onto, since G is acting almost-freely: hence $H^2(BG;\mathbb{Q}) \to H^2(X_G;\mathbb{Q})$ is zero. On the other hand $G \to X \times EG \to X_G$ is the pull-back of $G \to EG \to BG$ via $X_G \to BG$; in particular it is orientable with respect to H^*_Δ. Thus X (homotopy equivalent to $X \times EG$) has a K.S.-model of the form $M(X_G) \otimes \Lambda(s_1, \ldots, s_n)$, where $\deg(s_i) = 1$, $1 \le i \le n$, and $d(s_i) = 0$, $1 \le i \le n$ (since $H^2(BG;\mathbb{Q}) \to H^2(X_G;\mathbb{Q})$ is zero). Hence $H^*_\Delta(X;\mathbb{Q}) \cong H^*_\Delta(X_G;\mathbb{Q}) \otimes H^*_\Delta(G;\mathbb{Q})$ as algebras. So $H^*(X;\mathbb{Q}) \cong H^*(X/G;\mathbb{Q}) \otimes H^*(G;\mathbb{Q})$, as algebras, by the Vietoris-Begle mapping theorem.

(2.3) Remarks: (i) An argument similar to the above, applied to a simple closed connected subgroup, shows that no non-abelian compact connected Lie group can act almost-freely on a CKS: and only condition (i) of definition (2.1) is needed for this.

(ii) Again if we assume only condition (i) of definition (2.1), then we get $rk_0(X) \le \beta_1(X) := \dim_{\mathbb{Q}} H^1(X;\mathbb{Q})$.

(iii) Theorem (2.2) is "best possible", since T^{2m} is a CKS, and it can act freely on itself.

3. Let X be a simply connected reasonable space and let $L_*(X) = \pi_*(\Omega X) \otimes \mathbb{Q}$ denote its rational homotopy Lie algebra. The following result is proved in [2], Theorem (4.6)':

If $L_i(X) = 0$ for all odd i, then $rk_0(X) \le \dim_{\mathbb{Q}} Z\, L_*(X)$, where $Z\, L_*(X)$ denotes the centre of the Lie algebra $L_*(X)$.

This improves the bound given by $-\chi_\pi(X) = \dim L_*(X)$ for this type of spaces and it is clear that one does need some improvement in this direction to get an affirmative answer to (HQ) since $\dim_{\mathbb{Q}} H^*(X;\mathbb{Q}) \le 2^{\dim L_*(X)}$ in the case at hand and equality holds only if the minimal model of X has trivial boundary (compare [2],(4.5)). In fact, the minimal model for a space X with $L_{odd}(X) = 0$ is the exterior algebra $\Lambda^*(V)$ over the vector space $V = \mathrm{Hom}(L_*(X), \mathbb{Q})$ dual to $L_*(X)$ with a degree shift and a boundary which is a derivation on $\Lambda(V)$. The quadratic part of the boundary corresponds (under duality) to the Lie multiplication of $L_*(X)$. The next simplest case to having a trivial boundary on the minimal model M(X) of X would be to have the boundary completely determined by the Lie product of $L_*(X)$. These are the so-called π-formal (or co-formal) spaces. Their rational homotopy type is determined by $L_*(X)$ and in particular the cohomology $H^*(X;\mathbb{Q})$ is just the algebraically defined cohomology $H^*(L_*(X))$ of the Lie algebra $L_*(X)$. Together with the above bound on $rk_o(X)$ one would get an affirmative answer to (HQ) if $\dim H(L_*) \geq 2^{\dim Z\, L*}$ for graded, connected Lie algebras with $L_{odd} = 0$. We do not know whether this holds in general, but Deninger and Singhof (s.[9]) have given lower bounds for the dimension of the cohomology of a nilpotent Lie algebra which imply the above inequality if $L_*^3 = 0$ (i.e. all three fold Lie brackets are zero). Putting all this together we get:

(3.1) Proposition: Let X be a simply-connected, reasonable, π-formal space such that $L_{odd}(X) = 0$ and $L_*(X)^3 = 0$, then

$$\dim_{\mathbb{Q}} H^*(X;\mathbb{Q}) \geq 2^{\dim Z\, L_*(X)} \geq 2^{rk_o(X)} .$$

If X is not π-formal (i.e. the boundary of M(X) is not determined by the Lie product of $L_*(X)$), then - similar to the situation described above for $\chi_\pi(X)$ - the upper bound on $rk_o(X)$ given by $\dim Z\, L_*(X)$ will not suffice to provide the desired lower bound on $\dim_{\mathbb{Q}} H^*(X;\mathbb{Q})$. But non-vanishing higher order Whitehead products (i.e. non-trivial higher order terms in the boundary of M(X)) will reduce the torus rank of X below $\dim Z\, L_*(X)$ in general. This is illustrated by the following example:

(3.2) Example: Let X be a finite CW-complex such that $M(X) = \Lambda(y_1,y_2,y_3,y_4,y)$, where $\deg(y_i) = 3, 1 \leq i \leq 4$, $\deg(y) = 11$, $dy_i = 0$, $1 \leq i \leq 4$ and $dy = y_1 \wedge y_2 \wedge y_3 \wedge y_4$. Then $\dim_{\mathbb{Q}} ZL_*(X) = \dim_{\mathbb{Q}} L_*(X) = 5$. But, by [2], Theorem (4.1), $rk_o(X) \leq 1$.

One might ask whether the bound given by $\dim Z\, L_*(X)$ is "best possible" for reasonable, simply connected, π-formal spaces X with $L_{odd}(X) = \pi_{ev}(X) \otimes \mathbb{Q} = 0$, or more general ask for lower bounds (in terms of M(X)) on the torus rank of X . Since M(X) depends only on the rational homotopy type of X, "best possible" is to be interpreted as "within the rational homotopy type of X there exists a simply-connected, reasonable space $\tilde{X}$ which has the desired torus rank" (compare

[11], 4.3). The following propositions provide an answer to this question.

(3.3) Proposition: Let $M(X) = (\Lambda^*(V),\delta)$ be the minimal model (over $\mathbb{Q}$) of a simply-connected, reasonable space X with $L_{odd}(X) = \pi_{ev}(X)\otimes\mathbb{Q} = 0$ and $\dim_{\mathbb{Q}} H^*(X;\mathbb{Q}) < \infty$. If the boundary $\delta: V \to \Lambda^*(V)$ factors through $\Lambda^*(W)$, where $W \subset V$ is a (graded) linear subspace of the (graded) vector space V of codimension $n < \infty$, then there exists a simply connected, finite CW-complex $\tilde{X}$ which is rationally homotopy equivalent to X and carries a free action of an n-dimensional torus $G = T^n$ ($n = \dim(V/W)$).

Proof: In view of [11], 4.2 it suffices to define a twisting $\tilde{\delta}$ of δ on $\mathbb{Q}[t_1,\ldots,t_n] \tilde{\otimes} \Lambda^*(V)$ (which makes $(\mathbb{Q}[t_1,\ldots,t_n] \tilde{\otimes} \Lambda^*(V),\tilde{\delta})$ a $\mathbb{Q}[t_1,\ldots,t_n]$-cochain algebra) such that the cohomology of $(\mathbb{Q}[t_1,\ldots,t_n] \tilde{\otimes} \Lambda^*(V),\tilde{\delta})$ is finite dimensional. We choose a splitting of V into a direct sum of graded vector spaces $V = W \oplus Z$ and a homogenous basis $z_1,\ldots,z_n$ of Z. We now define:

$\tilde{\delta}(t_i) = 0$, $i = 1,\ldots,n$

$\tilde{\delta}(w) := \delta(w)$ for $w \in W$

$\tilde{\delta}(z_i) := \delta(z_i) + t_i^{\frac{|z_i|+1}{2}}$, $i = 1,\ldots,n$.

Since $\delta(V) \subset \Lambda^*(W)$ one gets $\tilde{\delta}\circ\tilde{\delta} = \delta\circ\delta \equiv 0$.

Hence $\tilde{\delta}|_V$ extends to a unique derivation on $\mathbb{Q}[t_1,\ldots,t_n] \tilde{\otimes} \Lambda^*(V)$ and is a twisting (n-parameter family of deformations) of δ .

It remains to show that $\dim_{\mathbb{Q}} H(\mathbb{Q}[t_1,\ldots,t_n] \tilde{\otimes} \Lambda^*(V),\tilde{\delta})$ is finite. Since $\Lambda^*(V)$ is an exterior algebra ($V_{ev} = 0$) one can find $k_i \in \mathbb{N}$ such that $(\delta(z_i))^{k_i} = \left(\tilde{\delta}(z_i) - t_i^{\frac{|z_i|+1}{2}}\right)^{k_i} = 0$ for $i = 1,\ldots,n$. Since the t_i's are cycles it follows (using the fact that $\tilde{\delta}$ is a derivation) that $\left(t_i^{\frac{|z_i|+1}{2}}\right)^{k_i}$ is a boundary in $\mathbb{Q}[t_1,\ldots,t_n] \tilde{\otimes} \Lambda^*(V)$ (compare [12], Chap VII, Lemma (1.1)). A well known spectral sequence argument (using the "Serre" spectral sequence of the "fibration" $\mathbb{Q}[t_1,\ldots,t_n] \hookrightarrow \mathbb{Q}[t_1,\ldots,t_n] \tilde{\otimes} \Lambda^*(V) \to \Lambda^*(V)$ and the fact that $\mathbb{Q}[t_1,\ldots,t_n]$ is noetherian) shows that $H(\mathbb{Q}[t_1,\ldots,t_n] \tilde{\otimes} \Lambda^*(V),\tilde{\delta})$ is finitely generated over $\mathbb{Q}[t_1,\ldots,t_n]$ (since $\dim_{\mathbb{Q}} H(\Lambda^*(V),\delta) < \infty$, i.e. $H(\Lambda^*(V),\delta)$ finitely generated over $\mathbb{Q}$). Together with the fact that sufficiently high powers of the t_i's are zero in $H(\mathbb{Q}[t_1,\dots,t_n] \tilde{\otimes} \Lambda^*(V), \tilde{\delta})$ we get that $\dim_{\mathbb{Q}} H(\mathbb{Q}[t_1,\dots,t_n] \tilde{\otimes} \Lambda^*(V), \tilde{\delta}) < \infty$.

(3.4) Proposition: Let $M(X) = (\Lambda^*(V),\delta)$ be a minimal model (over $\mathbb{Q}$) of a simply connected CW-complex X of finite $\mathbb{Q}$-type ($H^*(X;\mathbb{Q})$ of finite type).

a) If $\delta: V \to \Lambda^*(V)$ factors through $\Lambda^*(W)$, where $W \subset V$ is a graded linear subspace, then $\dim_{\mathbb{Q}} ZL_*(X) \geq \dim_{\mathbb{Q}}(V/W)$.

b) If $z(X) := \dim_{\mathbb{Q}} ZL_*(X)$ then there exists a graded linear subspace $W \subset V$ of codimension $z(X)$ such that the quadratic part q of the boundary $\delta|_V$ (i.e. the composition $q: V \to \Lambda^*(V) \to \Lambda^2(V)$) factors through $\Lambda^2(W)$ ($\Lambda^*V = \bigoplus_i \Lambda^i V$ and $\Lambda^*V \to \Lambda^2 V$ is the canonical projection).

Proof: a) Clearly $q: V \xrightarrow{\delta} \Lambda^*(V) \to \Lambda^2(V)$ factors through $\Lambda^2(W)$. For the dual multiplication on $\mathrm{Hom}(V,\mathbb{Q}) = \pi_*(X) \otimes \mathbb{Q}$ it follows immediately that all products where one of the factors is contained in $\mathrm{Hom}(V/W,\mathbb{Q})$ vanish. Hence $\mathrm{Hom}(V/W,\mathbb{Q})$ corresponds (after dimension shift) to a subspace of the centre $ZL_*(X)$ of the Lie algebra $L_*(X) = \pi_*(\Omega X) \otimes \mathbb{Q}$.
b) Let $Z \subset \mathrm{Hom}(V,\mathbb{Q})$ be the subspace of $\pi_*(X) \otimes \mathbb{Q} = \mathrm{Hom}(V,\mathbb{Q})$ which corresponds to $ZL_*(X)$ under the dimension shift. The Lie product $L_*(X) \otimes L_*(X) \xrightarrow{\mu} L_*(X)$ factors through $L_*(X)/ZL_*(X) \otimes L_*(X)/ZL_*(X)$ and therefore the dual map $\mu^*: V \to V\otimes V$ factors through $W\otimes W$, where W is defined such that $\mathrm{Hom}(W,\mathbb{Q}) = \mathrm{Hom}(V,\mathbb{Q})/Z$. Since μ is anti-commutative the dual μ^* actually factors through $\Lambda^2 W$ and the map $V \to \Lambda^2 W \to \Lambda^2 V$ obtained this way coincides with the quadratic part q of δ .

(3.5) Corollary: Let X be a simply connected, finite, π-formal CW-complex with $L_{odd}(X) = \pi_{ev}(X) \otimes \mathbb{Q} = 0$. Then the torus rank of the rational homotopy type of X (s.[11], 4.3) is equal to the dimension of the centre $ZL_*(X)$ of the rational homotopy Lie algebra $L_*(X) = \pi_*(\Omega X) \otimes \mathbb{Q}$ of X . In other words:

$$rk_o(X) \leq \dim_{\mathbb{Q}} ZL_*(X) \text{ and this bound is "best possible".}$$

References

[1] ALLDAY, C. and HALPERIN, S.: Lie group actions on spaces of finite rank. Quart. J. Math. Oxford (2) 29, 69-76 (1978)
[2] ALLDAY, C. and PUPPE, V.: On the localization theorem at the cochain level and free torus actions. (preprint)
[3] BLANCHARD, A.: Sur les variétés analytiques complexes. Annales Ec. Norm. Sup. 73, 157-202 (1957)
[4] BOREL, A.: Seminar on Transformation Groups. Annals of Math. Studies, No. 46, Princeton, New Jersey: Princeton Univ. Press 1960

[5] BREDON, G.E.: Introduction to Compact Transformation Groups. New York - London: Academic Press 1972

[6] BROWDER, W.: Cohomology and group actions. Invent Math. 71, 599-607 (1983)

[7] BROWN, E.H.: Twisted tensor products, I. Ann. of Math. 69, 223-246 (1959)

[8] CARLSSON, G.: On the homology of finite free $(\mathbb{Z}/_2)^n$-complexes, Invent Math. 74, 139-147 (1983)

[9] DENNINGER, C. and SINGHOF, W.: On the cohomology of nilpotent Lie algebras. (preprint)

[10] HALPERIN, S.: Lectures on Minimal Models. Memoirs de la Soc. Math. France (1984)

[11] HALPERIN, S.: Rational homotopy and torus actions. (preprint)

[12] HSIANG, W.Y.: Cohomology Theory of Topological Transformation Groups. Berlin-Heidelberg-New York: Springer 1975

[13] SULLIVAN, D.: Infinitesimal computations in topology. Inst. Hautes Etudes Sci. Publ. Math. No. 47, 269-331 (1977)

[14] TANRÉ, D.: Homotopie rationelle: Modèles de Chen, Quillen, Sullivan. Lect. Notes Math. 1025, Berlin, Heidelberg, New York: Springer 1983

THE EQUIVARIANT WALL FINITENESS OBSTRUCTION AND WHITEHEAD TORSION

Paweł Andrzejewski
Szczecin, Poland

Dedicated to the memory of
Andrzej Jankowski and Wojtek Pulikowski

Let G be a compact Lie group and X a G-CW-complex G-dominated by a finite one. Then it is natural to ask whether X has the G-homotopy type of a finite G-CW-complex. As in the non-equivariant case [16] one can expect that the answer to this question will depend on some algebraic invariants. The aim of this paper is to describe the equivariant version of the finiteness obstruction from the following two points of view.

The first one goes along the classical Wall's line. Namely, for any closed subgroup H of G and any component X_α^H of X^H one defines [10] the group $(WH)_\alpha = \{w \in WH : w(X_\alpha^H) = X_\alpha^H\}$ and its lifting $(WH)_\alpha^*$ which is a Lie group and acts on the universal covering $\widetilde{X_\alpha^H}$. These groups are used to define the family of elements $w_\alpha^H(X) \in \tilde{K}_o(Z[\pi_o(WH)_\alpha^*])$ and to show that the finitely dominated G-CW-complex X is G-homotopy equivalent to a finite G-CW-complex iff all $w_\alpha^H(X)$ are zero.

On the other hand, it is not difficult to generalize the construction of the finiteness obstruction given by S. Ferry [7] to the equivariant case. Precisely, under the above assumption on X there exists a single invariant $\sigma_G(X) \in Wh_G(X \times S^1)$, such that $\sigma_G(X) = 0$ iff X is G-finite (up to G-homotopy type).

Moreover, there is a natural relaion between these obstructions. By the results of Illman [10] and Bass-Heller-Swan [5] the equivariant Whitehead group $Wh_G(X \times S^1)$ maps onto the direct sum

$$\bigoplus_H \bigoplus_\alpha \tilde{K}_o(Z[\pi_o(WH)_\alpha^*])$$

of (reduced) projective class groups and we are able to prove that the image of $\sigma_G(X)$ decomposes exactly into the family of elements $w_\alpha^H(X)$.

In the case when G is a finite group and any fixed point set X^H is connected and non-empty, J. Baglivo [4] has defined an algebraic Wall-type obstruction to finiteness. The equivariant version of the finiteness obstruction for finite group actions was also established by D. Anderson [1]. (Unfortunately, the definition of the obstructions in [1] is not quite correct). The generalization of Ferry's work is due to S. Kwasik [12] and we briefly recall his results. Recently W. Lück [13] has presented another geometrical approach to the finiteness obstruction.

A short survey of the contents of the paper is as follows. Section 1 contains the description of the relative equivariant Wall-type obstruction $w_G(X,A)$ for relatively free actions which plays a crucial role in the next section that deals with the general construction of the invariants $w_\alpha^H(X)$. As stated above, we shortly recall the generalization of Ferry's results [12] and this is done in section 3 while section 4 contains the comparison of these obstructions via the Bass-Heller-Swan isomorphism. Finally, applying Illman's result [11] we obtain in section 5 a product formula for finiteness obstructions and its geometric application.

Our notations are the standard ones. For any closed subgroup H of G we define

$$X^{>H} = \{x \in X : G_x \supsetneq H\} .$$

Furthermore, we denote $X^{(H)} = GX^H$ and $X^{>(H)} = GX^{>H}$. We define a partial order in the set of all conjugacy classes by $(H) \geq (K)$ iff there exists $g \in G$ such that $gHg^{-1} \supset K$, and by $(H) > (K)$ we mean $(H) \geq (K)$ and $(H) \neq (K)$. We also assume familiarity with the first part of Illman's paper [10].

I wish to thank the referee for helpful suggestions which allowed to improve the final version of this paper.

1. The case of a relatively free action

Let G be a compact Lie group and X a G-space. The space X is <u>G-dominated</u> by the G-space K if there exist G-maps $\Phi : K \longrightarrow X$ and $s : X \longrightarrow K$ such that $\Phi \cdot s \overset{G}{\simeq} id_X$. Then Φ is called the <u>domination map and s its section</u>.

If now X is a connected G-CW-complex and $p : \tilde{X} \longrightarrow X$ denotes its universal covering then we can consider the lifting of the action of the group G on X to the covering action of a group G^* on $\tilde{X}$ (see sect. 5 in [10] for details). The group G^* is a Lie group and fits into the exact sequence

$$0 \longrightarrow \pi_1(X) \longrightarrow G^* \overset{\pi}{\longrightarrow} G \longrightarrow 0 .$$

Moreover, $\tilde{X}$ is a G^*-CW-complex ([10] th. 6.6). Let further A be a G-invariant subcomplex of X such that the inclusion induces an isomorphism $\pi_1(A,x_o) \cong \pi_1(X,x_o)$ of fundamental groups. Then one can define an action of $\pi_o(G^*)$ on homotopy and homology groups $\pi_n(X,A,x_o)$, $H_n(X,A)$ such that it makes them into modules over the group ring $Z[\pi_o(G^*)]$ (see sect. 7 in [10]). We say that the action of G on the pair (X,A) is <u>relatively free</u> if G acts freely on X-A. We say that the G-CW-pair (X,A) is <u>relatively finite</u> if X-A has finite number of G-cells. By a <u>relative G-domination,</u> we mean the G-map $\Phi : (K,L) \longrightarrow (X,A)$ along with its section $s : (X,A) \longrightarrow (K,L)$ such that $\Phi \cdot s \overset{G}{\simeq} id_{(X,A)}$.

Let now the relatively free G-CW-pair (X,A) be G-dominated by a relatively free, relatively finite G-CW-pair (K,L) via the map $\Phi : (K,L) \longrightarrow (X,A)$ and let

$q : \hat{K} \longrightarrow K$ be the pull-back of $p : \tilde{X} \longrightarrow X$ by Φ i. e.

$$\hat{K} = \{(\tilde{x},k) : p(\tilde{x}) = \Phi(k)\} .$$

The group G^* acts on $\hat{K}$ by the formula $g^*(\tilde{x},k) = (g^*(\tilde{x}),\pi(g^*)(k))$. Then $\hat{K}$ is a G^*-CW-complex and G^* acts freely on $\hat{K}-\hat{L} = \hat{K}-q^{-1}(\hat{L})$. Since $C_n(\tilde{X},\tilde{A}) \cong \oplus Z[\pi_o(G^*)]$ the cellular chain complexes $C_*(\hat{K},\hat{L})$ and $C_*(\tilde{X},\tilde{A})$ are complexes of free $Z[\pi_o(G^*)]$-modules and $C_*(\hat{K},\hat{L})$ is finite. Moreover the map Φ induces the domination map $C_*(\hat{K},\hat{L}) \longrightarrow C_*(\tilde{X},\tilde{A})$ of chain complexes. We define the <u>relative equivariant Wall finiteness obstruction</u> as

$$w_G(X,A) = w(C_*(\tilde{X},\tilde{A})) \in \tilde{K}_o(Z[\pi_o(G^*)]) ,$$

where $w(C_*)$ is the algebraic finiteness obstruction [17], [4] . The following property will serve as an inductive step in the next section.

<u>Proposition 1.1.</u> Let a relatively free G-CW-pair (X,A) be G-dominated by a relatively free, relatively finite G-CW-pair (K,L) and suppose that L is of finite type; then there exist a relatively free, relatively finite G-CW-pair (Y,A) and a G-homotopy equivalence $h : (Y,A) \longrightarrow (X,A)$ with $h|_A = id_A$ iff $w_G(X,A) = 0$ in $\tilde{K}_o(Z[\pi_o(G^*)])$.

<u>Remarks.</u>

1. The relative equivariant Wall finiteness obstruction was also defined in [2] by different (geometrical) methods.

2. A G-CW-complex is of finite type if it contains a finite number of G-cells in each dimension.

In order to prove the above proposition we need some auxiliary facts. Let K,X be connected G-CW-complexes and $\Phi : K \longrightarrow X$ a G-map such that

1) for any subgroup H of G Φ^H determines the bijection of the sets of components K^H_α and X^H_α and

2) for any H and corresponding components, $\Phi^H_\alpha : K^H_\alpha \longrightarrow X^H_\alpha$ induces the isomorphism of fundamental groups.

Let $M = M(\Phi)$ denote the mapping cylinder of Φ and $r : \tilde{M} \longrightarrow M$ its universal covering. We set $\pi_n(\Phi^H_\alpha) = \pi_n(M^H_\alpha,K^H_\alpha,x_o)$ and $H_n(\Phi^H_\alpha) = H_n(M^H_\alpha,K^H_\alpha)$. Let us fix the connected component X^H_α of X^H and take the elements $a_i \in \pi_n(\Phi^H_\alpha)$ $(i=1,2,\ldots k)$ represented by the pairs of maps (s_i,t_i), $s_i : D^n \longrightarrow X^H_\alpha$, $t_i : S^{n-1} \longrightarrow K^H_\alpha$ such that $\Phi^H_\alpha \cdot t_i = s_i|_{S^{n-1}}$. Let $L = K \cup e^n_1 \cup \ldots \cup e^n_k$ denote a G-CW-complex obtained from K by adding k G-n-cells of type (H) via the G-maps determined by t_i . Extend Φ to a G-map $\psi : L \longrightarrow X$ by means of the maps s_i . <u>Such pair</u> (L,ψ) <u>is said to be obtained from</u> (K,Φ) <u>by attaching G-n-cells of type</u> (H) <u>to</u> K <u>via</u> a_i . The following is an immediate consequence of the construction.

Lemma 1.2. If $i < n$ then $\pi_i(\Phi^H_\alpha) = \pi_i(\psi^H_\alpha)$. If $n > 1$ then in the exact sequence of $Z[\pi_o(WH)^*_\alpha]$-modules

$$\ldots \longrightarrow \pi_{n+1}(\psi^H_\alpha) \longrightarrow \pi_n(L^H_\alpha, K^H_\alpha, x_o) \longrightarrow \pi_n(\Phi^H_\alpha) \longrightarrow \pi_n(\psi^H_\alpha) \longrightarrow 0 \longrightarrow \ldots$$

$\pi_n(L^H_\alpha, K^H_\alpha, x_o)$ is a free $\mathbb{Z}[\pi_o(WH)^*_\alpha]$-module with generators b_i such that $d(b_i) = a_i$ $(i = 1,2,\ldots,k)$.

Now we apply the cell-attaching technique to generalize the results of [4].

Lemma 1.3. Let G be a compact Lie group and X a G-CW-complex. If X is G-dominated by a finite G-CW-complex then X is G-homotopy equivalent to a G-CW-complex Y of finite type.

Proof. We shall show inductively that:

If $\Phi : K \longrightarrow X$ is a G-(n-1)-connected domination map and K is finite then there exists a finite G-CW-complex L containing K and G-n-connected extension $\psi : L \longrightarrow X$ of Φ.

There are three cases to consider.

Case 1 $(n = 0)$. Since $\Phi^H_* : \pi_o(K^H) \longrightarrow \pi_o(X^H)$ is an epimorphism so let us suppose the components K^H_1, K^H_2 map to the component X^H_α. We attach a G-1-cell of type (H) to $K^{(H)}$ to obtain an extension $\psi : L \longrightarrow K$ of Φ^H which induces an isomorphism on the π_o-level.

Case 2 $(n = 1)$. Let $\Phi^H_\alpha : K^H_\alpha \longrightarrow X^H_\alpha$ be the restriction to corresponding components. The group $\pi_2(\Phi^H_\alpha)$ is finitely generated in $\pi_1(K^H_\alpha)$ so we can extend Φ^H_α to ψ by attaching G-2-cells of type (H) via the generators of the group $\pi_2(\Phi^2_\alpha)$.

Case 3 $(n > 1)$. By assumption $\pi_n(\Phi^H_\alpha) \cong H_n(\widetilde{\Phi^H_\alpha})$ is finitely generated $Z[\pi_o(WH)^*_\alpha]$-module, where $(WH)_\alpha = \{w \in WH : wX^H_\alpha = X^H_\alpha\}$. If L denotes a G-CW-complex obtained from K by attaching G-n-cells of type (H) via the generators of $\pi_n(\Phi^H_\alpha)$ then lemma 1.2 shows $\pi_n(\psi^H_\alpha) = 0$. Lemma 1.3 is now obvious (cf. [4] p. 312).

Now we are ready to prove the proposition 1.1.

If (X,A) is G-homotopy equivalent to relatively finite pair then $w_G(X,A) = 0$ by the homotopy type invariance of algebraic Wall obstruction.

Suppose now that (X,A) is G-dominated by (K,L) and that $w_G(X,A) = 0$. Let $\Phi : (K,L) \longrightarrow (X,A)$ be a domination map. It follows from the proof of lemma 1.3 that we can assume $\Phi|_L : L \longrightarrow A$ to be a G-homotopy equivalence and $\Phi : K \longrightarrow X$ to be G-n-connected where $n = \max(\dim(K-L),2)$. By lemma 2.1 in [16] $\pi_{n+1}(\Phi) \cong H_{n+1}(\tilde{\Phi})$ is projective and finitely generated $Z[\pi_o(G^*)]$-module and it represents $w_G(X,A)$ ([8] p. 340). By assumption there exist finitely generated, free $Z[\pi_o(G^*)]$-modules C, D such that $\pi_{n+1}(\Phi) \oplus C = D$. Let rank C = m and let $\Phi_1 : K_1 \longrightarrow X$ be a G-map obtained from Φ by attaching m free G-n-cells to K via trivial maps $a_i \in \pi_n(\Phi)$. Then lemma 1.2 shows that $\pi_{n+1}(\Phi_1) = \pi_{n+1} \oplus C = D$. Attach now to K_1 free G-(n+1)-cells via free generators of the module $\pi_{n+1}(\Phi_1)$ to obtain a G-map $\Phi_2 : (K_2,L) \longrightarrow (X,A)$ such that $\Phi_2 : K_2 \longrightarrow X$ is a homotopy equivalence and $\Phi_2|_L : L \longrightarrow A$ is a G-homotopy equivalence of pairs (cf. [3], prop. 1.2) with (K_2,L) relatively finite. Now, extending the G-homotopy inverse of $\Phi_2|_L$ one can obtain the required G-homotopy equivalence $h : (Y,A) \longrightarrow (X,A)$.

2. The equivariant Wall-type obstruction to finiteness

Throughout this section G will denote a compact Lie group and X a G-CW-complex. Suppose that X is G-dominated by a finite G-CW-complex K and let $\Phi : K \to X$ be a domination with the section $s : X \longrightarrow K$. In this section we will define the family of Wall obstructions which determine if the G-CW-complex X has the G-homotopy type of a finite one.

For any closed subgroup H of G the fixed point set X^H is an NH-space as well as a WH-space where WH = NH/H. If X is a G-CW-complex then Illman ([10] sect. 4) observed that X^H is a WH-CW-complex and it is finite if X is. We will need the following observation, the proof of which is completely straightforward.

Lemma 2.1. If X is an H-CW-complex then the twisted product $G \times_H X$ is a G-CW-complex and it is finite if X is.

Let further X^H_α be a connected component of X^H and denote $(NH)_\alpha = \{n \in NH : nX^H_\alpha = X^H_\alpha\}$ and $(WH)_\alpha = \{w \in WH : wX^H_\alpha = X^H_\alpha\}$. Both, $(NH)_\alpha$ and $(WH)_\alpha$, are compact Lie groups and X^H_α is a $(WH)_\alpha$-CW-complex. The set $(WH)X^H_\alpha = (NH)X^H_\alpha$ is called the WH-component of X^H_α.

Let now X^H_α be a connected component of X^H such that H occurs as an isotropy subgroup in X^H_α, i. e. $X^H_\alpha - X^{>H}_\alpha \neq \emptyset$. We define an equivalence relation ~ in the set of such components X^H_α, by setting $X^H_\alpha \sim X^K_\beta$ iff there exists an element $n \in G$ such that $nHn^{-1} = K$ and $nX^H_\alpha = X^K_\beta$. We denote the set of equivalence classes of this relation by CI(X). Note that CI(X) is a subset of the set C(X) introduced by Illman [10].

Lemma 2.2. Suppose that a G-CW-complex X is G-dominated by a G-CW-complex K and let $\Phi : K \longrightarrow X$ denote the domination map with the section $s : X \longrightarrow K$. Let X^H_α and K^H_β be components of X^H and K^H, respectively, such that $s(X^H_\alpha) \subset K^H_\beta$. Then $(WH)_\alpha = (WH)_\beta$ and K^H_β $(WH)_\alpha$-dominates X^H_α.

If X_α^H is a component of X^H which represents an element of the set $CI(X)$ then let K_β^H be a component of K^H such that $s(X_\alpha^H) \subset K_\beta^H$. The group $(WH)_\alpha$ acts on the pairs $(X_\alpha^H, X_\alpha^{>H})$ and $(H_\beta^H, K_\beta^{>H})$ in such a way that $(X_\alpha^H, X_\alpha^{>H})$ is relatively free and $(K_\beta^H, K_\beta^{>H})$ is relatively free and relatively finite. By the relative version of lemma 2.2 we have that $(K_\beta^H, K_\beta^{>H})$ $(WH)_\alpha$-dominates $(X_\alpha^H, X_\alpha^{>H})$. We define <u>an invariant</u> $w_\alpha^H(X)$ to be $w_\alpha^H(X) = w_{(WH)_\alpha}(X_\alpha^H, X_\alpha^{>H}) \in \tilde{K}_o(Z[\pi_o(WH)_\alpha^*]$. We wish to show that this is independent of the choice of representative X_α^H from the equivalence class $[X_\alpha^H]$ in $CI(X)$. Let X_β^K be a component of X^K such that $X_\alpha^H \sim X_\beta^K$. This means that there exists $n \in G$ such that $nHn^{-1} = K$ and $nX_\alpha^H = X_\beta^K$. The map $n: X_\alpha^H \to X_\beta^K$ is a $\gamma(n)$-isomorphism from the $(WH)_\alpha$-CW-complex X_α^H to the $(WK)_\beta$-CW-complex X_β^K. Here $\gamma(n) : (WH)_\alpha \longrightarrow (WK)_\beta$ is an isomorphism defined by $\gamma(n)(n_\alpha H) = (nn_\alpha n^{-1})K$. Furthermore, $\gamma(n)$ induces the canonical isomorphism

$$\Gamma : \tilde{K}_o(Z[\pi_o(WH)_\alpha^*]) \longrightarrow \tilde{K}_o(Z[\pi_o(WK)_\beta^*])$$

which is independent of n. The isomorphism $n : X_\alpha^H \longrightarrow X_\beta^K$ induces an isomorphism $\tilde{n}_* : C_*(\tilde{X}_\alpha^H, \tilde{X}_\alpha^{>H}) \longrightarrow C_*(\tilde{X}_\beta^K, \tilde{X}_\beta^{>K})$ of chain complexes and from this it follows that $\Gamma(w_\alpha^H(X)) = w_\beta^K(X)$.

We can now state the following result.

<u>Theorem 2.3.</u> Let a G-CW-complex X be G-dominated by a finite G-CW-complex K. Suppose X has a finite number of isotropy types. Then X has the G-homotopy type of a finite G-CW-complex iff all the invariants $w_\alpha^H(X)$ vanish.

<u>Proof.</u> Since the necessity part is clear, we only have to prove the sufficiency. Suppose that $w_\alpha^H(X) = 0$ for any equivalence class $[X_\alpha^H]$ in $CI(X)$. Note that the set $CI(X)$ consists of one connected component from each WH-component $(WH)X_\alpha^H$ for which $(WH)X_\alpha^H - (WH)X_\alpha^{>H} \neq \emptyset$, i. e. $X_\alpha^H - X_\alpha^{>H} \neq \emptyset$. Here H runs through a complete set of representatives for all the isotropy types (H) which occur in X.

By assumption on X the set $CI(X)$ is finite.

Let $(H_1), \ldots, (H_r)$ be isotropy types occurring on X ordered in such a way that if $(H_i) \geq (H_j)$ then $i \leq j$. Let $X_{\alpha_1}^{H_i}, \ldots, X_{\alpha_{s_i}}^{H_i}$ denote the representatives of WH_i-components of X^{H_i}. Order the set of pairs $\{(p,q) : 1 \leq p \leq r, 1 \leq q \leq s_p\}$ lexicographically.

The proof goes by induction. We shall construct for each pair (p,q) a G-CW-complex $Y_{p,q}$ and a G-homotopy equivalence $f_{p,q} : Y_{p,q} \longrightarrow X$ such that

1) $(Y_{p,q})^H$ is WH-finite for any subgroup H of G with $H \in (H_i)$ for some $1 \leq i < p$.

2) $G(Y_{p,q})^{H_p}_{\beta_j}$ is G-finite for any component $(Y_{p,q})^{H_p}_{\beta_j}$ of $(Y_{p,q})^{H_p}$ corresponding to $X^{H_p}_{\alpha_j}$ under $f_{p,q}$ for $1 \leq j \leq q$.

Then Y_{r,s_r} will be a finite G-CW-complex G-homotopy equivalent to X.

We begin with $Y_{0,0} = X$ and $f_{0,0} = id_X$. Suppose now that $Y_{p,q}$ satisfying the above conditions has been constructed. We will identify X with $Y_{p,q}$ via $f_{p,q}$, and will assume that X^H is WH-finite for any H conjugate to some H_i $(1 \leq i < p)$ and that $GX^{H_p}_{\alpha_j}$ is G-finite for $1 \leq j \leq q$.

There are two cases to consider.

Case 1 $(q < s_p)$. We simplify the notation by setting $H = H_p$ and $\alpha = \alpha_{q+1}$. Since $(X^H_\alpha, X^{>H}_\alpha)$ is $(NH)_\alpha$-dominated by a finite pair, $w^H_\alpha(X) = 0$ and $X^{>H}_\alpha$ is $(NH)_\alpha$-finite by inductive assumption, the proposition 1.1 implies that there exists $(NH)_\alpha$-homotopy equivalence of pairs $f : (Z, X^{>H}_\alpha) \longrightarrow (X^H_\alpha, X^{>H}_\alpha)$ with Z a finite $(NH)_\alpha$-CW-complex and $f|_{X^{>H}_\alpha}$ an identity. Now f induces a G-homotopy equivalence of twisted products

$$f' : (G \times_{(NH)_\alpha} Z, G \times_{(NH)_\alpha} X^{>H}_\alpha) \longrightarrow (G \times_{(NH)_\alpha} X^H_\alpha, G \times_{(NH)_\alpha} X^{>H}_\alpha) .$$

We define an equivalence relation $\sim$ on $G \times_{(NH)_\alpha} X^{>H}_\alpha$ by setting $[g,x] \sim [g',x']$ iff $gx = g'x'$ in X. We extend this relation to $G \times_{(NH)_\alpha} Z$ and $G \times_{(NH)_\alpha} X^H_\alpha$ by identifying no points outside $G \times_{(NH)_\alpha} X^{>H}_\alpha$. Let $Y = G \times_{(NH)_\alpha} Z/\sim$ and note that $G \times_{(NH)_\alpha} X^H_\alpha/\sim \; \simeq GX^H_\alpha \subset X$. By lemma 2.1 Y is a finite G-CW-complex and f' induces the G-homotopy equivalence $f^H : Y \longrightarrow GX^H_\alpha$. Now we can use the techniques of [9], section 4 to extend f'' to a G-homotopy equivalence $f_{p,q+1} : Y_{p,q+1} \longrightarrow X$.

Case 2 $(q = s_p)$. The proof of it is completely analogous to that of case 1.

3. The equivariant version of Ferry's construction

In [12] S. Kwasik has generalized the construction of the finiteness obstruction presented by Ferry to the equivariant case. We briefly recall his description especially as the proof the the theorem 3.4 in [12] is not totally clear. I am indebted to S. Illman and S. Kwasik for helpful remarks concerning the proof of the theorem 3.2 below.

Let $\Phi : K \longrightarrow X$ be a domination map with the section $s : X \longrightarrow K$. If $A = s \cdot \Phi : K \longrightarrow K$ then denote by $T(A)$ the mapping torus of A obtained from the mapping cylinder $M(A)$ by identification of the top and bottom of $M(A)$ by means of the identity map.

Proposition 3.1. [12] If the G-space X is G-dominated by a finite G-CW-complex K then the mapping torus $T(A)$ is a finite G-CW-complex and has the G-homotopy type of the G-space $X \times S^1$ (with trivial G-action on S^1).

Let $B : T(A) \longrightarrow X \times S^1$ denote the G-homotopy equivalence of proposition 3.1. The natural infinite cyclic covering of $X \times S^1$ induces an infinite cyclic covering $I(A)$ of $T(A)$. The G-space $I(A)$ is an infinite G-CW-complex with two ends ε_+, ε_- and the G-homotopy equivalence B gives rise to a G-homotopy equivalence between X and $I(A)$.

Let $u : S^1 \longrightarrow S^1$ be the homeomorphism given by the complex conjugation and B^{-1} the homotopy inverse to B. Denote by $\tau(h) \in WH_G(T(A))$ the torsion of the G-homotopy equivalence $h = B^{-1} \cdot (id_X \times u) \cdot B : T(A) \longrightarrow T(A)$. We define the equivariant obstruction to finiteness as $\sigma_G(X) = B_*(\tau(h)) \in Wh_G(X \times S^1)$.

One can show that this obstruction is well-defined (see [7] th. 2.3).

Theorem 3.2. The finitely dominated G-space X has the equivariant homotopy type of a finite G-CW-complex iff $\sigma_G(X) = 0$.

Proof. If X has the G-homotopy type of a finite G-CW-complex K we may assume that the domination map $\Phi : K \longrightarrow X$ is the G-homotopy equivalence and $A \overset{G}{\simeq} id_K$. Then $T(A)$ is G-homotopy equivalent to $K \times S^1$ and $B = \Phi \times id_{S^1}$. Hence $\tau(h) = \tau(id_K \times u) \in WH_G(K \times S^1)$ and we show that this torsion vanishes.

By the product formula for equivariant Whitehead torsion [11] we have for the $(H \times Q, \alpha \times \beta)$-component $(K \times S^1)^{H\times Q}_{\alpha\times\beta}$

$$\tau(id \times u)^{H\times Q}_{\alpha\times\beta} = \overline{\chi}^{H}_{\alpha}(K) \cdot j_*\tau(u)^{Q}_{\beta} \in Wh(\pi_o(WH)^*_\alpha \times \pi_o(WQ)^*_\beta) .$$

Since the action of G on S^1 is trivial we get $\tau(u)^Q_\beta = 0$ and $\sigma_G(X) = 0$.

If $\sigma_G(X) = 0$ then $\tau(h) = 0$ and it means that $h : T(A) \longrightarrow T(A)$ is an equivariant simple-homotopy equivalence. Making use of the equivariant version of [6], exercise 4. D., p. 16, we can find a finite G-CW-complex W and two equivariant collapses $f_i : W \longrightarrow T(A)$ such that the diagram

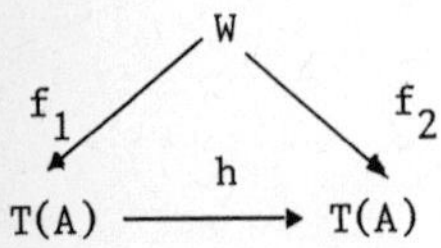

commutes up to G-homotopy.

Now, passing to infinite cyclic coverings we have a diagram

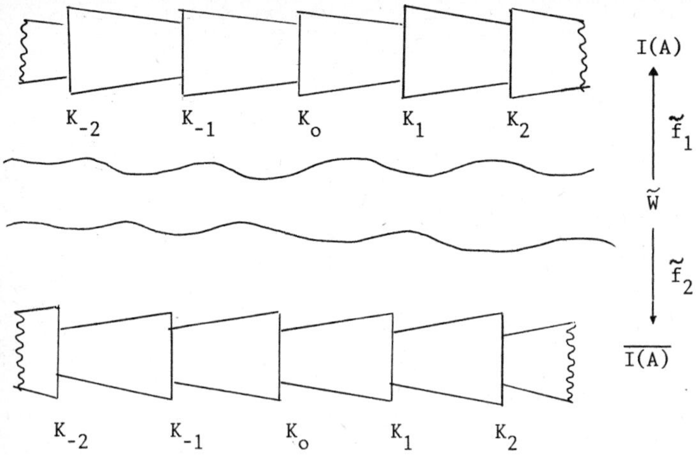

where $\overline{I(A)}$ denotes the reversed (with respect to the ends) copy of $\overline{I(A)}$ and $K_i=K$. For sufficiently large m the region of $\tilde{W}$ between $(\tilde{f}_1)^{-1}(K_{-m})$ and $(\tilde{f}_2)^{-1}(K_m)$ is an equivariant strong deformation retract of $\tilde{W}$ and therefore it has the G-homotopy type of X .

4. The relation between $\sigma_G(X)$ and $w_\alpha^H(X)$

Supose that a G-CW-complex X is G-dominated by a finite G-CW-complex K . In [10] Illman showed that there is a natural isomorphism

$$WH_G(X \times S^1) = \bigoplus_{C(X)} Wh(\pi_o(WH)_\alpha^* \times Z) .$$

Furthermore, for an arbitrary short exact sequence of groups

$$0 \longrightarrow R \longrightarrow P \longrightarrow Z \longrightarrow 0$$

we have the natural Bass-Heller-Swan decomposition of the Wh-functor [5]

$$Wh(P) = Wh(R) \oplus \tilde{K}_o(Z[R]) \oplus N .$$

In particular, we have an epimorphism

$$S : Wh(P) \longrightarrow \tilde{K}_o([R])$$

whose definition is given below. Hence we obtain the natural decomposition

$$Wh_G(X \times S^1) \cong \bigoplus_{C(X)} Wh(\pi_o(Wh)_\alpha^*) \oplus \bigoplus_{C(X)} \tilde{K}_o(Z[\pi_o(WH)_\alpha^*]) \oplus N$$

and the natural epimorphism

$$S: Wh_G(X \times S^1) \longrightarrow \tilde{K}_o(Z[\pi_o(WH)_\alpha^*]) .$$

The aim of this section is to prove the following result.

Theorem 4.1.. The equivariant finiteness obstruction $\sigma_G(X)$ decomposes into the family of obstructions $w^H_\alpha(X)$. Precisely, the image of the (H,α)-component $\sigma_G(X)^H_\alpha$ of the obstruction $\sigma_G(X)$ under epimorphism

$$S : Wh(\pi_o(WH)^*_\alpha \times Z) \longrightarrow \tilde{K}_o(Z[\pi_o(WH)^*_\alpha])$$

is equal to the equivariant Wall-type obstruction $w^H_\alpha(X)$.

We start with the definition of the homomorphism

$$S : Wh(P) \longrightarrow \tilde{K}_o(Z[R]) .$$

Decompose the group ring $Z[P]$ as $Z[P] = \bigoplus_{r\in Z} Z[R]t^r$ where $t\in P$ maps to $1\in Z$. Denote by $D = \bigoplus_{r\geq o} Z[R]t^r$ and $D' = \bigoplus_{r\leq o} Z[R]t^r$ and let $x\in Wh(P)$ be represented by a $Z[P]$-isomorphism

$$d : Z[P]^k \longrightarrow Z[P]^k .$$

Choose $s > 0$ so large that $d(D^k t^s) \subset D^k$ and $D'^k \subset d(D'^k t^s)$. Then a $Z[R]$-module $D^k/d(D^k t^s)$ is finitely generated and projective ([15] prop. 10.2) and, by definition, it represents $S(x)$. One can show that S is a well-defined group homomorphism ([15] th. 8.1).

Now the G-homotopy equivalence $h : T(A) \longrightarrow T(A)$ induces the G-homotopy equivalence $\tilde{h}$ between $I(A)$ and its reversed copy $\overline{I(A)}$. Taking the mapping cylinder of $\tilde{h}$ we may assume that $\tilde{h} : I(A) \longrightarrow I(A)$ is an equivariant strong deformation retraction of $I(A)$ (see prop. 1.3 in [9]). In $I(A)$ consider a G-invariant subcomplex L such that L is a neighborhood of ε_+ and $(I(A)-L) \cup \overline{I(A)}$ is a neighborhood of ε_- . Let $L_1 = L \cap \overline{I(A)}$. We will need the following observation.

Lemma 4.2: The pair (L,L_1) is G-dominated by a relatively finite pair $(L_o \cup L_1, L_1)$.

Proof. The G-homotopy equivalence $h : T(A) \longrightarrow T(A)$ induces proper homotopy equivalence $h' : T(A)/G \longrightarrow T(A)/G$ and $\tilde{h} : I(A) \longrightarrow I(A)$ induces the proper strong deformation retraction of $I(A)/G$, [14], lemma 4.7.

Let $h_t : I(A) \longrightarrow I(A)$ be the G-homotopy between id and $\tilde{h}$. Passing to the orbit spaces one can find a G-subcomplex L_2 of L such that $h_t(L_2) \subset L$ for all t . Extend now the G-homotopy $h_t : L_2 \cup L_1 \longrightarrow L$ to the G-homotopy $k_t : L \longrightarrow L$ constant on L_1 . The complex $L - (L_1 \cup L_2)$ is G-finite so there exists a G-finite subcomplex $L_o \subset L$ with $k_t(L-L_2) \subset L_o \cup L_1$. Now the inclusion $(L_o \cup L_1, L_1) \longrightarrow (L,L_1)$ is a G-domination map .

Hence by lemma 2.2 the pair $((L_o \cup L_1)^H_\alpha, (L_1)^H_\alpha \cup (L_o \cup L_1)^{>H}_\alpha)$ $(WH)_\alpha$-dominates the pair $(L^H_\alpha, (L_1)^H_\alpha \cup L^{>H}_\alpha)$ and we can define the obstruction

$$w_\alpha^H(I(A),\overline{I(A)},\varepsilon_+) = w(C_*(\widetilde{L_\alpha^H},\widetilde{(L_1)_\alpha^H \cup L_\alpha^{>H}})) \in \tilde{K}_o(Z[\pi_o(WH(I(A)))_\alpha^*]) .$$

This obstruction is independent of the choice of subcomplex L , because for another $L' \subset L$ there are only finitely many G-cells in $L-(L' \cup \overline{I(A)})$.

Choose now neighborhoods L_+ , L_- of ε_+ , ε_- so that $I(A)-L_+$, $I(A)-L_-$ are neighborhoods of ε_- and ε_+ , respectively, and $L_+ \cup L_- = I(A)$. Then the subcomplex $L_+ \cup L_-$ is G-finite and since $I(A)$ is G-dominated by K the Mayer-Vietoris sequence

$$0 \longrightarrow C_*(\widetilde{(L_+ \cap L^-)_\alpha^H},\widetilde{(L_+ \cap L_-)_\alpha^{>H}}) \longrightarrow C_*(\widetilde{(L_-)_\alpha^H},\widetilde{(L_-)_\alpha^{>H}}) \oplus C_*(\widetilde{(L_+)_\alpha^H},\widetilde{(L_+)_\alpha^{>H}}) \longrightarrow$$
$$\longrightarrow C_*(\widetilde{I(A)_\alpha^H},\widetilde{I(A)_\alpha^{>H}}) \longrightarrow 0$$

shows that $C_*(\widetilde{(L_+)_\alpha^H},\widetilde{(L_+)_\alpha^{>H}})$ and $C_*(\widetilde{(L_-)_\alpha^H},\widetilde{(L_-)_\alpha^{>H}})$ are dominated by finitely generated free complexes. Thus we can define the obstructions

$$w_\alpha^H(I(A),\varepsilon_+) = w(C_*(\widetilde{(L_+)_\alpha^H},\widetilde{(L_+)_\alpha^{>H}}))$$

$$w_\alpha^H(I(A),\varepsilon_-) = w(C_*(\widetilde{(L_-)_\alpha^H},\widetilde{(L_-)_\alpha^{>H}}))$$

which do not depend on the choice of L_+ and L_- . Similarly, the neighborhoods $L_1^+ = L_+ \cap \overline{I(A)}$, $L_1^- = L_- \cap \overline{I(A)}$ give the obstructions $w_\alpha^H(\overline{I(A)},\varepsilon_+)$, $w_\alpha^H(\overline{I(A)}\varepsilon_-)$ and and we have

$$w_\alpha^H(I(A),\varepsilon_+) = w_\alpha^H(\overline{I(A)},\varepsilon_+) + w_\alpha^H(I(A),\overline{I(A)},\varepsilon_+) .$$

In our situation L_1^+ and L_- have the G-homotopy type of K so

$$w_\alpha^H(I(A),\varepsilon_+) = w_\alpha^H(I(A),\varepsilon_-) = 0 ,$$

and again the Mayer-Vietoris sequence yields

$$w_\alpha^H(I(A)) = w_\alpha^H(I(A),\varepsilon_+) + w_\alpha^H(I(A),\varepsilon_-) = w_\alpha^H(I(A),\varepsilon_+)$$
$$= w_\alpha^H(I(A),\overline{I(A)},\varepsilon_+) .$$

The crucial step in the proof of the theorem 4.1 lies in the following.

Proposition 4.3. If $S_1 : Wh(\pi_o(WH(T(A)))_\alpha^*) \longrightarrow \tilde{K}_o(Z[\pi_o(WH(I(A)))_\alpha^*])$ is the B-H-S-epimorphism and $\tau(h)_\alpha^H$ denotes the (H,α)-component of the equivariant Whitehead torsion of h then

$$S_1(\tau(h)_\alpha^H) = w_\alpha^H(I(A),\overline{I(A)},\varepsilon_+) .$$

Proof. First of all one can observe that $w_\alpha^H(I(A),\overline{I(A)},\varepsilon_+)$ does not change under the equivariant formal deformations of the mapping cylinder $M(h)$ mod $T(A)$. Hence, by corollary 4.4 in [9] we may assume that the pair $(M(h),T(A))$ is in simplified form i.e.

$$M(h) = T(A) \cup \bigcup b_i^2 \cup \bigcup c_i^3 .$$

Let $V = M(\tilde{h})$ be the mapping cylinder of $\tilde{h} : I(A) \longrightarrow I(A)$ and $p : \tilde{V} \longrightarrow V$ its universal covering. Then we have

$$\tilde{V} = \widetilde{I(A)} \cup \bigcup \tilde{b}_i \cup \bigcup \tilde{c}_i .$$

By the second part of the corollary 4.4 of [9] the cellular chain complex

$$C_*(\widetilde{V^H_\alpha}, \widetilde{I(A)^H_\alpha \cup V^{>H}_\alpha})$$

has the form

$$\ldots \longrightarrow 0 \longrightarrow C_3 \xrightarrow{d} C_2 \longrightarrow 0 \longrightarrow \ldots$$

where $C_2 \cong C_3 \cong (Z[\pi_o(WH(T(A)))^*_\alpha])^k$ with preferred bases derived from the lifted equivariant 2- and 3-cells, respectively. Denote by T the generating covering translation of V over $M(h)$ and by t its lifting to $\tilde{V}$. Then $pt = Tp$.

Now we choose large $s > 0$ and let $L(s)$ be a G-subcomplex of V obtained from $I(A)$ by attaching G-2-cells $T^r p(\tilde{b}_i)$ and G-3-cells $T^r p(t^s \tilde{c}_i) = T^{r+s} p(\tilde{c}_i)$ for $r \geq o$ and all i.

Then $L(s)$ is a neighborhood of ε_+ and $(M(\tilde{h}) - L(s)) \cup I(A)$ is a neighborhood of ε_-, so by definition

$$w^H_\alpha(I(A), \overline{I(A)}, \varepsilon_+) = w(C_*(\widetilde{L(s)^H_\alpha}, \widetilde{I(A)^H_\alpha \cup L(s)^{>H}_\alpha}))$$

On the other hand, the cellular chain complex $C_*(\widetilde{L(s)^H_\alpha}, \widetilde{I(A)^H_\alpha \cup L(s)^{>H}_\alpha}$ is a complex of free $Z[\pi_o(WH(I(A)))^*_\alpha]$-modules and again by [9] corollary 4.4 we have

$$C_2(\widetilde{L(s)^H_\alpha}, \widetilde{I(A)^H_\alpha \cup L(s)^{>H}_\alpha}) = D^k \subset C_2$$

and

$$C_3(\widetilde{L(s)^H_\alpha}, \widetilde{I(A)^H_\alpha \cup L(s)^{>H}_\alpha}) = D^k t^s \subset C_3 .$$

For large s the quotient module $B_s = D^k/d(D^k t^s)$ is projective and by definition

$$S_1(\tau(h)^H_\alpha) = [B_s] \in \tilde{K}_o(Z[\pi_o(WH(I(A)))^*_\alpha]) .$$

The projectivity of B_s implies that $C_*(\widetilde{L(s)^H_\alpha}, \widetilde{I(A)^H_\alpha \cup L(s)^{>H}_\alpha})$ is chain homotopy equivalent to the complex of the form

$$\ldots \longrightarrow 0 \longrightarrow 0 \longrightarrow B_s \longrightarrow 0 \longrightarrow \ldots$$

with B_s in dimension 2. Thus $w^H_\alpha(I(A), \overline{I(A)}, \varepsilon_+) = [B_s]$.

Now we have the commutative diagram

$$\begin{array}{ccc} Wh(\pi_o(WH(T(A)))^*_\alpha) & \xrightarrow{B_*} & Wh(\pi_o(WH)^*_\alpha \times Z) \\ \downarrow S_1 & & \downarrow S \\ \tilde{K}_o(Z[\pi_o(WH(I(A)))^*_\alpha]) & \xrightarrow{B_*} & \tilde{K}_o(Z[\pi_o(WH)^*_\alpha]) \end{array}$$

which yields finally

$$S(\sigma_G(X)^H_\alpha) = SB_*(\tau(h)^H_\alpha) = B_*S_1(\tau(h)^H_\alpha) = B_*(w^H_\alpha(I(A),\overline{I(A)},\varepsilon_+))$$
$$= B_*(w^H_\alpha(I(A))) = w^H_\alpha(X) \quad .$$

5. A product formula for equivariant finiteness obstruction and its application

In this section G and P denote arbitrary compact Lie groups, unless otherwise is stated. Recently S. Illman [11] has given the product formula for the equivariant Whitehead torsion $\tau(f\times h)$ in terms of the equivariant Whitehead torsions of f and h and various Euler characteristics. We use his formula to derive the corresponding formula for the obstructions $\sigma_G(X)$ and $w^H_\alpha(X)$ and its geometric application.

Let X be a G-CW-complex G-dominated by a finite G-CW-complex K and L a finite P-CW-complex. Then the product $L\times X$ is finitely $(P\times G)$-dominated by $L\times K$ and we have the obstruction

$$\sigma_{P\times G}(L\times X) \in Wh_{P\times G}(L\times X\times S^1).$$

Now the domination map defines the $(P\times G)$-homotopy equivalence

$$\bar{B} : T(id_L\times A) \longrightarrow L\times X\times S^1 .$$

But we have $T(id\times A) = L\times T(A)$, $\bar{B} = id_L\times B$ and our finiteness obstruction is given by

$$\sigma_{P\times G}(L\times X) = (id\times B)_*(\tau(id\times h)) \in Wh_{P\times G}(L\times X\times S^{-1}) .$$

Since for the $(Q\times H,\beta\times\alpha)$-component of $\tau(id\times h)$ we have

$$\tau(id\times h)^{Q\times H}_{\beta\times\alpha} = \bar{\chi}^Q_\beta(L)i_*(\tau(h)^H_\alpha)$$

we obtain for the $(Q\times H,\beta\times\alpha)$-component $(L\times X)^{Q\times H}_{\beta\times\alpha}$

$$\sigma_{P\times G}(L\times X)^{Q\times H}_{\beta\times\alpha} = \bar{\chi}^Q_\beta(L)i_*(\sigma_G(X)^H_\alpha) \in Wh(\pi_o(WQ)^*_\beta \times \pi_o(WH)^*_\alpha) \qquad (1)$$

where

$$\bar{\chi}^Q_\beta(L) = \chi(L^{(Q)}_\beta/P, L^{>(Q)}_\beta/P) = \chi(L^Q_\beta/(WQ)_\beta, L^{>Q}_\beta/(WQ)_\beta)$$

and

$$i : \pi_o(WH)^*_\alpha \longrightarrow \pi_o(WQ)^*_\beta \times \pi_o(WH)^*_\alpha$$

denotes the inclusion.

By naturality of the B-H-S decomposition and theorem 4.1 we also obtain

$$w^{Q\times H}_{\beta\times\alpha}(L\times X) = \overline{\chi}^Q_\beta(L) \cdot i_*(w^H_\alpha(X)) \ . \tag{2}$$

Moreover, any obstruction $w^S_\gamma(L\times X)$ where (S,γ) is not of a product form, equals zero.

As an immediate corollary of the formula (1) or (2) we have the following geometric result (cf.[13] cor. 6.4)

Theorem 5.1. Let G be a finite group and X a G-CW-complex G-dominated by a finite one. Let V be any unitary complex representation of the group G and $S(V)$ its unit sphere. Then the product $X \times S(V)$ with the diagonal G-action has the G-homotopy type of a finite G-CW-complex.

Remark. The above theorem is not true for arbitrary compact Lie groups.

References.

[1] D.R. Anderson: Torsion invariants and actions of finite groups, Michigan Math. J. 29 (1982), 27-42.

[2] P. Andrzejewski: On the equivariant Wall finitenes obstruction, preprint.

[3] S. Araki, M. Muruyama: G-homotopy types of G-complexes and representation of G-homotopy theories, Publ. RIMS Kyoto Univ. 14 (1978), 203-222.

[4] J:A. Baglivo: An equivariant Wall obstruction theory, Trans. Amer. Math. Soc. 256 (1979), 305-324.

[5] H. Bass, A. Heller, R. Swan: The Whitehead group of a polynomial extension, Publ. IHES 22 (1964), 67-79.

[6] M.M. Cohen: A course in simple-homotopy theory, Graduate Texts in Math. Springer-Verlag, 1973.

[7] S. Ferry: A simple-homotopy approach to the finiteness obstruction Shape Theory and Geometric Topology, Lecture Notes in Math. 870 (1981), 73-81.

[8] S.M. Gersten: A product formula for Wall's obstruction, Amer. J. Math. 88(1966), 337-346.

[9] S. Illman: Whitehead torsion and group actions, Ann. Acad. Sci. Fennicae, Ser. AI 588 (1974), 1-44.

[10] S. Illman: Actions of compact Lie groups and equivariant Whitehead torsion, preprint, Purdue Univ. (1983).

[11] S. Illman: A product formula for equivariant Whitehead torsion and geometric applications, these proceedings.

[12] S. Kwasik: On equivariant finiteness, comp. Math. 48 (1983), 363-372

[13] W. Lück: The geometric finiteness obstruction, Mathematica Gottingensis, Heft 25 (1985).

[14] L.C. Siebenmann: On detecting Euclidean space homotpically among topological manifolds, Invent. Math. 6 (1968), 245-261.

[15] L.C. Siebenmann: A total Whitehead torsion obstruction to fibering over the circle, Comment. Math. Helv. 45 (1970), 1-48.

[16] C.T.C. Wall: Finiteness conditions for CW-complexes, Ann, Math. 81 (1965), 55-69.

[17] C.T.C. Wall: Finiteness conditions for CW-complexes, II, Proc. Royal Soc. London, Ser. A, 295 (1966), 129-139.

Homotopy Actions and Cohomology of Finite Groups

Amir H. Assadi *)
University of Virginia
Charlottesville, Virginia 22903
Max-Planck-Institut für Mathematik, Bonn

Introduction

Let X be a connected topological space, and let $\mathcal{H}(X)$ be the monoid of homotopy equivalences of X. The group of self-equivalences of X, $\mathcal{E}(X)$, is defined to be $\pi_o\mathcal{H}(X)$. A homomorphism $\alpha : G \to \mathcal{E}(X)$ is called a homotopy action of G on X. Equivalently, the assignment of a self-homotopy equivalence $\alpha(g) : X \to X$ to each $g \in G$ such that $\alpha(g_1g_2) \sim \alpha(g_1)\alpha(g_2)$ and $\alpha(1) \sim 1_X$ is also called a homotopy action. Since it is easier to construct self-homotopy equivalences rather than homeomorphisms of X, it is natural to consider the questions of existence of actions first on the homotopy level, (i.e. homotopy actions) and then try to find an equivalent topological action. A topological G-action φ on Y is said to be equivalent to a homotopy action α on X, if there exists a homotopy equivalence $f : Y \to X$ which commutes with φ and α up to homotopy, i.e. f is homotopy equivariant (for short, f is an h-G-map). This is the point of view taken in [16] and the motivation for G. Cooke's study of the question:

---

*) This work has been partially supported by an NSF grant, the Center for advanced Study of University of Virginia, the Danish National Science Foundation, Matematisk Institut of Aarhus University, and Forschungsinstitut für Mathematik of ETH, Zürich, and Max-Planck-Institut für Mathematik, Bonn, whose financial support and hospitality is gratefully acknowledged. It is a pleasure to thank W. Browder, N. Habegger, I. Madsen, G. Mislin, L. Scott, R. Strong, and A. Zabrodsky for helpful and informative conversations. Special thanks to Leonard Scott for explaining the results of [8] to me which inspired some of the algebraic results, and to Stefan Jackowski for his helpful and detailed comments on the first version of this paper.

<u>Question 1</u>. Given a homotopy action α on X, when is (X,α) equivalent to a topological action?

The problem is quickly and efficiently turned into a lifting problem: A homomorphism $\alpha : G \to E(X)$ yields a map $B\alpha : BG \to BE(X)$. On the other hand the exact sequence of monoids $H_1(X) \to H(X) \to E(X)$ yields a fibration $BH_1(X) \to BH(X) \to BE(X)$.

<u>Theorem (G. Cooke)</u> [16]. (X,α) is equivalent to a topological action if and only if $B\alpha : BG \to BE(X)$ lifts to $BH(X)$ in the fibration $BH(X) \to BE(X)$.

Note that if X does not have a "homotopically simple structure", e.g. if X is not a $K(\pi,n)$ and $\dim X<\infty$, then $\pi_i(BH_1(X))$ is exceedingly difficult to calculate, and the above lifting problem will have infinitely many a priori non-zero obstructions. However, if G is a finite group (and we will assume this throughout) and X is localized away from the prime divisors of $|G|$, e.g. if $\pi_1(X) = 1$ and X is rational, then all the obstructions vanish, and any such (X,α) is equivalent to a topological action. Algebraically, this can be interpreted by the fact that all the relevant RG-modules (where R is a ring of characteristic prime to $|G|$) are semi-simple and consequently cohomologically trivial. Thus the interest lies in the "modular case", (i.e. when a prime divisor of $|G|$ divides the characteristic of R) and the inetgral case $R = \mathbb{Z}$.

In comparison with topological actions, homotopy actions have very little structure in general. For instance, there are no analogues of "fixed point sets", "orbit spaces" or "isotropy groups". This makes a general study of homotopy actions a difficult task. Notwithstanding, there has been some applications to problems in homotopy theory and geometric (differential) topology (e.g. [5] [6] [16] [22] [34] [35] for a sample).

Given a homotopy functor h and a homotopy action of G, say (X,α), we obtain a "representation of G". E.g. if $X \cong K(\pi,n)$ and $h = \pi_n$, then $\pi_n(X) \cong \pi$ becomes a $\mathbb{Z}G$-module. In this case, any $\mathbb{Z}G$-module π also gives rise to a homotopy G-action on $X \simeq K(\pi,n)$, and in fact a topological G-action.

For spaces which are not homotopically easy to understand (such as most manifolds and finite dimensional spaces) homology and cohomology provide a more useful representation module. From this point of view, spaces with a single non-vanishing homology, known as Moore spaces, are the simplest to study. For simplicity, suppose we are given a $\mathbb{Z}G$-module M which is $\mathbb{Z}$-free. Then it is easy to see that there exists a homotopy action α of G on a bouquet of spheres X such that $\bar{H}_*(X) \cong M$ as $\mathbb{Z}G$-modules. We say that "(X,α) realizes M", or that M is realizable by (X,α). An obstruction theory argument shows that the question of realizability of $\mathbb{Z}G$-modules by homotopy G-actions on Moore spaces has a 2-torsion obstruction ([7] [22]) which can be identified with appropriate cohomological invariants of the $\mathbb{Z}G$-module M ([7] P. Vogel, unpublished). In relation with the question of how close these homotopy actions are to topological actions, one should mention the following well-known problem attributed to Steenrod [26]:

Question 2. Is an integral representation of G realizable by a G-action on a Moore space?

There has been some partial progress in answering the above question and we refer the reader to [3] [9] [13] [22] [30] [32] [33] and their references. In an attempt to understand homotopy actions, we will specialize and apply the methods of this paper to the above problem. Thus constructions and the study of the counterexamples for Question 2 in this paper should be regarded as a method of producing and investigating "invariants of homotopy actions" for more general spaces.

As mentioned above, the usual notion of transformation groups such as fixed points, isotropy groups, and orbit spaces do not carry over to homotopy actions as such. Therefore, we will try to attach other invariants, mostly of cohomological nature, to both G-spaces and homotopy G-actions, and compare them. For topological actions these invariants are naturally (and expectedly) related to fixed point sets and isotropy groups (whenever they are well-defined). Thus we have placed special emphasis on topological actions with some finiteness condition on the underlying space (e.g. finite cohomological dimension) as well as G-actions with collapsing spectral sequence in their Borel construction. On the algebraic side, our feeling is that the category of integral (modular) representations of G which arise as homology (cohomology) of G-spaces is an important part of the category of all representations, and its algebraic study is worthwhile in its own right. The projectivity criterion (Thm. 2.1) as well as the complexity criterions (Sec. 3) and their consequences are some steps in this direction.

In comparing homotopy and topological actions, we will study:

Question 3. When is a representation of G realizable by the homology of a G-space?

As we will see below, there are integral (and modular) representations of G which are not realizable via the homology of any G-space (we do not restrict ourselves to Moore spaces). On the other hand, there are representations which are not realizable by G-actions on Moore spaces but they can still be realized by G-actions on other spaces (Section 5). All these representations arise from homotopy actions. These examples show that, even for homologically simple spaces, such as bouquet of spheres, the collection of integral representation of G on $H_*(X)$ induced by a homotopy action $\alpha : G \to E(X)$ does not by itself decide whether (X,α) is equivalent to a topological action. It is the interrelationship of all $H_i(X)$ as $\mathbb{Z}G$-modules which determines the realizability in this case (Section 5). In the applications of homotopy actions to differential topological problems, one often needs to find finite dimensional G-spaces which realize a given homotopy action. The solution to the lifting problem mentioned earlier in the introduction, provides an infinite dimensional free G-space. In this context, the following problem is often necessary to answer:

Question 4. Suppose X is homotopy equivalent to a finite dimensional space and $\varphi : G \times X \to X$ is an action. When does there exist a finite dimensional G-space K and a G-map $f : X \to K$ inducing homotopy equivalence?

We study this problem and the related question Question 3 by "reduction to p-groups". This is the subject of a future paper. In particular, one has satisfactory characterizations for groups with periodic cohomology and some other classes of groups which includes nilpotent groups or some of the alternating groups.

Notation and conventions. All rings are commutative with unit. $\mathbb{F}_p$ is the field with p-element, where p always denotes a prime number, and k is a field of characteristic $p > 0$ (often an algebraic closure of $\mathbb{F}_p$). For a finite group G , H_G denotes the ring $\oplus_i H^{2i}(G;k)$ if p is odd and $H_G = \oplus_i H^i(G;k)$ if $p = 2$. $\hat{H}^*$ denotes Tate cohomology [14] and the terminologies in this context are in [14] and [28]. $\mathbb{Z}_p \equiv \mathbb{Z}/p\,\mathbb{Z} \equiv$ integers (mod p) . The localization of a ring R with respect to the multiplicative subset generated by an element $\gamma \in R$ is denoted by $R[\gamma^{-1}]$. For an ideal J in a ring R , rad(J) is the radical of J and if M is an R-module, Ann(x) is the annihilating ideal of $x \in M$. The dual of a k-algebra A is denoted by A* . For an RG-module M and a subgroup H , $M|RH$ denotes the restriction to H . The terminology and conventions in topological group actions are taken from [10] and [19] and those related to homotopy actions are to be found in [16]. For example E_G is the contractible free G-space and $E_G \times_G X$ is the Borel construction of a G-space X . If a G-space X needs to have a base point in the context, we replace X by its suspension ΣX and take $x \in X^G \neq \emptyset$, unless X is already endowed with a base point. Many of the statemets which are phrased in terms of cohomology have their counterparts in homology and we have avoided repeating this fact. The spaces X are not necessarily CW complexes unless otherwise specified. We may use sheaf cohomology for more general situations and the proofs are still valid (with some mild modification if necessary). The basic reference is [27] part I in particular its appendix, and we have used Quillen's terminology and notation when appropriate. E.g. $cd_p(X)$ means cohomological dimension of X (mod p) .

The bibliography contains the references which have been available to us, at least in some written form. Otherwise they have been mentioned in the context.

Section 1. Localization and Projectivity

In this section we present a variation on P.A. Smith's theorem as a consequence of Quillen's version of the localization theorem of Borel (cf. [19] or [27]). The statements are not as general as they could be because we will present different proofs when the cohomological finiteness of the G-spaces are not assumed. These finiteness assumptions are necessary when applying the localization theorem. There is an analogy between the finiteness assumptions of this section on the level of orbit spaces and the weaker finiteness assumptions for cohomology in the following sections. There is also a localization-type argument implicit in the arguments of sections 2 and 3 which are explicit in the context of this section. The special cases treated differently in this section will hopefully serve to give motivation and some insight into the more algebraic arguments of the following sections. The basic reference for some details of the assertions of this sections (as well as the terminology and the notation) is [10]. More general forms of the localization theorem are discussed in [19].

1.1 Proposition. Let G be a finite group and let X be a connected G-space which is either compact, or $cd_p(X/G) < \infty$ for a fixed prime p. Assume that for each subgroup $C \subset G$ in order p, $H^i(X;\mathbb{F}_p)$ is a cohomologically trivial $\mathbb{F}_p C$-module for all $i > 0$. Then the p-singular set of X, $S_p(X) \equiv \bigcup_P X^P$, where P ranges over non-trivial p-subgroups of G, satisfies $\bar{H}^*(S_p(X);\mathbb{F}_p) = 0$.

Proof: Let $C \subset G$ and $|C| = p$, and let $\gamma \in H^2(C;\mathbb{F}_p)$ be the polynominal generator. Without loss of generality, we may assume that $X^G \neq \emptyset$, hence $X^C \neq \emptyset$. Choose $x \in X^G \subset X^C$. The Serre spectral sequence of the Borel construction $(X,x) \to E_C \times_C (X,x) \to BC$ collapse since $H^i(BC;H^j(X,x;\mathbb{F}_p)) = 0$ for $i > 0$ and all j by cohomological triviality. Thus $H^*_C(X,x;\mathbb{F}_p) \cong H^0(BC;H^*(X,x;\mathbb{F}_p))$. Localization with respect to γ shows ([27]):

$$H^*_C(X,x;\mathbb{F}_p)[\gamma^{-1}] \cong H^*(BC;H^*(X,x))[\gamma^{-1}]$$

$$\cong \hat{H}^*(C;H^*(X,x)) = 0 ,$$

(by the hypothesis of cohomological triviality) where $\hat{H}^*$ denotes Tate cohomology. By the localization theorem

$$H^*_C(X^C,x;\mathbb{F}_p)[\gamma^{-1}] \cong H^*_C(X,x;\mathbb{F}_p)[\gamma^{-1}] = 0 .$$

Since $H^*_C(X^C,x;\mathbb{F}_p)[\gamma^{-1}] \cong H^*(X^C,x;\mathbb{F}_p) \otimes_{\mathbb{F}_p} \hat{H}^*(C;\mathbb{F}_p)$, it follows that $H^*(X^C,x;\mathbb{F}_p) = 0$.

For any subgroup $K \subseteq G$, such that $|K| = p^r$ and $K \supseteq C$, it follows that $X^K \neq \emptyset$ and $\bar{H}^*(X^K;\mathbb{F}_p) = 0$ by an induction. Since this holds for every cyclic p-subgroup $C \subseteq G$, one has $\bar{H}^*(X^K;\mathbb{F}_p) = 0$ for all subgroups $K \subseteq G$, $K \neq 1$. An inductive argument using Mayer-Vietoris sequences yields the desired conclusion. ∎

We will be particularly interested in the class of G-spaces for which the Serre spectral sequence of their Borel construction collapses. This is formulated as condition (DSBC) (degenerate spectral sequence of Borel construction) below.

<u>CONDITION (DSBC)</u>: Let X be a G-space and let $A \subset G$ be a subgroup. We say that X satisfies the condition (DSBC) for A if the Serre spectral sequence of the fibration $X \to E_A \times_A X \to BA$ (in the Borel construction of the A-space X) collapses.

<u>1.2 Proposition</u>. Let p be a prime divisor of order of G, and suppose that X is a connected G-space such that either X is compact or that $cd_p(X/G) < \infty$. Assume that:

(1) X satisfies condition (DSBC) for each maximal elementary abelian subgroup $A \subseteq G$.

(2) The p-singular set $S_p(X)$ satisfies: $S_p(X) \neq \emptyset$ and $\bar{H}^*(S(X);\mathbb{F}_p)=0$. Then $\bar{H}^*(X;\mathbb{F}_p)$ is cohomologically trivial as an $\mathbb{F}_pG$-module.

<u>Proof</u>: Let A be any p-elementary abelian rank t subgroup, and let $e_A \in H^{2t}(A;\mathbb{F}_p)$ be the product of the t 2-dimensional polynomial generators in $H^2(A;\mathbb{F}_p)$, (cf. [27] Part I). Since $S_p(X)^A = X^A$ and (2) implies that $\bar{H}^*(X^A,x;\mathbb{F}_p) = 0$ (where $x \in X^G \neq \emptyset$ is the base point), it follows that $H^*_A(X,x;\mathbb{F}_p)[e_A^{-1}] = 0$, by the localization theorem ([27] Part I). Since the Serre spectral sequence of $(X,x) \to E_A \times_A (X,x) \to BA$ collapses by (1), we may localize the E_2-term with respect to e_A and conclude that $H^*(BA;H^*(X,x;\mathbb{F}_p))[e_A^{-1}] = 0$. But $H^*(BA;H^*(X,x;\mathbb{F}_p))[e_A^{-1}] \cong \hat{H}^*(A;H^*(X,x;\mathbb{F}_p))$. Since this is true for all p-elementary abelian groups $A \subseteq G$, $|A|=p^r$, it follows that $H^*(X,x;\mathbb{F}_p)$ is cohomologically trivial over all p-elementary abelian subgroups of G. By Chouinard's theorem (cf. [15] and [20]) $H^*(X,x;\mathbb{F}_p)$ is cohomologically trivial over G (see the introduction to section 2). ∎

We obtain a special case of Theorem 2.1 as a corollary:

<u>1.3 Corollary</u>. Suppose that X is a connected G-space with the following properties:

(1) Either X is compact or $cd_p(X/G) < \infty$ for each p dividing order G.

(2) X satisfies condition (DSBC) for each p-elementary abelian subgroup $A \subseteq G$. Then $\bar{H}^*(X)$ is $\mathbb{Z}G$-projective if and only if $\bar{H}^*(X)|\mathbb{Z}C$ is $\mathbb{Z}C$-projective for each subgroup $C \subseteq G$ of prime order. In particular, this conclusion holds if X is a Moore space which satisfies (1).

<u>Proof</u>: By 1.1 and 1.2, the cohomological triviality of $\bar{H}^*(X)$ over G is equivalent to the cohomological triviality of $\bar{H}^*(X;\mathbb{F}_p)$ for all cyclic subgroups of order p. But a $\mathbb{Z}G$-module is $\mathbb{Z}G$-projective if and only if it is $\mathbb{Z}$-free and cohomologically trivial (cf. [28]). ∎

<u>Section 2</u>. <u>The Projectivity Criteria</u>

Let G be a finite group. Sylow(G) denotes the set of Sylow subgroups, and $G_p \in$ Sylow(G) denotes a p-Sylow subgroup. Let R be a ring and RG be the group algebra over R. In studying the cohomological properties of RG-modules, it is necessary to have a good understanding of projective modules. The following two theorems have played important roles in the "local-to-global" arguments.

(1) <u>Rim [28]</u>: A $\mathbb{Z}G$-module is $\mathbb{Z}G$-projective if and only if $M|\mathbb{Z}G_p$ is $\mathbb{Z}G_p$-projective for all $G_p \in$ Sylow(G).

(2) <u>Chouinard [15]</u> (See also Jackowski [20]): A $\mathbb{Z}G$-module M is $\mathbb{Z}G$-projective if and only if $M|\mathbb{Z}E$ is $\mathbb{Z}E$-projective for all p-elementary abelian groups.

Chouinard's theorem is particularly useful in the problems related to cohomological properties of M, since the cohomology of elementary abelian groups are well-understood, whereas the cohomology ring of a general p-group is far more complicated and has remained mysterious as yet.

Thus, the projectivity of a $\mathbb{Z}G$-module M is detected by its restrictions to the elementary abelian subgroups. Now suppose that M is a kE-module, where E is p-elementary of rank n (i.e. of order p^n), and where k is a field of characteristic p. (For simplicity, assume that k is algebraically closed, although for the most part this assumption is not used.)

It is tempting to look for a projectivity criterion for M in terms of a family of proper subgroups of E . In general there is no such criterion if we consider only subgroups of E . However, there is such a characterization if we include a certain family of well-behaved subgroups of kE . This is basically the content of a result due to Dade [17]. To describe this, let I be the augmentation ideal: $0 \to I \to kE \overset{\varepsilon}{\to} k \to 0$ and choose an $\mathbb{F}_p$-basis for E , say $\{e_1,\dots,e_n\} \subset E$. Let $A = (a_{ij})$ be a non-singular $n \times n$ matrix over k and define the homomorphism $\psi_A: kE \to kE$ by:

$$\psi_A(e_i) = 1 + \sum_{j=1}^{n} a_{ji}(e_j-1) .$$

Then ψ_A is an automorphism since A is non-singular. In [11] J. Carlson called subgroups of order p^m in kE , $m \leq n$, generated by $\{\psi_A(e_1),\dots,\psi_A(e_m)\}$, "shifted subgroups" of kE . Such subgroups are p-elementary abelian and for $m = n$, $\{\psi_A(e_1),\dots,\psi_A(e_n)\}$ generate kE as a k-algebra. A cyclic subgroup S of the shifted subgroup $\langle\psi_A(e_1),\dots,\psi_A(e_n)\rangle$ is called a "shifted cyclic subgroup" and any generator of S is called a "shifted unit". From now on we assume that all kE-modules are finite dimensional over k .

(3) Dade [17]: A kE-module M is kE-projective if and only if $M|kS$ is kS-projective for every shifted cyclic subgroup of kE .

(Since kE is a local ring, projective, injective, cohomologically trivial, and free modules coincide [28]). In fact, one can show that $M|kS$ is kS-projective if and only if $M|kS'$ is kS'-projective provided that the shifted units generating S and S' are congruent modulo I^2 . This leads to the following more intrinsic definition of shifted subgroups and units [11] [8]. Let L be an n-dimensional k-subspace of I such that $I = L \oplus I^2$. Then every element $\ell \in L$ satisfies $\ell^p = 0$, and a k-basis of L generates kE as a k-algebra. Consequently, for any $\ell \in L$, $1 + \ell$ is a shifted unit and for any k-basis of L , say $\{\ell_1,\dots,\ell_n\}$, the p-elementary subgroup generated by $\{1+\ell_1,\dots,1+\ell_n\}$ is a shifted subgroup. J. Carlson attached a global invariant to a kE-module M , by taking the set $V_L^r(M)$ consisting of all nonzero $\ell \in L$ for which $M|k\langle 1+\ell\rangle$ is not $k\langle 1+\ell\rangle$-free (where $\langle 1+\ell\rangle$ is the group generated by $1+\ell$) together with zero. He showed that this is an affine algebraic variety and exhibited many beautiful properties of $V_L^r(M)$, called "the rank variety of M " (cf. [11]). Carlson conjectured that $V_L^r(M)$ is isomorphic to the cohomology variety of M , $V_E(M)$ (called the Quillen variety and inspired by Quillen's ideas in [27]), and he showed that $V_L^r(M)$ injects into $V_E(M)$. The Quillen variety $V_E(M)$ is the affine variety in k^n defined by the ideal of elements in the commutative graded ring $H_E \equiv \oplus_i H^{2i}(E;k)$ which annihilate the H_E-module $H^*(E;M)$ ($H_E = \oplus_i H^i(E;k)$ when E is a 2-group. The conjecture of Carlson is proved by Avrunin-Scott [8], and as a corollary $V_L^r(M)$ is independent of L up to isomorphism. Thus the projectivity criterion of Dade which can be detected "locally" by shifted units, has the following "global formulation". From now on we drop the subscript L in $V_L^r(M)$.

(4) Carlson [11]: M is kE-free if and only if $V^r(M) = 0$.

This motivates the search for a projectivity criterion for $\mathbb{Z}G$-modules which appear as (reduced) homology of G-spaces. It turns out that the family of cyclic subgroups of order p of G detects the projectivity (and cohomological triviality). Thus "the geometry of M " is determined by a restricted class of subgroups of G in this case, and gives an idea of how restricted the category of realizable $\mathbb{Z}G$-modules is. This is not true for homology of all G-spaces, rather a special class which includes Moore spaces. The projectivity criterion for the homology of more general G-spaces should be described in terms of "global invariants" attached to a G-space. The specific nature of a G-action on a space X determines a certain interrealationship between $H_i(X)$ and $H_j(X)$ as $\mathbb{Z}G$-modules, and this fact is not detectable by simply considering the graded module $\oplus_i \bar{H}_i(X)$. The examples of the following sections will elaborate more on this point.

2.1 Theorem. Suppose X is a connected G-space which satisfies the condition (DSBC) for each p-elementary abelian subgroup $A \subseteq G$. Let M be the $\mathbb{Z}G$-module determined by the G-action on the total homology of X in positive dimensions. Then M is $\mathbb{Z}G$-projective if and only if $M|\mathbb{Z}C$ is $\mathbb{Z}C$-projective for each subgroup $C \subset G$ of prime order. (Similarly for cchomological triviality).

2.2 Corollary. Suppose the $\mathbb{Z}G$-module M appears as the homology of a Moore G-space. Then M is $\mathbb{Z}G$-projective if and only if M is $\mathbb{Z}C$-projective for each cyclic subgroup of G .

We will give two proofs of the above theorem. The first is in the spirit of transformation group theory and while it is quite elementary it reveals the topological nature of this criterion. The second proof is in a more general setting and hopefully will provide some motivation for introducing and emphasis on the global invariants of a G-space.

2.3 Corollary. Suppose X_1 and X_2 are connected G-spaces, both of which satisfy (DSBC) as in (2.1) and suppose $f : X_1 \to X_2$ is a G-map. Let M_1 and M_2 denote the total reduced homology of X_1 and X_2 as $\mathbb{Z}G$-modules and let $\varphi : M_1 \to M_2$ be the $\mathbb{Z}G$-homomorphism induced by f. Then there are $\mathbb{Z}G$-projective modules P_1 and P_2 such that $M_1 \oplus P_1 \cong M_2 \oplus P_2$ if and only if $\varphi_*:\hat{H}^i(C;M_1) \to \hat{H}^i(C;M_2)$ are isomorphisms for $i = 0,1$, and all cyclic subgroups $C \subseteq G$ of prime order.

Section 3. Varieties associated to a G-space

Let k be an algebraically closed field of characteristic $p>0$, and let G be a p-elementary abelian group of rank n. For a connected G-space X, we will assume $X^G \neq \emptyset$ (when needed) and $x \in X^G$ is the base point. As far as homological invariants of X are concerned at this point, this will be no restriction, since we acn always suspend the action. For a kG-module M, the rank variety $V^r(M)$ reveals much about its cohomological invariants. Thus, we are tempted to consider the rank variety $V^r(\oplus_i H^i(X,x;k)$ and investigate its influence on the topology of the G-space X. However, the more directly related variety, (when we have sufficient knowledge about the G-action) is the "support variety" $V_G(X)$.

In [27], Quillen studied cohomological varieties arising from equivariant cohomology rings $H^*_G(X;k)$ for a G-space X (cohomology with constant coefficients), and he proved his celebrated stratification theorem among other results. According to Quillen's stratification theorem, the cohomological variety of a G-space X for a general finite group G has a piecewise description in terms of varieties arising from elementary abelian subgroups of G. Inspired by this work of Quillen, Avrunin-Scott in [8] defined the cohomological variety $V_G(M)$ for a finitely generated kG-module M and proved an anlogous stratification theorem for $V_G(M)$ in terms of elementary abelian subgroups of G. Here, $V_G(M)$ is the largest support (in Max H_G) of the H_G-module $H^*(G,N\otimes M)$ where N ranges over all finitely generated kG-modules. Avrunin-Scott's stratification theorem may be regarded as generalizing the special case of Quillen's result for the G-space X=point to the equivariant cohomology with local coefficients H_G^* (point;M) (the kG-module M replacing the constant coefficients k of Quillen). The stratification of support varieties in the case of equivariant cohomology with local coeeficients $H_G^*(X;M)$ for a G-space X (whose orbit space X/G has finite cohomological dimension over k) is carried out by Stefan Jackowski in [21] under the extra hypothesis that M is a kG-algebra. Jackowski's theorem yields a topological proof of Avrunin-Scott theorem in the spirit of Quillen's original approach.

Such stratification theorems describe the above mentioned cohomological varieties of a general finite group G in terms of elementary abelian subgroups of G . When G is an elementary abelian group, $V_G(X)$ is the affine algebraic variety defined by the annihilator ideal in H_G of $H_G^*(X,x;k)$. For the rest of this section, we will assume that G is an elementary abelian group. The corresponding results and notions for the case of a general finite group is obtained from this basic case and the appropriate stratification theorem. Elaboration of these ideas will appear elsewhere.

While one hopes that $V_G(X) \cong V_G^r(\oplus_i H^i(X,x))$, this turns out to be true only for a restricted, but nevertheless important class of G-spaces. For a G-space with $H^i(X) \neq 0$ for only finitely many i (and some mildly more general class), it turns out that one can define a different, (but related) rank variety in a natural way. This is done by associating to X a $\mathbb{Z}G$-module defined up to a suitable stable equivalence. The $V_G^r(X)$ is defined to be the rank variety of this module (tensored with k). The isomorphism $V_G(X) = V_G^r(X)$ will show that the "cohomological support variety" is also a "rank variety" and as such, it will enjoy the properties of rank varieties.

Following [5], call two G-spaces X_1 and X_2 "freely equivalent", if there exists a G-space Y such that $X_i \subset Y$, and $Y-X_i$ are free G-spaces with $cd_p(Y-X_i) <\infty$ for $i = 1,2$. This defines an equivalence relation between G-spaces. We may also consider the case when Y/X_i is compact if $cd(y-X_i)=\infty$ with appropriate modifications.

<u>3.1 Lemma</u>. Suppose X_1 and X_2 are freely equivalent. Then $V_G(X_1) \cong V_G(X_2)$.

Proof: Compare the Leray spectral sequences for $E_G \times_G X_i \to X_i/G$ with $E_G \times_G Y \to Y/G$ where X_i and Y are as above, $Y-X_i$ = free G-space [27]. It follows that $V_G(X_i) \cong V_G(Y)$. ■

3.2 Proposition. Suppose $H^i(X;k) \neq 0$ for only finitely many i . Then $V_G(X) \subset V_G^r(\oplus_i H^i(X,x;k))$. If X satisfies the condition (DSBC) for G, then $V_G(X) \cong V_G^r(\oplus_i H^i(X,x;k))$.

Proof: Proceed by induction on $\nu(X) \overset{\text{def}}{\equiv} \text{number}\ \{i | H^i(X,x;k) \neq 0\}$. For $\nu(X) = 1$, X is a Moore space and the spectral sequence of $(X,x) \to E_G \times_G (X,x) \to BG$ degnerates to one line, which shows that $V_G(X) \cong V_G(\oplus_j H^j(X,x;k)$ ($\equiv$ its support variety). By Avrunin-Scott's proof of J. Carlson's conjecture [8], the latter is isomorphic to $V_G^r(\oplus_j H^j(X,x;k))$. Suppose the assertion is true whenever $\nu(X) < m$, $m > 1$. Given X_1 with $\nu(X_1) = m$, we add free G-cells to X_1 to obatin the G-space Y so that $Y-X$ is free, $\dim(Y-X) < \infty$, and $\nu(Y) < m$. For example, kill the first non-vanishing homology, say $H_\ell(X,x;k)$, using Serre's version of the Hurewicz theorem, (after suspending X , if needed). Then $V_G(X) \cong V_G(Y)$ since X and Y are freely equivalent (Lemma 3.1) and $V_G(Y) \cong V_G^r(\oplus_j H^j(Y,x;k))$ by induction. On the other hand, $V_G^r(Y) \subseteq V_G^r(X)$. This follows again because $(Y/X)^G$ = point and $\dim(Y/X) < \infty$. Alternatively, if we kill $H_\ell(X,x;k)$ (the first non-vanishing) to obtain Y , we have the exact sequence:

$$0 \to H_{\ell+1}(X;k) \to H_{\ell+1}(Y;k) \to F \to H_\ell(X;k) \to 0$$

where F is a free kG-module, and

$$H_i(X;k) \cong H_i(Y;k) \quad \text{for} \quad i > \ell+1 .$$

For every shifted cyclic subgroup S of kG for which $H^i(X,x;k)|kS$ is kS-free, $H^i(Y,x;k)|kS$ will also be kS-free by Schanuel's lemma. Hence $V_G^r(\oplus_j H^j(Y,x;k)) \subset V_G^r(\oplus_i H^i(X,x;k))$ as desired.

If X satisfies the condition (DSBC) for G , then in the Serre spectral sequence of $X \to E_G \times_G X \to BG$, $E_2^{p,q} = E_\infty^{p,q}$. Thus $\text{rad}(\text{Ann}\ H_G^*(X,x;k)) \cong \text{rad}(\text{Ann}\ H^*(G,H^*(X,x;k)))$ by a simple calculation and a filtration argument. Since $\text{rad}(\text{Ann}\ H^*(G,H^*(X,x;k))) \cong \bigcap_i \text{rad}(\text{Ann}\ H^*(G;H^i(X,x;k)))$, it follows that

$$V_G(X) \cong V_G(\oplus_i H^i(X,x;k)) \cong \bigcup_i V_G(H^i(X,x;k)) \cong \bigcup_i V_G^r(H^i(X,x;k)) \cong$$

$V_G^r(\oplus_i H^i(X,x;k))$

(where the isomorphism between V_G and V_G^r of $H^i(X,x;k)$ is due Avrunin-Scott's theorem again). ∎

The second assertion of 3.2 is not true in general. The examples in the following sections illustrate this point.

The above observations lead us to define a kG-module $M(X)$ for each G-space X with $H^i(X;k) \neq 0$ for only finitely many i, such that $V_G(X) \cong V_G^r(M(X))$. Since for Moore spaces X, $V_G(X) \cong V_G^r(H^*(X,x;k))$, we embed X in a "mod k" Moore G-space Y freely equivalent to it. This is possible since $H^i(X;k) = 0$ for large i and we can add free G-cells inductively using Serre's Hurewicz theorem. Let $M(X) \equiv H_*(Y,x;k)$. Although $M(X)$ is not well-defined, $H^*(G;M(X)^*)$ and $H_G^*(X,x;k)$ are isomorphic modulo H_G-torsion. Hence $V_G(X) \cong V_G(M(X)^*) \cong V_G^r(M(X)^*) \cong V_G^r(M(X))$ and $V_G(X)$ has a description as a rank variety.

The module $M(X)$ is well-defined only in a "stable sense". For a kG-module L, define $\omega^0(L) \cong L$, and $\omega^1(L) \equiv \omega(L)$ by the exact sequence $0 \to \omega(L) \to F \to L \to 0$, where F is kG-free, and $\omega^{i+1}(L) \equiv \omega(\omega^1(L))$. These modules are stably well-defined by Schanuel's lemma (cf. e.g. Swan's Springer-Verlag LNM 76).

<u>3.3 Proposition</u>. Suppose X is a G-space such that $H^i(X;k) \neq 0$ for finitely many i. Let Y_1 and Y_2 be two mod k Moore G-spaces freely equivalent to X. Then there are integers s and $t \geq 0$, such that $\omega^s(H^*(Y_1,x;k))$ is stably isomorphic to $\omega^t(H^*(Y_2,x;k))$. (Call this ω-stability for short.)

<u>Proof</u>: Choose a G-space Z freely equivalent to Y_1 and Y_2 and containing Y_1 and Y_2, and such that $H_i(Z,x;k) = 0$ for $i \neq \ell$, $\ell \gg$ nonzero dimensions in $H^*(Y_j;k)$ for $j = 1,2$. Then $C_*(Z/Y_i;k)$ are free kG-modules except for $* = 0$, where the base point naturally defines a split augmentation $C_0(Z/Y_i;k) \underset{\longleftarrow}{\overset{\varepsilon_i}{\longrightarrow}} k \to 0$. $C_*(Z/Y_i;k)$ has homology (mod k) nonzero only in two dimensions above 0, corresponding to $H_\ell(Z;k)$ and $H_*(Y_i,x;k)$. An appropriate application of the Schanuel's lemma shows that $\omega^t(H_*(Y_1,x;k)) \cong H_\ell(z;k) \cong \omega^s(H_*(Y_2,x;k))$ for some integers $t,s \geq 0$. ∎

<u>3.4 Corollary</u>.Given a G-space X with $H^i(X;k) = 0$ for sufficiently

large i, there exists a kG-module $M(X)$ which is well-defined up to ω-stability and $V_G(X) \cong V_G^r(M(X))$. □

The ω-stable class of $M(X)$ is in fact a "composite extension" of various $\omega^{s_i}(H_i(X;k))$ for all $i > 0$ and appropriate integers $s_i \geq 0$. This means that if $0 < i(1) < i(2) < \ldots < i(m)$ are the dimensions where $H_i(X;k) \neq 0$, then there are integers $s(1),\ldots,s(m)$ and extensions:

$0 \to H_{i(j+1)}(X;k) \to L_{i(j+1)} \to \omega^{s(j)}(L_{i(j)}) \to 0$ for $j = 1,\ldots,m$, and where $L_{i(1)} \equiv H_{i(1)}(X;k)$ and $M(X) \cong \omega^t L_{i(m)}$ for some $t \geq 0$. Let us refer to this construction as "an ω-composite extension".

We have the following formal corollary:

3.5 Corollary. Suppose that $H^i(X;k) = 0$ for all sufficiently large i, and suppose X has a homotopy G-action $\alpha : G \to E(X)$. Then (X,α) is equivalent to a topological G-action only if some ω-composite extension L of the kG-modules $H_i(X;k)$ (as given by α) is realizable by a mod k Moore G-space. ■

While this corollary seems to be a formal consequence of definitions, it does lead to the following theorem which will be proved in section 5.

3.6 Theorem. There exist decomposable kG-modules M which are realizable by homotopy G-actions, but they are not realizable by the homology of any G-space X.

Next, we apply the above results to give a proof of Theorem 2.1.

Proof of Theorem 2.1: Let $M = \bigoplus_{i>0} H_i(X)$. Then, if M is $\mathbb{Z}G$-projective, clearly M is $\mathbb{Z}C$-projective for any subgroup, in particular cyclic subgroups of G. Conversely, suppose M is $\mathbb{Z}C$-projective for all such $C \subseteq G$ as in the theorem. Let $M' = \bigoplus_{i>0} H_i(X;k)$. By Chouinard's theorem, it suffices to consider the case where G is p-elementary abelian, and we will assume this for the sequel. Since X satisfies the condition (DSBC) for G, one has $V_G(X) \cong V_G^r(M')$, by Proposition 3.2. At this point one has several (basically equivalent) ways of finishing the proof. The first is somewhat longer, but more illuminating, and we will refer to it in the applications.

First argument: $V_G(X)$ is defined via the radical of the annihilator of $H_G^*(X,x;k)$, say J, in H_G, which is the intersection of associated prime ideals $\mathrm{Ann}_{H_G}(\alpha)$, for $\alpha \in H_G^*(X,x;k)$. Since associated primes are closed under the Steenrod algebra, a theorem of Landweber [24] and [25] (generalizing a theorem of Serre [29]; see also [1]) shows that they are generated by two dimensional classes in $\bigoplus_{i>0} H^{2i}(G;\mathbb{F}_p) \subset H_G$. Landweber's proof is for $\mathbb{F}_p$-coefficients throughout, but one can easily check that his arguments goes through with k-coefficients and the same conclusion. (The invariance of associated primes under the Steenrod algebra has been observed by several authors [25] [31] [18]). Thus J is defined by linear equations with $\mathbb{F}_p$-coefficients. Consequently $V_G(X)$ as well as $V_G^r(M')$ are $\mathbb{F}_p$-rational, (i.e., a union of subvarieties defined by linear equations with $\mathbb{F}_p$-coefficients). For a shifted cyclic subgroup $S \subset kG$, $V_S^r(M'|kS) \cong V_G^r(M') \cap \mathrm{tr}_{S,G}(V_S^r(k))$ (cf. [8]) where $\mathrm{tr}_{S,G}$ is the transfer. It follows that for each shifted cyclic subgroup which is not a subgroup of G, $S \cap G = \{1\}$ and $\mathrm{tr}_{G,S}(V_G^r(k)) \cap V_G^r(M') = 0$. (Here we assume to have chosen a k-vector space L such that $I = L \oplus I^2$, I = augmentation ideal, as described in Section 2.) Hence $V_G^r(M')$ is detected by the shifted cyclic subgroups S such that $S \cap \{G\} \neq \{1\}$, i.e. cyclic subgroups of G. By the hypothesis, $M'|kS$ is kS-free for all such $S \subset G$. Thus, $V_G^r(M') = 0$ and M' is kG-free. Since $H^i(X,x)$ is $\mathbb{Z}C$-projective, it is $\mathbb{Z}$-free. The long exact sequence of cohomology associated to $0 \to \mathbb{Z} \to \mathbb{Z} \to \mathbb{F}_p \to 0$ breaks into short exact sequences:

$$0 \to H^i(X;\mathbb{Z}) \xrightarrow{xp} H^i(X;\mathbb{Z}) \to H^i(X;\mathbb{F}_p) \to 0 .$$

But for all $A \subseteq G$, $\hat{H}^*(A;H^*(X,x;\mathbb{F}_p)) = 0$ ($\hat{H}^*$ = Tate cohomology and kG-projectivity implies $\mathbb{F}_pG$-cohomological triviality [14]). Hence $\hat{H}^*(A,H^*(X,x))$ is p-divisible, which means that it vanishes for all $A \subseteq G$. Therefore $H^*(X,x)$ is $\mathbb{Z}G$-projective, being $\mathbb{Z}$-free and $\mathbb{Z}$-cohomologically trivial [28].

Second argument: An inductive argument using Cartan's formula shows that the annihilating ideal of $H_G^*(X,x;k)$ is invariant under the Steenrod algebra, as in G. Carlsson [13]. A theorem of Serre [29] then shows that the variety $V_G(X)$ is $\mathbb{F}_p$-rational. Hence $V_G^r(M')$ is rational using Proposition 3.2. The rest of the proof is as in the first argument and the details are left to the reader. ∎

3.7 Addendum. The examination of the proof shows that in fact the statement of Theorem 2.1 remains valid, if we replace $\mathbb{Z}$-coefficients by k-coefficients as well as $\mathbb{Z}G$- and $\mathbb{Z}C$-projective by kG- and kC-free respectively. Thus one needs that $H^i(X;k) = 0$ for all sufficiently large i, instead of the stronger statement with $\mathbb{Z}$-coefficients. ∎

The above proof also suggests that as in J. Carlson [12], one can determine the complexity of $H^*(X,x;k)$ by the dimension of the variety $V_G^r(\oplus_i H^i(X,x;k)) = V_G(X)$ for this particular case. This is the counterpart of Theorem 2.1 for non-projective modules.

Let p be a fixed prime and let k be a field of characteristic p, say algebraically closed for convenience sake. We denote by $cx_G(M)$ the complexity of the kG-module M (cf. [2] [23] [12]).

3.8 Theorem. Let X be a connected G-space which satisfies the condition (DSBC) for each maximal elementary abelian p-subgroup $A \subseteq G$ and $H^*(-;k)$. Let $M \equiv \bigoplus_{i>0} H_i(X;k)$ with the induced kG-module structure. Suppose $cx_G(M) = r$. Then there exists a p-elementary abelian subgroup $E \subseteq G$ of rank r such that $cx_E(M|kE) = r$.

Proof: By Alperin-Evens [2], $cx_G(M) = \max_A\{cx_A(M|kA) \mid A \subseteq G$ maximal p-elementary abelian$\}$. Thus we may assume that G is elementary abelian. Since $V_G^r(M) \cong V_G(X)$ is rational as in the proof of Theorem 2.1 above, $\dim V_G^r(M)$ is the maximum dimension of the rational linear subvarieties whose union is $V_G^r(M)$. Let V_0 be one such linear maximum dimensional subspace of $k^n \cong V_G^r(k)$, (where we assumed $n = \text{rank } G$) and let $E = G \cap V_0$ be the set of rational points of V_0. Then rank $E = \dim V_0$ since V_0 is rational. On the other hand, $tr_{E,G}(V_E^r(M|kE)) \cong V_0$ (cf. [8] and [11] for details) and $cx_E(M|kE) = \dim V_0 = \text{rank } E$. ∎

Let G be a p-elementary abelian group of rank n. In [23], Ove Kroll proves that if $cx_G(M) = t$ for a kG-module M, then there exists a shifted subgroup $\Gamma \subset kG$ of rank $n-t$ such that $M|k\Gamma$ is $k\Gamma$-free. J. Carlson's proof of Kroll's theorem [12] is in essence a "transversality argument" in the following sense. Since $cx_G(M) = t$, $\dim V_G^r(M) = t$, and it is always possible to find an $(n-t)$-dimensional linear subspace L of $k^n \cong V_G^r(k)$ which is in "transverse position" to $V_G^r(M)$, (i.e. it has intersection $\{0\}$.) Now restriction to the shifted subgroup Γ which is obtained from any k-basis of L yields $\dim V^r(M|k\Gamma) = \dim(L \cap V_G^r(M)) = 0$, which means that $M|k\Gamma$ is $k\Gamma$-free.

When $V_G^r(M)$ is rational, one would like to find a subgroup $\Gamma \subseteq G$ with the above property. But this is not possible in general as it can be seen from the following simple example:

<u>3.9 Example</u>. Let $M = \oplus_E (kG \otimes_{kE} k)$ where E runs over all cyclic subgroups of G. Then $cx_G(M) = 1$ and $M|kA$ is not kA-free for any non-trivial subgroup $A \subseteq G$.

However, the first argument of the proof of Theorem 2.1 above reveals that we can give a counterpart to Kroll's theorem in a particular case.

Call a G-space X "k-primary", if the radical of the annihilator ideal of $H_E^*(X,x;k)$ in H_E is prime for all maximal p-elementary abelian subgroups of G. (Here k is a field of characteristic p again.) Recall p-rank $(G) \overset{def}{\equiv} \max$ {rank of elementary abelian p-subgroup $E \subseteq G$}.

<u>3.10 Theorem</u>. Suppose p-rank $(G) = n$ and X is a connected k-primary G-space which satisfies the condition (DSBC) for all maximal p-elementary abelian subgroups and $H^*(-;k)$-coefficients. Also, assume that $H^i(X;k) = 0$ for all sufficiently large i. Then there exists a p-elementary abelian subgroup $E \subseteq G$ such that rank $E = n - \max_i \{cx_G(H_i(X,x;k))\}$ and $H_i(X,x;k)$ is kE-free for all i.

<u>Proof</u>: As in the preceding theorems, it suffices to assume that G is p-elementary abelian (Alperin-Evens [2]). By Proposition 3.2 $V_G^r(\oplus_i H^i(X,x;k)) \cong V_G(X)$. Since X is k-primary, the first argument in the proof of Theorem 2.1 shows that $V_G^r(\oplus_i H_i(X,x;k) \cong V_G^r(\oplus_i H^i(X,x;k))$ consists of one rational linear subvariety of $k^n \cong V_G^r(k)$, and its dimension equals to $cx_G(\oplus H_i(X,x;k)) = \max_{i>0} cx_G(H_i(X;k))$. Hence there is a rational linear subspace L transverse to $V_G^r(H_i(X;k))$, and we may choose $\dim L = n - \max_{i>0} cx_G(H_i(X;k))$. Let E be the subgroup of G whose $\mathbb{F}_p$-generators gives an $\mathbb{F}_p$-basis for L. This is the desired subgroup. ∎

<u>3.11 Remark</u>. One can modify the above argument to weaken the hypothesis that "X is k-primary" or that "X satisfies (DSBC)", etc. But these hypotheses cannot be removed altogether by the above example 3.9 and the example in Sections 4 and 5.

Section 4. Applications to Steenrod's problem

In this section we consider the special case of G-actions of Moore spaces. Suppose M is a finitely generated $\mathbb{Z}$-free $\mathbb{Z}G$-module. Then M is determined by a homomorphism $\rho : G \to GL(n,\mathbb{Z})$, where $n = \mathrm{rank}_{\mathbb{Z}}(M)$. Suppose that X is homotopy equivalent to a bouquet of spheres of dimension $k \geq 2$, and $H_k(X) \cong \mathbb{Z}^n$. Then $E(X) \equiv \pi_0 H(X) \cong GL(n,\mathbb{Z})$ by obstruction theory. Thus ρ induces a homomorphism $\alpha : G \to E(X)$ such that the homotopy action (X,α) realizes the $\mathbb{Z}G$-module M . More generally, if $\mathrm{Tor}_1^{\mathbb{Z}}(M,\mathbb{Z}_2) = 0$, or if G is of odd order, then an obstruction theory argument (cf. [22]) shows that any homomorphism $\rho : G \to GL(n,\mathbb{Z})$ (which induces the $\mathbb{Z}G$-module structure of M) can be lifted to a homomorphism $\alpha : G \to E(X)$. Thus the homotopy action (X,α) realizes M .

On the other hand, given M , we have the $\mathbb{Z}$-free $\mathbb{Z}G$-module M' from the exact sequence $0 \to M' \to F \to M \to 0$, where F is a free $\mathbb{Z}G$-module. It is not difficult to see that M is realizable by a Moore G-space, if and only if M' is realizable by a Moore G-space. Thus, as far as the question of realizability of $\mathbb{Z}G$-modules is concerned, one can consider $\mathbb{Z}$-free $\mathbb{Z}G$-modules with no loss of generality. Therefore, the realizability of modules by homotopy actions does not pose a difficult problem in the contexts where one is primarily interested in realizability by topological G-actions.

In passing, let us mention that the obstructions for realizability of a $\mathbb{Z}G$-module by a homotopy action on a Moore space has been studied by P. Vogel [7] (unpublished). Vogel has shown that for $G = \mathbb{Z}_2 \times \mathbb{Z}_2$, there is an $\mathbb{F}_2[G]$-module which is not realizable by a homotopy action on a Moore space:

<u>4.1 Example (P. Vogel) [7]</u>. Regard $\mathbb{Z}_2 \times \mathbb{Z}_2$ as the 2-Sylow subgroup of $GL(2,\mathbb{F}_4)$, i.e. as 2×2 upper triangular matrices of the form $\begin{pmatrix} 1 & x \\ 0 & 1 \end{pmatrix}$ where x belongs to the field with 4 elements. The natural action of $GL(2,\mathbb{F}_4)$ by left multiplication on the column vectors of $M = (\mathbb{F}_4)^2$ makes M into a $\mathbb{Z}[\mathbb{Z}_2 \times \mathbb{Z}_2]$-module. Vogel's obstruction theory shows that this modules is not realizable by a homotopy action of $\mathbb{Z}_2 \times \mathbb{Z}_2$ on a Moore space.

<u>4.2 Construction and Examples</u>. Let k be an algebraic closure of $\mathbb{F}_p$, and let $G = \mathbb{Z}_p \times \mathbb{Z}_p$ be generated by e_1 and e_2 . Let I be the augmentation ideal and choose the k-vector space L such that $I = L \oplus I^2$,

with $\{\ell_1,\ell_2\}$ a k-basis for L, (as in Section 2). Then for almost all choices of $\alpha = (\alpha_1,\alpha_2) \in k^2$, the shifted unit $u_\alpha = 1+\alpha_1\ell_1+\alpha_2\ell_2$ generates a shifted subgroup $S \equiv \langle u_\alpha \rangle$ of order p such that $S \cap G = \{1\}$. (Cf. Carlson [11] for details on shifted subgroups). More explicitly, for a (finite) Galois extension K of $\mathbb{F}_p$, choose $\alpha_1,\alpha_2 \in K$ such that $u_\alpha = 1+\alpha_1(e_1-1) + \alpha_2(e_2-1)$ satisfies $u_\alpha - 1 \notin I^2$ and a $u_\alpha \neq g \pmod{I^2}$ for any $g \in G$. The condition $1-u_\alpha \notin I^2$ ensures that kG is kS-free, and $S \equiv \langle u_\alpha \rangle \subset kG$ can be treated like an ordinary subgroup as far as induction and restriction is concerned [11]. In particular, Mackey's formula and Shapiro's Lemma are valid.

Recall that for the local ring kG, projective, injective, cohomologically trivial, and free modules coincide. First we need the following:

<u>4.3 Lemma</u>. (i) There exists an indecomposable kG-module M_0 such that M_0 is kC-projective for all cyclic subgroups $C \subset G$, but M_0 is not kG-projective.

(ii) There exists a finitely generated $\mathbb{Z}$-free $\mathbb{Z}G$-module M_1 which is $\mathbb{Z}C$-projective for all cyclic subgroups $C \subset G$, but M_1 is not $\mathbb{Z}G$-projective.

(iii) There exists an indecomposable $\mathbb{Z}G$-module M with the same properties as in (ii) above.

(iv) In above part (iii), one may choose M such that $k \otimes M \cong M' \oplus Q$, where M' is an indecomposable kG-module, and Q is kG-free.

<u>Proof</u>: (i) The above discussion, for (almost all) u_α chosen with $S = \langle u_\alpha \rangle$, one has $S \cap G = \{1\}$ and kG is a free kS-module. Let $M_0 = kG \otimes_{kS} k$ be the induced module. Then for each $C \subset G$, $|C| = p$, $C \cap S = \{1\}$. Hence $\hat{H}^*(C,M_0) = 0$ by Mackey's formula. But $\hat{H}(G,M_0) \cong \hat{H}^*(S;k) \neq 0$ by Shapiro's Lemma. Since kG is local, a cohomologically trivial kG-module is kG-free (= kG-projective). Thus (i) is proved.

(ii) One can choose u_α such that $u_\alpha = 1+\alpha_1(e_1-1) + \alpha_2(e_2-1)$, where α_1 and α_2 lie in a finite Galois extension of $\mathbb{F}_p$, say k_1, and $\langle u_\alpha \rangle = S$ still satisfies the same properties as in (i). Let $M_0 = k_1G \otimes_{k_1S} k_1$ be the k_1G-module which is k_1C-free for each $C \neq G$ but not k_1G-free as in (i). Consider the exact sequence $0 \to M_1 \to (\mathbb{Z}G)^t \to M_0 \to 0$. The long exact sequence of cohomology

$$\cdots\cdots \to \hat{H}^i(C,M_1) \to \hat{H}^i(C,(\mathbb{Z}G)^t) \to \hat{H}^i(C,M_0) \to \cdots\cdots$$

shows that M_1 is $\mathbb{Z}C$-projective for all $C \subset G$ $C \neq G$, and M_1 is not $\mathbb{Z}G$-projective.

(iii) Let $M_1 = M_1^1 \oplus \ldots\ldots \oplus M_1^r$ be a decomposition in terms of indecomposable $\mathbb{Z}G$-modules. Then all M_1^j are $\mathbb{Z}C$-projective, but at least one of them is not $\mathbb{Z}G$-projective, say M_1^1. Then M_1^1 satisfies (ii) and it is also indecomposable.

(iv) Tensor the exact sequence of (ii) by k:

$$0 \to k \otimes M_1 \to (kG)^t \to k \otimes M_0 \to 0 .$$

Note that we can choose M_0 so that $k \otimes M_0$ is indecomposable. (Briefly: $\dim_k k \otimes M_0 = \dim_k KG/kS = [G:S] = p$, and since $k \otimes M_0 | kC$ is projective, the dimension over k of each kC-indecomposable summand, and hence each kG-indecomposable summand must be divisible by p.) In the short sequence:

$$0 \to M' \to P \to k \otimes M_0 \to 0$$

where P is the projective cover of $k \otimes M_0$, M' is also indecomposable, since $k \otimes M_0$ is indecomposable. Hence Schanuel's Lemma shows that $k \otimes M_1 \cong M' \oplus$ (projective). ■

<u>4.4 Theorem</u>. Suppose G is a finite group such that $G \supset \mathbb{Z}_p \times \mathbb{Z}_p$. Then:

(I) there exists a kG-module M_0' which satisfies (i) of Lemma 4.3.

(II) There exists a $\mathbb{Z}G$-module M' which satisfies (iv) of Lemma 4.3. Further, it is not possible to find a Moore G-space X such that $\bar{H}_*(X;k) \cong M_0'$ as kG-modules.

Similarly, there does not exist a Moore G-space X such that $\bar{H}_*(X;\mathbb{Z}) \cong M'$ as $\mathbb{Z}G$-modules.

<u>Proof</u>: Let M_0 be the $k[\mathbb{Z}_p \times \mathbb{Z}_p]$-module of Lemma 4.3(i). Let $M_0' \equiv kG\otimes_{k[\mathbb{Z}p\times\mathbb{Z}p]} M_0$. Since $S \cap C = \{1\}$, Mackey's formula shows that for each $C \subset G$, $|C| =$ prime, M_0'/kC is kC-cohomologically trivial, hence kC-free. But M_0' is not kG-free since it is not $k[\mathbb{Z}_p \times \mathbb{Z}_p]$-free, as $M_0' | \mathbb{Z}_p \times \mathbb{Z}_p$ has M_0 as a direct summand, (or apply Shapiro's lemma).

(II) Let M be as in Lemma 4.3 (iv), and let $M' = \mathbb{Z}G\otimes_{\mathbb{Z}[\mathbb{Z}p\times\mathbb{Z}p]} M$. The assertion follows as in part (I). Now the non-existence of the Moore G-spaces realizing these G-modules is a consequence of the pro-

jectivity criterion Theorem 2.1. ∎
(Compare 4.4 with G. Carlsson's theorem [13].)

The case $G = Q_{2^n}$ = generalized quaternionic group of order 2^n is somewhat different, because the maximal elementary abelian subgroup of Q_{2^n} is the subgroup of order two generated by the central element $\tau \in Q_{2^n}$. Therefore kG-projectivity (or $\mathbb{Z}G$-projectivity) of a module is completely decided by the restriction to $k\langle\tau\rangle$ or $\mathbb{Z}\langle\tau\rangle$. Therefore Theorem 2.1 does not help directly in this situation. In the sequel, we present first a proof of non-realizability of a kG-module by Moore G-space (similarly for a $\mathbb{Z}G$-module) in the finite dimensional case, and we will use the geometric intuition of this case to remove the finite dimensionality restriction with a different proof.

<u>4.5 Proposition</u>. Let G be the quaternionic group of order 2^n, $n \geq 3$. Then there exists a $\mathbb{Z}G$-module M such that M is not $\mathbb{Z}G$-isomorphic to the (reduced) homology of a finite dimensional Moore G-space X. Similarly, $k \otimes M$ is not $\mathbb{Z}G$-isomorphic to $\bar{H}_*(X;k)$.

<u>Proof</u>: Let $\tau \in G$ be the central element of order 2 and let τ generate $T \cong \mathbb{Z}_2 \subset G$. Then $G/T \cong D_{2^{n-1}}$, the dihedral group of order 2^{n-1}. Let M be the module over $\mathbb{Z}_2 \times \mathbb{Z}_2$ constructed as in Lemma 4.3 (iv) above and let $N = \mathbb{Z}[D_{2^{n-1}}] \otimes_{\mathbb{Z}[\mathbb{Z}_2 \times \mathbb{Z}_2]} M$. Then M is not $\mathbb{Z}[D_{2^{n-1}}]$-isomorphic to $\bar{H}_*(X_0)$ for any Moore G-space X_0. In fact, $k_1 \otimes M$ is not $k_1[D_{2^{n-1}}]$-isomorphic to $\bar{H}_*(X_0;k_1)$ for any field k_1 of characteristic 2.

Consider N as a $\mathbb{Z}Q_{2^n}$-module, where T acts trivially on N. (To get a G-module on which all elements of G act non-trivially take $\mathbb{Z}G \oplus N$, or the group of n-cocycles in a minimal projective resolution of N over $\mathbb{Z}G$.). Suppose there exists a finite dimensional Moore G-space Y such that, $Y^G \neq \emptyset$ and $H_*(Y) \cong N$ as $\mathbb{Z}G$-module. Then Y^T is a $D_{2^{n-1}}$-space of finite dimension, and since the Serre spectral sequence of $Y \to E_T \times_T Y \to BT$ collapses, $\bar{H}^*(Y^T;k_1) \cong \hat{H}^*(T;N) \otimes_{\hat{H}^*(T;k_1)} k_1$ as in the proof of Proposition 1.1. Using this periodicity of $H^*(T;N)$, it follows that $\bar{H}^*(Y^T;k_1) \cong (N \otimes k_1)^\tau/(1+\tau)(N \otimes k_1) \cong N \otimes k_1$. But this means that $N \otimes k_1$ is realized by the $D_{2^{n-1}}$-space Y^T, i.e. $\bar{H}^*(Y^T;k_1) \cong N \otimes k_1$. By Proposition 1.2 or Theorem 4.4 this cannot happen. ∎

Alternatively, the ω-stable module $M(Y^T)$ up to ω-stability is k_1G-isomorphic to $N \oplus Q$ where N is the indecomposable factor and Q is kG-free. This is the case because $\bar{H}_*(Y^T;k_1)$ has only one decomposable k_1G-module N as a summand which is not k_1G-free. Thus the construction $M(X)$ and the definition of ω-composite extensions shows that any ω-composite extension of various $\bar{H}_i(Y^T;k_1)$ is of the form $N \oplus Q$ up to ω-stability. Now the Projectivity criterion Theorem 2.1 of Theorem 4.4 shows that $N \oplus Q$ of $\omega^j(N) \oplus Q$ cannot occur as $H_*(L;k_1)$ for any Moore $D_{2^{n-1}}$-space L. This contradiction shows that such a Moore G-space cannot exist.

The proof of the above implies the finite dimensional case of the following corollary. (The details are left to the reader).

<u>4.6 Corollary</u>. If $G \supseteq Q_{2^n}$, then there are $\mathbb{Z}G$-modules which are not $\mathbb{Z}G$-isomorphic to the reduced homology of a Moore G-space. □

Now we proceed to give a different proof which shows that such Moore G-spaces cannot exist regardless of their dimensions.

Since every quaternionic 2-group contains the quaternionic group Q_8 of order 8, we will prove the theorem for Q_8 and deduce the result for Q_{2^n}, $n \geq 3$ from it. Suppose that X is any Moore G-space, where $G = Q_8$, such that $H^*(X,x) \cong M$, $(x \in X^G \neq \emptyset)$. Let M be a $\mathbb{Z}Q_8$-module which is $\mathbb{Z}$-free, and $T \equiv \langle\tau\rangle \subset Q_8$ acts trivially on M, and let $A = Q_8/T \cong \mathbb{Z}_2 \times \mathbb{Z}_2$ induce a $\mathbb{Z}A$-module structure on M. Consider the Borel construction $(W,W_0) = E_G \times_T (X,x)$ which carries a free A-action. The Serre spectral sequence $(X,x) \to (W,W_0) \to BT$ collapses and $H^*(W,W_0;k_1) \cong H^*(T,M \otimes k_1)$. Denote $M \otimes k_1$ by M_1. Since T acts trivially on M_1, it follows that $H^*(T,M_1) \cong H^*(T,k_1) \otimes M \cong H^*(W,W_0,k_1)$.

Now consider the Borel construction $E_A \times_A (W,W_0) \to BA$. In the spectral sequence of this fibration, $E_2^{p,0} = 0$ for all p and $E_2^{p,1} \cong H^p(A;H^1(W,W_0)) \cong H^p(A;M_1)$. On the other hand, $E_A \times_A (W,W_0) \cong (W/A, W_0/A)$ since A acts freely, and $(W/A,W_0/A) = E_G \times_G (X,x)$. Hence $E_2^{1,1} \cong E_\infty^{1,1} \cong H_G^1(X,x;k_1) \cong H^1(G;M_1)$. The H_A-module structure of $E_G \times_G (X,x)$ is also related to the H_G-structure by the following commutative diagram:

$$\begin{array}{ccc}
E_G \times_G (X,x) & \longleftrightarrow & E \times_A (W,W_0) \\
\downarrow & & \downarrow \\
BG & \longrightarrow & BA\,.
\end{array}$$

At this point, let $M_1 \equiv k_1A \otimes_{k_1S} k_1$, and note that $H^*(A;M_1) \cong H^*(S;k_1) \cong k_1[g_\alpha]$ for $g_\alpha \in H^1(S;k_1)$. Let the corresponding generated be denoted by $\gamma \in H^1(A;M_1)$. Then $\operatorname{rad}(\operatorname{Ann}(\gamma))$ in H_A is the ideal $J = (\alpha_1 y+\alpha_2 x)$.

On the other hand, let C be the cyclic group of order 4 in $k_1[Q_8]$ given by the extension $T \to C \to S$. If we regard k_1 as a trivial module over kS on which T acts trivially also, it follows that $k_1A \otimes_{k_1S} k_1 | k_1C \cong k_1Q_8 \otimes_{k_1C} k_1 | k_1C$.

Thus, $H^*(Q_8;M_1) \cong H^*(C;k_1)$, and in the Lyndon-Hochschild-Serre spectral sequence of $T \to C \to S$, $H^1(S;k_1) \xrightarrow{\cong} H^1(C;k_1)$ while all other $H^1(S;k_1)$ map to zero in $H^i(C;k_1)$.

Since the diagram

$$\begin{array}{ccccc} T & \longrightarrow & C & \longrightarrow & S \\ \| & & \downarrow & & \downarrow \\ T & \longrightarrow & Q_8 & \longrightarrow & A \end{array}$$

commutes, we may identify $g_\alpha \in H^1(S;k_1)$ with a generator $g \in H^1(C;k_1) \cong H^1(Q_8;M_1)$. Under this identifiaction, $g \in H^1_{Q_8}(X,x;k_1) = H^1(E \times_A (W,W_0);k_1)$, is identified with $\gamma \in H^1(A;M_1) \cong H^1(S;k_1)$.

<u>4.7 Assertion</u>: $\operatorname{rad}(\operatorname{Ann}(g)) = J$ in H_A.

<u>Proof</u>: It suffices to show that $f = \alpha_1 y+\alpha_2 x$ belongs to $\operatorname{Ann}(g)$ since f generates J. But $f^t.\gamma = 0$ since $f \in \operatorname{Ann}(\gamma) = J$ for some $t \geq 0$. The naturality of all the identifications made above shows that $f^t.\gamma = 0 \iff f^t.g = 0 \iff f^t.g = 0 \Rightarrow (f) = \operatorname{rad}(\operatorname{Ann}(g))$.

On the other hand, $\operatorname{rad}(\operatorname{Ann}(g))$ must be invariant under Steenrod algebra, being an associated prime for the module $H^*_{Q_8}(X,x;k_1)$ over H_A. Hence its variety must be $\mathbb{F}_p$-rational by Serre's theorem [29], and J is not rational over $\mathbb{F}_p$ by the choice of α. This contradiction establishes the theorem.

<u>4.8 Remark</u>. An alternative proof using a complexity argument is briefly as follows. In the spectral sequence with $E_2^{p,q} = H^p(A;H^q(W,W_0))$ which converges to $H^*(E_G \times_G (X,x) \cong H^*(C;k_1)$, for $p+q =$ constant, $E_\infty^{p,q} \neq 0$ only for one pair (p,q). Thus multiplication by f^t shifts the filtration in E_∞. But since there is only one non-zero term, it

follows that an appropriate power of f^t kills the E_∞-term in this case. This shows that the radical of the annihilator of the module contains f . Hence the H_A-variety of X is the intersection of the line ℓ given by f with possible other lines. If this intersection does not include ℓ , then it must be zero dimensional, and one argues that M must be $\mathbb{Z}_2 \times \mathbb{Z}_2$ -projective accordingly, which is a contradiction again.

The above results show the following theorem, due to Carlsson for $G = \mathbb{Z}_p \times \mathbb{Z}_p$ [13] and to Vogel for $G \supset Q_8$ (to appear) using calculations with the Steenrod algebra. An exposition of Vogel's theorem can be found in [9].

4.9 Theorem. If all $\mathbb{Z}G$-modules are realizable by Moore G-spaces, then G is "metacyclic", i.e. all Sylow subgroups of G are cyclic.

4.10 Remark. Jackowski, Vogel and several others have observed that Carlsson's counterexample for $\mathbb{Z}_p \times \mathbb{Z}_p$ implies that for $G \supset \mathbb{Z}_p \times \mathbb{Z}_p$ the induced module is also a counterexample.

Section 5. Some Examples

We have seen how to construct examples of $\mathbb{Z}G$-modules which are not realizable by Moore G-spaces. These also give examples of homotopy actions on Moore spaces which are not equivalent to a topological action. The question arises whether these lead to criteria for homotopy actions on more general spaces to be equivalent to topological actions. It is helpful to consider the case of spaces which are bouquets of Moore spaces of different dimensions. We will briefly investigate the possibility of realizing a given $\mathbb{Z}G$-module M by a topological action on such a space. This module M arises from a homotopy action (X,α) and as a consequence our examples reveal some properties of homotopy actions on such spaces. Note that if a $\mathbb{Z}G$-module M is indecomposable, then M can be realized only by a Moore G-space. Thus to get new examples, we will consider decomposable modules.

By means of a simple construction using the modules of Section 4 and the theory of Sections 2 and 3, we will show that for $G \supset \mathbb{Z}_p \times \mathbb{Z}_p$ the following hold.

(5.1) There is a $\mathbb{Z}G$-module $M = M_1 \oplus M_2$, where $M_i \neq 0$ are indecomposable, such that neither M nor M_i are realizable by Moore G-spaces.

(5.2) There is an (n-1)-connected finite G-CW complex X of dimension $n+1$ such that $\oplus_i \bar{H}_i(X) = M$ as $\mathbb{Z}G$-module. Call this action $\varphi : G \times X \to X$.

(5.3) X is homotopy equivalent to a bouquet of spheres of dimension n and $n+1$, but (X,φ) is not G-homotopy equivalent to a bouquet of spheres, with a G-action.

(5.4) Let P be the projective cover of M_1 and $0 \to \Omega(M_1) \to P \to M_1 \to 0$ be an exact sequence of $\mathbb{Z}G$-modules. Then an extension of M_1 and $\Omega(M_1)$ is realizable by a finite dimensional Moore G-space. Similarly for M . This extension is non-trivial necessarily.

(5.5) We may choose $M_1 = M_2$ in the above.

(5.6) Since $\Omega(M_1)$ is not realizable by a Moore space either, we have also examples of modules M_1 and $M_1' = \Omega(M_1)$ such that $M_1 \oplus M_1'$ is not realizable by a topological action on a Moore space, but some non-trivial extension of M_1 and M_1' is realizable by a Moore G-space.

(5.7) We may construct examples where $M_1 = \Omega(M_1)$ in the above.

(5.8) There is a homotopy action of G , say α , on a finite bouquet of n-spheres L , such that (L,α) and any suspension of this h-action $(\Sigma^i L,\Sigma^i \alpha)$ are not equivalent to topological actions. But $(L \vee \Sigma L, \alpha \vee \Sigma\alpha)$ is equivalent to a topological action.

(5.9) $V_G(X) \neq V_G^r(X)$, thus the inclusion $V_G(X) \subset V_G^r(X)$ of Proposition 3.2 cannot be improved (even for finite dimensional spaces). Here the varieties are taken over kG . Here $V_G(X) = 0$ while $\oplus_i H_i(X,x;k)$ is not kG-free.

(5.10) Radicals of the annihilators in H_G of $H_G^*(X,x;k)$ and $H^*(G;H^*(X,x;k))$ are not equal.

(5.11) We may choose M_i such that the projectivity criterion 2.1 does not apply to X . This will follow because we will choose M_i such that $\oplus H_i(X,x)|\mathbb{Z}C$ is $\mathbb{Z}C$-projective for all $C \subset G$, $|C|$ = prime, but $\oplus_i H_i(X,x)$ is not $\mathbb{Z}G$-projective. Thus Theorem 2.1 cannot be extended to all G-spaces without additional hypotheses (even for finite dimensional G-spaces).

(5.12) For appropriate choices of M_1 and M_2 , $M = M_1 \oplus M_2$ will not be realizable by any G-space, $M_i \neq 0$, $i = 1,2$.

<u>5.13 Example</u>. It suffices to consider $G = \mathbb{Z}_p \times \mathbb{Z}_p$, and the above assertions (whenever applicable) hold for $G \supset \mathbb{Z}_p \times \mathbb{Z}_p$ or $G \supset Q_8$. Consider the $\mathbb{Z}G$-module M_1 constructed in Theorem 4.4. For some of the assertions such as (5.5), (5.6), and (5.7), let $p = 2$, otherwise p is any prime. We may choose M_1 to be $\mathbb{Z}$-free and $\mathbb{Z}G$-indecomposable. From the exact sequence:

$$(5.14) \qquad 0 \to M_2 \to (\mathbb{Z}G)^r \overset{\varphi}{\to} (\mathbb{Z}G)^s \to M_1 \to 0$$

it follows that $M_2|\mathbb{Z}C$ is $\mathbb{Z}C$-projective for all $C \subset G$, $|C| = $ prime while M_2 is not $\mathbb{Z}G$-projective, since M_1 is not $\mathbb{Z}G$-projective. Therefore M_2 is not realizable by a Moore G-space either. Let $M = M_1 \oplus M_2$. The same holds for M.

We may take bouquets of s and r free G-orbits of n-spheres,

i.e. $X_1 = \bigvee_{i=1}^{s} (G_+ \wedge S^n)_i$ and $X_2 = \bigvee_{j=1}^{r} (G_+ \wedge S^n)_j$.

There exists a G-map $f : X_2 \to X_1$ such that $f_*: H_n(X_2) \to H_n(X_1)$ can be identified with the $\mathbb{Z}G$-homomorphism $\varphi : (\mathbb{Z}G)^r \to (\mathbb{Z}G)^s$ after appropriate identifications $H_n(X_1) \cong (\mathbb{Z}G)^s$ and $H_n(X_2) \cong (\mathbb{Z}G)^r$. Then the mapping cone of f is a finite G-space X which satisfies (5.1) and (5.2) above, in view of the exact sequence (5.14), (5.1) and (5.2) imply (5.3).

The projective cover of M_1, namely P, satisfies $0 \to P \to F_1 \overset{\eta}{\to} F_2 \to 0$ where F_1 and F_2 are $\mathbb{Z}G$-free (not necessarily finitely generated). Thus P can be realized via the mapping cone X_0 of the G-map $g : \bigvee_i (G_+ \wedge S^{n-1})_i \to \bigvee_j (G_+ \wedge S^{n-1})_j$ corresponding to η (i.e. $g_* = \eta$ in $H_{n-1}(-;\mathbb{Z})$). X_0 is also free off the base point. In the exact sequence:

$$(5.15) \qquad 0 \to M_1' \to P \overset{\psi}{\to} M_1 \to 0$$

the homomorphism ψ can be realized by a G-map $f': X_0 \to X$ which induces $f'_* : H_n(X_0) \to H_n(X)$, $f'_* = \psi$, by equivariant obstruction theory (or see [3]). The mapping cone of f', say Y, is a Moore G-space and $H_{n+1}(Y)$ is the extension in the sequence:

$$(5.16) \qquad 0 \to M_2 \to H_{n+1}(Y) \to M_1' \to 0$$

Thus an extension of M_1' and M_2 is realizable by the Moore G-space Y. This proves (5.4). Since M_1 is a periodic module by construction, by taking $G = \mathbb{Z}_2 \times \mathbb{Z}_2$ we can fulfill (5.5) - (5.7). If we wish to choose $M_1 \cong M_2$ for odd p, just take the exact sequence

(5.17) $$0 \to M_1 \to P_1 \xrightarrow{\zeta} P_2 \to M_1 \to 0$$

where P_i are projective covers, and $\mathrm{Ker}\,\zeta = M_1$ since M_1 is chosen to be indecomposable. (5.17) exists due to periodicity of M_1. This is the analogue of (5.14) and we can use P_i instead of F_i, $i=1,2$.

Since all these modules are realizable by homotopy actions (obstruction theory), the assertion (5.8) follows easily from the previous ones.

To see (5.9), note that $V_G^r(k \otimes M_1)$ is not $\mathbb{F}_p$-rational by the construction (cf. Section 4). Thus $V_G^r(\oplus_i \bar{H}_i(X;k)) \equiv V_G^r(k \otimes (M_1 \oplus M_2)) = V_G^r(M_1)$ is not rational over $\mathbb{F}_p$. But $V_G(X)$ is rational over $\mathbb{F}_p$ (see the proof of 2.1). Thus $V_G(X) \neq V_G^r(\oplus_i H_i(X;k))$. Since $V_G^r(M_1)$ is only one line, in this case it follows that $V_G(X) = 0$ in fact. Except for (5.12) which will be proved below separately, the other assertions follow from the above discussion and elementary considerations.

Again, in the following $G \supset \mathbb{Z}_p \times \mathbb{Z}_p$ or Q_8.

<u>5.18 Theorem</u>. There exists a decomposable $\mathbb{Z}G$-module M which cannot be realized by the total reduced homology of any G-space. There are homotopy actions (X,α) realizing M, and all such (X,α) are not equivalent to topological actions.

<u>Proof</u>: As before, we may assume $G = \mathbb{Z}_p \times \mathbb{Z}_p$ and the general case follows from this case. Choose u_α and u_β as in Theorem 4.4, such that $u_\alpha \not\equiv u_\beta \pmod{I^2}$ and the lines in k^2 given by u_α and u_β are distinct. Corresponding to these choices we get indecomposable $\mathbb{Z}$-free $\mathbb{Z}G$-modules M_α and M_β whose rank varieties are the lines determined by u_α and u_β. Neither M_α nor M_β is realizable by a Moore G-space using the projectivity criterion 2.1. For the same reason, $\omega^t(M_\alpha)$, $\omega^s(M_\beta)$ or any direct sum of them are not realizable by Moore G-spaces (see Section 3). Any ω-composite of M_α and M_β is of the form:

(5.19) $$0 \to \omega^t(M_\alpha) \to U \to \omega^s(M_\beta) \to 0$$

and this extension is determined by a class $\eta \in \mathrm{Ext}^1_{\mathbb{Z}}(\omega^t(M_\alpha), \omega^s(M_\beta))$. By tensoring with k, we get

$$(5.20) \qquad 0 \to \omega^t(M_\alpha \otimes k) \to U \otimes k \to \omega^s(M_\beta \otimes k) \to 0$$

and a corresponding class $\eta' \in \mathrm{Ext}^1_{kG}(\omega^t(M_\alpha \otimes k), \omega^s(M_\beta \otimes k))$. We claim that this class vanishes, so that (5.20) is split and $U \otimes k \cong \omega(M_\beta \otimes k) \oplus \omega^t(M_\alpha \otimes k)$. But this follows from the fact that $\mathrm{Ext}^1_{kG}(\omega^t(M_\alpha \otimes k), \omega^s(M_\beta \otimes k)) \cong H^1(G, \omega^t(M_\alpha \otimes k)^* \otimes \omega^s(M_\beta \otimes k)) = 0$, where $*$ means dual with respect to k. The last assertion is a consequence of J. Carlson tensor product formula ([11] Theorem 5.6) as follows. The rank variety of $\omega^t(M_\alpha \otimes k)^*$ is seen to be the same as $V^r_G(\omega^t(M_\alpha \otimes k)) = V^r_G(M_\alpha \otimes k)$ by the definition of V^r, and $V^r_G(\omega^t(M_\alpha \otimes k)^* \otimes \omega^s(M_\beta \otimes k)) = V^r_G(M_\alpha \otimes k) \cap V^r_G(M_\alpha \otimes k) = 0$ by the choice of α and β. Hence $\omega^t(M_\alpha \otimes k)^* \otimes \omega^s(M_\beta \otimes k)$ is kG-free by (4) of Section 2, and $\eta' = 0$ as a consequence. Now suppose $M = M_\alpha \oplus M_\beta$ is realizable by a G-space. Then an ω-composite of $M_\alpha \otimes k$ and $M_\beta \otimes k$ is realizable by a Moore G-space by Corollary 3.5. By the above discussion, any such ω-composite is split and it cannot be realized by a Moore G-space since it does not satisfy the projectivity criterion (Theorem 2.1).

Since M is realizable by a homotopy action, the second assertion follows. ∎

References

[1] Adams, J.F. - Wilkerson, C.: "Finite H-spaces and algebras over the Steenrod algebra", Ann. Math. 111 (1980) 95-143.

[2] Alperin, J. - Evens, L.: "Varieties and elementary abelian groups", (to appear).

[3] Arnold, J.: "On Steenrod's problem for cyclic p-groups", Canad. J. Math. 29 (1977) 421-428.

[4] Assadi, A.: "Extensions libres des actions des groupes finis", Proc. Aarhus Top. Conf. 1982, Springer LNM 1052 (1984).

[5] Assadi, A.: "Finite group actions on simply-connected CW complexes and manifolds", Mem. AMS No. 259 (1982).

[6] Assadi, A. - Browder, W.: "Construction of finite group actions on simply-connected manifolds" (to appear).

[7] Assadi, A. - Vogel, P.: "Seminar on equivariant Moore spaces", (Informal seminar notes, Université de Genève, 1981).

[8] Avrunin, G. - Scott, L.: "Quillen stratification for modules", Invent. Math. 66 (1982), 277-286.

[9] Benson, D. - Habegger, N.: "Varieties for modules and a problem of Steenrod", (Preprint, May 1985).

[10] Bredon, G.: "Introduction to Compact Transformation Groups", Academic Press (1972).

[11] Carlson, J.: "The varieties and the cohomology ring of a module", J. Algebra 85 (1983), 104-143.

[12] Carlson, J.: "Complexity and varieties of modules", in Oberwolfach 1980, Springer-Verlag LNM 882, 62-67.

[13] Carlsson, G.: "A counterexample to a conjecture of Steenrod", Invent. Math. 64, 171-174 (1981).

[14] Cartan, H. - Eilenberg, S.: "Homological Algebra", Princeton Univ. Press, (1956).

[15] Chouinard, L.: "Projectivity and relative projectivity for group rings", J. Pure Appl. Alg. 7 (1976), 287-302.

[16] Cooke, G.: "Replacing homotopy actions by topological actions Trans. AMS 237 (1978), 391-406.

[17] Dade, E.: "Endo-permutation modules over p-groups II", Ann. of Math. 108 (1978), 317-346.

[18] Duflot, J.: "The associated primes of $H_G^*(X)$ ", J. Pure Appl. Alg. 30 (1983), 131.

[19] Hsiang, W.Y.: "Cohomology Theory of Topological Transformation Groups", Springer, Berlin (1975).

[20] Jackowski, S.: "The Euler class and periodicity of group cohomology", Comment. Math. Helv. 53 (1978), 643-650.

[21] Jackowski, S.: "Quillen decomposition for supports of equivariant cohomology with local coefficients", J. Pure Appl. Alg. 33 (1984), 49-58.

[22] Kahn, P.: "Steenrod's problem and k-inavriants of certain classifying spaces", Alg. K-theory, Proc. Oberwolfach (1980), Springer LNM 967 (1982).

[23] Kroll, O.: "Complexity and elementary abelian p-groups", J. Algebra (1984).

[24] Landweber, P. - Stong, R.: "The depth of rings of invariants over finite fields", (preprint) 1984. (To appear in Proceedings of Number Theory Conference, Columbia Univ., Springer-Verlag).

[25] Landweber, P.: "Dickson invariants and prime ideals invariant under Steenrod operations", (Talk given at Topology Conference, University of Virginia, April 1984).

[26] Lashof, R.: "Probelms in topology, Seattle Topology Conference", Ann. Math. (1961).

[27] Quillen, D.: "The spectrum of an equivariant cohomology ring I", and "II" Ann. of Math. 94 (1971), 549-572 and 573-602.

[28] Rim, D.S.: "Modules over finite groups", Ann. Math. 69 (1959), 700-712.

[29] Serre, J.-P.: "Sur la dimension cohomologique des groupes profinis", Topology 3 (1965), 413-420.

[30] Smith, J.: (to appear)

[31] Stong, R.: (Private Communication)

[32] Swan, R.: "Invariant rational functions and a problem of Steenrod", Inven. Math. 7 (1969), 148-158.

[33] Vogel, P.: "On Steenrod's problem for non-abelian finite groups", Proc. Alg. Top. Conf., Aarhus 1982, Springer LNM 1051 (1984).

[34] Zabrodsky, A.: "On G. Cooke's theory of homotopy actions", Proc. Topology Conf. London, Ontario 1981, in New Trends in Algebraic Topology.

[35] Zabrodsky, A.: "Homotopy actions of nilpotent groups", Proc. Topology Conf. Mexico 1981, Contemporary Math. 1984.

Normally Linear Poincaré Complexes And Equivariant Splittings

Amir H. ASSADI(*)
University of Virginia
Charlottesville, Virginia USA

INTRODUCTION: The study of a number of problems in group actions on manifolds calls for explicit constructions of actions. Successful applications of surgery theory in the non-equivariant problems has been a great motivation for various generalization of surgery to the equivariant set up. However, the variety of problems which may be approached via surgery in transformation groups is quite rich. The wide range of phenomena which are to be studied in some of the traditional problems (such as existence and classification problems) has limited the range of applicability of the existing equivariant theories. As a result, it seems appropriate to device specialized surgery theories which aim at different classes of more specific problems.

In the problems which arise in conjunction with the existence and classification of actions on manifolds, it is often useful (in agreement with the general philosophy of surgery) to divide roughly the constructions to two steps. In the first step, one uses methods of algebraic topology to study the problem in the homotopy category. In the second step one passes from the homotopy category to manifolds via surgery. The objects of interest in the first step are Poincaré complexes. Since the category of Poincaré complexes plays an important role in the study of smooth manifolds, it is natural, thus, to study homotopy related problems of G-manifolds on the level of equivariant analogues of this category.

The question arises, then, as to what extent a Poincaré complex with G-action should inherit the structure of a G-manifold. In this paper, we suggest a category of Poincaré complexes with G-actions, whose objects are called "normally linear Poincaré G-complexes" and the morphisms are "isovariant normally linear maps". This category inherits

---

(*) The author has been partially supported by an NSF grant, The Center For Advanced Study of the University of Virginia, and The Max-Planck-Institut Für Mathematik whose hospitality and financial support is gratefully acknowledged.

all the homotopy aspects of the category of Poincaré complexes without G-actions, while it has a certain amount of "manifold information" (from the category of G-manifolds) built into its objects and morphisms. This is in the form of a "suitable stratification" and a linearization of the Spivak normal sphere bundles of strata.

The range of applications and usefulness of this category, of course, depends on how successfully one is able to translate "the algebraic topology" of a problem into the kind of information which would allow one to construct "homotopy models" in this category. Constructions of objects in a category of Poincaré G-complexes becomes difficult if the candidate Poincaré G-complex is required to have "too much manifold information" built into it. On the other hand, imposing "insufficient manifold-like structure" on a Poincaré G-complex makes it difficult to construct equivariant surgery problems from such complexes, (mainly due to lack of equivariant transversality.)

Thus, it appears that the nature of the problem at hand should determine the extent of manifold-like data required from homotopy models. We will illustrate this point by studying the problem of equivariant splittings of closed G-manifolds in our category. Theorem II. 1 and II. 2 give necessary and sufficient conditions for the existence of splittings up to homotopy in terms of normally linear Poincaré G-complexes. Theorems IV. 1 , IV. 3 , and IV. 7 illustrate constructions and solutions of the relevant surgery problems, using the homotopy models of Theorem II. 1 . To give concrete examples, Theorem IV. 5 considers the problem for homotopy spheres and yields a generalization of Anderson-Hambleton's theorem ([1] Theorem A) while Theorem III. 1 illustrates a shorter and different proof of their theorem. Further applications of these ideas will appear in a subsequent paper.

The contents of this paper is as follows. In Section I the category of normally linear Poincaré G-complexes is introduced, and some relevant definitions and background information is mentioned. Section II contains the construction of objects of this category which will be used to study equivariant splittings. Section III illustrates the theory applied to the special case of homotopy spheres to give another proof of the Anderson-Hambleton theorem. This section serves to motivate the generalization of this theorem in Section IV. (Theorem IV. 5), and the solution of the splitting problem up to concordance (Theorem IV. 1 , IV. 3 and IV. 7) with varying degrees of generality. We conclude the paper by a brief discussion of the algebraic obstructions which arise in the general splitting problem.

Finally, we would like to point out a few remarks and mention some features which are implicit in this particular choice of application for normally linear Poincaré complexes. First, our methods does not require "general positionality" or the so called "Gap Hypotheses" which have been used by most authors. Here, the reader will find a discussion of the problem of relaxing "general positionality" in the equivariant surgery problems in Reinhard Schultz' survey article and collection of problems [20]. Thus, the theories which use general-position-type assumptions do not apply to our situation. Secondly, we have considered non-simply-connected manifolds, not only to achieve a greater degree of generality, but also to illustrate new applications for the algebraic K-theoretic functor Wh_1^T of [8], [9] which is the relevant functor to capture such obstructions. We have postponed explicit computations of these obstructions as well as certain other surgery obstructions to a forthcoming paper. The reader, however, will find some results in this direction in [9].

The third point concerns the notion of quasisimple actions and their constructions. The homological hypotheses which are necessary in the splitting problem and "the extension problem" of [9] use $\mathbf{Z}_q\pi$-coefficients (local coefficients) where $\mathbf{Z}_q = \mathbf{Z}/q\mathbf{Z}$. When π is an infinite group, one cannot replace $\mathbf{Z}_q\pi$-coefficients with $\mathbf{Z}_{(q)}\pi$-coefficients, where $\mathbf{Z}_{(q)}$ is the integers localized at q. While the constructions of [9] are given for $\mathbf{Z}_q\pi$ (in order to provide necessary and sufficient conditions for the constructions to exist), they work as well with $\mathbf{Z}_{(q)}\pi$ replacing $\mathbf{Z}_q\pi$ everywhere. Thus, in all the homological conditions in this paper, one can replace $\mathbf{Z}_q$ by $\mathbf{Z}_{(q)}$; but the sufficient conditions obtained in this form will not be necessary anymore. S. Weinberger has independently studied "unextended homologically trivial actions" [23] (which is the analogue of our quasisimple actions for the case of $\mathbf{Z}_{(q)}\pi$-coefficients) using "Zabrodsky mixing". Weinberger's survey article in [24] contains further ideas and developments in conjunction with construction of actions. We refer the reader to [20] for articles of Schultz and Weinberger and their references for discussions of related results and problems.

Finally, to study G-actions on Poincaré complexes which are not quasisimple, one encounters completely new phenomena. The methods of constructions which assume that G acts trivially on homology do not apply to non-quasisimple actions. An alternative is to study such problems via "homotopy actions". This is the point of view of [7] (see also [4]). Construction of non-quasisimple normally linear Poincaré

G-complexes (using homotopy actions) and further applications will be discussed in a forthcoming paper of the author.

REMARK: It appears to us that the constructions of normally linear Poincaré complexes, (e.g. as in Section II) may be combined with Browder-Quinn's paper in Manifolds, Tokyo, 1973, (University of Tokyo Press 1975) to give a general set up for classification theory of quasi-simple actions. Moreover, Browder-Quinn theory can be potentially useful to analyze the G-manifold structures on normally linear Poincaré G-complexes. In this fashion, one may try to refine the results of our Section IV by analyzing the relevant surgery obstructions in the Browder-Quinn theory (instead of passing to concordance to bypass possibly non-zero obstructions).

SECTION I. PRELIMINARY NOTIONS:

Throughout this paper G is a finite group of order q, and we will work in the category of G-CW complexes, while G-actions on smooth manifolds are assumed to be smooth. The smoothness assumption is made only for convenience sake and most of the results, when appropriate, are true about more general types of action with some regulariy conditions, e.g. locally smooth PL actions, etc.

An earlier definition for a Poincaré G-complex was suggested by Frank Connolly [11], where all the homotopy analogues of the ingredients involved in a G-manifold were built into the definition of a so called "G-Poincaré complex". For our purposes, however, it is appropriate to introduce G-complexes which have inherited some linear structure on the regular neighborhoods of various strata. This restriction, in this case, makes it possible to translate the homotopy problems involving (non-free) G-manifolds into questions which involve the homotopy structure of the fixed point sets without losing the linear information naturally given for their normal bundles. Furthermore, we will discuss methods of construction for such G-complexes with this richer structure, and obtain positive answers in a variety of circumstances.

Let $\mathcal{C}$ be a category of Poincaré complexes (pairs). $\mathcal{C}$ could be the category of simple Poincaré complexes, or the category of finite Poincaré complexes, or a more general category, for instance [22]. We will fix $\mathcal{C}$ during the following discussion and suppress any reference to it unless it is necessary. For the applications, the context will determine the category $\mathcal{C}$.

I.1. DEFINITION: A normally linear Poincaré G-complex (pair) with one orbit type is a Poincaré complex (pair) in $\mathcal{C}$ in the ordinary sense (not necessarily connected). A normally linear Poincaré G-pair (X,Y) with $(k+1)$ orbit-types is defined inductively as follows. Let H be a maximal isotropy subgroup. Then $(G \cdot X^H, G \cdot Y^H)$ is required to be a Poincaré G-pair with one orbit type which has an equivariant regular neighborhood pair $(R,\partial_1 R)$ in (X,Y) such that:

(1) there exists a G-bundle ν over $G \cdot X^H$ such that $(R,\partial_1 R)$ is G-homeomorphic to $(D(\nu), D(\nu|G \cdot Y^H))$;

(2) there is a normally linear Poincaré G-pair $(C,\partial C)$ with k orbit types and a G-homeomorphism $f : S(\nu) \longrightarrow \partial_+ C \subset \partial C$ such that $X = C \cup_f D(\nu)$ and $Y = \partial_- C \cup_{f'} D(\nu|G \cdot Y^H)$ where $\partial_- C = \overline{\partial C - \partial_+ C}$ and $f' = f|S(\nu|G \cdot Y^H)$.

REMARK: Normally linear Poincaré G-complexes defined above are different from Conolly's [11] G-Poincaré complexes in at least two different points. First, the Spivak normal fibre space of one stratum in the next is already given a linear structure. Second, the Poincaré embeddings of our definition are more manifold like in that the complement of one stratum in the next is also prescribed (subject to the appropriate identifications coming with the structure). As we shall illustrate in Section IV, this results in a great simplification of the construction of surgery problems.

ν is called the equivariant normal bundle of $G \cdot X^H$. *An isovariant normally linear map* is an equivariant map which preserves the isotropy types and the normal bundles (after the identification of regular neighborhoods and disk bundles). The G-homeomorphisms f and f' above are (G-cellular) isovariant normally linear maps of Poincaré pairs with k orbit types. It is possible to show inductively that for each subgroup $K \subseteq G$, (X^K, Y^K) is a Poincaré complex which is Poincaré embedded in (X,Y) , and its Spivak normal fibre space has a $N(K)$-linear structure. ($N(K)$ = normalizer of K in G).

Normally linear Poincaré G-complexes are constructed in [4],[5], [7],[9] in the semifree case. Smooth G-manifolds are normally linear

Poincaré G-complexes in a natural manner. We drop the prefix G whenever the context allows us to do so.

I.2. CONVENTION: All Poincaré complexes with G-actions are assumed to be normally linear Poincaré complexes. If $L \underset{\neq}{\subseteq} K \subseteq G$, $\dim X^K - \dim X^L < 2$. If X is connected, we assume that X^K is connected for all $K \subseteq G$. All manifolds are compact and all Poincaré complexes are finite.

We will study first the case of semifree actions which serve as a model for the inductive proofs of similar results for actions with several isotropy groups. However, the generalization of the results of the semifree case is not immediate, even in the case of actions on spheres (or disks) due to the fact that the fixed point sets of isotropy subgroups of composite order satisfy very little homological restrictions in general. In fact, Oliver's work [17] shows that in the case of disks, only certain Euler characteristic relationships are necessary (and sufficient). Therefore, it is inevitable to consider some restricted classes of actions where some minimal homological conditions are imposed on the fixed point sets of various isotropy subgroups.

A convenient category of G-complexes is the category of quasisimple actions.

I.2. DEFINITION: An action $\phi : G \times X \to X$ is called quasisimple if for each isotropy subgroup $K \subseteq G$, the action of $N(K)/K$ on the fundamental group of each component of X^K_α and, subsequently, $H_*(X^K_\alpha ; \mathbf{Z}_q\pi_1(X^K_\alpha))$ are trivial. Note that the triviality of the action of $N(K)/K$ on $\pi_1(X^K_\alpha)$ makes it possible to define unambiguously the action on the homology of X^K_α with local coefficients $\mathbf{Z}_q\pi_1(X^K_\alpha)$. (Recall that $\mathbf{Z}_q = \mathbf{Z}/q\,\mathbf{Z}$. One may also use $\mathbf{Z}_{(q)}$ systematically).

REMARKS: (1) Quasisimple actions were introduced and studied in [9].

(2) Replacing $\mathbf{Z}_q$ by $\mathbf{Z}_{(q)}$ in the above definition, for a free G-space X, quasisimplicity means that $\pi_1(X/G) \cong \pi_1(X) \times G$ and G acts trivially on the homology. This notion has been called "an unextended action" by S. Weinberger and studied in [23] independently.

I.3. DEFINITION: Let X be a connected G-CW complex, where G is a finite group of order q. X is called a simple G-space (and the action is called simple) if $(E_G \times_G X)_q$ is fibre homotopy equivalent to $(BG \times X)_q$. Here X_q denotes the localization of X which preserves $\pi_1(X)$ and localizes $\pi_i(X)$ for $i > 1$ at $\mathbf{Z}/q\,\mathbf{Z}$. Cf. [10] and [9] Section II.

In dealing with non-simply-connected complexes, it is necessary to consider simple homotopy types and simple homotopy equivalences. The equivariant generalizations of the Whitehead torsion are studied in [18],[15],[14]. To construct a G-action on a simply-connected finite complex X (up to homotopy type in the category of finite complexes) the projective class group $K_0(\mathbf{Z}G)$ and certain subgroups or subquotients play an important role (cf. [21],[17],[2],[1] etc.). If $\pi_1(X) \neq 1$, then the analogue of $K_0(\mathbf{Z}G)$ is an abelian group $Wh_1^T(\pi \subset \Gamma)$ where $\pi = \pi_1(X)$ and Γ is the extension $1 \longrightarrow \pi \longrightarrow \Gamma \longrightarrow G \longrightarrow 1$ obtained from the action of G on $\pi_1(X)$ (whenever defined). $Wh_1^T(\pi \subset \Gamma)$ and its algebraic properties and topological applications are treated in [9], and an alternative definition in terms of the fibre of a transfer map between Whitehead spaces is given in [8].

We will briefly recall the definition and some properties of $Wh_1^T(\pi \subset \Gamma)$ when $\Gamma = \pi \times G$ (the case of quasisimple actions). Let $\mathcal{A}$ be the category whose objects are pairs $(M,\mathcal{B})$ where M is a finitely generated $\mathbf{Z}\Gamma$-projective module which is free over π and $\mathcal{B}$ is a π-basis for M. Two objects are equivalent $(M,\mathcal{B}) \sim (M',\mathcal{B}')$ if there is a π-simple isomorphism $f : (M,\mathcal{B}) \longrightarrow (M',\mathcal{B}')$. Let $\mathcal{A}' = \mathcal{A}/\sim$ and consider the monoid structure on $\mathcal{A}'$ induced by direct sums (and disjoint union), taking $(0,\emptyset)$ as the neutral element. Then $(\mathbf{Z}\Gamma,G)$ generates the monoid of trivial elements $\mathcal{T}$, and we define $Wh_1^T(\pi \subset \Gamma) = \mathcal{A}'/\mathcal{T}$. It is an abelian group which fits into a 5-term transfer exact sequence $Wh_1(\Gamma) \xrightarrow{tr} Wh_1(\pi) \xrightarrow{\beta} Wh_1^T(\pi \subset \Gamma) \xrightarrow{\alpha} \tilde{K}_0(\mathbf{Z}\Gamma) \xrightarrow{tr} \tilde{K}_0(\mathbf{Z}\pi)$.
The homomorphism α is induced by the forgetful functor $(M,\mathcal{B}) \longrightarrow M$. Furthermore, let $Wh_1^T(\pi;\mathbf{Z}_q) = K_1(\mathbf{Z}_q\pi)/\{\pm\pi\}$. Then one has a commutative diagram

$$\begin{array}{ccccc}
Wh_1(\pi) & \xrightarrow{\beta} & Wh_1^T(\pi \subset \Gamma) & \xrightarrow{\alpha} & \tilde{K}_0(\mathbf{Z}\Gamma) \\
 & \searrow^{\text{canon.}} & & \nearrow_{\gamma} & \\
 & & Wh_1(\pi;\mathbf{Z}_q) & &
\end{array}$$

where $\alpha \circ \gamma$ is a generalization of the Swan homomorphism (cf. [21]) $\sigma_G: (\mathbf{Z}_q)^{\times} \longrightarrow \widetilde{K}_0(\mathbf{Z}G)$ (when $\pi = 1$).

A topological application of $Wh_1^T(\pi \subset \Gamma)$ is as follows. Suppose (X,Y) is a pair, $\pi_1(X) = \pi$, and X is a finite G-complex. Let $\phi : G \times Y \longrightarrow Y$ be a free quasisimple action, and let $H_*(X,Y;Z_q\pi) = 0$. Then there exists a free finite G-complex X' such that Y is an invariant subcomplex, and there exists a π-simple homotopy equivalence $f : X' \longrightarrow X$ rel Y if and only if $\gamma\tau(X,Y) = 0$ in $Wh_1^T(\pi \subset \pi \times G)$, where $\tau(X,Y)$ is the Reidemeister torsion of the pair (X,Y) (well-defined in $Wh_1(\pi;\mathbf{Z}_q)$ due to the homological hypothesis). Cf. [9] Section I for further details.

<u>I.3. LEMMA</u>: Suppose X is a finite semifree simple G-complex. Then $H_*(X,X^G;Z_q\pi) = 0$ and $\gamma\tau(X,X^G) \in Wh_1^T(\pi \subset \pi \times G)$ vanishes, where $\pi = \pi_1(X)$.

<u>PROOF</u>: Cf. [9] Proposition II.3.

We extend the notion of admissible splittings of [1] to non-simply connected closed manifolds (Poincaré complexes). Let M^n be a closed manifold and let $M^n = M_1^n \cup M_2^n$ be a splitting so that $M_1 \cap M_2 = \partial M_2$. It is an "admissible splitting" if $\pi_1(\partial M_1) \cong \pi_1(M_i) \cong \pi_1(M) = \pi$ (similarly for Poincaré complexes).

<u>I.4. LEMMA</u>: Let $\phi : G \times M^n \longrightarrow M^n$ be semifree and suppose that $M = M_1 \cup M_2$ is an equivariant admissible decomposition of (M,ϕ) such that M_i are simple. Let $M^G = F$ and $M_i \cap F = F_i$. Then $H_*(M_i,F_i;Z_q\pi) = 0$ and $\gamma\tau(M_i,F_i) = 0$ for $i = 1,2, \pi = \pi_1(M)$.

<u>PROOF</u>: This follows from I.3.

■

In the next section we will show how to construct normally linear Poincaré complexes to solve the equivariant splitting problem for closed G-manifolds on the level of homotopy.

SECTION II. SPLITTING UP TO HOMOTOPY:

As before, G is a finite group of order q . Let $\varphi : G \times \Sigma^n \to \Sigma^n$ be a smooth, semifree action on a homotopy sphere. In [1] Anderson and Hambleton studied criteria for the existance of equivariant homological symmetry of (Σ^n,ϕ) , i.e. $\Sigma^n = D_1^n \cup D_2^n$ where each D_i^n is an invariant disk and $H_j((D_1^n)^G) \cong H_j((D_2^n)^G)$ for all j . Roughly speaking, vanishing of a semi-characteristic type invariant characterizes (Σ^n,ϕ) which are homologically double in the above sense, provided that $n > 2\dim \Sigma^G$. Anderson and Hambleton call this structure a (strong) balanced splitting.

Since any homotopy sphere is a twisted double, the results of [1] may be interpreted as finding obstructions to make a (given) "non-equivariant symmetry" into an equivariant one. Besides leading to the discovery of a new and interesting invariant of such semifree actions, this equivariant symmetry may be regarded as a homological regularity condition (i.e. similarity to the linear actions). From this perspective, it is natural to ask if such equivariant splittings exist for more general actions. In this section, we propose to study this question for closed manifolds under some homological restrictions which impose P.A. Smith Theoretic conditions on the fixed point sets of isotropy subgroups. Our approach is to find invariants which characterize the existence of equivariant splittings on the level of normally linear Poincaré complexes, thus reducting the problem to an equivariant surgery problem. Since the fixed-point sets of non-trivial subgroups are, in general, non-simply connected, we will study the problem with special attention to the fundamental group. The following theorem gives necessary and sufficient conditions for the existence of equivariant splittings in the category of normally linear Poincaré complexes, with semifree actions. The general case is stated separately and its proof is an elaboration of the arguments for the semifree case.

II.1. THEOREM: Suppose $\phi : G \times X \longrightarrow X$ is a quasisimple semifree action such that (X,ϕ) is a normally linear finite Poincaré complex with $(X,\phi)^G = F$, $\nu(F \subset X) = \nu$, and a (non-equivariant) admissible splitting $X = X_1 \cup X_2$, $F \cap X_i = F_i$. Suppose (1) $H_*(X_i,F_i;\mathbf{Z}_q\pi) = 0$ and (2) $\gamma\tau(X_i,F_i) \in Wh_1^T(\pi \subset \pi \times G)$ vanishes. Then there exists a quasi-simple semifree normally linear finite Poincaré G-complex X' with the following properties: (a) $(X')^G = F$, $\nu(F \subset X') = \nu$; (b) X' has an equivariant admissible splitting $X' = X_1' \cup X_2'$, $X_i \cap F = F_i$ and X_i' are simple; (c) there exists a normally linear isovariant map $f : X' \longrightarrow X$ which induces a π-simple homotopy equivalence; (d) X_i' and $\partial X_i'$ are π-simple homotopy equivalent to X_i and ∂X_i rel F_i and ∂F_i respectively. Furthermore, the hypotheses (1) and (2) above are necessary for the existence of such X' .

PROOF: Since X is normally linear, there exists a Poincaré pair $(C, \partial C)$ with a free G-action, such that $\partial C = S(\nu)$ and $X = D(\nu) \cup C$ (after appropriate identifications.) Let $C_i = C \cap X_i$, and let $\partial_- C_i = \partial C \cap X_i$, $\partial_+ C_i = C \cap \partial X_i$, $\partial_0 C_i = \partial_+ C_i \cap \partial_- C_i$. Note that $\partial_0 C_1 = \partial_0 C_2$ and $\partial_+ C_1 = \partial_+ C_2$; denote them by ∂_0 and ∂_+ respectively. Thus we have the following diagram

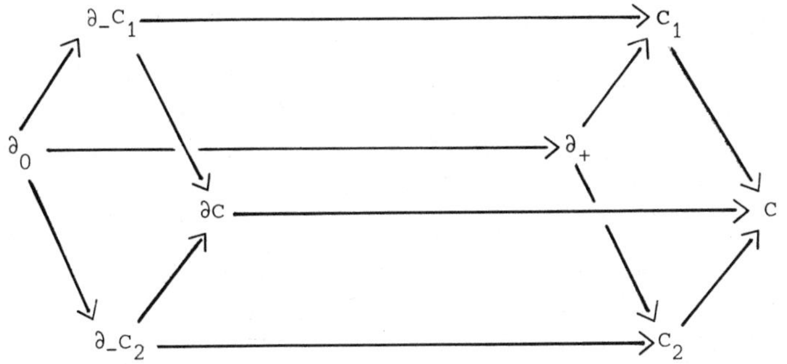

DIAGRAM (D)

in which not all maps are equivariant. If X' exists with the desired properties, we can write $X' = C' \cup D(\nu)$ and obtain a diagram (D') involving C', C_i' and the analoguous boundary decompositions in which all maps are equivariant. Furthermore, we will get a map of diagrams (D') $\longrightarrow$ (D) with the induced map $\partial_0' \longrightarrow \partial_0$, $\partial_\pm' \longrightarrow \partial_\pm$, $\partial C' \longrightarrow \partial C$, and $C' \longrightarrow C$ being the identity or an equivariant π-simple homotopy equivalence, as it is clear from the context and the requirements (a) - (d) above.

Let us use an asterisks to denote orbit spaces (e.g. $X^* = X/G$) and a bar to denote a covering with the deck transformation group G (e.g. $\overline{C}^* = C$ in the above situation). Thus we look for a diagram (D'^*) of orbit spaces in which the spaces $C_i'^*$ and $\partial_+'^*$ as well as the dotted arrows are to be determined:

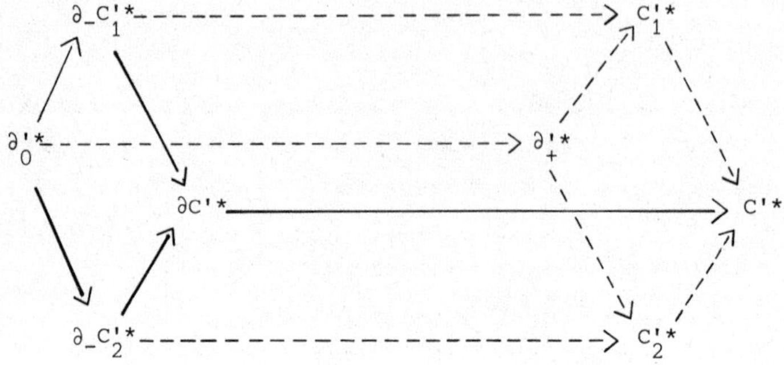

DIAGRAM (D'*)

The left side and the right side faces of the parallelograms in (D), (D') and (D'*) are push-outs with respective push-out maps, and we denote them by (LD), (RD), (LD'), etc. Moreover, in (D'*) we have the following equalities up to homotopy: $C'^* = C^*$, $\partial_0'^* = \partial_0^*$, $\partial_-C_1'^* = \partial_-C_i^*$ and $\partial C'^* = \partial C^*$ and the appropriate maps are induced by the corresponding maps in (D).

In the terminology of [9] Theorem V.1, we wish "to push forward" the free action from the push out diagram of free G-spaces (LD) to the corresponding diagram (RD) after possibly replacing (RD) by homotopy equivalent complexes. Since the constructions of [9] are sufficiently functorial, they apply to this situation. Briefly, note that $H_*(C_i,\partial_+C_i;\mathbf{Z}_q\pi) = H_*(C_i,\partial_-C_i;\mathbf{Z}_q\pi) = H_*(X_i,F_i;\mathbf{Z}_q\pi) = 0$ by Poincaré duality, excision and hypothesis (1) of the Theorem. Further, the quasisimplicity condition ensures that the scheme of [9] applies to construct the appropriate localizations of the diagrams

$$\begin{array}{ccc} \partial_-C_1 & \longrightarrow & C_i \\ \uparrow & & \uparrow \\ \partial_0 & \longrightarrow & \partial_+ \end{array}$$

and then take push-outs. The finiteness obstruction as well as the Whitehead torsion obstruction for choosing C_i' to be finite and π-simple homotopy equivalent to C_i, is the image of the Reidemeister torsion $\tau(X_i, F_i)$ in $Wh_1^T(\pi \subset \pi \times G)$, and it vanishes by hypothesis (2). It follows from the duality of the Reidemeister torsion [16] that the corresponding obstructions for choosing ∂_+' to be finite and (equivariantly) π-simple homotopy equivalent to ∂_+ vanishes as well (cf. [9] Theorem 1.13). The existence of an equivariant π-simple homotopy equivalence $(C', \partial C') \longrightarrow (C, \partial C)$ and π-simple homotopy equivalence of C_i and C_i' follows from the constructions and the functoriality of push-outs.

The necessity of conditions (1) and (2) of the Theorem for existence of X' together with the appropriate equivariant splitting follows as in [9] Section II.

∎

Equivariant splittings of actions with two isotropy types and semifree actions can be treated in a similar fashion. This observation allows one to generalize Theorem II.1 to actions with several isotropy types, provided that the fixed point sets of adjacent strata are related to each other in the same manner that the stationary point set of G and the free stratum are related in the semifree case. The condition of quasisimplicity as in Definition I.2 ensures that this is the case. (The hypotheses of the following theorem may be relaxed at the expense of introducing more complicated notions and longer statements, but we will not do this). The proof of this theorem uses an inductive argument similar to II.1 and we will omit it.

<u>II.2. THEOREM</u>: Let (X, ϕ) be a finite G-Poincaré complex. Suppose $X = X_1 \cup X_2$, and denote $F_i(K) = X^K \cap X_i$. Assume that the splittings $F_1(K) \cup F_2(K) = X^K$ are admissible for all isotropy subgroups $K \subseteq G$ such that $H_*(F_i(K), F_i(L); \mathbf{Z}_q \pi_1(F_i(K))) = 0$ and $\gamma\tau(F_i(K), F_i(L)) \in Wh_1^T(\pi_1(F_1(K)) \subset \pi_1(F_i(K)) \times G)$ vanish. Then there exists a finite G-Poincaré complex (X', ψ) with $(X', \psi)^G = (X, \phi)^G$ and an equivariant admissible splitting $X' = X_1' \cup X_2'$ such that (a) there exists a normally linear isovariant simple homotopy equivalence $f : X' \longrightarrow X$ extending the inclusion $X'^G = X^G \subset X$; (B) X_i and X' and $X_i' \cap X'^K$ are simple homotopy equivalent to X_i and $X_i \cap X^K$ repectively and $X'^G \cap X_i' = X^G \cap X_i$.

SECTION III: A SPECIAL CASE

In the special case where M^n is a homotopy sphere, an equivariant splitting is obtained as an application of II.1, or by a direct argument. This yields another proof for a Theorem of Anderson-Hambleton [1]. We will mention this special case separately to illustrate the theory in a concrete case.

III.1. THEOREM: Let Σ^n be a homotopy sphere, $\phi : G \times \Sigma^n \longrightarrow \Sigma^n$ a semifree action and $F^k = \Sigma^G$ where $\nu = \nu(F \subset \Sigma)$ and $\dim \nu > k$. Given a splitting $F = F_1 \cup F_2$, there exists a corresponding equivariant splitting $\Sigma^n = D_1^n \cup D_2^n$ into disks such that $D_i^n \cap F = F_i$ if and only if $\overline{H}_*(F_i;\mathbf{Z}_q) = 0$ and $\mathcal{O}(F_i) = 0$ in $\widetilde{K}_0(\mathbf{Z}G)$, where $\mathcal{O}(F_i) = \sum_{j>0} (-1)^i \sigma_G(H_j(F_i))$ and σ_G is the Swan map of Section I.

PROOF: Choose $X_1^n \subset \Sigma^n$ to be diffeomorphic to D^n and $X_1 \cap F = F_1$ and $\partial X_1 \cap F = \partial F_1$. This follows easily from handle body theory and general positionality, since $n > 2k$, and we are working non-equivariantly. By Theorem II.1 we have a normally linear finite Poincaré complex X_1' such that $(X_1', \partial X_1')^G = (F_1, \partial F_1)$, and $\nu(F_1 \subset X_1') = \nu|F_1$, and there is an isovariant map $f_1 : X_1' \longrightarrow \Sigma$ which extends the inclusion on $D(\nu|F_1)$. Since $H_i(F_1) = 0$ for $i \geq k-1$, it follows that X_1' is obtained from $D(\nu)$ by adding free G-cells of dimension at most k. Thus, f_1 can be deformed into an isovariant embedding extending the inclusion of $D(\nu|F_1)$. Let R be an equivariant regular neighborhood of $f_1(X_1') \cup D(\nu)$ in Σ^n. Then closure $(R - D(\nu|F_2))$ is diffeomorphic to D_1^n and Σ^n is equivariantly split as $D_1^n \cup D_2^n$ where $D_2^n = \Sigma^n - \mathrm{int}(D_1^n)$. The necessity of these conditions follows easily as in [1] or [2] Section II.

III.2. REMARK: The existence of X_1' follows from a direct argument, by attaching free G-cells of dimension $\leq k$ to $S(\nu|F_1)$ as in [2] II.V or [3] Section II. Then the equivariant map $f_1 : X_1' \longrightarrow \Sigma^n$ extending the inclusion $D(\nu|F_1) \longrightarrow \Sigma^n$ is a direct consequence of obstruction theory, and it can be deformed rel $D(\nu|F_1)$ to an isovariant map using general positionality of F.

SECTION IV. SPLITTING UP TO CONCORDANCE:

In this Section we use the existence of equivariant splittings of Section II to find equivariant splittings of a G-manifold (M,ϕ) based on a given non-equivariant splitting. This illustrates the construction of surgery problems from a given normally linear Poincaré G-complex.

When the appropriate obstructions for the existence of an equivariant splitting in the category of normally linear Poincaré complexes vanish, we obtain (X',ϕ') which is isovariantly π-simple homotopy equivalent to (M,ϕ). Next, we return to the category of G-manifolds by smoothing (X',ϕ') equivariantly, while preserving the splitting up to equivariant homotopy. The result will be (M',ψ) which is isovariantly π-simple homotopy equivalent to (M,ϕ) (relative to an equivariant regular neighborhood of $M^G = M'^G$). Rather than a detailed analysis of the relevant surgery exact sequence (leading to the surgery obstructions in order to arrange (M',ψ) to be G-diffeomorphic to (M,ϕ) and inherit the desired splitting from (X',ϕ')), we pass to a restricted concordance in order to get a positive answer. Namely, we change the action on the free part of (M,ϕ) rel $S(\nu(M^G))$ to get (M,ψ) concordant to $(M,\phi)(\text{rel } M^G)$ such that (M,ψ) is equivariantly split as desired.

If $\pi_1(M) = 1$, then this change in action is merely taking the equivariant connected sum of (M,ϕ) and an "almost linear" sphere (S^n,σ). Thus in this case, the G-homeomorphism type of (M,ϕ) is not changed in order to be equivariantly split. Again we give the proof in the case of semifree actions and only state the general case.

IV.1. THEOREM: Let $\phi : G \times M^n \longrightarrow M^n$ be a quasisimple smooth semifree action with $(M,\phi)^G = F^k$, $\nu(F \subset M) = \nu$ and a (non-equivariant) admissible splitting of the closed manifold $M = M_1 \cup M_2$, $M_i \cap F = F_i$. Assume that (1) $H_*(M_i,F_i;\mathbf{Z}_q\pi) = 0$, and (2) $\gamma\tau(M_i,F_i) \in Wh_1^T(\pi \subset \pi \times G)$ vanishes. Then there exists a quasisimple semifree G-action on M, say $\psi : G \times M \longrightarrow M$, such that: (a) (M,ψ) is concordant to (M,ϕ) relative to F; (b) (M,ψ) has an equivariant splitting $M = M_1' \cup M_2'$ where M_i' are simple, $M_i' \cap F = F_i$ and M_i' and $\partial M_i'$ are π-simple homotopy equivalent to M_i and ∂M_i respectively. Furthermore, conditions (1) and (2) are necessary for the existence of (M,ψ).

PROOF: Since by Theorem II.1 the conditions (1) and (2) are necessary for the existence of equivariant splittings in the category of normally linear Poincaré complexes, (cf. II.1) we need to show only their sufficiency.

First, we construct the concordance on the level of normally linear Poincaré complexes. Thus we have an equivariantly split G-complex X' which satisfies all the stated properties if we replace X by M in Theorem II.1.

IV.2. PROPOSITION: Under the hypotheses of IV.1, there exists a normally linear G-Poincaré pair $(Y,\partial Y)$ such that: (1) $Y^G = F \times [0,1]$ and $\nu(Y^G \subset Y) = \nu \times [0,1]$; (2) $\partial Y = M \cup X'$ where the induced action on M is ϕ and on X' is the action given by Theorem II.1.

PROOF: Let $f : X \longrightarrow M$ be the isovariant map of II.1, and let Y be the mapping cylinder of f.

■

We continue the proof of IV.1 by finding a normal invariant for $(Y,\partial Y)$ which restricts to the natural one given on $M \subset \partial Y$. Using the normal linearity, let $Y = D(\nu \times [0,1]) \cup Y'$ where $\partial Y' = C \cup S(\nu \times [0,1]) \cup C'$ using the notation of II.1, and Y' has a free quasisimple G-action.

Let BG be Stasheff's classifying space for stable spherical fibrations. As before, we denote the orbit space by an asterisk: $X^* \equiv X/G$. Let $\alpha : Y'^* \longrightarrow BG$ be the classifying map for the Spivak spherical fibration of Y'^*. Then $\alpha \mid C^*$ lifts to BO since C^* is a manifold. Also this lift extends over $S(\nu \times [0,1])^*$. The obstruction to extending this to a lift of α to BO is an element $\chi \in h^*(Y'^*, C^* \cup S(\nu \times [0,1])^*)$, where h^* = generalized cohomology theory of G/O. Since $H^*(Y', C \cup S(\nu \times [0,1]); \mathbf{Z}_q) = 0$ by excision, the Cartan-Leray spectral sequence for the covering pair $(Y',C) \longrightarrow (Y'^*,C^*)$ collapses and $H^0(B\ ,h^*(Y',C)) \cong h^*(Y'^*,C^*)$. From the hypothesis of quasisimplicity, it follows that G acts trivially on $h^*(Y',C)$ (cf. [9] II.6 and Lemma II.10) and $h^*(Y'^*,C^*) \cong h^*(Y',C)$. Thus χ is q-divisible. On the other hand the transfer $tr(\chi) \in h^*(Y',C)$ vanishes, since Y' is (non-equivariantly) homotopy equivalent to $C \times [0,1]$.

Therefore $\chi = 0$ and α lifts to BO .

This yields the desired normal invariant, say $f : (W^{n+1}, \partial W) \longrightarrow (Y'^*, \partial Y'^*)$ such that $\partial W = C^* \cup S(\nu \times [0,1])^* \cup V^n$ and $f \mid C^* \cup S(\nu \times [0,1])^*$ is the inclusion. The splitting $C'^* = C_1'^* \cup C_2'^*$ (as given in II.1) induces an equivariant decomposition $V = V_1 \cup V_2$, $V_1 \cap V_2 = V_0 = \partial V_1 = \partial V_2$. Let $f_i \mid V_i$, $i = 0,1,2$. The surgery obstruction to making $f_1 : (V_1, \partial V_1) \longrightarrow (C_1'^*, \partial C_1'^*)$ into a homotopy equivalence rel $S(\nu \times 1)^*$ such that $\bar{f}_1 : (\bar{V}_1, \partial \bar{V}_1) \longrightarrow (C_1', \partial C_1')$ is a π-simple homotopy equivalence rel $S(\nu \times 1)$ vanishes by [22] Theorem 3.3 (cf. [9] Theorem II.7). Let N_1^{n+1} be this normal cobordism, and add N_1^{n+1} to V^{n+1} along V_1 to obtain a new normal map (after smoothing corners, etc.). Then $f' : W' \longrightarrow Y'^*$ with $\partial W' = C^* \cup S(\nu \times [0,1])^* \cup V'$, $V' = V_1' \cup V_2'$, and $f' \mid V_1 : (V_1', \partial V_1') \longrightarrow (C_1'^*, \partial C_1'^*)$ is a homotopy equivalence rel $S(\nu \times 1)^* \cap \partial C_1'$ (and the induced map on the G-coverings is π-simple). Next, we can do surgery on f' rel $C^* \cup S(\nu \times [0,1])^* \cup V_1'$ to make it into a homotopy equivalence of pairs, applying again Wall's Theorem ([22] Theorem 3.3) since $\pi_1(C_2') \cong \pi_1(Y') \cong \pi$. Call the new map $f'' : W'' \longrightarrow V'$, where $\partial W'' = C^* \cup S(\nu \times [0,1])^* \cup V''$ and $V'' = V_1'' \cup V_2''$, $V_1'' = V_1'$ and $f'' \mid V_2''$ is also a homotopy equivalence (and $f'' : \bar{V}'' \longrightarrow C'$ and $f'' \mid \bar{V}_2''$ are π-simple equivalences). Adding $D(\nu \times [0,1])$ back to $\bar{W}''$ along $S(\nu \times [0,1])$ yields the desired concordance. (The reader can easily verify that $\bar{W}''$ is an s-cobordism with a free G-action, and $M_1' = V_1' \cup D(\nu \times 1 \mid F_1)$ and $M_2' = \bar{V}_2'' \cup D(\nu \times 1 \mid F_2)$ yield the equivariant splitting required by the Theorem).

■

<u>IV.3.3 THEOREM</u>: Suppose $\pi_1(M) = 1$ in IV.1. Then there exists an almost linear sphere (S^n, σ) such that the equivariant connected sum $(M, \psi) = (M, \phi) \# (S^n, \sigma)$ admits an equivariant splitting as in the conclusion of Theorem IV.1.

<u>PROOF</u>: Let $\tau \in Wh_1(G)$ be the torsion of the relative h-cobordism W'' with respect to C/G . Choose $\chi \in M^G$ and the linear sphere $S(T_\chi M \oplus \mathbf{R}) = S^n$ where the tangent space $T_\chi M$ has the linear representation induced by ϕ . Let K^{n+1} be the concordance $S^n \times [0,1]$ obtained by adding free 2-handles and 3-handles to the free stratum of S^n so that the resulting G-h-cobordism has torsion $-\tau$. The new

equivariant concordance $M \times [0,1] \# S^n \times [0,1]$ (where the connected sum is along an arc $\{\chi\} \times [0,1]$ in the stationary point sets) is actually an equivariant s-cobordism, and hence a product. But $\partial(M \times [0,1] \# S^n \times [0,1]$ with the induced action is G-diffeomorphic to $(M,\phi) \cup (M,\phi \# \sigma)$ where σ is the "alomost linear" action induced on $S^n \times \{1\}$ in the concordance $S^n \times [0,1]$. (See [5]).

IV.4. COROLLARY: Given (M,ϕ) as in IV.1, and so that $\pi_1(M) = 1$, there exists a smooth action $\psi : G \times M \longrightarrow M$ such that (M,ψ) has an equivariant splitting as in the conclusion of Theorem IV.1, and (M,ϕ) is G-homeomorphic to (M,ψ) .

∎

If M^n is a homotopy sphere, then we get an equivariant decomposition into disks, thus generalizing the Anderson-Hambleton Theorem [1] Theorem A. Note that the methods of [1] which are based on general position arguments do not apply here, since codimensions could be quite small.

IV.5. THEOREM: Let $\phi : G \times \Sigma^n \longrightarrow \Sigma^n$ be a semifree action with $(\Sigma^n,\phi)^G = F$ and $\nu(F \subset \Sigma) = \nu$, $\dim \nu > 2$. Assume that $\Sigma^n = D_1^n \cup D_2^n$ with $D_i^n \cap F = F_i$ is a non-equivariant splitting. Then there exists a smooth semifree G-sphere (Σ,ψ) such that (Σ,ψ) is G-homeomorphic to (Σ,ϕ) , $(\Sigma,\psi)^G = F$, $\psi \mid G \times \nu = \phi \mid G \times \nu$, and (Σ,ψ) has an equivariant splitting into disks $\Sigma = D^n \cup D^n$ with $(D^n)^G = F_i$ if and only if $\sum_{j>0} (-1)^j \sigma_G (H_j(F_i)) = 0$ in $\widetilde{K}_0(\mathbb{Z}G)$, $i = 1$ or 2 .

∎

IV.6. REMARK: Suppose F^k is a $\bmod q$ homology sphere. Then Anderson and Hambleton prove that the necessary and sufficient conditions for existence of a "balanced splitting" of F^k (i.e. F is homologically a double) is that a certain semicharacteristic type invariant vanishes (cf. [1] Theorem B). Thus Theorem IV.5 can be applied to generalize this result of Anderson-Hambleton and improve their dimension hypothesis in Theorem B of [1] from $\dim \nu \geq k+2$ to $\dim \nu > 2$.

As in Section II, we can generalize the above results to actions with many isotropy subgroups. The proofs of the semifree cases can be adapted to serve as the inductive step of the following theorem. The normally linear Poincaré G-complex which is the homotopy model in this case is provided by Theorem II.2. We omit the details.

IV.7. THEOREM: Let (X^n,ϕ) be a smooth closed G-manifold with an admissible splitting $X = X_1 \cup X_2$, satisfying all the hypotheses of Theorem II.2. Then there exists a smooth G-action $\psi : G \times X \longrightarrow X$ such that (X,ψ) is concordant to (X,ϕ) rel X^G and (X,ψ) has an equivariant splitting $X = X_1' \cup X_2'$ which satisfies the conclusions (a) and (b) of Theorem II.2.

SECTION V. REALIZATION OF OBSTRUCTIONS:

One may use normally linear Poincaré complexes to construct actions with admissible splittings which do not admit necessarily equivariant splittings. Again, the results of this section may be specialized to the situation considered by Anderson-Hambleton [1] to give an alternative proof of their Theorem C. The important algebraic calculations of the hyperbolic map in the Rothenberg-Ranicki exact sequence for the quaternionic groups are due to Anderson-Hambleton ([1] Proposition 5.2 and [13] Lemma 6.1) who applied it in their examples of actions on spheres without balanced splittings. These calculations are used to take care of the case where the 2-Sylow subgroup is the quaternion group of order 8, denoted by $\mathcal{Q}_8$.

V.1. THEOREM: Let M^n be a simply-connected closed manifold, and $M^n = M_1^n \cup M_2^n$ be an admissible splitting. Suppose that $F^k \subset M$ is a closed submanifold with normal bundle ν which admits a G-bundle structure with a free representation on each fibre, where G is a subgroup of SU(2) whose 2-Sylow subgroup is either (i) cyclic or (ii) $\mathcal{Q}_8$ and $K \not\equiv 1 \bmod 4$. Assume that (M_0^{n+1},F_0^{k+1}) is a manifold pair such that $\partial(M_0,F_0) = (M,F)$, $F_i = M_i \cap F$ satisfying the hypotheses:
(1) for $i = 1,2$ $\pi_1(M_i) = 1$ and $H_*(M_i,F_i;\mathbb{Z}_q) = 0$, where $i = 0,1,2,$.
(2) The G-bundle structure of ν extends to the normal bundle of F_0 in M_0. Then there exists a quasisimple semifree action $\phi : G \times M' \longrightarrow M'$ such that $M'^G = F$, where M' is homotopy equivalent to M. Further, (M',ϕ) has an equivalent splitting $M_1' \cup M_2'$, $M_i' \cap F = F_i$ if and only if $\Sigma(-1)^j\sigma_G H(M_1,F_1) = 0$ in $\widetilde{K}_0(\mathbb{Z}G)$.

The idea of the proof of this theorem is the following. Using the hypotheses (1) and (2) in this context, we construct a normally linear Poincaré pair $(X,\partial X)$ with semifree G-action such that $X^G = F_0$ and $(X,\partial X) \simeq (M_0,\partial M_0)$. This pair is not necessarily finite, however, one shows that the finiteness obstruction for the boundary vanishes, so that ∂X is a finite Poincaré G-complex. Then a surgery problem is set us as in [9] and in the spirit of section IV of the present paper. To realize the obstructions for equivariant splittings, one may choose (M_0,F_0) such that for any choice of an admissible splitting, the cohomology class in $\hat{H}(\mathbf{Z}_2;\tilde{K}_0(\mathbf{Z}G)$ represented by the finiteness obstruction $\Sigma(-1)^j\sigma_G(H_j(M_1,F_1))$ be non-zero. E.g. when $G = Q_8$ $\tilde{K}_0(ZQ_8) \cong \mathbf{Z}_2$, and there are such pairs (M,F) with non-zero obstructions. One instance of this is Anderson-Hambleton's example using thickerings of Moore spaces with appropriate homology. The crucial algebraic fact is that this non-zero element contributes non-trivally only to the surgery obstructions which arise in the process of equivariant splittings i.e. (M_1,F_1). This contribution is zero when the surgery problem is considered over all of ∂X. This is reflected in the algebraic calculations of Anderson-Hambleton [1] of the hyperbolic map in the Ranicki-Rothenberg exact sequence. In fact, the approach of constructing the normally linear Poincaré model of this problem simplifies and shortens only the geometric part of the proof of Theorem C of [1]. The more delicate algebraic computations are already treated in [1], and we use them almost in the same way as in [1] (only at the last stage to complete the surgery and produce M' which is G-homotopy equivalent to ∂X rel F.) We comment that the cobounding surgery problem (X,F) is only auxiliary and simplifies the study of the surgery obstruction on ∂X.

This theorem may be generalized to actions with several isotropy subgroups. The full proof of this theorem and further applications of normally linear Poincaré complexes will appear elsewhere.

REFERENCES

[1] *D. Anderson and I. Hambleton:* "Balanced splittings of semi-free actions of finite groups on homotopy spheres", Com. Math. Helv. 55 (1980) 130-158.

[2] *A. Assadi:* "Finite Group Actions on Simply-connected Manifolds and CW Complexes", Memoirs AMS, No. 257 (1982).

[3] *A. Assadi:* "Extensions of finite group actions from submanifolds of a disk", Proc. of London Top. Conf. (Current Trends in Algebraic Topology) AMS (1982).

[4] *A. Assadi:* "Extensions Libres des Actions des Groupes Finis dans les Variétés Simplement Connexes", (Proc. Aarhus Top. Conf. Aug. 1982) Springer-Verlag LNM 1051.

[5] *A. Assadi:* "Concordance of group actions on spheres", Proc. AMS Conf. Transformation Groups, Boulder, Colorado (June 1983) Editor, R. Schultz, AMS Pub. (1985).

[6] *A. Assadi and W. Browder:* "On the existence and classification of extensions of actions of finite groups on submanifolds of disks and spheres", (to appear in Trans. AMS).

[7] *A. Assadi and W. Browder:* "Construction of free finite group Actions on Simply-connected Bounded Manifolds", (in preparation).

[8] *A. Assadi and D. Burghelea:* "Remarks on transfer in Whitehead-Theory", Max-Planck-Institut, Preprint 86-6 (1986).

[9] *A. Assadi and P. Vogel:* "Finite group actions on compact manifolds", (Preprint). A shorter version has been published in Proceedings of Rutgers conference on surgery and L-Theory, 1983, Springer-Verlag LNM 1126 (1985).

[10] *A.K. Bousfield and D.M. Kan:* Springer-Verlag LNM, No. 304 (1972).

[11] *F. Connolly:* (Talk in Oberwolfach meeting in Transformation groups, August 1982).

[12] *A. Fröhlich, M. Keating and S. Wilson:* "The class groups of quaternion and dihedral 2-groups", Mathematika 21 (1974) 64-71.

[13] *I. Hambleton and J. Milgram:* "The surgery obstruction groups of finite 2-groups", Inv. Math. 61 (1980) 33-52.

[14] *H. Hauschild:* "Äquivariante Whitehead torsion", Manus. Math. 26 (1978) 63-82.

[15] *S. Illman:* "Whitehead torsion and group actions", Ann. Acad. Sci. Fenn. Ser. Al. 588 (1974) 1-44.

[16] *J. Milnor:* "Whitehead torsion", Bull. AMS. 72 (1966) 358-426.

[17] *R. Oliver:* "Fixed-point sets of group actions on finite acyclic complexes", Comm. Math. Helv. 50 (1975) 155-177.

[18] *M. Rothenberg*: "Torsion invariants and finite transformation groups", Proc. Symp. Pur Math. vol. 32, Part I, AMS. (1978).

[19] *A. Ranicki*: "Algebraic L-theory I: Foundations", Proc. Lon. Math. Soc. (3) 27 (1973) 101-125.

[20] *R. Schultz*, Editor, Proceeding of AMS summer conference in transformation groups, Boulder Colorado 1982, AMS. R.I. (1985).

[21] *R. Swan*: "Periodic resolutions and projective modules", Ann. Math. 72 (1960) 552-578.

[22] *C.T.C. Wall*: "Surgery on compact manifolds", Academic Press, New York 1970.

[23] *S. Weinberger*: "Homologically trivial actions I" and "II", (preprint), Princeton University (1983).

[24] *S. Weinberger*: "Constructions of group actions", Proceedings of AMS summer conference in transformation groups, Boulder Colorado 1982, AMS. R.I. (1985).

FREE $(\mathbb{Z}/_2)^k$-ACTIONS AND A PROBLEM IN COMMUTATIVE ALGEBRA

Gunnar Carlsson*

(I) INTRODUCTION. In [1,2], the following theorem is proved.

THEOREM I.1. Suppose $G = (\mathbb{Z}/_p\mathbb{Z})^k$ acts freely on a finite complex X, where X is homotopy equivalent to $(S^n)^\ell$, and suppose that G acts trivially on n-dimensional mod-p homology. Then $\ell \geq k$.

The analogous theorem for $G = (S^1)^k$ is proved in [6]. In fact, for this case, the theorem is proved for $S^{n_1} \times \ldots \times S^{n_\ell}$, where the n_i's may be distinct. The proofs of these theorems rely heavily on the special homological properties of the spaces involved, in particular on the non-vanishing of cup-products in $H^*(X;\mathbb{Z}/_p\mathbb{Z})$ or $H^*(X;\mathbb{Q})$. One's initial reaction is to attempt to remove the hypothesis of trivial action on homology in Theorem I.1, to extend the result to $S^{n_1} \times \ldots \times S^{n_\ell}$. However, in attempting this, one is still utilizing the special properties of the spaces involved; a more appealing approach is to try to find a priori homological properties which must be satisfied by spaces which admit free $(\mathbb{Z}/_p\mathbb{Z})^k$ or $(S^1)^k$-actions, and which apply in a wide family of examples.

Such general properties are hard to come by; an example is:

THEOREM I.2 [3]. Let X be a finite free G-complex, $G = (\mathbb{Z}/_2\mathbb{Z})^k$ or $(S^1)^k$, and suppose G acts trivially on $H^*(X;\mathbb{Z}/_2\mathbb{Z})$, if $G = (\mathbb{Z}/_2\mathbb{Z})^k$. Then X has at least k non-trivial homology groups.

We now propose as a conjecture the following much more striking a priori restriction.

CONJECTURE I.3. Suppose $G = (\mathbb{Z}/_p\mathbb{Z})^k$ or $(S^1)^k$, and suppose X is a finite free G-complex. Then $\sum_i \mathrm{rk}_{\mathbb{Z}/_p\mathbb{Z}} H_i(X,\mathbb{Z}/_p\mathbb{Z})$ or $\sum_i \mathrm{rk}_{\mathbb{Q}} H_i(X,\mathbb{Q})$, respectively, is $\geq 2^k$.

REMARK: The rational version of this conjecture has also been proposed by S. Halperin.

* The author is an Alfred P. Sloan Fellow, and is supported in part by N.S.F. Grant 82-01125.

So far, this conjecture can be proved for $(\mathbb{Z}/_{2\mathbb{Z}})^k$ and $(S^1)^k$, with $k \leq 3$ (see [4], where the case of $(\mathbb{Z}/_{2\mathbb{Z}})^k$ is handled. The proof for $(S^1)^k$ is entirely similar.) The case $k=4$ can probably also be carried through with these techniques.

In this paper, we'll formulate the algebraic analogue of the conjecture for $G = (\mathbb{Z}/_{2\mathbb{Z}})^k$, and prove its equivalence with a question concerning differential graded modules over polynomial rings. We'll also briefly discuss its relationship with commutative algebraic conjectures of Horrocks, related to the study of algebraic vector bundles on projective spaces.

The author wishes to thank L- Avramov, S. Halperin, and J.E. Roos for stimulating discussions concerning this subject.

(II) THE ALGEBRAIC FORMULATION. We consider $\mathbb{F}_2[G]$, $G = (\mathbb{Z}/_{2\mathbb{Z}})^k$, and let $\Lambda_k = \mathbb{F}_2[G]$. As an algebra, Λ_k is isomorphic to the exterior algebra $E(y_1,\ldots,y_k)$, $y_i = T_i+1$, where $\{T_1,\ldots,T_k\}$ is a basis for $(\mathbb{Z}/_{2\mathbb{Z}})^k$. We view Λ_k as a graded ring by assigning the grading 0 to all elements of Λ_k.

Let A_* be a graded ring.

DEFINITION II.1. A DG (Differential Graded) A_*-module is a free, graded A_*-module M with a graded A_*-module homomorphism $d : M \to M$ of degree (-1) so that $d^2 = 0$.

A DG A_*-module is said to be finitely generated, bounded above, or bounded below if its underlying graded module is. The homology of M, H_*M is defined in the usual way; H_*M is itself a graded A_*-module. The notions of homomorphism, chain homotopies, and chain equivalences of DG A_*-modules are the evident ones.

Now, for a graded ring A_*, we let $\mathcal{D}(A_*)$ denote the category of finitely generated DG A_*-modules, and if A_* is bounded above, we let $\mathcal{D}_\infty(A_*)$ denote the category of bounded above DG A_*-modules. $\mathcal{D}(A_*)$ is of course a subcategory of $\mathcal{D}_\infty(A_*)$.

The algebraic formulation of our Conjecture I.3 is the following.

CONJECTURE II.2. Let $M \in \mathcal{D}(\Lambda_k)$. Then $\mathrm{rk}_{\mathbb{F}_2} H_*M \geq 2^k$.

We observe that Conjecture II.2 implies Conjecture I.3. For if X is any finite G-complex, then the cellular chains $\tilde{C}_*(X;\mathbb{F}_2)$ are a finitely generated chain complex of free $\mathbb{F}_2[G] = \Lambda_k$-modules, which is the same as an object of $\mathcal{D}(\Lambda_k)$, and $H_*(X;\mathbb{F}_2)=H_*(\tilde{C}_*(X;\mathbb{F}_2))$.

Suppose that the ring A_* is an augmented algebra over a field k, so that k is a module over A_*. If $M \in \mathcal{D}_\infty(M)$, we denote by $H_*(M,k)$ the homology of the DG k-module $k \otimes_A M$. Λ_k is of course an augmented ring over $\mathbb{F}_2$ via the augmentation $\Lambda_k \to \mathbb{F}_2$, $T_i \to 1$.

Now, let P_k denote the polynomial ring $\mathbb{F}_2[x_1,\dots,x_k]$, which we grade by assigning each variable the grading (-1). P_k is also augmented over $\mathbb{F}_2$; the augmentation is determined by the requirement that $x_i \to 0$ for all i. Recall from [3] that there is a functor $\beta : \mathcal{D}(\Lambda_k) \to \mathcal{D}(P_k)$ defined as follows. For a DG Λ_k-module (M,∂) the underlying module of $\beta(M,\partial)$ is $M \otimes_{\mathbb{F}_2} P_k$, and the differential δ on $\beta(M)$ is defined by $\delta(m \otimes f) = \partial m \otimes f + \sum_{i=1}^{k} y_i m \otimes x_i f$. The P_k action is on the right hand factor. We also have

PROPOSITION II.3. [3; Propositions II.1 and II.2]. There are natural isomorphisms $H_*M \to H_*(\beta M;\mathbb{F}_2)$ and $H_*(M;\mathbb{F}_2) \to H_*\beta M$.

An immediate consequence is

COROLLARY II.4. [3; Corollary II.3]. For any $M \in \mathrm{ob}\,\mathcal{D}(\Lambda_k)$, $H_*\beta M$ is finitely generated as an $\mathbb{F}_2$-vector space.

For any graded ring A_*, bounded above, we let $h\mathcal{D}(A_*)$ and $h\mathcal{D}_\infty(A_*)$ denote the "homotopy categories" of $\mathcal{D}(A_*)$ and $\mathcal{D}_\infty(A_*)$. These are obtained from $\mathcal{D}(A_*)$ and $\mathcal{D}_\infty(A_*)$ by inverting all chain equivalences. Let $\mathcal{D}^o(P_k)$ and $\mathcal{D}^o_\infty(P_k)$ denote the full subcategories of $\mathcal{D}(P_k)$ and $\mathcal{D}_\infty(P_k)$, respectively, whose objects are the DG-P_k-modules (M,∂) for which H_*M is a finitedimensional $\mathbb{F}_2$-vector space. We also let $h\mathcal{D}^o(P_k)$ and $h\mathcal{D}^o_\infty(P_k)$ denote the corresponding homotopy categories. Finally, let $\mathcal{D}^o_\infty(\Lambda_k)$ denote the full subcategory of $\mathcal{D}_\infty(\Lambda_k)$ whose objects are chain requivalent to objects in $\mathcal{D}(\Lambda_k)$ and let $h\mathcal{D}^o_\infty(\Lambda_k)$ denote the corresponding homotopy category. Let $h\beta : h\mathcal{D}(\Lambda_k) \to h\mathcal{D}(P_k)$ be the induced map on homotopy categories. Then Corollary II.4 shows that $h\beta$ factors through $h\mathcal{D}^o(P_k)$. Moreover, it is easy to check that it extends to a functor $H : h\mathcal{D}^o_\infty(\Lambda_k) \to h\mathcal{D}^o_\infty(P_k)$.

DEFINITION II.5. Let $(M,\partial) \in \mathrm{ob}\,\mathcal{D}_\infty(A_*)$, where $A_* = 0$ for $* > 0$, and where A_* is augmented over a field k. We say that (M,∂) is minimal if the map $\partial \otimes \mathrm{id} : M \otimes_A k \to M \otimes_A k$ is the zero map.

PROPOSITION II.6. For every $(M,\partial) \in \mathrm{ob}(\mathcal{D}_\infty(A_*))$, there exists $(\bar{M},\bar{\partial}) \in \mathrm{ob}(\mathcal{D}_\infty(A_*))$, where $(\bar{M},\bar{\partial})$ is minimal and is chain equivalent to (M,∂).

PROOF. This is Proposition I.7 of [4].

We now prove our main theorem.

THEOREM II.7. $H : h\mathcal{D}^o_\infty(\Lambda_k) \to h\mathcal{D}^o_\infty(P_k)$ is an equivalence of categories.

PROOF. We first construct a functor $G : \mathcal{D}^o_\infty(P_k) \to \mathcal{D}^o_\infty(\Lambda_k)$ as follows. Given a DG P_k-module (M,∂), the underlying module is of $G(M,\partial)$ is $M \otimes_{\mathbb{F}_2} \Lambda_k$, and the differential δ on $G(M,\partial)$ is defined by $\delta(m \otimes \alpha) = \partial m \otimes \alpha + \sum_{i=1}^{k} x_i m \otimes y_i \alpha$. One proves, by arguments identical to those in the proofs of Propositions II.1 and II.2, that $H_*(G(M)) \cong H_*(M,\mathbb{F}_2)$ and $H_*M \cong H_*(G(M);\mathbb{F}_2)$. To see that $G(M) \in \mathrm{ob}\,\mathcal{D}^o_\infty(\Lambda_k)$, we note that since $M \in \mathcal{D}^o_\infty(P_k)$, $\dim_{\mathbb{F}} H_*M < +\infty$. Therefore $\dim_{\mathbb{F}_2} H_*(G(M);\mathbb{F}_2) < +\infty$. Let $\overline{G(M)}$ be any minimal DG Λ_k-module, chain equivalent to $G(M)$. Then $\overline{G(M)} \otimes_{\Lambda_k} \mathbb{F}_2 \cong H_*(\overline{G(M)};\mathbb{F}_2) \cong H_*(G(M);\mathbb{F}_2)$ so $\overline{G(M)}$ is finitely generated which was to be shown.

We now construct a natural transformation $N : G \circ H \to \mathrm{Id}$ as follows. The underlying module of $G \circ H(M)$ is $M \otimes_{\mathbb{F}_2} P_k \otimes_{\mathbb{F}_2} \Lambda_k$; let $\epsilon : P_k \to \mathbb{F}_2$ be the augmentation, and let $\mu : M \otimes \Lambda_k \to M$ be the structure map for M as a Λ_k-module. Then we define $N(M)$ to be the map $M \otimes_{\mathbb{F}_2} P_k \otimes_{\mathbb{F}_2} \Lambda_k \xrightarrow{\mathrm{id} \otimes \epsilon \otimes \mathrm{id}} M \otimes_{\mathbb{F}_2} \Lambda_k \xrightarrow{\mu} M$; it is easily checked to be a chain map, and a chain equivalence. Similarly, we define $N'(M) : H \circ G(M) \to M$ to be the composite $M \otimes_{\mathbb{F}_2} \Lambda_k \otimes_{\mathbb{F}_2} P_k \to M \otimes_{\mathbb{F}_2} P_k \to M$. This is also easily checked to be a chain equivalence, which proves the theorem. ∎

This equivalence of categories leads us to propose the following:

CONJECTURE II.8. Let $M \in \mathrm{ob}\,\mathcal{D}^o(P_k)$. Then $\mathrm{rk}_{P_k} M \geq 2^k$.

Finally, we prove

PROPOSITION II.9. Conjecture II.8 is equivalent to Conjecture II.2.

PROOF. By Theorem II.7 and Proposition II.3, Conjecture II.2 is equivalent to the conjecture that for all $M \in \mathrm{ob}\,\mathcal{D}^o(P_k)$, $\mathrm{rk}_{\mathbb{F}_2} H_*(M;\mathbb{F}_2) \geq 2^k$. But Proposition II.6 shows that M is equivalent to a minimal DG P_k-module $\overline{M}$. $\mathrm{rk}_{P_k} M \geq \mathrm{rk}_{P_k} \overline{M} = \mathrm{rk}_{\mathbb{F}_2} \overline{M} \otimes_{P_k} \mathbb{F}_2 = \mathrm{rk}_{\mathbb{F}_2} H_*(\overline{M};\mathbb{F}_2) \geq 2^k$. ∎

(III) THE RELATION WITH HORROCKS' CONJECTURE. G. Horrocks' has conjectured the following (see [5] for discussion of related material.)

CONJECTURE III.1. Let M be an Artinian graded module over the polynomial ring $R = F[x_1,\ldots,x_k]$, where F is a field. Then $rk_F Tor_i^R(M,F) \geq \binom{k}{i}$.

We may weaken this slightly to

CONJECTURE III.2. Let M be an Artinian graded module over the polynomial ring $R = F[x_1,\ldots,x_k]$, where F is a field. Then $\sum_i rk_F Tor_i^R(M,F) \geq 2^k$.

The relationship between our conjectures and this one is now given by the following.

PROPOSITION III.3. Conjecture II.8 implies Conjecture III.2 for $F = \mathbb{F}_2$.

PROOF. Let M be any Artinian graded module over P_k, and let $R(M)$ denote a minimal graded resolution of M. Then $rk_{P_k} R(M) = \sum_i rk_{\mathbb{F}_2} Tor_i^{P_k}(M;\mathbb{F}_2)$, and $R(M)$ may certainly be viewed as an object of $\mathcal{D}(P_k)$. Since $H_*R(M) \cong M$, and M is $\mathbb{F}_2$-finite dimensional (since it is Artinian), $R(M)$ is in fact an object of $\mathcal{D}^o(P_k)$. Thus, if Conjecture II.8 holds, then

$$\sum_i rk_{\mathbb{F}_2} Tor_i^{P_k}(M,\mathbb{F}_2) = rk_{P_k} R(M) \geq 2^k .$$

REFERENCES

[1] CARLSSON, G.: On the non-existence of free actions of elementary abelian groups on products of spheres, Am. Journal of Math., 102, No. 6, (1980), pp. 1147-1157.

[2] CARLSSON, G.: On the rank of abelian groups acting freely on $(S^n)^k$, Inventiones Math., 69, (1982), pp. 393-400.

[3] CARLSSON, G.: On the homology of finite free $(\mathbb{Z}/2\mathbb{Z})^n$-complexes, Inventiones Math., 74, (1983), pp. 139-147.

[4] CARLSSON, G.: Free $(\mathbb{Z}/2)^3$-actions on finite complexes, to appear, Proceedings of a Conference in honor of John Moore.

[5] HARTSHORNE, R.: Algebraic vector bundles on projective spaces: a problem list. Topology, 18 (1979), pp. 117-128.

[6] HSIANG, W. Y.: Cohomology Theory of Topological Transformation Groups, Springer Verlag, 1975.

Department of Mathematics, University of California, San Diego
La Jolla, CA 92093

Verschlingungszahlen von Fixpunktmengen in Darstellungsformen. II

Tammo tom Dieck und Peter Löffler

Abstract: Let $G = H_0 \times H_1$ be a product of two cyclic groups of odd order. Let $j_i : S^{n(i)} \longrightarrow S^{n(0)+n(1)+1}$, i=0,1, be any two imbeddings of standard spheres into the standard sphere. Suppose

a) The integers n(0) and n(1) are both odd and greater or equal to 5.

b) The normal bundles ν_i of the imbeddings j_i, i=0,1, are both trivial.

c) The linking number k of $j_0(S^{n(0)})$ and $j_1(S^{n(1)})$ is a unit in $\mathbb{Z}/|G|$ and lies in the kernel of the Swan homomorphism $s_G : \mathbb{Z}/|G|^* \longrightarrow K(\mathbb{Z}G)$.

Then there is a smooth action of G on $X = S^{n(0)+n(1)+1}$ such that

1) the isotropy groups are 1, H_0, H_1,

2) the fixed point sets X^{H_i} are the spheres $j_i(S^{n(i)})$, i=0,1.

Ziel dieser Note ist es, den folgenden Satz zu beweisen:

Satz 1: Sei $G = H_0 \times H_1$ ein Produkt von zwei zyklischen Gruppen ungerader Ordnung. Für i=0,1 seien $j_i : S^{n(i)} \longrightarrow S^{n(0)+n(1)+1}$ disjunkte Einbettungen von Standardsphären in die Standardsphäre. Es gelte

a) Die Zahlen n(0) und n(1) sind beide ungerade und größer oder gleich fünf.

b) Die Normalenbündel ν_i der Einbettungen j_i, i=0,1, sind trivial.

c) Die Verschlingungszahl k von $j_0(S^{n(0)})$ und von $j_1(S^{n(1)})$ in $S^{n(0)+n(1)+1}$ sei eine Einheit in $\mathbb{Z}/|G|$ und liege im Kern des Swan Homomorphismus' $s_G : \mathbb{Z}/|G|^* \longrightarrow K(\mathbb{Z}G)$.

Dann gibt es eine glatte Operation von G auf $X = S^{n(0)+n(1)+1}$, so daß

1) die Isotropiegruppen 1, H_0, H_1 sind,

2) die Fixpunktmengen X^{H_i} die Sphären $j_i(S^{n(i)})$ sind, i=0,1.

Dieser Satz verallgemeinert den Hauptsatz aus [tDL], wo die Existenz einer solchen Verschlingunskonfiguration der Fixpunktmengen bewiesen worden war. In [Le] wird allerdings gezeigt, daß man mehrere solcher Konfigurationen bei festem n(0) und n(1) vorgeben kann. Der hier angegebene Beweis differiert auch erheblich von dem aus [tDL] und kann vermutlich - mit gewissen Einschränkungen - auf einfach zusammenhängende rationale Homologiesphären mit komplexen Strukturen auf den Normalenbündeln der Einbettungen erweitert werden.(Für die Einschränkungen vergleiche etwa [Sch]) Die hier benutzten Methode der Erweiterung von Gruppenoperationen, die auf dem Rand einer Mannigfaltigkeit vorgegeben sind, haben sich schon an anderer Stelle bewährt.

Wir setzen $n = n(0)+n(1)+1$. Ohne Beschränkung der Allgemeinheit dürfen wir $n(1) \geqslant n(0) \geqslant 5$ voraussetzen. Bekanntlich gilt dann, daß $j_0 : S^{n(0)} \longrightarrow S^n$ bis auf Isotopie die Standardeinbettung ist([Le]).

Nach Voraussetzung b) können die Einbettungen j_i zu disjunkten Einbettungen

$$\tilde{j}_i : S^{n(i)} \times B^{n-n(i)} \longrightarrow S^n$$

verdickt werden.

Wir setzen

$$X = S^n - \tilde{j}_0(S^{n(0)} \times \mathring{B}^{n-n(0)}) - \tilde{j}_1(S^{n(1)} \times \mathring{B}^{n-n(1)}).$$

X ist eine Mannigfaltigkeit mit Rand δX.

Es gilt $\quad \delta X = \delta_0 X \cup \delta_1 X$

mit $\quad \delta_i X \cong S^{n(0)} \times S^{n(1)} \quad , i=0,1,$

wobei die Diffeomorphismen durch die $\tilde{j}_i$ induziert werden.

Man errechnet

$$H_i(X,\delta_1 X) \cong \begin{cases} \mathbb{Z}/k & i = n(1) \\ 0 & \text{sonst.} \end{cases}$$

Bekanntlich gibt es Inklusionen

$$i_0 : S^{n(0)} \longrightarrow S^n - j_1(S^{n(1)})$$

und

$$i_1 : S^{n(1)} \longrightarrow S^n - j_0(S^{n(0)}),$$

die Homotopieäquivalenzen sind [M].Seien

$$\phi_0 : S^n - j_1(S^{n(1)}) \longrightarrow S^{n(0)}$$

$$\phi_1 : S^n - j_0(S^{n(0)}) \longrightarrow S^{n(1)}$$

Homotopieinverse zu diesen Inklusionen.

Wir betrachten nun das folgende Diagramm

$$
(*)\qquad
\begin{array}{ccccc}
\delta_0 X & \longrightarrow & X & \longleftarrow & \delta_1 X \\
\Big\downarrow{\scriptstyle l_0} & \swarrow{\scriptstyle k_0} & & \searrow{\scriptstyle k_1} & \Big\downarrow{\scriptstyle l_1} \\
S^n - j_1(S^{n(1)}) & & & & S^n - j_0(S^{n(0)}) \\
\Big\downarrow{\scriptstyle \phi_0} & & & & \Big\downarrow{\scriptstyle \phi_1} \\
S^{n(0)} & & & & S^{n(1)}
\end{array}
$$

wobei $l_i = k_i \circ \tilde{j}_i$, $i = 0,1$, gesetzt ist

und k_i die offensichtliche Inklusion ist. Wir definieren schließlich

$$\alpha' : X \longrightarrow S^{n(0)} \times S^{n(1)}$$

durch $\alpha' = (\phi_0 \circ k_0) \times (\phi_1 \circ k_1)$.

Ist $r \in \mathbb{Z}$, so bezeichne $[r]: S^a \longrightarrow S^a$ eine Abbildung vom Grad r.

<u>Lemma 1</u>: Wir haben ein homotopiekommutatives Diagramm

$$
\begin{array}{ccccc}
\delta_0 X & \longrightarrow & X & \longleftarrow & \delta_1 X \\
\Vert & & & & \Vert \\
S^{n(0)} \times S^{n(1)} & & \Big\downarrow{\scriptstyle \alpha'} & & S^{n(0)} \times S^{n(1)} \\
& {\scriptstyle [k] \times [1]} \searrow & & \swarrow {\scriptstyle [1] \times [k]} & \\
& & S^{n(0)} \times S^{n(1)} & &
\end{array}
$$

<u>Beweis</u>: Dies folgt leicht aus dem Diagramm (*) und der Definition und den Eigenschaften der Verschlingungszahl.

Wir benötigen nun den folgenden Satz:

<u>Satz 2</u>: Sei G eine endliche Gruppe der Ordnung g. Sei X^n eine kompakte Mannigfaltigkeit, $n \geq 6$, mit $\delta X = \delta_0 X \cup \delta_1 X$ und $\delta_0 X \cap \delta_1 X = \emptyset$. Es gelte

1) $\pi_i(X) = \pi_i(\delta_0 X) = \pi_i(\delta_1 X) = 0 \quad i = 0,1.$

2) $H_*(X, \delta_0 X) \otimes \mathbb{Z}_{(g)} = 0$

3) G operiere frei auf $\delta_0 X$ und die auf $H_*(\delta_0 X) \otimes \mathbb{Z}[g^{-1}]$ induzierte Operation sei trivial.

4) Es bezeichne h_i die Ordnung von $H_i(X, \delta_0 X)$. Setze $P(X, \delta_0 X) = \prod h_{2i} / \prod h_{2i+1}$. Wegen 2) definiert $P(X, \delta_0 X)$ durch Reduktion ein

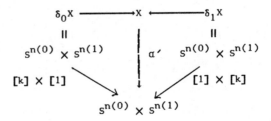

Element in $(\mathbb{Z}/g)^*$. Es sei $s_G(P(X,\delta_0 X)) = 0$ (s_G = Swan Homomorphismus von G).

Unter diesen Voraussetzungen gibt es auf X eine freie G-Operation, die die auf $\delta_0 X$ gegebene erweitert. Diese induziert auf $H_*(X) \otimes \mathbb{Z}[g^{-1}]$ wieder die triviale G-Operation.

Dieser Satz wurde von mehreren Autoren unabhängig voneinander bewiesen [AB], [W], um nur zwei Quellen zu nennen.

Der Satz wurde so zitiert, daß er genau auf unseren Sachverhalt paßt. Jede G-Operation auf $\delta_0 X$ (die der Bedingung 3) genügt) kann auf X erweitert werden und induziert eine auf $\delta_1 X$.

Wir wählen nun für i = 0, 1 freie H_i-Darstellungen V_i mit $\dim_{\mathbb{R}} V_i = n(i)+1$. Seien $S(V_i)$ die zugehörigen Einheitssphären mit induzierter freier G-Operation. Wie in [tDL] 2.2 zeigt man, daß die Normalenabbildung

$k \cdot id : k \cdot S(V_i) \longrightarrow S(V_i)$ vom Grad k durch Umhenkeln (n(i)-1)-zusammenhängend gemacht werden kann. So erhält man eine Sphäre $\Sigma(k,V_i)$ mit freier G-Operation. Wir nennen $\Sigma(k,V_i)$ (mit der Grad-k Normalenabbildung) ein k-faches von $S(V_i)$. Sicher ist $\Sigma(k,V_i)$ nicht eindeutig. Aber da $L_n^h(G)$ verschwindet [B], werden je zwei Vertreter von $\Sigma(k,V_0) \times S(V_1)$ bzw. $S(V_0) \times \Sigma(k,V_1)$ h-kobordant. Nun gilt:

<u>Satz 3</u>: Versehen wir $\delta_0 X$ mit der G-Operation $\Sigma(k,V_0) \times S(V_1)$, und versehen wir X mit der durch Satz 2 garantierten freien G-Operation, so wird auf $\delta_1 X$ gerade $S(V_0) \times \Sigma(k,V_1)$ induziert.

<u>Folgerung</u>: Satz 1 ist richtig.

<u>Beweis</u>: Man betrachte

$$\Sigma(k,V_0) \times B(V_1) \cup X \cup B(V_0) \times \Sigma(k,V_1).$$

Daß man nach Vergessen der G-Operation wieder das Objekt erhält, mit dem man anfing, liegt daran, daß der Diffeomorphismus

$\delta_1 X \cong S(V_0) \times \Sigma(k,V_1)$ nach Vergessen der G-Operation die Identität ist (vergleiche den Beweis nach Lemma 3).

Es bleibt Satz 3 zu zeigen:

Dazu versehen wir X mit der durch Satz 2 garantierten freien G-Operation. Man betrachte nun das folgende Diagramm:

$$\begin{array}{ccccc}
\delta_0 X = \Sigma(k,V_0) \times S(V_1) & \xrightarrow{\ i_0\ } & X & \longleftarrow & \delta_1 X \\
 & \searrow^{\alpha_0 = [k]\times[1]} & \downarrow \alpha & \swarrow_{\alpha_1} & \\
 & & S(V_0) \times S(V_1) & &
\end{array}$$

Lemma 2: Die G-äquivariante Normalenabbildung α_0 kann (bis auf Homotopie) eindeutig zu einer G-äquivariante Normalenabbildung α erweitert werden.

Beweis:a) Existenz einer G-Abbildung.
Wegen Lemma 1 gibt es α', eine nicht äquivariante Abbildung. Invertieren wir die Gruppenordnung g, so hat man ein homotopiekommutatives Diagramm

$$\begin{array}{ccc}
(\delta_0 X/G)_{(1/g)} & \longrightarrow & (X/G)_{(1/g)} \\
\wr\| & & \wr\| \\
|\delta_0 X|_{(1/g)} \times BG & \longrightarrow & |X|_{(1/g)} \times BG
\end{array}$$

($|\delta_0 X|$ bzw $|X|$ bezeichnet den Raum $\delta_0 X$ bzw X mit trivialer G-Operation) und wir setzen $\alpha[1/g] = \alpha'[1/g] \times id_{BG}$. (Für die Bezeichnungen und den Hintergrund über Lokalisierungen vergleiche man etwa [ELP]). Lokalisieren wir an der Gruppenordnung, so wird $i_{0(g)}$ eine Homotopieäquivalenz. Wir setzen $\alpha_{(g)} = \alpha_{(0)} \circ i^{-1}_{0(g)}$. Die Abbildungen $\alpha[1/g]$ und $\alpha_{(g)}$ passen rational zusammen und definieren α.

b)Behauptung: Ist T(X/G) das Tangentialbündel von X/G, so gilt als Gleichung in KO(X/G)

$$(V_0 \oplus V_1) \times_G X \cong T(X/G).$$

Beweis: Betrachte hierzu

$$\begin{array}{ccccc} KO(X,\delta_0 X) & \longrightarrow & KO(X) & \longrightarrow & KO(\delta_0 X) \\ \uparrow{\scriptstyle \pi_1^*} & & \uparrow{\scriptstyle \pi_2^*} & & \uparrow{\scriptstyle \pi_3^*} \\ KO(X/G,\delta_0 X/G) & \longrightarrow & KO(X/G) & \longrightarrow & KO(\delta_0 X/G) \end{array}$$

(hierbei sind die π^* durch Projektionen induziert).

Aus der Atiyah-Hirzebruch Spektralfolge ergibt sich:

1) π_1^* ist ein Isomorphismus;

2) $\pi_2^* \otimes \mathbb{Z}[1/g]$ ist ein Isomorphismus.

Setze $\Delta = T(X/G) - (V_0 \oplus V_1) \times_G X$.

Man hat nun

$$\pi_2^*(\Delta) = T(X) = 0$$

sowie $i_0^*(\Delta) = 0,$

weil die geforderte Gleichheit sicher über $\delta_0 X/G$ gilt. Wegen 2) ist deshalb Δ nur g-Torsion. Wegen 1) und der Struktur von $H_*(X,\delta_0 X)$ besteht $KO(X/G,\delta_0 X/G)$ aber nur aus k-Torsion.

c) Behauptung: Die Erweiterung α von α_0 kann eindeutig (bis auf Homotopie) als G-äquivariante Normalenabbildung gewählt werden.

Beweis: Wähle einen Isomorphismus

$$\phi\colon T(X/G) \oplus \epsilon^N \longrightarrow \alpha^*(T(S(V_0) \times S(V_1))/G) \oplus \epsilon^{N+1},$$

den es wegen b) gibt. Man betrachte

$$\begin{array}{ccccccc} KO^{-1}(X,\delta_0 X) & \longrightarrow & KO^{-1}(X) & \longrightarrow & KO^{-1}(\delta_0 X) & \longrightarrow & KO(X,\delta_0 X) \\ \uparrow{\scriptstyle \pi_1^*} & & \uparrow{\scriptstyle \pi_2^*} & & \uparrow{\scriptstyle \pi_3^*} & & \uparrow{\scriptstyle \pi_4^*} \\ KO^{-1}(X/G,\delta_0 X/G) & \longrightarrow & KO^{-1}(X/G) & \longrightarrow & KO^{-1}(\delta_0 X/G) & \longrightarrow & KO(X/G,\delta_0 X/G) \end{array}$$

Wir müssen zeigen, daß es einen Automorphismus ψ des stabilen Bündels $T(X/G)$ gibt, der die folgenden Eigenschaften hat:

1) $\psi \circ \phi|_{\delta_0 X}$ ist die gegebene Normalenabbildung.

2) ψ ist eindeutig bestimmt.

Nun entsprechen stabile Automorphismen von $T(X/G)$ gerade $\widetilde{KO}^{-1}(X/G)$. Schränken wir ϕ auf $\delta_0 X/G$ ein, so gibt es einen Automorphismus ψ_1 über $\delta_0 X/G$ mit den geforderten Eigenschaften. Eine Diagrammjagd zeigt, daß man ein ψ mit den geforderten Eigenschaften finden kann. Torsionsbetrachtungen wie unter b) zeigen die Eindeutigkeit.

Bemerkung: Eigentlich ist Lemma 2 ein Teil eines ausführlichen Beweises von Satz 2.

Wie in [HM] bezeichne $N_k(((S(V_0) \times S(V_1))/G)$ die Menge der Normaleninvarianten vom Grad k. Wir wählen $(\delta_0 X/G, \alpha_0)$ als ausgezeichnetes Element.

Seien W_i, i = 0,1, freie H_i-Darstellungen, so daß die gegebenen Normalenabbildung vom Grad k $\Sigma(k,V_i) \longrightarrow S(V_i)$ zu Normalenabbildungen $\Sigma(k,V_i) \longrightarrow S(W_i)$ vom Grad 1 hochgehoben werden kann (vergleiche den Beweis von [tDL] 2.2).

Man betrachte jetzt

$$N_1(((S(W_0) \times S(V_1))/G) \xrightarrow[{[k] \times [1]}]{} N_k(((S(V_0) \times S(V_1))/G) \longleftarrow$$

$$\xleftarrow[{[1] \times [k]}]{} N_1(((S(V_0) \times S(W_1))/G).$$

Es definiert $(\delta_0 X, \alpha_0)$ das ausgezeichnete Element auf der linken Seite und in der Mitte. Offenbar besagt Lemma 2

$$[\delta_0 X, \alpha_0] = [\delta_1 X, \alpha_1] \in N_k.$$

Andererseits gilt (siehe [tDL] 2.3)

$$[\Sigma(k,V_0) \times S(V_1)] = [S(V_0) \times \Sigma(k,V_1)] \in N_k.$$

Lemma 3: Die äquivariante Normalenabbildung

$$\alpha_1 : \delta_1 X \longrightarrow S(V_0) \times S(V_1)$$

vom Grad k kann zu einer äquivariante Normalenabbildung

$$\tilde{\alpha}_1 : \delta_1 X \longrightarrow S(V_0) \times S(W_1)$$

vom Grad 1 hochgehoben werden.

Beweis: Mit Lokalisierungen beweist man dies analog zu Teil a) aus Lemma 2.

Damit haben wir die Gleichheit

$$([1] \times [k])[\delta_1 X/G, \tilde{\alpha}_1] = ([1] \times [k])[(S(V_0) \times \Sigma(k,V_1))/G] \in N_k$$

Aus [BM] Proposition 4.6 folgt, daß der Kern von [1] × [k] nur aus k-Torsion besteht. Beachten wir andererseits, daß die Projektion

$$\pi : S^{n(0)} \times S^{n(1)} \longrightarrow (S(V_0) \times S(V_1))/G$$

einen k-lokalen Isomorphismus

$$\pi^*_{(k)} : [(S(V_0) \times S(V_1))/G,\ QS^0/Cat]_{(k)} \longrightarrow [S^{n(0)} \times S^{n(1)},\ QS^0/Cat]_{(k)}$$

induziert (g ist jetzt invertierbar) und folgern daraus

$$\pi^*_{(k)} : (\delta_1 X/G,\ \tilde{\alpha}_1) = \pi^*_{(k)}\ ((S(V_0) \times \Sigma(k,V_1))/G),$$

so ergibt sich, daß die Normaleninvarianten von $(\delta_1 X,\ \tilde{\alpha}_1)$ und $(S(V_0) \times \Sigma(k,V_1))/G$ übereinstimmen müssen. Da $L^h_n(G)$ verschwindet [B], müssen beide h-kobordant sein.

Literatur

[AB] A. Assadi-W. Browder: In preparation.

[B] A. Bak: Odd dimension surgery groups of odd torsion groups vanish. Topology 14(1975), 367-374.

[BM] G. Brumfiel-I. Madsen: Evaluation of the transfer and the universal surgery class. Inv. math. 32(1976), 133-169.

[tDL] T. tom Dieck-P. Löffler: Verschlingungen von Fixpunktmengen in Darstellungsformen.I, Math. Gottingensis 1 (1985) und Alg. Top. Gött. 1984, Proc. LNM 1172(1985), 167 - 187.

[ELP] J. Ewing-P.Löffler-E. Pedersen: A Local Approach to the Finiteness Obstruction, Math. Gott. 40(1985).

[HM] I. Hambleton-I. Madsen: Local surgery obstructions and space forms, preprint 1984.

[Le] J. Levine: A classification of differentiable knots. Ann. of Math. 82(1965), 15-50.

[M] W. Massey: On the normal bundle of a sphere imbedded in Euclidean space. Proc. AMS 10,(1959), 959-964.

[Sch] R. Schultz: Differentiability and the P. A. Smith theorems for spheres: I. Actions of prime order groups. Conf. on Alg. Top., London, Ont., 1981, Can. Math. Soc. Conf. Proc. Vol. 2, Pt. 2 (1982), 235-273.

[W] S. Weinberger: Homologically trivial group actions, preprint 1983.

An algebraic approach to the generalized Whitehead group

Karl Heinz Dovermann* and Melvin Rothenberg**

Department of Mathematics
Purdue University and
Unversity of Hawaii at Manoa

Department of Mathematics
University of Chicago

Abstract: The notions of simple homotopy theory and Whitehead torsion have generalizations in the theory of transformation groups. One does not have to consider free actions. A geometric description of a generalized Whitehead group was given by Illman. The approach resembles that of Cohen. An algebraic approach was pursued by Rothenberg. This approach has been developed only under certain assumptions. In this paper we generalize the approach to give an algebraic description of the generalized Whitehead group for a finite group. In particular we put no restrictions on the component structure of the action and we do not assume that H fixed point components are 1-connected. We prove that our and Illman's approach lead to the same group.

*Partially supported by NSF Grant MCS 8100751 and 8514551

**Partially supported by NSF Grant MCS 7701623

0. Introduction

Simple homotopy theory was introduced by J.H.C. Whitehead [13] attempting to find a computational approach to homotopy theory. This notion turned out to be different from homotopy theory. Two standard references for simple homotopy theory and the related notion of Whitehead torsion are Milnor [11] and Cohen [3]. These references also include many geometric applications. Applications to the theory of free transformation groups are obtained by passing to quotient spaces.

The notions of simple homotopy theory and Whitehead torsion have generalizations in the theory of transformation groups. One does not have to consider free actions. A geometric description of a generalized Whitehead group was given by Illman [7]. The approach resembles that of Cohen. An algebraic approach was pursued by Rothenberg [12]. Which approach is preferable depends on the particular application one has in mind. Rothenbergs approach has been developed only under certain assumptions. In this paper we generalize this approach to give an algebraic description of the generalized Whitehead group for a finite group G. In particular we put no restrictions on the component structure of the action and we do not assume that H fixed point components are 1-connected. As one may expect, the groups defined by Illman and by us are related to each other. We prove that our and Illman's approach lead to the same group.

The paper is organized as follows:

In the first nine sections we introduce the basic categorical notation, K_0, K_1, and the Whitehead group Wh. In sections 10-14 we define the generalized torsion of a G-homology equivalence. The concepts required are strictly algebraic. Theorem A states that our algebraically defined group coincides with Illman's geometrically defined one. This result is based on Theorem B which describes the generalized Whitehead group as a sum of classical Whitehead groups. Finally we state the basic geometric properties of the generalized Whitehead torsion as well as the most important geometric conclusions.

The generalized Whitehead torsion has been considered in several other articles by Illman, Hauschild, Anderson, and ourselves [9,10,6,1,5] but the formalism and generality of our present approach is new. For some more recent articles by Araki, Araki-Kawakubo, and Steinberger-West see also [14,15,16].

1. Basic categories

A generic category will be denoted by M. All categories considered in the next seven sections will be assumed to have unique initial-terminal objects ∞, and all functors will be assumed to preserve them. For such M, any two objects are connected by a uniquely defined trivial map, denoted by 0. A morphism a is an epimorphism if $ba = 0$ implies $b = 0$. Projective objects are defined through the common universal property. The category $C(M)$ of finite chain complexes over M is defined in the obvious manner. An object of $C(M)$ will be denoted by (C_j, d_j), where $j \in \mathbb{Z}$. All categories we consider will be small, so that the usual set theoretic operations can be performed. We will systematically surpress mentioning that fact.

2. Exact sequences

An ES structure (ES = exact sequence) on M is a collection $ES(M) = \{(p,i)\}$ of pairs of morphisms, where domain p = range i, such that for isomorphisms α,γ,ϕ of M, $(p,i) \in ES(M)$ if and only if $(\alpha p\gamma^{-1}, \gamma i\phi^{-1}) \in ES(M)$. We further assume that for the initial-terminal object ∞ the pairs $(0_1, Id)$ and $(Id, 0_2)$ are in $ES(M)$, where $0_1\colon \infty \to A$ and $0_2\colon A \to \infty$. Subcategories always inherit ES structures, as does $C(M)$, if M has one. For abelian categories we always use the usual ES structure.

3. $K_0(M)$

For the category M with an ES structure $K_0(M)$ is well defined. It is the free abelian group generated by isomorphism classes of objects M subject to the relations $\infty = 0$, and if $(p,i) \in ES(M)$ then domain p = range i = domain i + range p. If $d\colon M_1 \to M_2$ is an exact functor, i.e., d preserves ES structures, then d induces a homomorphism $d_*\colon K_0(M_1) \to K_0(M_2)$. The inclusion of a subcategory is an example of such an exact functor.

4. Category of F chain complexes

If $F\colon M_1 \to M_2$ is a functor we define $C(F)$, the category of finite F chain complexes, as follows. An object of $C(F)$ is a sequence (C_j, d_j) with $C_j \in M_1$, $d_j \in M_2(F(C_j), F(C_{j-1}))$, and with $d_{j-1}d_j = 0$. We assume all but a finite number of the C_j's are ∞. A map $\alpha\colon (C_j,d_j) \to (\overline{C}_j, \overline{d}_j)$ in $C(F)$ is a sequence $\alpha_j\colon C_j \to \overline{C}_j$, where $\alpha_j \in M_1(C_j, \overline{C}_j)$ and $\overline{d}_j F(\alpha_j) = F(\alpha_{j-1})d_j$. When M_1 has an ES structure, $C(F)$ inherits one and the natural functor $j_1\colon C(M_1) \to C(F)$ is exact. If $M_1 \to M_2$ is exact then the natural functor $j_2\colon C(F) \to C(M_2)$ is also exact.

5. Categories of acyclic complexes

Let $F\colon M_1 \to M_2$ be as in 4. If M_2 is an Abelian category, we denote by $C_\alpha(M_2) \subset C(M_2)$ the full subcategory of acyclic complexes, $C_\alpha(F) \subset C(F)$ the full subcategory whose objects are in $j_2^{-1}(C_\alpha(M_2))$, and $C_\alpha(M_1,F)$ the full subcategory of $C(M_1)$ whose objects are in $j_1^{-1}j_2^{-1}(C_\alpha(M_2))$.

6. $K_1(F)$ and $\overline{K}_1(F)$.

Let M_1 be an ES category, M_2 an Abelian category and $F\colon M_1 \to M_2$ a functor. We <u>define</u>

$$K_1(F) = K_0(C_\alpha(F))/j_{1*}(K_0(C_\alpha(M_1,F)).$$

Consider elements in $C_\alpha(F)$ of the form

$$\infty \to \infty \to \cdots \to \infty \to A \xrightarrow{-Id} A \to \infty \to \cdots$$

These sequences generate a subgroup I in $K_1(F)$. We define

$$\overline{K}_1(F) = K_1(F)/I.$$

7. K_1 of a ring

We now specialize to the categories we are interested in. R will be a ring with identity and $\overline{R}$ will be the category of left R-modules. Let S be the category of base pointed sets and base point preserving maps. To assure that S satisfies the assumptions of (1), we assume base point of A = base point of $B = \#$, for A,B in S and that $\infty = \{\#\}$. Let $f\colon \overline{R} \to S$ be the forgetful functor, and $F\colon S \to \overline{R}$ the left adjoint of f. That is, $F(A)$ is the free R-module on $A - \{\#\}$. For A, B in S there exists a coproduct unique up to isomorphism, any of whose representatives will be denoted by $A \vee B$. If $A \cap B = \{\#\}$ we can take $A \vee B = A \cup B$. The ES structure on S is given by pairs (p,i), $i: A \to A \vee B$ and $p: A \vee B \to B$, the injection and projection of the coproduct. The category $\overline{R}$ is an Abelian category and we take the natural ES structure from exact sequences. The functor F, but not f, is exact. $K_1(F)$ is not interesting since we have not yet imposed a finiteness condition. However, if we let $F_0 = F|S_0$, where S_0 is the subcategory of S consisting of finite sets, then $\overline{K}_1(F_0)$ is the usual $\overline{K}_1(R)$. This motivates the notation of (6).

8. Categories of functors

To get the Whitehead groups we proceed as follows. For categories M_1 and M_2 we consider the functor category $C(M_1,M_2)$ whose objects are functors from M_1 to M_2 and whose morphisms are natural transformations. Note that $C(M_1,M_2)$ will have an initial-terminal object if M_2 does. We need no such assumption on M_1. If M_2 has an ES structure then $C(M_1,M_2)$ does by setting $\alpha_1 \xrightarrow{i} \alpha_2 \xrightarrow{p} \alpha_3$ to be in ES of $C(M_1,M_2)$ if and only if for each $A \in M_1$, $\alpha_1(A) \xrightarrow{i(A)} \alpha_2(A) \xrightarrow{p(A)} \alpha_3(A)$ is in ES of M_2. If $G: M_2 \to M_3$ is a functor, the composite yields $G_*: C(M_1,M_2) \to C(M_1,M_3)$. If G is exact so is G_*. With the notation from (7) we now set $M_2 = S$, $M_3 = \overline{R}$, and $G = F$. Again, $K_1(F_*)$ is not yet interesting since we have not yet imposed finiteness or projectivity conditions.

9. The Whitehead group of a category

A functor $\alpha\colon M \to S$ is of <u>finite type</u> if for each $A \in M$, $\alpha(A)/(\mathrm{Iso}(A))$ is finite. Here $\mathrm{Iso}(A)$ denotes the invertible elements of $M(A,A)$ and $\mathrm{Iso}(A)$ acts on $\alpha(A)$ via the functor α. We let $C_0(M,S) \subset C(M,S)$ be the full subcategory consisting of projective functors of finite type. From (7) and (8) we have $F_*\colon C(M,S) \to C(M,\overline{R})$. We let F_0 be the restriction of F_* to $C_0(M,S)$. Finally, we set

$$\mathrm{Wh}(M,R) = \overline{K}_1(F_0).$$

We repeat that for this M need not have an initial or terminal object. This definition is related to the classical one in the following example. Let G be a group, and $\underline{G}$ the category with one object whose morphisms are the elements of G. Then $\mathrm{Wh}(\underline{G},R)$ is the classical Whitehead group of G with coefficients in R.

10. The generalized Whitehead group.

Consider the following category $\mathcal{O}(G)$, crucial in transformation groups. The objects of $\mathcal{O}(G)$ are the subgroups of G. The morphisms $\mathcal{O}(G)(H_1,H_2)$ are G maps from G/H_2 to G/H_1. Alternatively, $\mathcal{O}(G)(H_1,H_2) = \{g \in G \mid H_2 \subset gH_1g^{-1}\}/H_1$. H_1 acts by right multiplication on $\{g \in G \mid H_2 \subset gH_1g^{-1}\}$. This category is sometimes described as the orbit category, but this is deceptive since G/H and G/gHg^{-1} are indistinguishable as orbits but represent different, although isomorphic, objects of $\mathcal{O}(G)$.

This category is central because for a G CW complex X the map which assigns to each $H \subset G$ the n cells of X^H determines a functor $\overline{X}: \mathcal{O}(G) \to S$ which encodes the G cell structure of X. To continue our examples, we have $Wh(\mathcal{O}(G),R) = Wh(G;R)$, the generalized Whitehead group of G defined in [12].

11. Partially ordered G sets

To continue our setup we need to digress and consider G posets. This notion is helpful in the study of the combinatorial structure of a G action, and it has been discussed in much detail in [4]. Suppose Π is a partially ordered set and G acts on Π preserving the partial order on Π. Then we call Π a partially ordered G set. As example consider $S(G) = \{H \subset G | H$ is a subgroup of $G\}$. A partial ordering is given by $H \leq K$ if and only if $H \supseteq K$. The G action is given by conjugation. Suppose $\rho: \Pi \to S(G)$ is an order preserving equivariant map. For any $\alpha \in \Pi$ we set

11.1
$$\Pi_\alpha = \{\beta \in \Pi | \beta \geq \alpha\},$$
$$\Pi_{(\alpha)} = \{\beta \in \Pi | g\beta \geq \alpha \text{ for some } g \in G\},$$
$$G_\alpha = \{g \in G | g\alpha = \alpha\}.$$

Throughout we assume that (an assumption satisfied in 11.4).

11.2 $\rho(\alpha) \triangleleft G_\alpha$ and $\rho: \Pi_\alpha \to S(G)_{\rho(\alpha)}$ is injective.

Note that $S(G)_{\rho(\alpha)} = S(\rho(\alpha)) \subset S(G_\alpha)$ and $G_\alpha \subset G_{\rho(\alpha)} = N_G(\rho(\alpha))$, the normalizer of $\rho(\alpha)$ in G. A pair (Π,ρ) as we just discussed it is called <u>G poset.</u> As example of a G poset consider $(S(G),Id)$. If H is a subgroup of G, then $S(G)_H = \{K \in S(G) | K \subset H\}$ and

$S(G)_{(H)} = \{K \in S(G) | gKg^{-1} \subset H$ for some $g \in G\}$. In general, Π_α is a G_α poset and $\Pi_{(\alpha)}$ is a G poset.

A G poset (Π,ρ) is called <u>complete</u> if

11.3 $\rho: \Pi_\alpha \to S(G)_{\rho(\alpha)}$ is bijective for all $\alpha \in \Pi$.

To any G space X we associate a G poset $(\Pi(X),\rho_X)$. Set

11.4
$$\Pi(X) = \coprod_{H \in S(G)} \pi_0(X^H)$$

Here $\coprod$ denotes the disjoint union. The action of G on X provides an action of G on $\Pi(X)$. If $\alpha \in \pi_0(X^H)$ we set $\rho_X(\alpha) = H$. If $\alpha \in \pi_0(X^H)$ then α is the name of a path component of X^H. We denote this subspace of X by either X_α or $|\alpha|$. The partial ordering on $\Pi(X)$ is given by: $\alpha \le \beta$ if $X_\alpha \subseteq X_\beta$ and $\rho(\alpha) \le \rho(\beta)$. Often we abbreviate $(\Pi(X),\rho_X)$ by $(\Pi(X),\rho)$. Notice that $(\Pi(X),\rho)$ is always complete.

<u>Definition 11.5</u> Let (Π,ρ) be a G poset. A (Π,ρ) space is a G space X with a collection of distinguished subspaces $\{X_\alpha | \alpha \in \Pi\}$, X_α could be empty, such that

(i) $X_{g\alpha} = gX_\alpha$ for all $g \in G$ and $\alpha \in \Pi$

(ii) $X_\alpha \subseteq X_\beta$ if $\alpha, \beta \in \Pi$ and $\alpha \le \beta$

(iii) $$X^H = \coprod_{\rho(\alpha) = H} X_\alpha$$

If X is a G CW complex and the X_α's are subcomplexes we call $(X,\{X_\alpha | \alpha \in \Pi\})$ also a (Π,ρ) complex. We say that X is a

(Π,ρ) space or complex if the X_α's are understood.

The obvious example is as follows. Let X be a G CW complex and $\Pi = \Pi(X)$ as in 11.4. The natural choice for the subcomplex in 11.5 are the spaces X_α distinguished in the paragraph before 11.5.

Let (Π,ρ) be a G poset. To each $\alpha \in \Pi$ we associate

11.6 $$W(\alpha) = G_\alpha/\rho(\alpha).$$

Suppose (Π,ρ) and (Π',ρ') are G posets. A G poset map $a\colon (\Pi,\rho) \to (\Pi',\rho')$ is an equivariant order preserving map $a\colon \Pi \to \Pi'$ such that $\rho(\alpha) = \rho'(a(\alpha))$. Suppose $f\colon X \to Y$ is an equivariant map of G spaces. This map f induces a map $\tilde{f}\colon (\Pi(X),\rho_X) \to (\Pi(Y),\rho_Y)$ by setting $\tilde{f}(\alpha) = \beta$ where β is defined by $\rho_X(\alpha) = \rho_Y(\beta)$ and $f(X_\alpha) \subseteq Y_\beta$. By restriction f induces the map

11.7 $$f_\alpha\colon X_\alpha \to Y_{\tilde{f}(\alpha)} .$$

12. The Whitehead group for 1-connected fixed point components

Let (Π,ρ) be a G poset. We define an associated category $\overline{(\Pi,\rho)}$. The objects are the elements of Π. For $\alpha,\gamma \in \Pi$, define $N(\alpha,\gamma) = \{g \in G \mid g\alpha < \gamma\}$. Then $\rho(\alpha)$ acts on $N(\alpha,\gamma)$ by right multiplication. The morphisms of the category are $\overline{(\Pi,\rho)}(\alpha,\gamma) = N(\alpha,\gamma)/\rho(\alpha)$. Composition of morphisms is defined by multiplication in G. The Whitehead group $Wh(\overline{(\Pi,\rho)};R)$ is a generalization of the Whitehead group $Wh(\mathcal{O}(G);R)$ from (10). It is appropriate for the study of actions of G where, for subgroups H of G, the H fixed point set need not be connected but each component is simply connected. The case of an empty H fixed point set is included.

13. The Whitehead group for non 1-connected fixed point components.

Next, we wish to describe algebraically the Whitehead groups for a G complex X where the fixed point sets need not be simply connected. We define a category $\mathcal{O}(X)$. The objects will be the components of X^H, as H runs over the subgroups of G with $X^H \neq \phi$. So, the objects are the elements of $\Pi(X)$. For each component $\alpha \in \Pi(X)$, we select a base point $x(\alpha)$ in X_α (11.4). A morphism from α to γ will be an element of $N(\alpha,\gamma)/\rho(\alpha)$ where $N(\alpha,\gamma)$ consists of pairs $(g,\lambda), g \in G$, $g\alpha \leq \gamma$, and λ is a homotopy class of paths in X_γ joining $gx(\alpha)$ to $x(\gamma)$. The subgroup $\rho(\alpha)$ of G acts on the pair by acting on the first factor on the right. Notice that if each component of X^H is simply connected we are exactly back in the category of (12). The product of elements of G, along with the composition of paths, describes a composition law for morphisms in $\mathcal{O}(X)$. Strictly speaking, the category depends on the choices of base points. However, choosing paths connecting two different sets of base points, defines an isomorphism from the category with one set of base points to the category with another set. This isomorphism is not canonical but it depends on the choices of paths. We now define the group $Wh(X;R) = Wh(\mathcal{O}(X);R)$ and see that, at least as abstract group, its isomorphism class is independent of the choices of base points. We claim that this is the algebraic description of Illman's group $Wh(X)$ when $R = \mathbb{Z}$, see Theorem A below.

14. The Whitehead torsion of a G homology equivalence mod R.

Suppose we are given two finite G CW complexes X and Y and a G map $f: X \to Y$ such that f maps components bijectively

($\tilde{f}$: $\Pi(X) \to \Pi(Y)$ is a bijection) and on each component f is a mod R homology isomorphism ($(f_\alpha)_*$: $H_*(X_\alpha,R) \to H_*(Y_{\tilde{f}(\alpha)},R)$ is an isomorphism). Naturally, we suppose that we selected base points for X_α and $Y_{\tilde{f}(\alpha)}$ and that f preserves them. In addition, we suppose that f_α <u>induces an isomorphism from</u> $\pi_1(X_\alpha)$ <u>to</u> $\pi_1(Y_{\tilde{f}(\alpha)})$. For notation see (11). So, our notion of an R homology equivalence is tied to the category and makes stronger assumptions than usual. We will see how to get an invariant $\tau(f)$ in Wh(X;R) from f.

The two functors, which assign the cellular chains of $\tilde{X}_\alpha$ and of $\tilde{Y}_{\tilde{f}(\alpha)}$ to α are finite chain complex functors on $\mathcal{O}(Y)$. They can be checked to be projective. If f is a G cellular map, it induces a transformation of chain complexes which induces an isomorphism on homology mod R by assumption. The mapping cone of this transformation is then an R valued acyclic functor from $\mathcal{O}(Y)$ to finite R complexes, which is projective and thus defines an element of Wh(Y,R). The argument of [12, p. 285] shows that this element depends only on the G homotopy class of f and thus the invariant is well-defined.

Let IWh(X) denote Illman's Whitehead group [7]. The construction of this paragraph defines a homomorphism

$$\alpha: \ \mathrm{IWh}(X) \to \mathrm{Wh}(X, \mathbb{Z}).$$

We then have

<u>Theorem A.</u> α is an isomorphism.

The proof will be carried out in the next few sections. In particular it will follow from Theorem B of the next section.

15. Computation of $Wh(X,R)$.

The proof of Theorem A follows from a calculation. Illman calculated his group; we shall calculate ours and see that one gets the same result. For each $\alpha \in \Pi(X)$ we defined

$$G_\alpha = \{g \in G | g\alpha = \alpha\} = \{g \in G | gX_\alpha = X_\alpha\} \cap N_G\rho(\alpha) \quad \text{and}$$

$$W(\alpha) = G_\alpha/\rho(\alpha).$$

Let $\tilde{X}_\alpha$ be the universal covering space of X_α. Recall that we have selected base point so that this canonical. Let $\overline{W}(\alpha)$ be the group of homeomorphisms of $\tilde{X}_\alpha$ which cover the action of G_α on X_α. We then have the exact sequence

$$0 \to \pi_1(X_\alpha) \to \overline{W}(\alpha) \to W(\alpha) \to 0.$$

In each G orbit of $\Pi(X)$ pick one representative. Call the set of components so constructed A. Then

<u>Theorem B.</u>

$$Wh(X,R) = \sum_{\alpha\in A} Wh(\overline{W}(\alpha),R).$$

In the theorem $Wh(\overline{W}(\alpha),R)$ is the classical Whitehead group with coefficients in R.

Theorem A follows from Theorem B since it is easily seen that α, composed with the isomorphism of Theorem B, is just Illman's isomorphism. We shall prove Theorem B in the next few sections.

16. Extending functors.

<u>Proposition.</u> Let $V \subset W$ be a full subcategory. Given any functor $T\colon V \to S$ there exists a unique (up to natural isomorphisms preserving T) minimal extension $\overline{T}\colon W \to S$ satisfying the following property. Given a natural transformation $\alpha\colon T \to F|V$, where F is any functor from W to S, there exists a unique natural transformation $\overline{\alpha}\colon \overline{T} \to F$ extending α.

It follows easily from the universal property that if T is projective so is $\overline{T}$. In the cases that we are interested in, we can check that, if T is of finite type so is $\overline{T}$. This is in particular true for each pair of objects A and B of W the morphism set $W(A,B)$ is finite.

<u>Proof of the Proposition.</u> We construct $\overline{T}$ as follows. For $A \in W$ set:

$$\overline{T}(A) = \{(f,x) \mid f \in W(C,A) \text{ for some } C \in V \text{ and } x \in T(C)\}/\sim \cup \infty .$$

The relation $\sim$ is the smallest equivalence relation such that $(f_1,x_1) \sim (f_2,x_2)$ if there exists f_3, C_3, j_1, j_2 and a commutative diagram

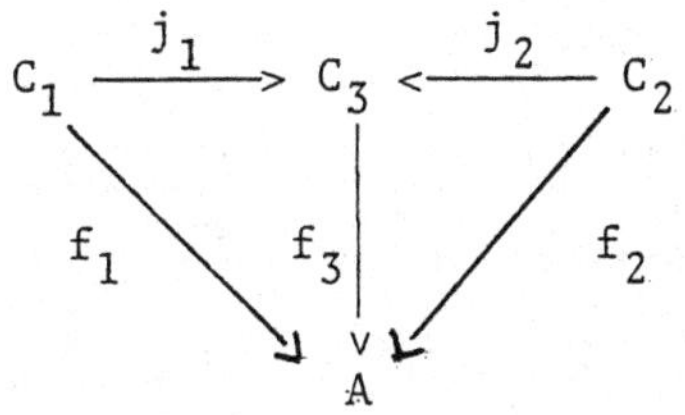

and $T(j_1)(x_1) = T(j_2)(x_2)$. If $\lambda \in W(A,B)$ set $\overline{T}(\lambda)(f,x) = (\lambda f,x)$. For $A \in V$ we have a natural indentification of $\overline{T}(A)$ and $T(A)$. Futhermore, if α is a natural transformation $T \to F|V$, we define $\overline{\alpha}\colon \overline{T}(A) \to F(A)$ by $\overline{\alpha}(f,x) = F(f)(\alpha(x))$. The properties of $\overline{T}$ asserted above follow immediately from the construction.

17. Restricting projective functors

Proposition. Let W be any category, $V \subset W$ a full subcategory such that for A in $ob(W) - ob(V)$ and $B \in ob(V)$ then $W(A,B) = \phi$. If $J\colon W \to S$ is projective, then $J_1 = J|V$ is projective.

Proof. Let $\alpha\colon F \twoheadrightarrow J_1$ be a surjective natural transformation where F is any functor form V to S. We may extend α to $\overline{\alpha}\colon \overline{F} \to J$ by (16), however $\overline{\alpha}$ may not be epi. Let $\phi\colon W \to S$ be a functor, with $\phi(A) = \infty$ if $A \in V$ and $\phi(A) = J(A)$ if $A \in W - V$. By the assumptions of the proposition ϕ is a functor and $\overline{\alpha}$ extends naturally to $\overline{\overline{\alpha}}\colon \overline{F} \vee \phi \twoheadrightarrow J$. Since J is projective there exists $\gamma\colon J \to \overline{F} \vee \phi$ with $\overline{\overline{\alpha}}\gamma = Id$. By construction $\gamma_1 = \gamma|J_1$ factors through F and $\alpha\gamma_1 = Id$. Hence J_1 is projective.

18. Quotients of natural transformation

Let $T_1,T_2\colon W \to S$ be functors as above and let $\alpha\colon T_1 \to T_2$ be a natural transformation. We can form the functor $T_2/\alpha(T_1)$, where $T_2/\alpha(T_1)(A) = T_2(A)/\sim$, and $\sim$ is the equivalence relation which identifies points of $\alpha(T_1(A))$ with #. The functor $T_2/\alpha(T_1)$ applied to a morphism has the obvious meaning, namely, it is the induced morphism on quotients. With this notation we have

<u>Propositon.</u> Suppose for each $A \in W$ such that $T_2(A) \neq \alpha(T_1(A))$ and $f \in W(A,B)$ we have that f has a two sided inverse. If T_2 is projective so is $T_2/\alpha(T_1)$.

<u>Proof.</u> There exists a natural transformation $\pi\colon T_2 \to T_2/\alpha(T_1)$. The functor $T_2/\alpha(T_1)$ is projective if and only if π splits, i.e., there is $j\colon T_2/\alpha(T_1) \to T_2$ with $\pi j = \mathrm{Id}$. There is an obvious unique base pointed splitting of $\pi(A)$, $j(A)\colon T_2/\alpha(T_1)(A) \to T_2(A)$, and we must show that this is functorial in A. This is true if and only if $x \in T_2(A)$, $x \neq \alpha(y)$ implies that for all $f \in W(A,B)$, $T_2(f)(x) \neq \alpha(z)$, $z \neq \infty$. By assumption, if there exists an x in $T_2(A)$ and $x \neq a(y)$ then f in $W(A,B)$ is invertible. But if $T_2(f)(x) = \alpha(z)$, then

$$x = T_2(f^{-1})T_2(f)(x) = T_2(f^{-1})\alpha(z) = \alpha(T_1(f^{-1})(z)),$$

which is a contradiction.

19. Consequence of (17) and (18)

<u>Corollary of (17) and (18).</u> Let V be a full subcategory of W such that if $A \in W - V$, then $W(A,B) = \phi$ for $B \in V$ and $a \in W(A,C)$ is invertible for $C \in W - V$. If $T\colon W \to S$ is projective, then $T_1 = T|V$ is projective and $T/j(T_1)$ is projective, where $j\colon T_1 \to T$ is the natural transformation of 16 extending the identity transformation.

20. Decomposition of the Whitehead group

Let V be a full subcategory of W as in 19. Then

$$Wh(W;R) = Wh(V;R) \oplus Wh(W,V;R)$$

where $Wh(W,V;R) = \overline{K}(F_{00})$, $F_{00} = F_0|C_{00}(W,V)$, and $C_{00}(W,V)$ is the full subcategory of $C(W,S)$ of projective functors of finite type, γ, such that $\gamma(A) = \infty$ for all A in V.

Proof. To see this we need only show that if for the functor T_n, d_n

$$\cdots \to T_n(A) \xrightarrow{d_n} T_{n-1}(A) \to \cdots$$

is acyclic for each A in V, then

$$\cdots \quad \overline{T}_n(A) \xrightarrow{\overline{d}_n} \overline{T}_{n-1}(A) \quad \cdots$$

is acyclic for all $A \in W$. Here $\overline{T}_n, \overline{d}_n$ is a minimal extension of T_n, d_n. This follows by an easy Meyer-Victoris argument. So we have a splitting of the natural map $p\colon Wh(W,R) \to Wh(V,R)$. It follows also from 19 that the kernel of p is generated by complexes whose terms are of the form $T/j(\overline{T|V})$, that is, by elements which come from $Wh(W,V;R)$.

21. Proof of Theorem B

Let X be a finite G CW complex. For the conjugacy class $[H]$ of the subgroup H of G let

$$X^{[H]} = \bigcup_{K \in [H]} X^K \text{ and}$$

$$X_s^{[H]} = \{x \in X^{[H]} \mid G_x \underset{\neq}{\supset} gHg^{-1} \text{ for some } g \in G\}.$$

By repeated application of (20) we have

$$\mathrm{Wh}(\mathcal{O}(X);R) = \sum_{[H]} \mathrm{Wh}(\mathcal{O}(X^{[H]}),\ \mathcal{O}(X_s^{[H]});R)$$

where the summation runs over conjugacy classes of subgroups of G. Now, an easy calculation as in [12, p. 274] shows that

$$\mathrm{Wh}(\mathcal{O}(X^{[H]}),\ \mathcal{O}(X_s^{[H]});R) = \sum \mathrm{Wh}(\overline{W}(\alpha);R).$$

Here α runs over elements in $\Pi(X)$ such that $\rho(\alpha) = H$ and we have to pick one such α in each G orbit. This proves Theorem B which was stated in (15).

22. Geometric properties

As a first step we discuss <u>induced maps between generalized Whitehead groups</u>. Let $s\colon (\Pi,\rho) \to (\Pi',\rho')$ be a poset map such that $W(\alpha) = W(s(\alpha))$ for all $\alpha \in \Pi$. Taking direct sums of the chain complex functors we obtain an induced map $s_*\colon \mathrm{Wh}(\overline{\Pi,\rho}) \to \mathrm{Wh}(\overline{\Pi',\rho'})$. More generally, suppose $g\colon B \to Y$ is an equivariant map. Pick collections of base points in B and Y (see 13) and suppose that g preserves them. Using an appropriate definition of universal coverings we have base points $\tilde{x}(\alpha)$ in $|\tilde{\alpha}|$. Now suppose that for all $\alpha \in \Pi(B)$ $(f_\alpha)_\#\colon \pi_1(B_\alpha) \to \pi_1(Y_{\tilde{f}(\alpha)})$ is an isomorphism and that $\overline{W}(\alpha) = \overline{W}(\tilde{f}(\alpha))$. Then we have the induced maps $\overline{W}(\alpha) \to \overline{W}(\tilde{f}(\alpha))$ (compare (compare [2, p. 65]) which are isomorphisms by the Five Lemma. We can continue as above and take direct sums of chain complex functors to obtain an induced map $g_*\colon \mathrm{Wh}(B,R) \to \mathrm{Wh}(Y,R)$. Obviously, g_* can be naturally defined if g is a G homotopy equivalence.

We give the geometric interpretation of the process we just described. Suppose $s\colon (\Pi,\rho) \to (\Pi',\rho')$ is a G poset map and X is a (Π,ρ) space. Then X can be understood as (Π',ρ') space by setting $X_\beta = \bigcup_{\alpha \in s^{-1}(\beta)} X_\alpha$. Suppose $f: A \to B$ is a G homotopy equivalence of finite G CW complexes, or, more generally, f is just a mod R G homology equivalence as in (14). So $\tau(f) \in Wh(B,R)$ is defined. Setting $\Pi(B) = \Pi$, and assuming s or g as above, we obtain an element $s_*(\tau(f)) \in Wh(\overline{\Pi',\rho'})$ or $g_*(\tau(f)) \in Wh(Y,R)$.

We generalize the definition given in (14). Suppose

$$\begin{array}{ccc} A & \xrightarrow{f_1} & B \\ \Big\downarrow{\scriptstyle g_1} & & \Big\downarrow{\scriptstyle g_2} \\ X & \xrightarrow{f_2} & Y \end{array}$$

is a square of equivariant maps of finite G CW complexes. Suppose the square is G homotopy commutative and $h: g_2 f_1 \simeq f_2 g_1$ is a G homotopy. This data determines an induced map

$$F: (M_{g_1}, A) \to (M_{g_2}, B)$$

which defines

$$f_3: M_{g_1}/A \to M_{g_2}/B$$

Suppose, g_2 and $p: Y \xrightarrow{\text{incl.}} M_{g_2} \xrightarrow{\text{proj.}} M_{g_2}/B$ induce maps on the level of Whitehead groups:

$$Wh(B,R) \xrightarrow{(g_2)_*} Wh(Y,R) \xrightarrow{p_*} Wh(M_{g_2}/B,R).$$

If F is a $G-R$ homology equivalence we set

$$\tau(F) = \tau(f_3) \in Wh(M_{g_2}/B,R).$$

If $\tau(f_1) \in Wh(B,R)$ and $\tau(f_2) \in Wh(Y,R)$ are defined

$$(22.1) \qquad p_*\tau(f_2) = \tau(f_3) + p_*(g_2)_*\tau(f_1) \in Wh(M_{g_2}/B,R).$$

It follows from this formula that $\tau(f_3)$ will not depend on the choice of the particular homotopy h in this case.

There is an interesting special case. Suppose the H fixed point set of B and Y are 1-connected (possibly empty) for each $H \subset G$. In this case we call B and Y G simply connected. In a natural way $Wh(B,R)$ and $Wh(Y,R)$ are subgroups of $Wh(point,R)$, and so is $Wh(M_{g_2}/B,R)$. Remember that $Wh(point,R)$ is $Wh(G,R)$ from [12]. This was pointed out in (10). With this understood, 22.1 simplifies to

$$22.2 \qquad \tau(f_2) = \tau(f_3) + \tau(f_1).$$

If $i: Z \to W$ is an inclusion we set $\tau(i) = \tau(W,Z)$. If f_i and g_i are inclusions 22.1 can be reformulated as

$$22.3 \qquad \tau(Y,X) = \tau(B,A) + \tau(Y,X \cup B) = \tau(B,A) + \tau(Y/B,X/A)$$

Let $f: X \to Y$ and $g: Y \to Z$ be G-R homology equivalences. A standard proof [12, 1.32] based on the exact sequence of the appropriate chain complexes of mapping cones implies

22.4 $$\tau(g\circ f) = g_*\tau(f) + \tau(g)$$

Here are some properties of the generalized Whitehead torsion. A generalization of [12, 2.5] is

Proposition 22.5 The generalized Whitehead torsion of an equivariant subdivision is zero.

The existence and uniqueness of a smooth equivariant triangulation of a smooth G manifold has been shown in [8]. From this follows

Proposition 22.6 The torsion of a G homotopy equivalence (mod R G homology equivalence) between smooth compact G manifolds is well defined and vanishes for diffeomorphisms.

Independently, this has also been shown by Illman [9, Theorem 3.1 and Corollary 3.2]. We use the notation of a G h-cobordism as it has been defined in [12, 3.1]. If X is a G space and $H \subset G$ we set $X^{[H]} = \{x \in X | G_x$ is conjugate to $H.\}$. Let $i: X \to \partial Y$ be a G imbedding of compact G manifolds with $\dim X = \dim Y - 1$. We call i a G cobordism of X to $\overline{\partial Y - X}$. We call i a G h-cobordism if for each $H \subset G$ the induced maps $i^{[H]}: X^{[H]} \to Y^{[H]}$ and $i'^{[H]}: \overline{\partial Y - X}^{[H]} \to Y^{[H]}$ are homotopy equivalences. Here i' is the inclusion $\overline{\partial Y - X} \to \partial Y$.

We restate some results from [12, section 3] in our language. The results are more general but the proofs are similar.

Equivariant s-Cobordism Theorem: Let $i: X \to Y$ be a G h-cobordism such that $Y_\alpha = X_\alpha \times I$ if $\dim X_\alpha \leq 4$. The pair (Y,X) is G diffeomorphic to $(X \times I, X \times 0)$ if and only if $\tau(i) = 0$ in $Wh(Y, \mathbb{Z})$.

Let τ represent a class in $Wh(Y,R)$ for some finite

G CW complex Y. The direct sum decomposition of Theorem B determines components $\tau_\alpha \in Wh(\overline{W}(\alpha), R)$.

<u>Realization Theorem</u>: Let X be a compact G manifold. Let τ be any element of $Wh(X, \mathbb{Z})$ such that τ_α vanishes if $\dim X_\alpha \leq 4$. Then there exists a G h-cobordism $i: X \to Y$ such that $\tau(i) = \tau$ and $Y_\alpha = X_\alpha \times I$ whenever $\dim X_\alpha \leq 4$.

<u>Classification of h-cobordism Theorem</u>: Let $i_j: X \to Y_j$ be a G h-cobordism, $j = 1,2$. Suppose that $(Y_j)_\alpha = X_\alpha \times I$ if $\dim X_\alpha \leq 4$, $\alpha \in \Pi(X)$, and $\tau(i_1) = \tau(i_2)$. Then there exists a diffeomorphism $\lambda: Y_1 \to Y_2$ with $i_2 = \lambda i_1$.

The notions of elementary expansions and collapses (briefly deformations) discussed in [7] and [12, p. 288] have natural generalizations in the category of (Π, ρ) complexes. Such deformations have vanishing torsion. We use the symbol $A \nearrow\!\searrow B$ to denote that A and B are connected through a sequence of equivariant elementary deformations. More precisely, we are given a sequence of spaces C_i, $1 \leq i \leq \ell$, and maps $k_i: C_i \to C_{i+1}$, $1 \leq i \leq \ell-1$ such that $A = C_1$, $B = C_\ell$, and k_i is either an equivariant collapse, or k_i is an inclusion and the G homotopy inverse of an equivariant collapse. So it makes sense to consider maps which are G homotopy equivalent to a sequence of elementary deformations. Finally, we have

<u>Proposition 22.7</u> Suppose $f: A \to B$ is a G homotopy equivalence of finite G CW complexes. Then f is G homotopic to a sequence of elementary deformations if and only if $\tau(f)$ vanishes in $Wh(B, \mathbb{Z})$.

The proof is standard based on [7, Theorem 3.6'] and our Theorem A. For a special case this was observed in [12, 2.3].

References

1. D. Anderson, Torsion invariants and actions of finite groups. Michigan Math. J. 29(1982), 27-42.

2. G. Bredon, Introduction to compact transformation groups, Academic Press, 1972.

3. M.M. Cohen, A course in simple homotopy theory. Springer Verlag, Berlin-Heidelberg-New York, 1970.

4. K.H. Dovermann and T. Petrie, G surgery II. Memoirs of the AMS, Vol. 260, (1982).

5. K.H. Dovermann and M. Rothenberg, The equivariant Whitehead torsion of a G fibre homotopy equivalence, preprint (1984).

6. H. Hauschild, Äquivariante Whitehead torsion. Manuscripta Math. 26(1978), 63-82.

7. S. Illman, Whitehead torsion and group actions. Annales Academiae Scientiarum Fennicae, Vol. 588(1974).

8. ________, Smooth equivariant triangulations of G manifolds for a finite group. Math. Ann. 233(1978), 199-220.

9. ________, Equivariant Whitehead torsion and actions of compact Lie groups, Group action on manifolds, Contemporary Mathematics Vol. 36(1985), 91-106.

10. ________, A product formula for equivariant Whitehead torsion, preprint (1985), ETH Zurich.

11. J. Milnor, Whitehead torsion. Bull AMS 72(1966), 358-426.

12. M. Rothenberg, Torsion invariants and finite transformation groups. Proc. of Symp. in Pure Math., AMS, Vol. XXXII, (1978), 267-311.

13. J.H.C. Whitehead, Simplicial spaces, nuclei and m-groups, Proc. London Math. Soc. (2), 45(1939), 243-327.

14. S. Araki, Equivariant Whitehead groups and G expansion categories, preprint.

15. S. Araki and K. Kawakubo, Equivariant s-cobordism theorem, preprint.

16. M. Steinberger and J. West, Approximation by equivariant homeomorphisms, preprint.

Almost complex S^1-actions on cohomology complex projective spaces

To the memory of A. Jankowski and W. Pulikowski

Akio Hattori

1. Introduction.

Let X be a closed C^∞ manifold which has the same cohomology ring as the complex projective space $\mathbb{C}P^n$. Such a manifold will be called cohomology complex projective space or, briefly, cohomology $\mathbb{C}P^n$. There is a conjecture due to T. Petrie [P] to the effect that if the group S^1 acts non-trivially on X then X has the same total Pontrjagin class $p(X)$ as $\mathbb{C}P^n$. The conjecture was partially solved in various special cases; cf. [D], [H_1], [M], [P], [Y_1] and [Y_2]. In [P] Petrie presented an interesting example of exotic S^1-action on a cohomology complex projective space X whose normal representations at fixed points are different from the linear S^1-actions on $\mathbb{C}P^n$ but, nevertheless, the Pontrjagin class of X is the same as $p(\mathbb{C}P^n)$.

On the other hand the author investigated certain almost complex S^1-actions in [H_2]. The results there suggest the following conjecture.

Conjecture A. Let X be an almost complex manifold which is a cohomology $\mathbb{C}P^n$ such that $c_1(X)^n[X] > 0$ and $T[X] \neq 0$ where $T[X]$ denotes the Todd genus of X. If X admits an almost complex S^1-action with only isolated fixed points then the normal representations of S^1 at fixed points are the same as those of a linear action on $\mathbb{C}P^n$ (the precise statement will be given in the statement of conjecture B). In particular, the total Chern class $c(X)$ is the same as $c(\mathbb{C}P^n)$, i.e.

$$c(X) = (1+x)^{n+1}$$

where x is a generator of $H^2(X; \mathbb{Z})$.

In the present paper we shall present a proof of the above conjecture for $n \leq 3$. In fact we shall formulate a more general conjecture concerning certain almost complex S^1-actions and prove it affirmatively when the complex dimension of the manifold is less than 4.

2. Main results.

First we recall some of the results in [H_2]. Let X be a compact connected almost complex manifold with an S^1-action which preserves the almost complex structure. Our basic assumptions in the sequel are the following:

(2.1) The fixed points are all isolated.

(2.2) The Euler number χ of X is equal to $n+1$. This together with (2.1) implies that there are exactly $n+1$ fixed points $P_0, P_1, \ldots, P_n$.

(2.3) $c_1(X)^n[X] > 0$ where $c_1(X)$ is the first Chern class of X and $[X] \in H_{2n}(X;\mathbb{Z})$ is the fundamental class of X. Moreover there exists $x \in H^2(X;\mathbb{Z})$ such that

$$x^n[X] = 1 \quad \text{and} \quad c_1(X) = kx \quad \text{with} \quad k > 0.$$

(2.4) $T[X] \neq 0$.

Let ξ be a complex vector bundle such that $c_1(\xi) = x$.

Lemma 2.5. The action of S^1 on X can be lifted to ξ.

This follows from [H_2, Corollary 3.3] since $\xi^k = \Lambda^n \tau(X)$ admits a lifting where $\tau(X)$ denotes the complex tangent bundle of X.

If we restrict ξ to each fixed point P_i we get an S^1-module $\xi|P_i$ which is of the form t^{a_i} where t is the standard 1-dimensional S^1-module and $a_i \in \mathbb{Z}$.

Lemma 2.6. The integers $\{a_i\}$ are all distinct.

This follows from [H_2, Corollary 3.8] in view of (2.2) and (2.3).

On the other hand the normal representation at each P_i takes the form

$$\tau(X)|P_i = \sum t^{m_{ij}}.$$

Lemma 2.7. The integers $\{m_{ij}\}$ are related to a_i by the formula

$$\sum_j m_{ij} = ka_i + d \tag{2.8}$$

where d is a fixed integer.

This is an easy consequence of the identity $\Lambda^n \tau(X) = \xi^k$; cf. [$H_2$, Corollary 3.15] for some related results.

The problem is to determine the possible values of k. In [H_2, Theorem 5.1] it is proved that

$$k \leq n+1 \tag{2.9}$$

under more general assumption than (2.1), (2.2), (2.3) and (2.4).

Conjecture B. Under the assumptions (2.1), (2.2), (2.3) and (2.4) the only possible value of k is $n+1$.

In [H_2, Corollaries 5.8 and 5.9] it is proved Conjecture B implies the following

Consequence 2.10. Under the assumptions (2.1), (2.2), (2.3) and (2.4) the weights $\{m_i\}$ at each fixed point P_i are given by

$$\{m_{i\nu}\} = \{a_i - a_j\}_{j\neq i},$$

that is, the weights are the same as those of a linear action on $\mathbb{C}P^n$. If moreover X is a cohomology CP^n then

$$c(X) = (1+x)^{n+1}.$$

Conjecture B combined with Consequence 2.10 reduces to Conjecture A when X is a cohomology $\mathbb{C}P^n$.

Theorem 2.11. Conjecture B and hence Conjecture A is true for $n \leq 3$.

Remark. Conjecture B is also true for $n = 4$. We can give a proof similar to the one which will be given in the next section. However it is too cumbersome to be reproduced here.

Also the condition $c_1(X)^n[X] > 0$ can be relaxed to $c_1(X)^n[X] \neq 0$ when n is even. But we do not insist on this point here.

3. Proof of Theorem 2.11.

As in [H_2] we set

$$p_i = \text{number of } \nu \text{ such that } m_{i\nu} > 0$$

and

$$\rho_q = \text{number of } i \text{ such that } p_i = q.$$

By [H_2, Proposition 2.6 and Remark 2.10] we have

(3.1) $$\rho_{n-q} = \rho_q \text{ for all } q \text{ and } \rho_0 = \rho_n = T[X].$$

For each i we set

$$\varphi_i(t) = \frac{\prod_{j\neq i}(1-t^{a_i-a_j})}{\prod_\nu(1-t^{m_{i\nu}})}, \quad 0 \leq i \leq n.$$

The $\varphi_i(t)$ is a Laurent polynomial of t, and by [H_2, Proposition 3.7] we know that

$$(3.2) \qquad \varphi_i(1) = \frac{\prod_{j\neq i}(a_i - a_j)}{\prod_\nu m_{i\nu}} = 1.$$

From (3.2) it follows easily that

$$(3.3) \qquad p_i \equiv i \mod 2 \quad \text{for} \quad i = 0, 1, \ldots, n.$$

Moreover [H_2, Theorem 4.2] implies the following

Proposition 3.4. If we set $\ell = n + 1 - k$ then there are Laurent polynomials $r_0(t), \ldots, r_\ell(t)$ such that

$$\varphi_i(t) = r_0(t) + r_1(t)t^{a_i} + \cdots + r_\ell(t)t^{\ell a_i} \quad \text{for all } i,$$

$$r_0(t) = T[X] = \rho_0 = \rho_n$$

and

$$r_{\ell-s}(1) = r_s(1), \quad 0 \leq s \leq \ell.$$

As a consequence of Proposition 3.4 we deduce the following

Lemma 3.5. ℓ must be even, i.e.

$$k \equiv n+1 \mod 2.$$

In fact if $\ell = 2s + 1$ then

$$\varphi_i(1) = 2(r_0(1) + \cdots + r_s(1)).$$

But this contradicts (3.2).

Proposition 3.6. Let p be a prime. For each i let x_j and x'_ν be the exponents of p in the prime factor decomposition of $a_i - a_j$, $j \neq i$, and $m_{i\nu}$ respectively. Then $\{x_j\}_{j\neq 1}$ and $\{x'_\nu\}_\nu$ coincide up to permutations.

Proof. The following argument is essentially due to [P]. $\varphi_i(t)$ can be expressed as a product of cyclotomic polynomials $\psi_d(t)$ (up to multiplication by some unit $\pm t^N$) where d ranges over those integers such that the number of $\{j;\ j\neq i,\ d$ divides $a_i - a_j\}$ is strictly larger than the number of $\{\nu;\ d$ divides $m_{i\nu}\}$.

On the other hand it is well known that $\psi_d(1) = q$ if d is a power of prime q and $\psi_d(1) = 1$ otherwise. Since $\varphi_i(1) = 1$ the conclusion follows easily.

Corollary 3.7. Let $m > 1$ be an integer. Let Y be a component of the fixed point set of the restricted $\mathbb{Z}/m$-action on X. If P_i and P_j both belong to Y then m divides $a_i - a_j$. Conversely if P_i

belongs to Y and m divides $a_i - a_j$ then P_j also belongs to Y provided m is a power of a prime p. In this case the Euler number $\chi(Y)$ of Y is equal to $\dim_{\mathbb{C}} Y + 1$.

Proof. The first statement is easy; see e.g. [H_2]. The remaining part follows from Proposition 3.6 and the observation that $\chi(Y)$ equals the number of j such that P_j belongs to Y and $\dim_{\mathbb{C}} Y$ is equal to the number of ν such that m divides $m_{i\nu}$ once we choose a fixed point P_i in Y.

With these preliminaries we can now proceed to the proof of Theorem 2.11.

First we consider the case $n = 1$. Since $0 < k \leq 2$ and k is even by (2.9) and Lemma 3.5, k must equal 2.

Remark 3.8. It is known that the only Riemann surface admitting an S^1-action with non-empty fixed point set consisting of only isolated fixed points is CP^1 and thus the first Chern class c_1 evaluated on the fundamental class is equal to 2. It follows that the conclusion $k = 2$ holds without the assumption (2.3).

In the sequel we shall assume $n \geq 2$. We also suppose that the fixed points $\{P_i\}$ are indexed so that $a_0 < a_1 < \cdots < a_n$.

Suppose $n = 2$. Let p be a prime number dividing $a_2 - a_0$. Let Y be the component of the fixed point set of the restricted $\mathbb{Z}/p$-action containing P_2. By Corollary 3.7, Y also contains P_0. We may assume the action on X is effective. Then P_1 can not be contained in Y. Therefore $\dim_{\mathbb{C}} Y = 1$, and by Remark 3.8 and Consequence 2.10 applied to the case $n = 1$ we see that $a_2 - a_0$ is the weight of Y at P_2. Hence from (3.2) it follows that the remaining weight of X at P_2 must be equal to $a_2 - a_1$. Similarly the weights at P_0 are precisely $a_0 - a_2$ and $a_0 - a_1$. Then from (2.8) for $i = 0$ and $i = 2$ we deduce that k must be equal to 3.

Remark 3.9. We have proved Theorem 2.11 and Consequence 2.10 under the following milder condition (2.3)' instead of (2.3). Let ξ be a complex line bundle on which the action can be lifted and such that $x^n[X] = 1$ where $x = c_1(\xi)$. We define the integers $\{a_i\}$ as before. Then under (2.1) and (2.2), the integers $\{a_i\}$ are mutually distinct and the equality (2.8) holds for some integers k and d as was proved in [H_2, Corollary 3.8, Corollary 3.15 and (3.17)]. Now we state the condition (2.3)'

(2.3)' There exists a complex line bundle ξ as above with $x^n[X] = 1$ and $k \geq 0$.

Note. It was shown in [H_2, Theorem 4.2] that if we also assume (2.4), i.e. $T[X] \neq 0$, then $k > 0$ in (2.3)'.

We now proceed to the case $n = 3$.

The possibilities are $k = 2$ or $k = 4$. Assuming $k = 2$ we shall deduce a contradiction. It is known that the numbers $\{a_i\}$ are altered to $\{a_i + a\}$ for some a if we take another lifting of the action to ξ; see e.g. [H_2]. Therefore we may assume that $0 \leq d \leq 1$ in the equality when $k = 2$. We divide into three subcases.

Subcase 1: $a_0 < a_1 < a_2 \leq 0 < a_3$. Evidently we have $p_3 = 3$, i.e. all the $m_{3\nu}$ are positive. Therefore

$$\varphi_3(1) = \frac{\prod_{j\neq 3}(a_3-a_j)}{\prod_\nu m_{3\nu}} > \frac{a_3^3}{\left(\frac{2a_3+1}{3}\right)^3} \geq 1.$$

This contradicts the assumption $\varphi_3(1) = 1$.

Subcase 2: $a_0 < 0 \leq a_1 < a_2 < a_3$. Similarly to Subcase 1 we deduce $\varphi_1(1) > 1$ which is a contradiction.

Subcase 3: $a_0 < a_1 < 0 < a_2 < a_3$. First assume $a_3 - a_2 > 1$ and let p be a prime integer dividing $a_3 - a_2$. Let Y be the component of the fixed point set of the restricted $\mathbb{Z}/p$ action containing P_3. We may assume the given S^1-action is effective. Thus $\dim_{\mathbb{C}} Y$ is equal to 2 or 1.

Assertion. If $\dim_{\mathbb{C}} Y$ is equal to 2 then k must equal 4.

Proof of Assertion. We first show that $K = (x|Y)^2[Y] = 1$. In fact assume $k > 1$. There is a unique P_i not contained in Y where $i = 0$ or 1. If m is the weight at P_3 normal to Y then from (3.2) we get

$$m = K(a_3 - a_i). \tag{3.10}$$

Let Y' be the component of the fixed point set of the restricted Z/m action containing P_3. There exists $P_j \in Y'$, $j \neq 3$. Then, from the equality (cf. [H_2, (3.20)])

$$\sum a_s = -\frac{n+1}{k}d = -2d$$

and the assumption $0 \leq d \leq 1$ it follows that $|a_j| \leq 2a_3$ and hence

$$a_3 - a_j \leq 3a_3 \tag{3.11}$$

On the other hand we see that

$$m = K(a_3 - a_i) \geq 2a_3 \tag{3.12}$$

since $i = 0$ or 1 and we assumed $K > 1$.

Now m divides $a_3 - a_j$ by Corollary 3.7. Hence by virtue of (3.11) and (3.12), m must equal $a_3 - a_j$, i.e.

$$a_3 - a_j = K(a_3 - a_i) = m$$

where $i = 0$ or 1 and $j \neq 3$. If $j = i$ then $K = 1$. If $j \neq i$ then $P_j \in Y$ so that p divides $a_3 - a_j = m$; but this can not happen since we have assumed the action is effective from the first. In any case we have proved $K = 1$.

If $K = 1$ then the weight of X at P_3 normal to Y is equal to $a_3 - a_i$ as above, and similarly the weight at P_0 normal to Y is $a_0 - a_i$. Putting this in (2.8) we get

$$\sum_{\nu=1}^{2} m_{s\nu} = a_s + (d - a_i)$$

for $s = 0$ and $s = 3$ where m_{s1} and m_{s2} are the weights of Y at P_s. But

$$\sum_{\nu=1}^{2} m_{s\nu} = k'a_s + d'$$

for all s, and k' must equal 1. This contradicts Remark 3.9 and completes the proof of Assertion.

If $a_3 - a_0$ has a common prime divisor with $a_3 - a_1$ or $a_3 - a_2$ then we can apply the above procedure and we have $k = 4$ by Assertion.

Similarly if $a_0 - a_1$ has a common divisor with $a_0 - a_2$ or $a_0 - a_3$ then $k = 4$.

Thus we are left with the case where $a_3 - a_2$ is prime to both $a_3 - a_1$ and $a_3 - a_0$ and $a_0 - a_1$ is prime to both $a_0 - a_2$ and $a_0 - a_3$. Then $a_3 - a_2$, $a_3 - a_1$, $a_3 - a_0$ are prime to each other. Let q_0 and q_1 be prime integers dividing $a_3 - a_0$ and $a_3 - a_1$ respectively and let Y_i be the component of the fixed point set of the restricted $\mathbb{Z}/q_i$ action containing P_3 for $i = 0, 1$. We see easily that $\dim_{\mathbb{C}} Y_i = 1$. Hence by Remark 3.8 and Consequence 2.10 that the weight of Y_i at P_3 is $a_3 - a_i$, $i = 0, 1$. Thus the weights of X at P_3 are precisely $\{a_3 - a_j\}_{j \neq 3}$. A similar argument shows that the weights of X at P_0 are precisely $\{a_0 - a_j\}_{j \neq 0}$. Putting these in (2.8) for $i = 3$ and $i = 0$ we get $k = 4$.

This completes the proof of Theorem 2.11 for the case $n = 3$.

References

[D] I.J. Dejter, Smooth S^1-manifolds in the homotopy type of $\mathbb{C}P^3$, Michigan Math. J. 23(1976), 83-95.

[H_1] A. Hattori, $Spin^c$-structures and S^1-actions, Invent. math. 48(1978), 7-13.

[H_2] A. Hattori, S^1-actions on unitary manifolds and quasi-ample line bundles, J. Fac. Sci. Univ. Tokyo, Sect IA, 31(1985), 433-486.

[M] M. Masuda, On smooth S^1-actions on cohomology complex projective spaces. The case where the fixed point set consists of four connected components, J. Fac. Sci. Univ. Tokyo, Sect IA, 28(1981), 127-167.

[P_1] T. Petrie, Smooth S^1-actions on homotopy complex projective spaces and related topics, Bull. Amer. Math. Soc. 78(1972), 105-153.

[Y_1] T. Yoshida, On smooth semi-free S^1-actions on cohomology complex projective spaces, Publ. Res. Inst. Math. Sci. 11(1976), 483-496.

[Y_2] T. Yoshida, S^1-actions on cohomology complex projective spaces, Sûgaku 29(1977), 154-164(in Japanese).

Department of Mathematics
University of Tokyo

A PRODUCT FORMULA FOR EQUIVARIANT WHITEHEAD TORSION AND GEOMETRIC APPLICATIONS

By Sören Illman

Dedicated to the memory of
Andrzej Jankowski and Wojtek Pulikowski

In the following, G and P denote arbitrary compact Lie groups, unless otherwise is specifically stated. Let $f: X \to X'$ be a G-homotopy equivalence between finite G - CW complexes and let $h: Y \to Y'$ be a P-homotopy equivalence between finite P - CW complexes. In this paper we shall give a formula which determines the equivariant Whitehead torsion $t(f \times h) \in Wh_{G\times P}(X \times Y)$ of the $(G \times P)$-homotopy equivalence

$$f \times h: X \times Y \longrightarrow X' \times Y' \tag{1}$$

in terms of the equivariant Whitehead torsions of f and h, and various Euler characteristics derived from the G-space X and the P-space Y. We are here concerned with equivariant simple-homotopy theory and the corresponding notion of equivariant Whitehead torsion as defined in [7]. We wish to point out that even in the case when $G = P$ our formula for the equivariant Whitehead torsion of (1) deals with the situation where (1) is considered as a $(G \times G)$-homotopy equivalence between finite $(G \times G)$-complexes. Nevertheless we are able to give in Corollary B, for G a finite group, a geometric application in which we are dealing with the diagonal G-action on $X \times Y$ and $X' \times Y'$, see also Corollaries D and G.

In the case when P is a finite group we obtain as a corollary of the product formula the geometric result given in Theorem A. Specializing further we obtain in the case when $G = P$, a finite group, the application given in Corollary B.

<u>THEOREM A</u>. Let G be a compact Lie group and let $f: X \to Y$ be a G-homotopy equivalence between finite G - CW complexes. Assume that P is a finite group and that B is a finite P - CW complex, such that $\chi(B^Q_\beta) = 0$ for each component B^Q_β of any fixed point set B^Q. Then

$$f \times id: X \times B \longrightarrow Y \times B$$

is a simple $(G \times P)$-homotopy equivalence.

COROLLARY B. Let G be a finite group and $f\colon X \to Y$ a G-homotopy equivalence between finite G - CW complexes. Assume that V is a unitary complex representation of G. Then

$$f \times id\colon X \times S(V) \longrightarrow Y \times S(V)$$

is a simple G-homotopy equivalence, where G acts diagonally on $X \times S(V)$ and $Y \times S(V)$.

Theorem A does not hold in general if P is a non-finite compact Lie group and Corollary B does not either hold for a non-finite compact Lie group G, see section 8.

Recall that in the case of ordinary simple-homotopy theory we have the following. Let $f\colon X \to X'$ and $h\colon Y \to Y'$ be homotopy equivalences between finite connected CW complexes. Then the Whitehead torsion of $f \times h\colon X \times Y \to X' \times Y'$ is given by

$$\tau(f \times h) = \chi(Y)i_*\tau(f) + \chi(X)j_*\tau(h). \tag{2}$$

Here $i\colon X \to X \times Y$ and $j\colon Y \to X \times Y$ denote inclusions given by $i(x) = (x,y_o)$ and $j(y) = (x_o,y)$, for some fixed $y_o \in Y$ and $x_o \in X$, and χ denotes the Euler characteristic. (See e.g., 23.2 in [1].) In particular we have that the map

$$f \times id\colon X \times S^{2n-1} \longrightarrow X' \times S^{2n-1} \tag{3}$$

has zero Whitehead torsion and hence is a simple-homotopy equivalence for each $n \geq 1$. The fact that (3) is a simple-homotopy equivalence is an important result in geometric topology. Our Corollary B establishes, for any finite group G, the corresponding result in equivariant simple-homotopy theory. Our formula for the equivariant Whitehead torsion of (1), valid for arbitrary compact Lie groups G and P, is a generalization of the classical formula (2).

This paper also contains some other results than those already mentioned and a quick survey of the contents of the paper is as follows. Section 1 contains a review of the algebraic description of the equivariant Whitehead group $Wh_G(X)$, where G denotes an arbitrary compact Lie group and X is a finite G - CW complex. In Section 2 we define the Euler characteristics that we will use. The statement of the product formula for equivariant Whitehead torsion is given in Section 3 and the proof of the product formula is given in Section 4. In Section 5 we prove Theorem A and Corollary B. Section 6 gives a formula for the equivariant Whitehead torsion of the join of two equivariant homotopy equivalences, and corresponding formulae in the case of the smash product and reduced join are given in section 7. In Section 8 we give an example which shows that equivariant Whitehead torsion in the case of a compact Lie group G is <u>not</u> determined by the restrictions to all finite subgroups of G. This example also shows that Theorem A does not hold when P is a non-finite compact Lie group and that Corollary B does not hold for G a non-finite compact Lie group.

In the case of a finite group G and with the additional assumption that each component of any fixed point set X^H and Y^K is simply connected, a product formula is given in Dovermann and Rothenberg [4], see the Corollary on p. 3 of [4]. They consider the product spaces $X \times Y$ and $X' \times Y'$ as G-spaces through the diagonal action of G. In fact they are mainly concerned with the more general situation of a G fiber homotopy equivalence. They work with the generalized Whitehead torsion as defined in Rothenberg [14], and they also establish formulae for the generalized Whitehead torsion of joins and smash products. There is also some unpublished work by Shôrô Araki on product formulae for equivariant Whitehead torsion. For product formulae for equivariant finiteness obstructions see tom Dieck [2], tom Dieck and Petrie [3] and Lück [11], [12].

1. Review of the componentwise formula for $Wh_G(X)$

We will need to recall the algebraic determination of $Wh_G(X)$ as given in [8], see also [9]. (The first algebraic determination of $Wh_G(X)$, for G a compact Lie group, is due to H. Hauschild [5].) We have an isomorphism

$$\Phi\colon Wh_G(X) \xrightarrow{\cong} \sum_{\underline{C}(X)} Wh(\pi_0(WK)^*_\alpha). \tag{1}$$

The direct sum is over the set $\underline{C}(X)$ of equivalence classes of connected (non-empty) components X^K_α of arbitrary fixed point sets X^K, for all closed subgroups K of G. Two components X^K_α and X^L_β of the fixed point sets X^K and X^L, respectively, are in relation, denoted $X^K_\alpha \sim X^L_\beta$, if there exists $n \in G$ such that $nKn^{-1} = L$ and $n(X^K_\alpha) = X^L_\beta$. Given a component X^K_α of X^K we define

$$(WK)_\alpha = \{w \in WK \mid wX^K_\alpha = X^K_\alpha\}.$$

Here $WK = NK/K$. There is a short exact sequence of topological groups

$$e \longrightarrow \Delta \longrightarrow (WK)^*_\alpha \longrightarrow (WK)_\alpha \longrightarrow e$$

where Δ denotes the group of deck transformations of $\widetilde{X^K_\alpha}$, and hence $\Delta \cong \pi_1(X^K_\alpha)$. The group $(WK)^*_\alpha$ is a Lie group (not necessarily compact) which acts on the universal covering $\widetilde{X^K_\alpha}$ of X^K_α by an action which covers the action of $(WK)_\alpha$ on X^K_α. For the details of the construction of $(WK)^*_\alpha$ we refer to Section 5 of [8]. Observe that the groups $(WK)_\alpha$

and $(WK)^*_\alpha$ in fact depend on the actual geometry of the G-space. When we find it necessary to emphasize this fact we will use the following more complete notation:

$$(WK)_\alpha = W(X^K_\alpha), \text{ and}$$

$$(WK)^*_\alpha = W^*(\widetilde{X^K_\alpha}).$$

Using the more complete notation we may write the above exact sequence as

$$e \to \Delta(\widetilde{X^K_\alpha}) \to W^*(\widetilde{X^K_\alpha}) \to W(X^K_\alpha) \to e.$$

By $\pi_o(WK)^*_\alpha$ we denote the group of components of $(WK)^*_\alpha$ and $Wh(\pi_o(WK)^*_\alpha)$ is the Whitehead group of the discrete group $\pi_o(WK)^*_\alpha$.

We may also think of the direct sum over $\underline{C}(X)$ as a double direct sum

$$\sum_{(K)} \sum_{\alpha} Wh(\pi_o(WK)^*_\alpha) \tag{2}$$

where the first direct sum is over all conjugacy classes (K), of closed subgroups of G, for which $X^K \neq \emptyset$, and the second direct sum is, for a fixed K representing the conjugacy class (K), over the set of NK-components of X^K, with one connected component X^K_α representing the NK-component $(NK)X^K_\alpha$.

The isomorphism Φ is defined as follows. Let $s(V,X) \in Wh_G(X)$ be an arbitrary element in $Wh_G(X)$. Thus (V,X) is a finite G - CW pair with $i: X \to V$ a G-homotopy equivalence. Let K be a closed subgroup of G and X^K_α a connected component of X^K, and let V^K_α be the corresponding component of V^K. We denote

$$V^{>K}_\alpha = \{v \in V^K_\alpha | K \subsetneq G_v\}.$$

Then $(V^K_\alpha, X^K_\alpha \cup V^{>K}_\alpha)$ is a finite $(WK)_\alpha$ - CW pair, such that $(WK)_\alpha$ acts freely on $V^K_\alpha - (X^K_\alpha \cup V^{>K}_\alpha)$, and the inclusion

$$j: X^K_\alpha \cup V^{>K}_\alpha \longrightarrow V^K_\alpha$$

is a $(WK)_\alpha$-homotopy equivalence, see [8], Corollary 4.5 and Corollary 8.5b. Let $\widetilde{V^K_\alpha}$ be a universal covering space of V^K_α and let $\widetilde{X^K_\alpha \cup V^{>K}_\alpha}$ be the induced universal covering space of $X^K_\alpha \cup V^{>K}_\alpha$. Now $(\widetilde{V^K_\alpha}, \widetilde{X^K_\alpha \cup V^{>K}_\alpha})$ is a finite $(WK)^*_\alpha$ - CW pair, where $(WK)^*_\alpha$ acts freely on $\widetilde{V^K_\alpha} - (\widetilde{X^K_\alpha \cup V^{>K}_\alpha})$, and the inclusion

$$\tilde{j}: \widetilde{X^K_\alpha \cup V^{>K}_\alpha} \longrightarrow \widetilde{V^K_\alpha}$$

is a $(WK)^*_\alpha$-homotopy equivalence, see [8], Theorem 6.6 and Corollary 8.6.

We now consider the chain complex

$$C(\widetilde{V^K_\alpha},\widetilde{X^K_\alpha \cup V^{>K}_\alpha}) \tag{3}$$

where $C_n(A,B) = H_n(A^n \cup B, A^{n-1} \cup B;Z)$ and A^n denotes the equivariant n-skeleton of A, (here, the $(WK)^*_\alpha$-equivariant n-skeleton), and $H_n(\ ;Z)$ is ordinary singular homology with integer coefficients. We have that (3) is a finite acyclic chain complex of finitely generated free based $Z[\pi_o(WK)^*_\alpha]$-modules, see [8], Section 9. Hence (3) determines an element in the Whitehead group of $\pi_o(WK)^*_\alpha$, which we denote by

$$\tau(V,X)^K_\alpha = \tau(C(\widetilde{V^K_\alpha},\widetilde{X^K_\alpha \cup V^{>K}_\alpha})) \in Wh(\pi_o(WK)^*_\alpha).$$

The isomorphism Φ is given by

$$\Phi(s(V,X))_{K,\alpha} = \tau(V,X)^K_\alpha.$$

Here we think of the right hand side of (1) as given in the form (2), and $\Phi(s(V,X))_{K,\alpha}$ denotes the (K,α)-coordinate of $\Phi(s(V,X))$. We shall also denote

$$\Phi(s(V,X)) = \tau(V,X).$$

Observe that we have $\tau(V,X)^K_\alpha = 0$ unless

$$V^K_\alpha - (X^K_\alpha \cup V^{>K}_\alpha) \neq \emptyset.$$

2. Euler characteristics

Let X be a finite G - CW complex and let K be a closed subgroup of G. Then we have

$$X^{(K)} = \{x \in X \mid (K) \leq (G_x)\} = GX^K$$

and

$$X^{>(K)} = \{x \in X \mid (K) \lneq (G_x)\} = GX^{>K}$$

where $X^{>K} = \{x \in X \mid K \subsetneq G_x\}$. Now let X^K_α be a connected component of X^K. We then define

$$X_\alpha^{(K)} = \{x \in X^{(K)} | Gx \cap X_\alpha^K \neq \emptyset\}.$$

Then we have

$$X_\alpha^{(K)} = GX_\alpha^K.$$

Furthermore we define

$$X_\alpha^{>(K)} = X_\alpha^{(K)} \cap X^{>(K)}$$

and it then follows that

$$X_\alpha^{>(K)} = GX_\alpha^{>K}$$

where $X_\alpha^{>K} = X_\alpha^K \cap X^{>K}$.

For any $n \geq 0$ we set

$$\nu_{n,K,\alpha}(X) = \#\{\text{G-n-cells of type } G/K \text{ in } X_\alpha^{(K)}\}.$$

Another way to express this is that $\nu_{n,K,\alpha}(X)$ equals the number of G-equivariant n-cells in $X_\alpha^{(K)} - X_\alpha^{>(K)}$, and hence $\nu_{n,K,\alpha}(X)$ equals the number of ordinary n-cells in

$$(X_\alpha^{(K)} - X_\alpha^{>(K)})/G = X_\alpha^{(K)}/G - X_\alpha^{>(K)}/G.$$

We also have that $\nu_{n,K,\alpha}(X)$ equals the number of $(WK)_\alpha$-equivariant n-cells in $X_\alpha^K - X_\alpha^{>K}$, i.e., the number of ordinary n-cells in

$$(X_\alpha^K - X_\alpha^{>K})/(WK)_\alpha = X_\alpha^K/(WK)_\alpha - X_\alpha^{>K}/(WK)_\alpha.$$

We now define

$$\bar{\chi}_\alpha^K(X) = \sum_{n=0}^{s} (-1)^n \nu_{n,K,\alpha}(X)$$

where $s = \dim X$. It follows from the above discussion that we in fact have

$$\bar{\chi}_\alpha^K(X) = \chi(X_\alpha^{(K)}/G, X_\alpha^{>(K)}/G) = \chi(X_\alpha^K/(WK)_\alpha, X_\alpha^{>K}/(WK)_\alpha). \tag{1}$$

It is immediate that the following holds.

LEMMA. Let $f\colon X \to Y$ be a G-homotopy equivalence. Then $\bar{\chi}_{\alpha}^{K}(X) = \bar{\chi}_{f(\alpha)}^{K}(Y)$ for all (K,α). (Here $Y_{f(\alpha)}^{K}$ denotes the component of Y^K that contains $f(X_{\alpha}^{K})$.)

3. Statement of the product formula

Let G and P be compact Lie groups. Let $f\colon X \to X'$ be a G-homotopy equivalence between finite G - CW complexes and let $h\colon Y \to Y'$ be a P-homotopy equivalence between finite P - CW complexes. Then the equivariant Whitehead torsion $\tau(f \times h)$ of the $(G \times P)$-homotopy equivalence

$$f \times h\colon X \times Y \longrightarrow X' \times Y'$$

is given as follows. Given a connected component X_{α}^{K} of a fixed point set X^K and a connected component Y_{β}^{Q} of Y^Q, where K and Q are closed subgroups of G and P, respectively, we have the connected component $X_{\alpha}^{K} \times Y_{\beta}^{Q} = (X \times Y)_{\alpha\times\beta}^{K\times Q}$ of $(X \times Y)^{K\times Q}$. The $(K\times Q,\alpha\times\beta)$-coordinate of $\tau(f \times h)$ is given by

$$\tau(f \times h)_{\alpha\times\beta}^{K\times Q} = \bar{\chi}_{\beta}^{Q}(Y)i_{*}\tau(f)_{\alpha}^{K} + \bar{\chi}_{\alpha}^{K}(X)j_{*}\tau(h)_{\beta}^{Q}. \tag{1}$$

Here $i\colon \pi_o(WK)_{\alpha}^{*} \to \pi_o(WK)_{\alpha}^{*} \times \pi_o(WQ)_{\beta}^{*}$ and $j\colon \pi_o(WQ)_{\beta}^{*} \to \pi_o(WK)_{\alpha}^{*} \times \pi_o(WQ)_{\beta}^{*}$ denote the natural inclusions. Furthermore any coordinate $\tau(f \times h)_{\gamma}^{S}$ of $\tau(f \times h)$, where (S,γ) is not of a product form as above, equals zero.

4. Proof of the product formula

We shall begin by proving the following fact. Given any element $s(V,X) \in Wh_G(X)$ and any finite P - CW complex B the equivariant Whitehead torsion $\tau(V \times B, X \times B)$ of the $(G \times P)$-pair $(V \times B, X \times B)$ is given as follows: If X_{α}^{K} and B_{β}^{Q} are connected components of X^K and B^Q, respectively, we have that

$$\tau(V \times B, X \times B)_{\alpha\times\beta}^{K\times Q} = \bar{\chi}_{\beta}^{Q}(B)i_{*}\tau(V,X)_{\alpha}^{K} \tag{1}$$

and $\tau(V \times B, X \times B)_{\gamma}^{S} = 0$ whenever (S,γ) is not of a product form.

The very last statement is easily seen to be true for the following reason. Every isotropy subgroup occurring in $V \times B$ is of the product form $G_v \times P_b$, and therefore $(V \times B)^S - ((X \times B)^S \cup (V \times B)^{>S}) = \emptyset$ and consequently $\tau(V \times B, X \times B)_{\gamma}^{S} = 0$, for each component $(X \times B)_{\gamma}^{S}$ of $(X \times B)^S$, if S is not a product of a closed subgroup of G and a closed subgroup of P.

Now consider a subgroup of $G \times P$ of the form $K \times Q$, where K and Q are closed subgroups of G and P, respectively. Then we have

$$(X \times B)^{K\times Q} = X^K \times B^Q.$$

Moreover any connected component $(X \times B)^{K\times Q}_{\gamma}$ of $(X \times B)^{K\times Q}$ is of the form

$$X^K_\alpha \times B^Q_\beta = (X \times B)^{K\times Q}_{\alpha\times\beta},$$

where X^K_α and B^Q_β are connected components of X^K and B^Q, respectively. It now remains to prove that (1) holds.

By definition

$$\tau(V \times B, X \times B)^{K\times Q}_{\alpha\times\beta} \in Wh(\pi_o(W(K \times Q))^*_{\alpha\times\beta}) \tag{2}$$

is the torsion of the chain complex

$$C(\widetilde{(V \times B)^{K\times Q}_{\alpha\times\beta}}, \widetilde{(X \times B)^{K\times Q}_{\alpha\times\beta} \cup (V \times B)^{>(K\times Q)}_{\alpha\times\beta}}) \tag{3}$$

which is a finite acyclic complex of finitely generated free based $Z[\pi_o(W(K \times Q))^*_{\alpha\times\beta}]$-modules. Observe that we have

$$((V \times B)^{K\times Q}_{\alpha\times\beta}, (X \times B)^{K\times Q}_{\alpha\times\beta} \cup (V \times B)^{>(K\times Q)}_{\alpha\times\beta})$$

$$= (V^K_\alpha \times B^Q_\beta, X^K_\alpha \times B^Q_\beta \cup V^{>K}_\alpha \times B^Q_\beta \cup V^K_\alpha \times B^{>Q}_\beta)$$

$$= (V^K_\alpha, X^K_\alpha \cup V^{>K}_\alpha) \times (B^Q_\beta, B^{>Q}_\beta).$$

It follows that the chain complex (3) equals the chain complex

$$C((\widetilde{V^K_\alpha}, \widetilde{X^K_\alpha \cup V^{>K}_\alpha}) \times (\widetilde{B^Q_\beta}, \widetilde{B^{>Q}_\beta})) \tag{4}$$

which is isomorphic to the chain complex

$$C(\widetilde{V^K_\alpha}, \widetilde{X^K_\alpha \cup V^{>K}_\alpha}) \otimes_Z C(\widetilde{B^Q_\beta}, \widetilde{B^{>Q}_\beta}). \tag{5}$$

It is easy to see that $(W(K \times Q))^*_{\alpha\times\beta} = (WK)^*_\alpha \times (WQ)^*_\beta$, and moreover we have a canonical isomorphism of rings

$$Z[\pi_o(WK)^*_\alpha \times \pi_o(WQ)^*_\beta] \cong Z[\pi_o(WK)^*_\alpha] \otimes_Z Z[\pi_o(WQ)^*_\beta]. \tag{6}$$

Taking into account the canonical ring isomorphism (6) we have that the chain complexes (4) and (5) are isomorphic as based chain complexes over the ring (6).

All in all it follows that the torsion of the chain complex (3) equals the torsion of the chain complex (5).

For simplicity we denote

$$C = C(\widetilde{V^K_\alpha}, \widetilde{X^K_\alpha \cup V^{>K}_\alpha})$$

$$C' = C(\widetilde{B^Q_\beta}, \widetilde{B^{>Q}_\beta})$$

and set $\pi = \pi_o(WK)^*_\alpha$, $\pi' = \pi_o(WQ)^*_\beta$, $R = Z[\pi]$ and $R' = Z[\pi']$. It follows by the Product Theorem in [10] that the torsion of the chain complex (5), i.e., the $R \otimes_Z R'$ complex $C \otimes_Z C'$, is given by

$$\tau(C \otimes_Z C') = \chi_{R'}(C') i_* \tau(C) \tag{7}$$

where $i_*\colon Wh(\pi) \to Wh(\pi \times \pi')$ is induced by the natural inclusion $i\colon \pi \to \pi \times \pi'$, and $\chi_{R'}(C')$ denotes the Euler characteristic of C' as an R'-complex. But we have that

$$\begin{aligned} \chi_{R'}(C') &= \chi_{R'}(C(\widetilde{B^Q_\beta}, \widetilde{B^{>Q}_\beta})) \\ &= \chi(\widetilde{B^Q_\beta}/(WQ)^*_\beta, \widetilde{B^{>Q}_\beta}/(WQ)^*_\beta) \\ &= \chi(B^Q_\beta/(WQ)_\beta, B^{>Q}_\beta/(WQ)_\beta) \\ &= \bar{\chi}^Q_\beta(B) \end{aligned}$$

where the last equality is given by (2.1). Since $\tau(C) = \tau(V,X)^K_\alpha$ we have that (7) shows that the formula (1) holds as claimed.

Now let $f\colon X \to X'$ be a G-homotopy equivalence between finite G - CW complexes. By the equivariant skeletal approximation theorem (see Theorem 4.4 in [13] or Proposition 2.4 in [6]) we may assume that f is skeletal. The geometric equivariant Whitehead torsion of f is then by definition

$$t(f) = s(M_f, X) \in Wh_G(X),$$

where M_f denotes the mapping cylinder of f. (In [7], Section 3 the element $t(f)$ is denoted by $\tau_g(f)$.) On the algebraic side we use the notation

$$\tau(f) = \tau(M_f, X) = \Phi(t(f))$$

for the equivariant Whitehead torsion of f. Let B be any finite P - CW complex and consider the (G × P)-homotopy equivalence $f \times id_B: X \times B \to X' \times B$. The mapping cylinder of $f \times id_B$ equals $M_f \times B$ and hence

$$\tau(f \times id_B) = \tau(M_f \times B, X \times B).$$

Therefore we obtain from (1) that

$$\tau(f \times id_B)^{K\times Q}_{\alpha\times\beta} = \overline{\chi}^Q_\beta(B) i_* \tau(f)^K_\alpha . \tag{8}$$

We are now ready to complete the proof of the general product formula. We write the (G × P)-homotopy equivalence $f \times h: X \times Y \to X' \times Y'$ as a composite

$$f \times h = (id_{X'} \times h) \circ (f \times id_Y)$$

and use the formula for the geometric equivariant Whitehead torsion of a composite ([7], Proposition 3.8) to obtain

$$t(f \times h) = t(f \times id_Y) + (f \times id_Y)^{-1}_* t(id_{X'} \times h). \tag{9}$$

Applying the isomorphism Φ to (9) and considering the $(K \times Q, \alpha \times \beta)$-coordinate of $\Phi(t(f \times h)) = \tau(f \times h)$ we obtain

$$\tau(f \times h)^{K\times Q}_{\alpha\times\beta} = \tau(f \times id)^{K\times Q}_{\alpha\times\beta} + (\phi(f \times id)^{-1}_* t(id \times h))^{K\times Q}_{\alpha\times\beta}. \tag{10}$$

Using a naturality property of the isomorphism Φ we now obtain

$$\tau(f \times h)^{K\times Q}_{\alpha\times\beta} = \tau(f \times id)^{K\times Q}_{\alpha\times\beta} + (f^K_\alpha \times id)^{-1}_* \tau(id \times h)^{K\times Q}_{f(\alpha)\times\beta}. \tag{11}$$

Here $(f^K_\alpha \times id)_*: Wh(\pi_o W^*(\widetilde{X^K_\alpha}) \times \pi_o W^*(\widetilde{Y^Q_\beta})) \to Wh(\pi_o W^*(\widetilde{(X')^K_{f(\alpha)}}) \times \pi_o W^*(\widetilde{Y^Q_\beta}))$ is induced by the map $f^K_\alpha \times id: X^K_\alpha \times Y^Q_\beta \to (X')^K_{f(\alpha)} \times Y^Q_\beta$, where $(X')^K_{f(\alpha)}$ denotes the component of $(X')^K$ that contains $f(X^K_\alpha)$. By (8) we have that

$$\tau(id \times h)^{K\times Q}_{f(\alpha)\times\beta} = \overline{\chi}^K_{f(\alpha)}(X') j'_* \tau(h)^Q_\beta$$

where $j': \pi_o W^*(\widetilde{Y^Q_\beta}) \to \pi_o W^*(\widetilde{(X')^K_{f(\alpha)}}) \times \pi_o W^*(\widetilde{Y^Q_\beta})$ denotes the natural inclusion. Since $(f^K_\alpha \times id)_* \circ j_* = j'_*$, and $\overline{\chi}^K_{f(\alpha)}(X') = \overline{\chi}^K_\alpha(X)$ by the Lemma in Section 2 we now obtain that

$$(f^K_\alpha \times id)^{-1}_* \tau(id \times h)^{K\times Q}_{f(\alpha)\times\beta} = \overline{\chi}^K_\alpha(X) j_* \tau(h)^Q_\beta. \tag{12}$$

Applying the basic formula (8) to the first term on the right hand side of (11) and using (12) we see that the formula (11) establishes the product formula. □

5. Proof of Theorem A and Corollary B

Assume that P is a finite group and let B be a finite P - CW complex such that $\chi(B^Q_\beta) = 0$, for each component B^Q_β of any fixed point set B^Q. It then follows that also $\chi(B^{>Q}_\beta) = 0$ and hence $\chi(B^Q_\beta, B^{>Q}_\beta) = 0$. Using (2.1) and the fact that $(WQ)_\beta$ acts freely on $B^Q_\beta - B^{>Q}_\beta$ we now obtain

$$\bar{\chi}^Q_\beta(B) = \chi(B^Q_\beta/(WQ)_\beta, B^{>Q}_\beta/(WQ)_\beta)$$

$$= \frac{1}{|(WQ)_\beta|}\chi(B^Q_\beta, B^{>Q}_\beta) = 0$$

for each component B^Q_β of any fixed point set B^Q.

Now let $f\colon X \to Y$ be a G-homotopy equivalence between finite G - CW complexes, where G is a compact Lie group. It then follows by the product formula (3.1) (or in fact by the simpler formula (4.8)) that the $(G \times P)$-homotopy equivalence $f \times id\colon X \times B \to Y \times B$ has algebraic equivariant Whitehead torsion equal to zero, i.e., $\tau(f \times id) = 0$. Since Φ in (1.1) is an isomorphism we also have that $t(f \times id) = 0 \in Wh_{G\times P}(X \times B)$, and therefore $f \times id\colon X \times B \to Y \times B$ is a simple $(G \times P)$-homotopy equivalence, by Theorem II.3.6' in [7]. This completes the proof of Theorem A.

In the case when $G = P$, a finite group, we thus have that $f \times id\colon X \times B \to Y \times B$ is a simple $(G \times G)$-homotopy equivalence. It is an easily established geometric fact that if one restricts the transformation group to any subgroup H of $G \times G$ one still has that the H-map $f \times id\colon X \times B \to Y \times B$ is a simple H-homotopy equivalence. In particular this applies to the case when H is the diagonal subgroup of $G \times G$, i.e., in the case when we are considering $X \times B$ and $Y \times B$ as G-spaces through the diagonal G-action on them. Taking B to be the unit sphere S(V) in a complex unitary representation space V of G we see that Corollary B holds. □

6. Equivariant Whitehead torsion of the join of two equivariant homotopy equivalences

In this section we denote $I = [-1,1]$. The join of X and Y is by definition

$$X * Y = (X \times Y \times I)/\sim$$

where $\sim$ stands for the identifications $(x,y,-1) \sim (x,y',-1)$ for any $x \in X$ and all $y,y' \in Y$ and $(x,y,1) \sim (x',y,1)$ for any $y \in Y$ and all $x,x' \in X$. The join $X * Y$ has the quotient topology induced from the natural projection $p: X \times Y \times I \to X * Y$, and we denote $p(x,y,t) = [x,y,t]$. If X is a finite G - CW complex and Y is a finite P - CW complex, where G and P are compact Lie groups, then $X * Y$ is a finite $(G \times P)$ - CW complex. We have the natural imbeddings

$$i_-: X \longrightarrow X * Y$$

$$i_+: Y \longrightarrow X * Y$$

$$j_o: X \times Y \longrightarrow X * Y$$

defined by $i_-(x) = [x,y_o,-1]$ and $i_+(y) = [x_o,y,1]$, where $y_o \in Y$ and $x_o \in X$ are arbitrary, and $j_o(x,y) = [x,y,0]$, for all $x \in X$ and $y \in Y$. Let π_1: $G \times P \to G$ denote the projection onto the first factor. Then i_- is a skeletal co-π_1-map from the G - CW complex X into the $(G \times P)$ - CW complex $X * Y$, i.e., we have $i_-(\pi_1(g,p)x) = (g,p)i_-(x)$ for all $(g,p) \in G \times P$ and $x \in X$. Hence i_- induces a homomorphism

$$i_{-*}: Wh_G(X) \to Wh_{G\times P}(X * Y).$$

This homomorphism is defined as follows. By changing X into a $(G \times P)$-space through $\pi_1: G \times P \to G$ we obtain a homomorphism

$$\pi_1^!: Wh_G(X) \to Wh_{G\times P}(X).$$

(It is not difficult to see that $\pi_1^!$ is a monomorphism.) Let us denote by $1_-: X \to X * Y$ the inclusion i_- when X is considered as a $(G \times P)$-space through π_1. Then 1_- is a $(G \times P)$-map and induces a homomorphism 1_{-*}: $Wh_{G\times P}(X) \to Wh_{G\times P}(X * Y)$. We now define

$$i_{-*} = 1_{-*} \circ \pi_1^!.$$

Similarly i_+ is a co-π_2-map from the P-space Y into the $(G \times P)$-space $X * Y$, where $\pi_2: G \times P \to P$ is the projection onto the second factor, and the induced homomorphism

$$i_{+*}: Wh_P(Y) \to Wh_{G\times P}(X * Y)$$

is defined in complete analogy with the above definition of i_{-*}. Finally the

$(G \times P)$-imbedding j_o induces a homomorphism

$$j_{o*}: Wh_{G\times P}(X \times Y) \longrightarrow WH_{G\times P}(X * Y).$$

Now let $f: X \to X'$ be a G-homotopy equivalence and $h: Y \to Y'$ a P-homotopy equivalence, where X' and Y' also denote finite G and P, respectively, CW complexes. Then we have.

<u>PROPOSITION C</u>. The equivariant Whitehead torsion $t(f * h) \in W_{G\times P}(X * Y)$ of the $(G \times P)$-homotopy equivalence $f * h: X * Y \to X' * Y'$ is given by

$$t(f * h) = i_{-*}t(f) + i_{+*}t(h) - j_{o*}t(f \times h). \tag{1}$$

<u>Proof</u>. Let us denote $Z = X*Y$ and $Z' = X'*Y'$, and define

$$Z_- = \{[x,y,t] \in Z \mid -1 \leq t \leq 0\}$$

$$Z_+ = \{[x,y,t] \in Z \mid 0 \leq t \leq 1\}.$$

The spaces Z'_- and Z'_+ are defined similarly. Then we have $Z = Z_- \cup Z_+$ and $Z_- \cap Z_+ = X \times Y \times \{0\} = X \times Y$, and Z' has an analogous decomposition. By the sum theorem for equivariant Whitehead torsion (see [7], Theorem II.3.12) we have

$$t(f * h) = j_{-*}t((f * h)_-) + j_{+*}t((f * h)_+) - j_{o*}t(f \times h). \tag{2}$$

Here $(f * h)_-: Z_- \to Z'_-$ and $(f * h)_+: Z_+ \to Z'_+$ are the $(G \times P)$-maps induced by $f * h: Z \to Z'$, and $j_-: Z_- \to Z$ and $j_+: Z_+ \to Z$ denote the inclusions. Now let $k_-: X \to Z_-$ be the natural $(G \times P)$-inclusion defined by $k_-(x) = [x,y_o,-1]$, where $y_o \in Y$ is any element in Y, and define a $(G \times P)$-retraction $r'_-: Z'_- \to X'$ by $r'[x',y',t] = x'$, where $-1 \leq t \leq 0$ and $x' \in X'$, $y' \in Y'$. Then we have

$$f = r'_- \circ (f * h)_- \circ k_-: X \longrightarrow X' \tag{3}$$

where f is considered as a $(G\times P)$-homotopy equivalence. (The factor P of $G \times P$ acts trivially on X and X'.) The torsion of $f: X \to X'$ when considered as a $(G \times P)$-map equals $\pi_1^!(t(f)) \in Wh_{G\times P}(X)$, where $t(f) \in Wh_G(X)$ is the torsion of the G-homotopy equivalence f. Applying the formula for the torsion of a composite map (see [7], Proposition 3.8) to (3) we obtain

$$\pi_1^! t(f) = t(k_-) + k_{-*}^{-1}t((f * h)_-) + ((f * h)_- \circ k_-)_*^{-1}t(r'_-). \tag{4}$$

But Z_- collapses $(G \times P)$-equivariantly to X. (This follows for example from

Corollary II.1.9 in [7].) Thus $k_-: X \to Z_-$ is a simple $(G \times P)$-homotopy equivalence and hence $t(k_-) = 0$. Similarly $t(r'_-) = 0$. Since $j_- \circ k_- = 1_-: X \to Z$ we now obtain from (4) that

$$j_{-*}t((f * h)_-) = j_{-*}k_{-*}\pi_1^! t(f) = 1_{-*}\pi_1^! t(f) = i_{-*}t(f).$$

Similarly we also obtain $j_{+*}t((f * h)_+) = i_{+*}t(f)$. Making these substitutions in (2) we see that we have proved (1). □

COROLLARY D. Let the assumptions be the same as in Theorem A. Then the equivariant Whitehead torsion of the $(G \times P)$-homotopy equivalence $f*id: X * B \to X' * B$ is given by

$$t(f * id) = i_* t(f) \tag{5}$$

where $i_*: Wh_G(X) \to Wh_{G\times P}(X * B)$ is induced by the natural inclusion $i: X \to X * B$. In case $G = P$, a finite group, and we consider $f*id: X * B \to X' * B$ as a G-homotopy equivalence, where $X * B$ and $X' * B$ have the diagonal G-action, the same formula (5) holds but now as an equality in $Wh_G(X * B)$.

7. Torsion of smash products and reduced joins

In this section we give simple explicit formulae for the equivariant Whitehead torsion of the smash product and the reduced join of two equivariant homotopy equivalences. First we need the following geometric result.

LEMMA E. Let (X,X_o) be a finite G - CW pair such that X_o collapses equivariantly to $\{x_o\}$, where $x_o \in (X_o^o)^G$. Then the natural projection $p: X \to X/X_o$ is a simple G-homotopy equivalence.

Proof. Clearly $p: X \to X/X_o$ is a skeletal G-map. In order to prove that p is a simple G-homotopy equivalence we need to show that $s(M_p,X) = 0 \in Wh_G(X)$. We shall show that M_p in fact collapses equivariantly to X. Let $c_1,\ldots,c_k$ be all the equivariant cells of $X - X_o$ ordered in such a way that $\dim c_i < \dim c_j$ implies $i < j$. Let us denote

$$X_i = X_o \cup c_1 \cup \ldots \cup c_i, \qquad 1 \leq i \leq k,$$

and $p_i = p|X_i: X_i \to X_i/X_o$, $0 \leq i \leq k$. By a direct use of the definition of an equivariant elementary collapse we now have that for each i, $1 \leq i \leq k$, there is an equivariant elementary collapse

$$M_{p_i} \cup X \searrow M_{p_{i-1}} \cup X.$$

Thus M_p collapses equivariantly to $M_{p_o} \cup X = CX_o \cup X$. But since X_o collapses equivariantly to $\{x_o\}$ it follows that $CX_o \cup X$ collapses equivariantly to $C\{x_o\} \cup X$, see Lemma 3.1 in [7]. Since $C\{x_o\} \cup X$ collapses equivariantly to X we have completed the proof. □

Using the above Lemma and the sum theorem for equivariant Whitehead torsion we can now prove the following.

<u>PROPOSITION F</u>. Let $f: (X,A) \to (Y,B)$ be a G-map between finite G - CW pairs such that $f: X \to Y$ and $f|A: A \to B$ are G-homotopy equivalences. Then the equivariant Whitehead torsion of the induced G-homotopy equivalence $\bar{f}: X/A \to Y/B$ is given by

$$t(\bar{f}) = p_* t(f) - q_* t(f|A)$$

where $p: X \to X/A$ denotes the natural projection and $q = p|A: A \to X/A$ is the constant map $q(a) = \{A\} \in X/A$ for every $a \in A$.

<u>Proof</u>. Let $X \cup CA$ be the union along A of X and the cone CA on A, and define $Y \cup CB$ analogously. Then f induces a G-homotopy equivalence $\tilde{f}: X \cup CA \to Y \cup CB$. Since both CA and CB collapse equivariantly to their respective vertices v_A and v_B it follows that $C(f|A): CA \to CB$ is a simple G-homotopy equivalence and hence $t(C(f|A)) = 0$. Thus we have by the sum theorem for equivariant Whitehead torsion (see [7], Theorem II.3.12) that

$$t(\tilde{f}) = i_{1*} t(f) - i_{o*} t(f|A) \tag{1}$$

where $i_1: X \to X \cup CA$ and $i_o: A \to X \cup CA$ denote the inclusions. Since $(X \cup CA)/CA = X/A$ and $(Y \cup CB)/CB = Y/B$ we now have the commutative diagram

$$\begin{array}{ccc} X \cup CA & \xrightarrow{\ \tilde{f}\ } & Y \cup CB \\ \Big\downarrow{\scriptstyle \tilde{p}} & & \Big\downarrow{\scriptstyle \tilde{p}'} \\ X/A & \xrightarrow{\ \bar{f}\ } & Y/B \end{array}$$

where $\tilde{p}$ and $\tilde{p}'$ denote the natural projections collapsing CA and CB, respectively, to a point. The maps $\tilde{p}$ and $\tilde{p}'$ are simple G-homotopy equivalences by Lemma E and hence $t(\tilde{p}) = 0$ and $t(\tilde{p}') = 0$. Therefore the above commutative diagram and the formula for the equivariant Whitehead torsion of a composite map (see [7], Proposition II.3.8) together with (1) give us that

$$t(\bar{f}) = \tilde{p}_* t(\tilde{f}) = \tilde{p}_* i_{1*} t(f) - \tilde{p}_* i_{o*} t(f|A) = p_* t(f) - q_* t(f|A)$$

where $p = \tilde{p} \circ i_1: X \to X/A$ is the natural projection and $q = \tilde{p} \circ i_o = p|A$:

$A \to X/A$ equals the constant map $q(a) = \{A\} \in X/A$ for every $a \in A$. □

Now assume that $X = (X,x_o)$ and $X' = (X',x_o')$ are two finite pointed G - CW complexes and let $f: X \to X'$ be a G-homotopy equivalence such that $f(x_o) = x_o'$. Similarly let $Y = (Y,y_o)$ and $Y' = (Y',y_o')$ be finite pointed P - CW complexes and let $h: Y \to Y'$ be a P-homotopy equivalence such that $h(y_o) = y_o'$. It then follows that f and h are equivariant homotopy equivalences between pointed equivariant CW complexes. Therefore we have that the smash product $f \wedge h$: $X \wedge Y \to X' \wedge Y'$ is a $(G \times P)$-homotopy equivalence between two finite $(G \times P)$ - CW complexes. Since $X \wedge Y = (X \times Y)/(X \vee Y)$ we have by Proposition F that

$$t(f \wedge h) = p_*t(f \times h) - q_*t(f \vee h)$$

where $p: X \times Y \to X \wedge Y$ denotes the natural projection and $q: X \vee Y \to X \wedge Y$ is the constant map onto the point $[x_o,y_o] \in X \wedge Y$. By the sum theorem the torsion of the $(G \times P)$-homotopy equivalence $f \vee h: X \vee Y \to X' \vee Y'$ equals $t(f \vee h) = i_{1*}t(f) + i_{2*}t(h)$. Here $X \vee Y = X \times \{y_o\} \cup \{x_o\} \times Y$, and $i_1: X \to X \vee Y$ and $i_2: Y \to X \vee Y$ denote the natural inclusions. Thus the equivariant Whitehead torsion of the smash product $f \wedge h: X \wedge Y \to X' \wedge Y'$ is given by the formula

$$t(f \wedge h) = p_*t(f \times h) - q_{1*}t(f) - q_{2*}t(h). \tag{2}$$

Here $q_1: X \to X \wedge Y$ and $q_2: Y \to X \wedge Y$ denote the constant maps onto the point $[x_o,y_o] \in X \wedge Y$, and $p: X \times Y \to X \wedge Y$ is the natural projection as above.

In particular we have the following.

<u>COROLLARY G</u>. Let $f: (X,x_o) \to (X',x_o')$ be a G-homotopy equivalence between finite pointed G - CW complexes, where G is a compact Lie group. Assume that P is a finite group and that $B = (B,b_o)$ is a finite pointed P - CW complex such that $\chi(B^Q_\beta) = 0$ for each component B^Q_β of any fixed point set B^Q. Then the equivariant torsion of the $(G \times P)$-homotopy equivalence $f \wedge id: X \wedge B \to X' \wedge B$ is given by

$$t(f \wedge id) = -q_*t(f) \tag{3}$$

where $q: X \to X \wedge B$ is the constant map $q(x) = [x_o,b_o]$, for every $x \in X$. In case $G = P$, a finite group, and we consider $f \wedge id: X \wedge B \to X' \wedge B$ as a G-homotopy equivalence the same formula (3) holds, now as an equality in $Wh_G(X \wedge B)$.

<u>Proof</u>. By Theorem A we have that $t(f \times id) = 0 \in Wh_{G\times P}(X \times B)$. Therefore we obtain from (2) that

$$t(f \wedge id) = -q_*t(f) \in Wh_{G\times P}(X \wedge P).$$

Now the second assertion follows directly when one recalls that $q_*: \mathrm{Wh}_G(X) \to \mathrm{Wh}_{G\times G}(X \wedge B)$ is defined by $q_* = \hat{q}_* \circ \pi_1^!$, where $\hat{q}: X \to X \wedge B$ equals the map q considered as a $(G \times G)$-map, and observes that $\mathrm{res} \circ \pi_1^! = \mathrm{id}$. Here $\mathrm{res}: \mathrm{Wh}_{G\times G}(X) \to \mathrm{Wh}_G(X)$ denotes the map induced by restriction to the diagonal subgroup of $G \times G$, and $G \times G$ acts on X by having the second factor G act trivially. □

Let us now return to the general case where G and P are compact Lie groups. We shall consider the reduced join of f and h. The reduced join of X and Y is by definition

$$X \tilde{*} Y = (X * Y)/(X * \{y_o\} \cup \{x_o\} * Y).$$

We claim that the $(G \times P)$-subcomplex $X * \{y_o\} \cup \{x_o\} * Y$ collapses equvariantly to $\{x_o\}$. This is seen as follows. Since $X * \{y_o\}$ equals the cone on X we have that $X * \{y_o\}$ collapses $(G \times P)$-equivariantly to $\{x_o\} * \{y_o\}$, by Lemma II. 1.8 in [7]. Likewise $\{x_o\} * Y$ collapses equivariantly to $\{x_o\} * \{y_o\}$, which then collapses equivariantly to $\{x_o\}$. Thus we have a commutative diagram

$$\begin{array}{ccc} X * Y & \xrightarrow{f * h} & X' * Y' \\ \downarrow{\scriptstyle \pi} & & \downarrow{\scriptstyle \pi'} \\ X \tilde{*} Y & \xrightarrow{f \tilde{*} h} & X' \tilde{*} Y' \end{array}$$

where the natural projections π and π' are simple $(G \times P)$-homotopy equivalences by Lemma E. Hence we obtain, in the same way as in the proof of Proposition F, that

$$t(f \tilde{*} h) = \pi_* t(f * h).$$

Applying Proposition C we now obtain

$$t(f * h) = \bar{q}_{1*}t(f) + \bar{q}_{2*}t(f) - \bar{p}_{o*}t(f \times h) \tag{4}$$

where $\bar{q}_1: X \to X \tilde{*} Y$ and $\bar{q}_2: Y \to X \tilde{*} Y$ are constant maps to the base point $\{*\} \in X \tilde{*} Y$ and $\bar{p}_o: X \times Y \to X \tilde{*} Y$ is given by $\bar{p}_o(x,y) = [x,y,0] \in X \tilde{*} Y$. But observe that $\bar{p}_o$ is $(G \times P)$-homotopic to the map sending (x,y) to $[x,y,1] = [x_o,y,1] = \{*\}$, i.e., to the constant map $\bar{q}_o: X \times Y \to X \tilde{*} Y$ onto the base point $\{*\} \in X \tilde{*} Y$. By Lemma II.2.1 in [7] we have that $\bar{p}_{o*} = \bar{q}_{o*}$ and hence we may write (4) in the form

$$t(f \tilde{*} h) = \bar{q}_{1*}t(f) + \bar{q}_{2*}t(f) - \bar{q}_{o*}t(f \times h). \tag{5}$$

Having established this formula (5) for the equivariant Whitehead torsion of the reduced join of two equivariant based homotopy equivalences let us point out that (5) is in fact already contained in the formula (2) for the equivariant Whitehead torsion of the smash product of two equivariant based homotopy equivalences. Namely, there is a natural $(G \times P)$-homeomorphism $X \tilde{*} Y \cong X \wedge Y \wedge S^1$, and a double application of the formula (2) gives us exactly the formula (5). Thus we have that the basic formulae are: the product formula, the formula (6.1) for the (unreduced) join, and in the based case the formula (7.2) for the smash product.

8. Restricting equivariant Whitehead torsion to finite subgroups, an example

We shall give an example which shows that equivariant Whitehead torsion in the case of a compact Lie group G is not determined by the restrictions to all finite subgroups of G. Examples of this kind are suggested by our product formula for equivariant Whitehead torsion, but we are in fact able to give a very simple example which only involves the use of the product formula for ordinary Whitehead torsion.

Let $f: X \to X'$ be an ordinary homotopy equivalence between ordinary finite CW complexes such that $t(f) \neq 0 \in Wh(X)$. Let S^1 act on S^1 by multiplication and consider the S^1-homotopy equivalence

$$f \times id: X \times S^1 \to X' \times S^1. \tag{1}$$

Since the action of S^1 on $X \times S^1$ is free there is a natural isomorphism

$$\phi_*: Wh_{S^1}(X \times S^1) \xrightarrow{\cong} Wh((X \times S^1)/S^1) = Wh(X)$$

see [7], Theorem 2.7. We have $\phi_*(t(f \times id)) = t(f) \neq 0$ and hence $t(f \times id) \neq 0 \in Wh_{S^1}(X \times S^1)$, and consequently (1) is not a simple S^1-homotopy equivalence. Before we continue let us observe that this fact already shows that Theorem A does not hold when P is a non-finite compact Lie group and that Corollary B does not hold when G is not finite.

We now claim that for any finite subgroup K of S^1 the K-homotopy equivalence $f \times id: X \times S^1 \to X' \times S^1$ has equivariant Whitehead torsion

$$t(f \times id) = 0 \in Wh_K(X \times S^1). \tag{2}$$

In order to prove this claim we note that in this case we have $\phi_*(t(f \times id)) = t(f \times id_{S^1/K})$, where

$$\phi_*: Wh_K(X \times S^1) \xrightarrow{\cong} Wh(X \times (S^1/K)) \qquad (3)$$

is an isomorphism. Since S^1/K is a circle we have $\chi(S^1/K) = 0$, and hence it follows by the product formula for ordinary Whitehead torsion (see, e.g. [1], Theorem 23.2) that $f \times id: X \times (S^1/K) \to Y \times (S^1/K)$ has Whitehead torsion $t(f \times id_{S^1/K}) = 0 \in Wh(X \times (S^1/K))$. Since ϕ_* in (3) is an isomorphism we also have that $t(f \times id) = 0 \in Wh_K(X \times S^1)$, which proves our claim that (2) holds. Thus we have that (1) is a non-simple S^1-homotopy equivalence which is such that when the action is restricted to any finite subgroup K of S^1 becomes a simple K-homotopy equivalence.

References

[1] M. Cohen, A course in simple-homotopy theory, Graduate Texts in Math. 10, Springer-Verlag, 1973.

[2] T. tom Dieck, Über projektive Moduln und Endlichkeitshindernisse bei Transformationsgruppen, Manuscripta Math. 34 (1981), 135-155.

[3] T. tom Dieck and T. Petrie, Homotopy representations of finite groups, Inst. Hautes Études Sci. Publ. Math. No. 56 (1983), 337-377.

[4] K.H. Dovermann and M. Rothenberg, The generalized Whitehead torsion of a G fibre homotopy equivalence. (Preprint, 1984).

[5] H. Hauschild, Äquivariante Whiteheadtorsion, Manuscripta Math. 26 (1978), 63-82.

[6] S. Illman, Equivariant singular homology and cohomology for actions of compact Lie groups, in Proceedings of the Second Conference on Compact Transformation Groups (Univ. of Massachusetts, Amherst, 1971), Lecture Notes in Math., Vol. 298, Springer-Verlag, 1972, pp. 403-415.

[7] S. Illman, Whitehead torsion and group actions, Ann. Acad. Sci. Fenn. Ser. A I 588 (1974), 1-44.

[8] S. Illman, Actions of compact Lie groups and the equivariant Whitehead group, to appear in Osaka J. Math. (Almost identical with the preprint: Actions of compact Lie groups and equivariant Whitehead torsion, Purdue University 1983.)

[9] S. Illman, Equivariant Whitehead torsion and actions of compact Lie groups, in Group Actions on Manifolds, Contemp. Math. Amer. Math. Soc. 36 (1985), pp. 91-106.

[10] K.W. Kwun and R.H. Szczarba, Product and sum theorems for Whitehead torsion, Ann. of Math. 82 (1965), 183-190.

[11] W. Lück, Seminarbericht "Transformationsgruppen und algebraische K-Theorie", Göttingen, 1982/83.

[12] W. Lück, The Geometric Finiteness Obstruction, Mathematica Göttingensis, Heft 25 (1985).

[13] T. Matumoto, On G - CW complexes and a theorem of J.H.C. Whitehead, J. Fac. Sci. Univ. Tokyo Sect. I A Math. Vol. 18 (1971), 363-374.

[14] M. Rothenberg, Torsion invariants and finite transformation groups, in Proc. Symp. Pure Math., Vol. 32, Part 1 (Algebraic and Geometric Topology), Amer. Math. Soc., 1978, pp. 267-311.

Department of Mathematics
University of Helsinki
Hallituskatu 15
00100 Helsinki
Finland

Balanced orbits for fibre preserving maps of S^1 and S^3 actions

Jan Jaworowski

Abstract. Let $G = S^1$ or $G = S^3$, and let $p : Z \to X$ be a bundle with a fibre preserving action of G. Let $q : V \to Y$ be a vector space bundle with a fibre preserving action of G. Let $f : Z \to V$ be a fibre preserving map. The paper studies the size of the subset A_f made up of the orbits over which the average of f is zero. The size of A_f depends on the cohomology index of the action on Z and on the type of the action on V which can be described in terms of a Euler number. The result can be viewed as an extension of a continuous version of the Borsuk-Ulam theorem.

1. The average of a map.

Let G be a compact Lie group, let Z be a G-space, let V be a finite dimensional representation space for G and let $f : Z \to V$ be a map. The average of f is the map $\mathrm{Av}\, f : Z \to V$ defined by

$$(\mathrm{Av}\, f)x := \int g^{-1}(fgx)dg ,$$

where $\int$ denotes the Haar integral on G. The classical version of the Borsuk-Ulam theorem says that for any map $f : S^n \to \mathbb{R}^n$ there is an orbit $\{x,-x\}$ over which the average of f (with respect to the antipodal $\mathbb{Z}_2$-actions) is zero. In [8], Liulevicius proved an extension of the Borsuk-Ulam theorem for arbitrary free compact Lie group actions on the sphere using the averaging construction. In [7] we studied the set of points where the average of a map $f : Z \to V$ from a single G-space Z to a representation space V is zero. In this note we study such a set in a fibre bundle setting : a single map f is replaced by a fibre preserving map of a bundle $p : Z \to X$ over X whose fibre is a sphere (or, more generally, a suitable manifold) with a free, fibre preserving action of G ; and the vector space V is also replaced by a vector space bundle. An alalogous extension of the Borsuk-Ulam theorem for $\mathbb{Z}_2$-actions was done in [5], [6] and [9].

The average can be defined in the same way for a fibre preserving map $f : Z \to V$ of a bundle $p : Z \to X$ with an action of G to a bundle $q : V \to Y$ of representations of G. It has the following properties:

(1.1) For any map f, $\mathrm{Av}\, f$ is an equivariant map.

(1.2) If f is equivariant then $\mathrm{Av}\, f = f$.

We say that f is <u>balanced</u> at x if $(\mathrm{Av}\, f)x = 0$. If f is balanced at x then it is balanced over the entire orbit of x. Let A_f denone the set of points where f is balanced. It is an invariant closed subset of Z and we have

(1.3) $$A_f = A_{(\mathrm{Av}\, f)} = (\mathrm{Av}\, f)^{-1}0 ,$$

where 0 is the zero section of $q : V \to Y$.

If $f : S^n \to \mathbb{R}^k$, then $A_f = \{x \in S^n \quad fx = f(-x)\}$ (with respect to the antipodal involution).

2. The index and the characteristic homomorphism

A free action of the groups $G = S^0$, $G = S^1$ or $G = S^3$ is well described by its characteristic class, or by the index of the action. Let $d = 1$, $d = 2$ or $d = 4$ according to whether G is the unit spherein the field $\mathbb{F}$ of real numbers, $\mathbb{R}$, complex numbers, $\mathbb{C}$ or quaternions, $\mathbb{H}$. The universal space EG for these groups is the infinite dimensional sphere and the classifying space $EG/G = BG$ is the infinite projective space $P_\infty\mathbb{F}$. The cohomology of BG is a polynomial algebra on one generator. In the case $\mathbb{F} = \mathbb{R}$, it is $\mathbb{Z}_2[c_\mathbb{R}]$ generated by $c_\mathbb{R} \in H^1(P_\infty\mathbb{R}; Z_2)$; for $\mathbb{F} = \mathbb{C}$ or $\mathbb{F} = \mathbb{H}$, the cohomology with integer coefficients is $\mathbb{Z}[c_\mathbb{F}]$ generated by $c_\mathbb{F} \in H^d P_\infty\mathbb{F}$. For simplicity, we will drop the subscript $\mathbb{F}$ and write $c := c_\mathbb{F}$. If G acts freely on a space Z then the characteristic class of Z is the image $c_\mathbb{F}(Z) = c(Z) := (\varphi/G)^*c \in H^d(Z/G)$ of c under a classifying map $\varphi : Z \to EG$. The case of $\mathbb{F} = \mathbb{R}$, from our point of view, was studied in [6] ; in this note we will deal with the cases $\mathbb{F} = \mathbb{C}$ and $\mathbb{F} = \mathbb{H}$.

We will write $\overline{Z} := Z/G$ for the orbit space of the action; and if $f : Z \to X$ is a map, then $\overline{f} : \overline{Z} \to X$ will denote the induced map of the orbit spaces.

The index for $G = \mathbb{Z}_2$ was defined by Yang [10] and Conner and Floyd [1]. Fadell, Husseini and Rabinowitz [2, 3] defined and studied the index for compact Lie groups other than $\mathbb{Z}_2$, including non-free actions. For $G = S^1$ or $G = S^3$ we will define it as follows: If Z is a free G-space, then $\mathrm{Ind}_{\mathbb{F}}(Z) = \mathrm{Ind}\,(Z)$ is the largest n such that $c^n(Z)$ is an element of infinite order in $H^{dn}\overline{Z}$. The following property of the index is often used:

(2.1) Proposition. If Z and Z' are two free G-spaces and $f : Z \to Z'$ is an equivariant map then $\mathrm{Ind}(Z) \leq \mathrm{Ind}(Z')$.

To study fibre preserving actions, the characteristic class, or the index alone, are not sufficient: one must take into account the action of the cohomology of the base space on the cohomology of the total space of the bundle. Just as it was done in [6] for $G = \mathbb{Z}_2$, one can define a "characteristic homomorphism" associated with the action.

(2.2) Definition. Let $G = S^1$ or $S = S^3$; let Z be a free G-space; and let $p : Z \quad X$ be a map such that the action of G on Z is fibre preserving with repect to p. The characteristic homomorphism for p is the map

$$\hat{p}_j = \hat{p}_j(Z) : H^i X \to H^{i+j}\overline{Z}$$

defined by $\hat{p}_j x := (\overline{p}^* x) \circ c^j(Z)$.

3. Equivariant cohomology

We will be using the Alexander-Spanier cohomology with integer coefficients and the Borel equivariant cohomology. If Z is a G-space then $Z_G := EG \times_G Z$, where EG is the universal space for G, G acts on $EG \times Z$ by $g(e,z) := (ge,gz)$ and $EG \times_G Z := \overline{EG \times Z}$. The map $Z_G \to \overline{EG} = BG$ induced by the first projection $EG \times Z \to EG$ is a bundle with fibre Z. If G act trivially on Z then $Z_G \cong BG \times Z$.

The equivariant cohomology of Z is $H^*_G Z := H^* Z_G$. If G acts freely on Z then the map $Z_G \to \overline{Z}$ induced by the second projection $EG \times Z \to Z$ is a bundle with a contractible fibre EG; hence $H^*_G Z \cong H^* \overline{Z}$.

If $\cdot$ denotes a one-point space then the constant map $EG \to \cdot$ induces an isomorphism $H^*_G(\cdot) \cong H^* BG$. We will be identifying the groups $H^*{}_G EG = H^* BG$ and $H^*_G(\cdot)$ under this isomorphism. In our case of $G = S^1$ and $G = S^3$, this ring is a polynomial algebra on the generator $c \in H^d P_\infty \mathbb{F}$, the universal characteristic class.

Suppose that W is a representation space for G with $\dim_{\mathbb{R}} = m$ and let $W_o := W - (0)$. The map $\pi : W_G \to BG$ induced by the first projection is an orientable bundle with fibre W and it has its Thom class $U(W) \in H^m(W_G, W_{oG}) \in H^m_G(W, W_o)$. The restriction $U'(W)$ of $U(W)$

to W , $U'(W) \in H_G^*W$, corresponds to the Euler class of π under the isomorphism $\pi^* : H^m(\cdot) \cong H_G^m W$. The Euler class of π will also be called *the Euler class of W* and denoted by $e(W)$. Of course, $e(W) = 0$ unless k is a multiple of $d = \dim \mathbb{F}$. In our case of $G = S^{d-1}$, if $m = dk$, $H^m(\cdot)$ is freely generated by c^k and $e(W)$ can be characterized by an integer $\chi(W)$ such that $U'(W) = \chi(W) \cdot c^m(W)$. This integer, $\chi(W)$, will be called *the Euler number of W* . Here c is viewed as a class in $H_G^m(\cdot) \cong H_G^m EG$.

(3.1) Lemma. Let Z be a free G-space $(G = S^1$ or $S^3)$ and let W be a representation space for G with $\dim_{\mathbb{F}} W = k$ and with the Euler number $\chi(W)$. Let $f : X \to W$ be an equivariant map. then

$$f_G^* U'(W) = \chi(W) \cdot c^k(Z) .$$

Proof. Let $\varphi : Z \to EG$ be a classifying map for Z . Let $c \in H_G^d EG = H_G^d(\cdot)$ be the universal characteristic class. Because of the isomorphism $H^*\overline{Z} \cong H_G^* Z$, we can consider $c(Z) = \varphi_G^* c$. Let $\gamma : W \to \cdot$ be the constant map. Since W is equivariantly contractible, the diagram

$$\begin{array}{ccc} H_G^* W & \xleftarrow{\varphi_G^*} & H_G^*(\cdot) \\ {\scriptstyle f_G^*} \nwarrow & & \swarrow {\scriptstyle \gamma_G^*} \\ & H_G^* W & \end{array}$$

is commutative, $\varphi_G^* = f_G^* \gamma_G^*$. Hence $f_G^* U'(W) = f_G^*(\chi(W) \cdot c^k(W))$
$= \chi(W) \cdot f_G^*(\gamma_G^* c^k(\cdot)) = \chi(W) \cdot (f_G^* \gamma_G^* c)^k = \chi(W) \cdot (\varphi_G^* c)^k = \chi(W) \cdot c^k(Z)$.

Suppose now that $q : V \to Y$ is an orientable vector space bundle with a fibre preserving linear action of G ; i.e., a bundle of representations of G . If $y \in Y$, let $V^y := q^{-1} y$ be the fibre of q over y . Let $\dim_{\mathbb{R}} V^y = m$. The map $q_G : V_G \to Y$ induced by q is a bundle whose fibre over $y \in Y$ is $q_G^{-1} y = V_G^y$ and $q^y : V_G^y \to BG$ is a bundle with fibre V^y.

Let V_o be the complement of the zero section in V and let $U_G(V) \in H_G^m(V, V_o)$ be the equivariant Thom class of q_G . It is characterized by the fact that for each $y \in Y$ it restricts to the Thom class $U(V^y) \in H_G^m(V^y, V_o^y)$ of the bundle $q^y : V_G^y \to BG$ which, in turn, restricts to the orientation class in $H^m(V^y, V_o^y)$ in every fibre V^y . The restriction $U_G'(V)$ of $U_G(V)$ to V will be called *the equivariant Euler class of V* . For each fibre V^y it restricts to the Euler class $U'(V^y)$. The Euler class is locally constant on Y . If Y is connected, it is constant and, as for a single representation space, is characterized by an integer, the Euler number.

(3.2) Proposition. If W is a representation space for G ($= S^1$ or S^3) free outside the origin the the Euler class of W is zero.

This proposition was proved in [7; (5.2) and (5.3)] . For the standard (scalar multiplication) representation $W = \mathbb{F}^k$, the restriction homomorphism $H_G^{dk}(\mathbb{F}^k, \mathbb{F}_o^k) \to H^{dk}\mathbb{F}^k$ is an isomorphism, hence $e(\mathbb{F}^k) = c^k(\mathbb{F}^k)$. Now if W is any representation for G free outside the origin, with $\dim_{\mathbb{R}} W = dk$, then there is an equivariant map $\psi : \mathbb{F}^k \to W$ and $\psi^* U'(W) = U'(\mathbb{F}^k)$; therefore $U'(W)$ is non-zero.

(3.2) Remark. If $p : Z \to X$ is a bundle with a fibre preserving action of $G = S^1$ or S^3 , then $\mathrm{Ind}(Z^x)$ is a locally constant function of $x \in X$; and $\mathrm{Ind}(Z) \geq \mathrm{Ind}(Z^x)$ because a fibre inclusion is an equivariant map. If X is connected, $\mathrm{Ind}(Z^x)$ is constant; it will be called the fibre index of Z .

4. Main Result.

As usual, $G = S^1$ or $G = S^3$, and $\mathbb{F} = \mathbb{C}$ or $\mathbb{F} = \mathbb{H}$, respectively, with $d = \dim_{\mathbb{R}}\mathbb{F}$.

(4.1) Theorem. Let $p : Z \to X$ be a bundle with a free fibre preserving action of G over a connected base space X , and let F be a fibre of p . Suppose that $H^{dn}\overline{F} \cong \mathbb{Z}$ is freely generated by $c^n(F)$ and $H^i\overline{F} = 0$ for $i > dn$. Suppose that $\pi_1(X)$ operates trivially on $H^*\overline{F}$. Let $q : V \to Y$ be an orientable vector space bundle of orthogonal representations of G whose Euler class is non-zero and with $\dim_{\mathbb{F}} V = k$. Let $f : Z \to V$ be a fibre preserving map. Then the characteristic homomorphism

$$\hat{p}_{n-k}(A_f) \ : \ H^i X \ \cong \ H^{i+d(n-k)}\overline{A}_f$$

is injective for every i .

(4.2) Remark. This theorem applies, for instance, if the fibre F is the unit sphere $S^{d(n+1)-1}$ in $\mathbb{F}^{n+1}$ and the action of G on Z and V is the standard scalar multiplication. It is a parametrized (or fibrewise) extension of a theorem proved in [7] in the same sense as the results of [5] and [6] are parametrized extensions of the classical theorems of Borsuk-Ulam and Yang.

(4.3) Corollary. The covering dimension of A_f is at least $\dim X + d(n-k) + d - 1$.

This is so because $H^{\dim X + d(n-k)}\overline{A}_f \neq 0$ and the orbit map $A_f \to \overline{A}_f$ is a bundle with fibre $G = S^{d-1}$.

(4.4) The kernel of a linear map. Suppose that $p : W \to X$ and $q : V \to Y$ are vector space bundles over $\mathbb{F}$ with fibre dimensions $\dim_{\mathbb{F}} W = n$, $\dim_{\mathbb{F}} V = k$, and suppose that $f : W \to V$ is a fibre preserving linear map. If f is of a constant rank, then the kernel of f is a subbundle of W of a fibre dimension (over $\mathbb{F}$) at least $n-k$ and hence its total space is of a covering dimension at least $\dim X + d(n-k)$. Corollary (4.3) can be used to obtain the same number as a lower bound for $\dim \operatorname{Ker} f$, even if the rank of f is not constant, as follows. Suppose that the bundle is furnished with a norm. With the standard scalar multiplication action, f is equivariant, and a non-zero vector w if W is in the kernel of f if and only if $\frac{x}{|x|}$ belongs to $A_{f|S^W}$, where S^W is the unit shphere bundle in W. Then $\operatorname{Ker} f - (0\text{-section}) \cong A_{f|S^W} \times \mathbb{R}$ and by (4.3) the covering dimension of $\operatorname{Ker} f$ is at least $\dim X + d(n-k)$. This lower bound can also be obtained more directly, without using (4.1).

5. Proof of (4.1).

We can assume that f is equivariant; otherwise we can replace f by $\operatorname{Av} f$, as in Section 1. Thus $A_f = f^{-1}0$, where 0 is the zero section in V.

Let $\hat{p} = \hat{p}_{n-k}(\bar{A}_f) : H^iX \to H^{i+d(n-k)}\bar{A}_f$. We will construct a transfer homomorphism, $t : H^q\bar{A}_f \to H^{q-d(n-k)}X$, and show that $tp = \chi$, where is the Euler number of V; by the assumption, $\chi \neq 0$. The construction of t will be similar to that used in [9].

By the continuity of the cohomology theory we are using, it suffices to show that for any invariant neighborhood N of A_f in Z there is a transfer map $t_N : H^*\bar{A}_f \to H^*X$ such that $t_N p_N = \chi$, where $\hat{p}_N = p(N) : H^*X \to H^*N$.

Consider the equivariant map of pairs $f : (Z, Z-A_f) \to (V, V_o)$ and the excision map $e : (N, N-A_f) \to (Z, Z-A_f)$. Denote by j the inclusion map $Z \to (Z, Z-A_f)$ or $V \to (V, V_o)$. Define t_N to be the composite map

$$t_N : H^{i+d(n-k)}\bar{N} \cong H_G^{i+d(n-k)}N \xrightarrow{\cup f_N^* U_G(V)} H_G^{i+dn}(N, N-A_f)$$

$$\xleftarrow{e^*} H_G^{i+dn}(Z, Z-A\ ) \xrightarrow{j^*} H_G^{i+dn}Z \cong H^{i+dn}\bar{Z} \xrightarrow{\bar{p}_\#} H^iX .$$

In this sequence, $f_N : (N, N-A_f) \to (V, V_o)$ is the restriction of f to N. We identify H_G^*Z with $H^*\bar{Z}$ (since Z is free) and thus consider $c(Z)$ as either a class in H_G^* or in $H^*\bar{Z}$. Accordingly, we replace p_G^* by $\bar{p}^*$. The map $p_\#$ is the Gottlieb "integration along a fibre"; we quote some of its properties:

(5.1) Proposition [4, p. 40]. Let $p : E \to X$ be a bundle with a fibre F such that $H^{i+m}F \cong H^i(\cdot)$ for $i \geq 0$. Then there exists a natural homomorphism $p_\# : H^{p+m}E \to H^pX$ such that:

(5.2) If $X = \cdot$ then $p_\#$ is the given isomorphism $H^m F \cong H^o(\cdot)$;

(5.3) If $x \in H^*X$ and $z \in H^*E$ then $p_\#((p^*x)\cdot z) = x\cdot(p_\# z)$.

We continue with the proof of (4.1). Let $x \in H^*X$. Then

$$t_N\hat{p}_N x = p_\# j^* e^{*-1}((p^*_{GN}x)\cdot c^{n-k}(N) \cdot f^*_N U_G(V)) =$$
$$= p_\# j^*((p^*_G x)\cdot c^{n-k}(Z)\cdot f^* U_G(V)) = p_\#((p^*x)\cdot c^{n-k}(Z)\cdot f^* U'_G(V))$$
$$= x\cdot p_\#(c^{n-k}(Z)\cdot f^* U'_G(V)) .$$

We claim that $f^*U'(V) = \chi c^k(Z)$. It suffices to check this against every fibre Z^x over $x \in X$. Let $y = fx$, let $i^x : Z^x \to Z$ and $i^y : V^y \to V$ be the fibre inclusions and let $f^x : Z^x \to V^x$ be the induced map. Then

$$i^{x*} f^* U'_G(V) = f^{x*} i^{y*} U'_G(V) = f^{x*} U'(V^y) = c^k(Z^x)$$
$$= i^{x*} c^k(Z) = i^{x*} c^k(Z) .$$

Therefore $t_N p_N x = x\cdot p_\#(c^{n-k}(Z)\cdot \chi c^k(Z)) = \chi x\cdot \bar{p}_\# c^n(Z)$. Since $H^{dn}\bar{F}$ is freely generated by $c^n(\bar{F})$, for each fibre Z^x , $\bar{p}_\# c^n(Z^x) = 1$ by (5.2). Hence $\bar{p}_\# c^n(Z) = 1$ and $t_N\hat{p}_N x = \chi x$.

This completes the proof.

References

1. Conner, P. E. and Floyd, E. E.: Fixed point free involutions and equivariant maps. Bull. Amer. Math. Soc. 66 (1960), 416-441.

2. Fadell, E. R. and Rabinowitz, P. H.: Generalized cohomological index theories for Lie group actions with an application to bifurcation questions for Hamiltonian systems. Invent. Math. 45 (1978), 139-174.

3. Fadell, E. R., Husseini, S. and Rabinowitz, P. H.: Borsuk-Ulam theorems for arbitrary S^1 actions and applications. Trans. Amer. Math. Soc. 275 (1982), 345-360.

4. Gottlieb, D. H.: Fibre bundles and the Euler characteristic. J. Differential Geometry 10 (1975), 39-48.

5. Jaworowski, J.: A continuous version of the Borsuk-Ulam theorem. Proc. Amer. Math. Soc. 82 (1981), 112-114.

6. Jaworowski, J.: Fibre preserving maps of sphere bundles into vector space bundles. Proc. of the Fixed Point Theory Workshop, Sherbrooke, 1980; Lecture Notes in Mathematics, vol. 886, Springer-Verlag, 1981.

7. Jaworowski, J.: The set of balanced orbits of maps of S^1 and S^3 actions. To be published in Proc. Amer. Math. Soc.

8. Liulevicius, A.: Borsuk-Ulam theorems for spherical space forms. Proceedings of the Northwestern Homotopy Theory Conference (Evanston, Ill., 1982). Contemp. Math. 19 (1983), 189-192.

9. Nakaoka, M.: Equivariant point theorems for fibre-preserving maps. Preprint.

10. Yang, C. T.: On theorems of Borsuk-Ulam, Kakutani-Yamabe-Yujobô and Dyson, I. Ann. of Math. 70 (1954), 262-282.

Jan Jaworowski
Department of Mathematics
Indiana University
Bloomington, IN 47405
U. S. A.

INVOLUTIONS ON 2-HANDLEBODIES

Joanna Kania-Bartoszynska
Mathematical Institute
Polish Academy of Sciences
Sniadeckich 8, P.O. Box 137
00-950 Warsaw, Poland

In this paper we classify actions of Z_2 on orientable and nonorientable handlebodies of genus 2. We use a method of splitting involutions on 2-handlebodies to involutions on handlebodies of lower genus. All of the considered objects and morphisms are from the PL (piecewise linear) category.
A 2-handlebody is a 3-manifold H which contains 2 disjoint, properly embedded 2-cells D_1, D_2 such that the result of cutting H along $D_1 \cup D_2$ is a 3-cell.

Involutions (i.e. homeomorphisms of period 2) are classified up to conjugation by a homeomorphism; i.e. two involutions h,g of a 2-handlebody H are conjugate if there exists a homeomorphism $f : H \rightarrow H$ such that

$$h = f \circ g \circ f^{-1}$$

It turns out that the involutions on 2-handlebodies are classified by their fixed-points sets together with their position in a handlebody. More precisely:

<u>Theorem</u>

Two involutions h_1 and h_2 on 2-handlebodies H_1 and H_2 respectively are conjugate if and only if there exists a homeomorphism of pairs

$$(H_1, \text{Fix } h_1) \cong (H_2, \text{Fix } h_2) \quad .$$

Possible fixed-points sets of Z_2-actions on 2-handlebodies can be found using Smith theory. It turns out that for every such set there is an involution of 2-handlebody realizing it. Using the constructions described in this paper we can verify that there are 17 conjugacy classes of involutions on an orientable 2-handlebody and 28 classes on a nonorientable handlebody.
The involutions are listed in the appendix.
The above theorem was already proved for 0-dimensional fixed-points sets by J.H. Przytycki (see [P-1], thm.2.1); for orientation-preserving involutions with homogeneously 2-dimensional fixed-points sets it was proved by R.B. Nelson ([N-1] and [N-2]).
For the rest of this paper let H denote a 2-handlebody (both orientable

or not), H_{or}-orientable 2-handlebody, H_{non}-nonorientable 2-handlebody, D^n-n-disk (i.e. n-cell), T-solid torus, Ks-solid Klein bottle, Mb-Möbius strip. (M,h) denotes involution h on a manifold M. Fix h denotes a fixed-points set of a map h.
To prove the classification theorem we shall split involutions on 2-handlebodies to involutions on 3-disks, solid tori and solid Klein bottles. To do this we look for a 2-disk D in H which is either preserved by an involution or is disjoint with its image. Then we analyse the situation obtained by removing that disk and its image from H. The existence of such disk has been proved by P.K. Kim and J.L. Tollefson in the following lemma (see [K-T], lemma 3).

Lemma

Let h be an involution on a compact manifold M. Suppose that there exists a 2-disk D in M such that ∂D lies in a given component $\partial_1 M$ of a boundary ∂M and ∂D does not bound a disk in $\partial_1 M$. Then there exists a disk S properly embedded in M with the properties:

(1) $\partial S \subset \partial_1 M$,

(2) ∂ S does not bound a disk in $\partial_1 M$,

(3) either $h(S) \cap S = \emptyset$ or $h(S) = S$ and S is in general position with respect to Fix h.

The proof can be found in [K-T] or in [G-L]. It is worth mentioning that this lemma was generalized by W.H. Meeks III and S.T. Yau for actions of any finite group of homeomorphisms; they used minimal surface techniques. The purely topological proof was given by A.L. Edmonds. Obviously, 2-handlebodies satisfy the assumptions of the Kim-Tollefson lemma. Let h be an involution on H. There exists a properly embedded 2-disk S with properties (1)-(3) of the lemma. Clearly, involution h acts on $H - (U \cup h(U))$, where U is a small regular neighborhood of S in H which is either h-invariant (in the case of $h(S) = S$) or disjoint with h(U) (in the other case).
It follows that any involution $h : H \to H$ is obtained in one of the 5 constructions described below.
Assume first that $h(S) \cap S = \emptyset$. We have to consider 3 cases depending on the number of components of $H - (S \cup h(S))$.

I.

The result of cutting H along $S \cup h(S)$ is a ball D^3.

Then h is obtained from an involution i of a ball D^3 by an identification of two pairs of 2-disks on the boundary of D^3 : D_1 with D_2 and $i(D_1)$ with $i(D_2)$, where

$$D_1 \cap D_2 = \emptyset \ , \ D_j \cap \text{Fix } i = \emptyset \quad \text{for } j = 1,2 \ ,$$

$$i(D_1) \cap D_2 = \emptyset = D_1 \cap i(D_2) \ .$$

To identify D_1 with D_2 and $i(D_1)$ with $i(D_2)$ we use a homeomorphism

$$f : D_1 \cup i(D_1) \to D_2 \cup i(D_2)$$

which commutes with the involution h.
Denote the result of this construction by

$$(D^3,i)_{D_1 \overset{f}{\equiv} D_2}$$

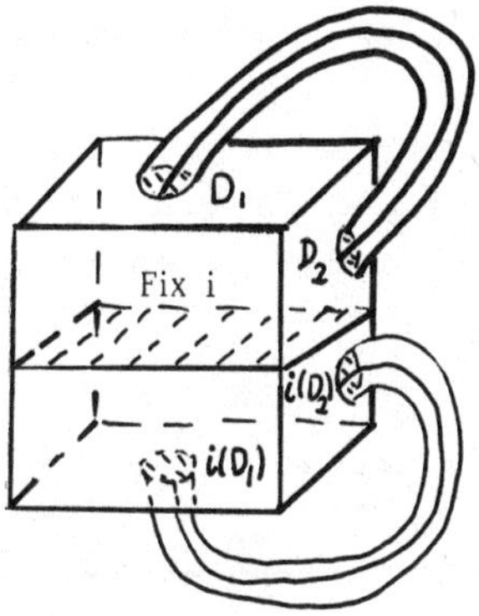

Fig. 1

Notice that if f changes orientation we obtain an involution of an orientable handlebody. If f preserves orientation we obtain an involution of H_{non} .
Observe also that conjugacy class of the involution $(D^3,i)_{D_1 \overset{f}{\equiv} D_2}$, remains the same for the different choice of D_1,D_2 and f as long as D_j' lies on the same side of Fix i as D_j (j = 1,2), and f,f' are in the same orientation class.

II.

The result of cutting H along $S \cup h(S)$ has two components i.e.

$$H - (S \cup h(S)) = D^3 \cup M \ ,$$

where M is a solid torus T or a solid Klein bottle Ks.
In this case h is obtained from an involution $i : D^3 \to D^3$ and an involution $j : M \to M$ by identifying a 2-disk $D_1 \subset \partial D^3$ with a 2-disk

$D_2 \subset \partial M$ and by identifying their images $i(D_1)$, $j(D_2)$ using a homeomorphism $f : D_1 \cup i(D_1) \rightarrow D_2 \cup j(D_2)$ such that

$$f \circ i = j \circ f \quad .$$

D_1 and D_2 have to satisfy

$$D_1 \cap \mathrm{Fix}\, i = \emptyset, \; D_2 \cap \mathrm{Fix}\, j = \emptyset \quad .$$

Denote the involution obtained in such way by

$$(D^3, i) \underbrace{}_{D_1 \overset{f}{\equiv} D_2} (M, j) \quad .$$

Notice that if $M = Ks$ we obtain an involution of a nonorientable 2-handlebody as well as in the case when $M = T$ and one of the involutions i,j preserves and the other changes orientation. If $M = T$ and either both involutions i,j preserve or both change orientation $(D^3,i) \underbrace{}_{D_1 \overset{f}{\equiv} D_2}$ (M,j) is an involution of H_{or} .

Observe also that the conjugacy class of an involution obtained in this construction does not depend on the choice of D_1, D_2 and f.

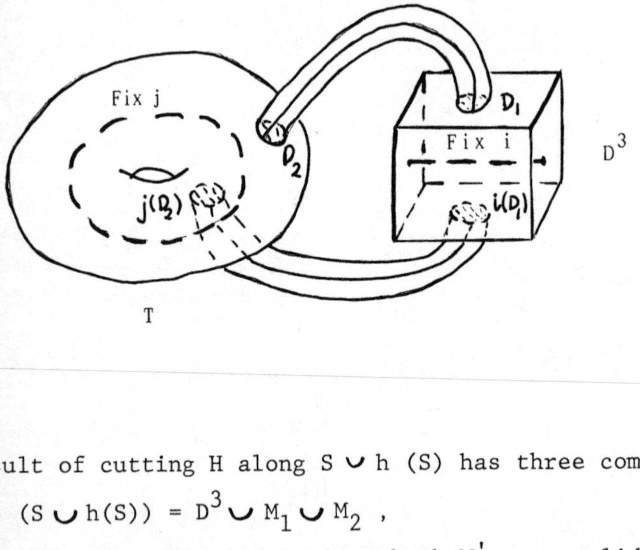

Fig. 2

III.

The result of cutting H along $S \cup h(S)$ has three components:

$$H - (S \cup h(S)) = D^3 \cup M_1 \cup M_2 \;,$$

where both M_1, M_2 are solid tori or both M_i's are solid Klein bottles. It is easy to see that in this case Fix h is equal to the fixed-points set of an involution i on D^3. Involution h is conjugate either to a central symmetry, to a line symmetry or to a plane symmetry in Fix h.

Figure 3

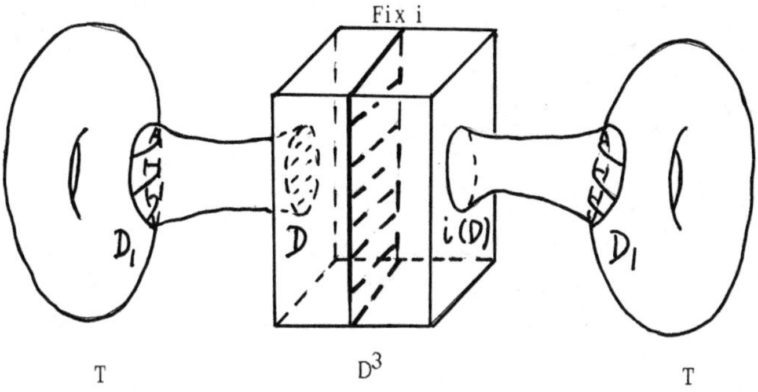

Now let $h(S) = S$.

We can assume that h does not exchange sides of S. If it does then for U - an h-invariant regular neighborhood of S in H we have

$$U \cong [-1,1] \times S, \text{ where } S = \{0\} \times S.$$

If we put $S_0 = \{-1\} \times S$ then $h(S_0) = \{1\} \times S$ so $h(S_0) \cap S_0 = \emptyset$ and h could be obtained by one of the constructions described above. Again we have to consider two cases depending on the number of components of H - S .

IV.

The result of cutting H along S is connected. Then

$$H - S = M ,$$

where M is a solid torus T or a solid Klein bottle Ks.

In this case h is obtained from an involution $j : M \to M$ by an identification of a 2-disk $D_1 \subset \partial M$ with a 2-disk $D_2 \subset \partial M$ using a homeomorphism $f : D_1 \to D_2$ commuting with j. Disks D_1, D_2 are chosen in such way that

$$D_1 \cap D_2 = \emptyset , \; j(D_i) = D_i \quad \text{for } i=1,2.$$

Denote this involution by

$$(M,j)_{D_1 \overset{f}{\equiv} D_2}$$

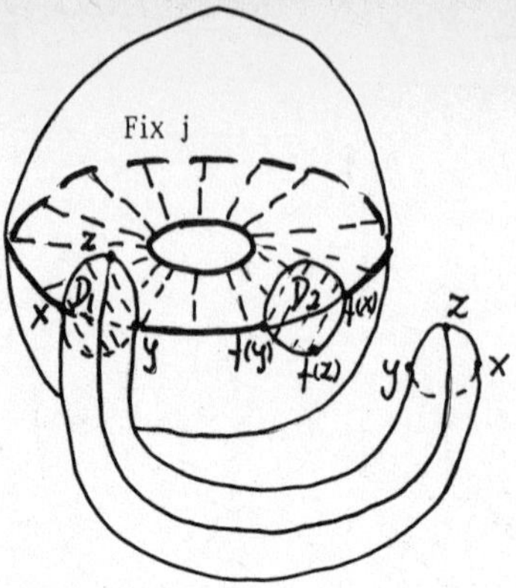

Fig.4

Observe that for M = T the conjugacy class of $(M,j)_{D_1 \overset{f}{=} D_2}$ depends only on the orientation class of f and $f|\mathrm{Fix}\, j$. For M - Ks the conjugacy class of the involutions obtained in this construction remains the same for a different choice of f as long as both f,f' either preserve or change local orientation on D_i's and either both preserve or change local orientation on $D_i \cap \mathrm{Fix}\, j$.

Let $D_1', D_2' \subset \partial M$ be the different choices of 2-disks such that

$$D_i' \cap D_2' = \emptyset , \; j(D_i') = D_i' \quad \text{for } i=1,2 .$$

Observe that if there exists an isotopy Φ of M taking D_i to D_i' (i=1,2) which is also an isotopy of Fix j (and thus takes $D_i \cap \mathrm{Fix}\, j$ to $D_i' \cap \mathrm{Fix}\, j$) then $(M,j)_{D_1 \overset{f}{=} D_2}$ and $(M,j)_{D_1' \overset{\Phi_1 \circ f}{=} D_2'}$ are conjugate.

V.

S disconnects H i.e.

$$H - S = M_1 \cup M_2 ,$$

where M_i is either a solid torus T or a solid Klein bottle Ks.
Then h is obtained from involutions $j_1 : M_1 \rightarrow M_1$, $j_2 : M_2 \rightarrow M_2$ by an identification of a 2-disk $D_1 \subset \partial M_1$ with a 2-disk $D_2 \subset \partial M_2$ using a homeomorphism $f : D_1 \rightarrow D_2$ such that

$$f \circ j_1 = j_2 \circ f \quad .$$

2-disks $D_i \subset \partial M_i$, i=1,2 have to be chosen so that

$j_i(D_i) = D_i$ for $i=1,2$.

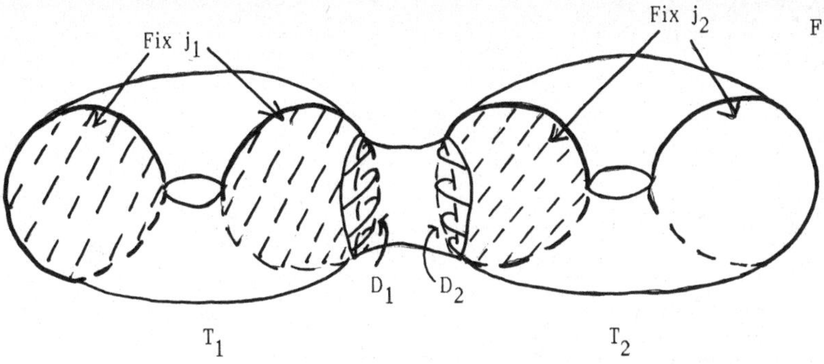

Fig.5

Clearly the conjugacy class of the obtained involution does not change for a different choice of 2-disks $D_1' \subset \partial M_1$, $D_2' \subset \partial M_2$ if there exist isotopies Φ_i of $(M_i, \mathrm{Fix}\, j_i)$ taking $(D_i, D_i \cap \mathrm{Fix}\, j_i)$ to $(D_i', D_i' \cap \mathrm{Fix}\, j_i)$ for $i=1,2$.

So we have reduced our problem to pasting together involutions of 3-disks, solid Klein bottles and solid tori, and to checking which constructions give us involutions from the same conjugacy class. Fortunately, involutions on handlebodies of lower genus are already classified.

Theorem

Involutions of D^3 are orthogonal up to conjugation.
Proof follows from C.R. Livesay theorem (see [Li]) and Smith Hypothesis proved by F. Waldhausen (see [Wa]). $\square$
Thus the only involutions of D^3 are central symmetry, line symmetry and plane symmetry. Denote them by i_1, i_2 and i_3 respectively.
Let solid torus T be represented as

$$T = S^1 \times D^2 = \mathbb{R} \times D^2/\sim \quad , \text{ where}$$

$$D^2 = \{ z \in \mathbb{C} \; : \; |z| \leqslant 1 \} \, ,$$

$$(t,y) \sim (t+1,y) \; .$$

Solid torus T can be also described as

$$T_A = \mathbb{R} \times D^2/\sim_A \quad , \text{ where}$$

$$(t,y) \sim_A (t+1,-y) \; .$$

Denote $T^* = T/\beta$, where $x \,\beta\, y$ iff ($x = y$ or $x = j(y)$).

Theorem

Every involution of the solid torus has one of the following forms (up to conjugation:

1) Involutions preserving orientation.

a) $j_{1a} : T \rightarrow T$, $j_{1a}(t,y) = (t,-y)$,
Fix $j_{1a} = S^1$, $T^* = S^1 \times D^2$

b) $j_{1b} : T \rightarrow T$, $j_{1b}(t,y) = (t+\frac{1}{2},-y)$
Fix $j_{1b} = \emptyset$, $T^* = S^1 \times D^2$

c) $j_{1c} : T \rightarrow T$, $j_{1c}(t,y) = (1-t,\bar{y})$
Fix $j_{1c} = D^1 \sqcup D^1$, $T^* = D^3$ ($\sqcup$ denotes disjoint sum)

2) Involutions changing orientation.

a) $j_{2a} : T \rightarrow T$, $j_{2a}(t,y) = (t+\frac{1}{2},\bar{y})$
Fix $j_{2a} = \emptyset$, $T^* = Ks$

b) $j_{2b} : T \rightarrow T$, $j_{2b}(t,y) = (t,\bar{y})$
Fix $j_{2b} = S^1 \times D^1$ $T^* = D^2 \times S^1$

c) $j_{2c} : T \rightarrow T$, $j_{2c}(t,y) = (1-t,y)$
Fix $j_{2c} = D^2 \sqcup D^2$, $T^* = D^1 \times D^2$

d) $j_{2d} : T \rightarrow T$, $j_{2d}(t,y) = (1-t,-y)$
Fix j_{2d} = two points

e) $j_{2e} : T_A \rightarrow T_A$, $j_{2e}(t,y) = (t+1,-\bar{y})$
Fix $j_{2e} = Mb$, $T^* = Ks$

f) $j_{2f} : T_A \rightarrow T_A$, $j_{2f}(t,y) = (1-t,-y)$
Fix j_{2f} = point $\sqcup D^2$

Proof : see [P-2] , theorem 6.5, I . □

Let solid Klein bottle be represented as

$$Ks = \mathbb{R} \times D^2/\approx \quad , \text{ where } (t,y) \approx (t+1,\bar{y}) .$$

Theorem

Every involution of solid Klein bottle has one of the following forms (up to conjugation):

1) $K_1 : Ks \rightarrow Ks$, $K_1(t,y) = (t+1,-\bar{y})$
Fix $K_1 = S^1$, $Ks^* = Ks$

2) $K_2 : Ks \to Ks$, $K_2(t,y) = (t+1,y)$

Fix $K_1 = S^1 \times D^1$, $Ks^* = S^1 \times D^2$

3) $K_3 : Ks \to Ks$, $K_3(t,y) = (t+1,-y)$

Fix $K_3 = Mb$, $Ks^* = Ks$

4) $K_4 : Ks \to Ks$, $K_4(t,y) = (1-t,-\bar{y})$

Fix $K_4 = D^1 \sqcup$ point

5) $K_5 : Ks \to Ks$, $K_5(t,y) = (1-t,\bar{y})$

Proof : see [P-2] , theorem 6.5 , II . □

Possible fixed-points sets of involutions on 2-handlebodies can be found using Smith theory (see [F1], thm. 4.3 and 4.4). If we denote a fixed-points set of an involution on 2-handlebody H by F then the following have to be satisfied

$$\sum_{i \geq n} \operatorname{rk} H^i(F;Z_2) \leq \sum_{i \geq n} \operatorname{rk} H^i(H;Z_2) \quad \text{for any integer } n$$

$$\chi(H;Z_2) \equiv \chi(F;Z_2) \mod 2 \quad .$$

Thus only the following cases may occur:

1) $\operatorname{rk} H^0(F;Z_2) = 1$, $\operatorname{rk} H^1(F;Z_2) = 0$

2) $\operatorname{rk} H^0(F;Z_2) = 1$, $\operatorname{rk} H^1(F;Z_2) = 2$

3) $\operatorname{rk} H^0(F;Z_2) = 2$, $\operatorname{rk} H^1(F;Z_2) = 1$

4) $\operatorname{rk} H^0(F;Z_2) = 3$, $\operatorname{rk} H^1(F;Z_2) = 0$

It is easy now to list all possible fixed-points sets of involutions on 2-handlebodies.
For each of these sets we check which constructions give us an involution with such fixed-points set. In all cases it is seen immediately that the results of different constructions of an involution for a given pair (H,Fix h) are conjugate. To show it we use a technique of cutting H along some suitably chosen properly embedded 2-disk and along its image.

Example

Consider the case of (H, Fix h) = (H_{or}, point $\sqcup$ annulus).
Involutions with such fixed-points set can be obtained only by the following constructions:

1) Construction II , for (D^3, i_1) , (T, j_{2b}) i.e.

$$(H,h) = (D^3, i_1) \underset{D_1 \overset{f}{\equiv} D_2}{\smile} (T, j_{2b})$$

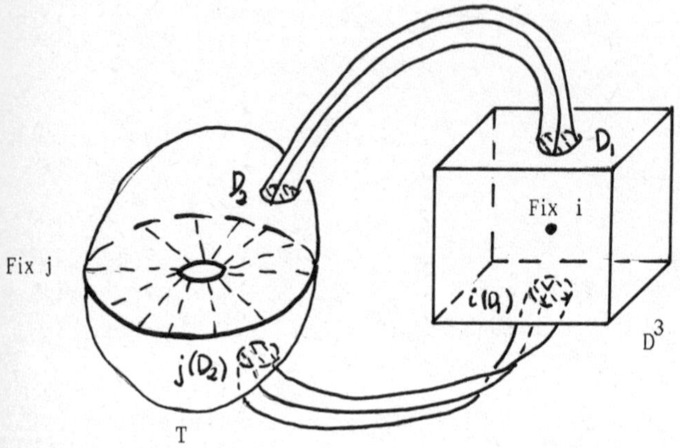

Fig.6

2) Construction IV , for $M = T$, $j = j_{2f}$.

$(H,h) = (T, j_{2f})_{D_1 \overset{f}{\equiv} D_2}$, where $f : D_1 \to D_2$ changes orientation on D_1 but locally preserves it on $D_1 \cap \mathrm{Fix}\, j_{2f}$.

3) Construction V, for $M_1 = T$, $j_1 = j_{2f}$ and $M_2 = T$, $j_2 = j_{2b}$, i.e.

$$(H,h) = (T, j_{2f}) \underset{D_1 \overset{f}{\equiv} D_2}{\smile} (T, j_{2b}) .$$

We will prove that the three involutions described above are conjugate. To show that $(T, j_{2f})_{D_1 \overset{f}{\equiv} D_2} \sim (D^3, i_1) \underset{D_1 \overset{f}{\equiv} D_2}{\smile} (T, j_{2b})$ it suffices to find a properly embedded 2-disk $D \subset T$ disjoint with D_1 and D_2, disjoint with $\mathrm{Fix}\, j_{2f}$ and disjoint with its image $j_{2f}(D)$.

The result of cutting $(H_{or}, h) = T, j_{2f})_{D_1 \overset{f}{\equiv} D_2}$ along $D \cup j_{2f}(D)$ is a disjoint sum of a solid torus with the involution which has an annulus as a fixed-points set and a 3-disk with central symmetry.

Thus $(T, j_{2f})_{D_1 \overset{f}{\equiv} D_2}$ could have been obtained by 1) (see figure 7).

Figure 7

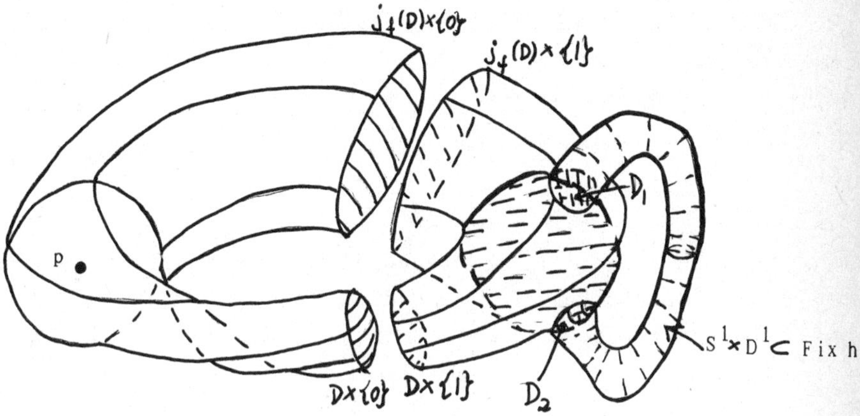

The proof that 3) $\sim$ 1) is analogous : we cut $(T,j_{2f}) \underset{D_1 \overset{f}{=} D_2}{\frown} (T,j_{2b})$ along $D \cup j_{2f}(D)$, where D is a 2-disk properly embedded in (T,j_{2f}) and such that $D \cap D_1 = \emptyset$, $D \cap \mathrm{Fix}\ j_{2f} = \emptyset$, $D \cap j_{2f}(D) = \emptyset$.

Fig. 8

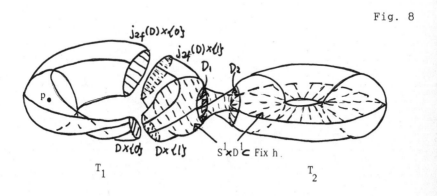

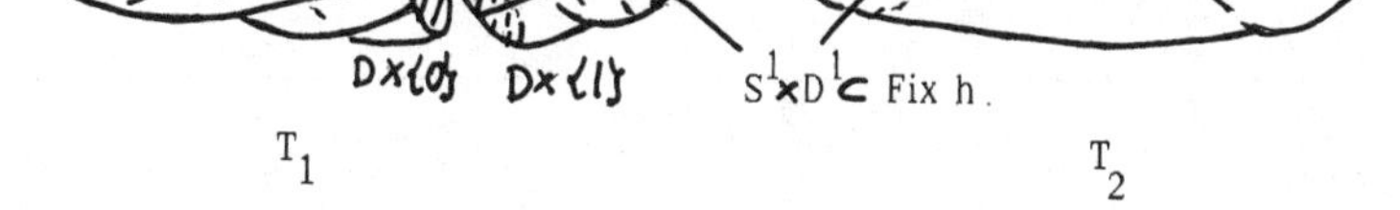

For all the other fixed-points sets we proceed in the same way.
It turns out that there are 17 conjugacy classes of involutions on an orientable 2-handlebody and 28 conjugacy classes of involutions on a

nonorientable 2-handlebody.
The involutions with their fixed-points sets are listed in the appendix.

This paper is based on my Master's thesis. I would like to express my deepest gratitude to my advisor Stefan Jackowski and to Józek Przytycki for their invaluable help.

APPENDIX

Observe first that there is no fixed-points free involution on H since the Euler characteristics of H is odd.
Denote by U(Fix h) a regular neighborhood of Fix h in H. p_i denotes a point, $\sqcup$ denotes disjoint sum.

Fix h	(H,h)	Description of (H,h)
point p	(H_{or}, h_1)	central symmetry in Fix h_1
	$(H_{non}, \tilde{h}_1)$	central symmetry in Fix $\tilde{h}_1$
$p_1 \sqcup p_2 \sqcup p_3$	(H_{or}, h_2)	$(D^3, i_1) \underset{D_1 \overset{f}{\equiv} D_2}{\smile} (T, j_{2d})$
D^1	(H_{or}, h_3)	line symmetry in Fix h_3
	$(H_{non}, \tilde{h}_2)$	line symmetry in Fix $\tilde{h}_2$
$D^1 \sqcup S^1$	(H_{or}, h_4)	$(D^3, i_2) \underset{D_1 \overset{f}{\equiv} D_2}{\smile} (T, j_{1a})$
	$(H_{non}, \tilde{h}_3)$	$(D^3, i_2) \underset{D_1 \overset{f}{\equiv} D_2}{\smile} (Ks, K_1)$
$D^1 \sqcup D^1 \sqcup D^1$	(H_{or}, h_5)	$(D^3, i_2) \underset{D_1 \overset{f}{\equiv} D_2}{\smile} (T, j_{1c})$
D^2 , H - D^2 is	(H_{or}, h_6)	plane symmetry in Fix h_6
not connected	$(H_{non}, \tilde{h}_4)$	plane symmetry in Fix $\tilde{h}_4$

D^2 , $H - D^2$ is connected	(H_{or}, h_7) $(H_{non}, \widetilde{h}_5)$	$(D^3, i_3) \underset{D_1 \overset{f}{\equiv} D_2}{\smile} (T, j_{2a})$ $(D^3, i_3) \underset{D_1 \overset{f}{\equiv} D_2}{\smile} (T, j_{1b})$
trinion (i.e. D^2 with two holes) H - Fix h orientable	(H_{or}, h_8) $(H_{non}, \widetilde{h}_6)$	$(T, j_{2b})_{D_1 \overset{f}{\equiv} D_2}$, where f changes orientation but locally preserves it on $D_1 \cap$ Fix j_{2b} $(Ks, K_2)_{D_1 \overset{f}{\equiv} D_2}$, where f changes local orientation on D_1, preserves local orientation on $D_1 \cap$ Fix K_2
trinion H - Fix h nonorientable	$(H_{non}, \widetilde{h}_7)$	$(Ks, K_2)_{D_1 \overset{f}{\equiv} D_2}$, where f preserves local orientation on D_1 and on $D_1 \cap$ Fix K_2
Klein bottle with a hole, H - Fix h connected	(H_{or}, h_9) $(H_{non}, \widetilde{h}_8)$	$(T_A, j_{2e}) \underset{D_1 \overset{f}{\equiv} D_2}{\smile} (T_A, j_{2e})$ $(Ks, K_3)_{D_1 \overset{f}{\equiv} D_2}$, where f locally preserves orientation on D_1 and on $D_1 \cap$ Fix K_3
Klein bottle with a hole, H - Fix h is not connected	$(H_{non}, \widetilde{h}_9)$	$(Ks, K_3)_{D_1 \overset{f}{\equiv} D_2}$, where f locally changes orientation on D_1 but preserves it on $D_1 \cap$ Fix K_3

Möbius strip with a hole, H - Fix h is connected	(H_{or}, h_{10}) $(H_{non}, \widetilde{h}_{10})$	$(T_A, j_{2e}) \underset{D_1 \overset{f}{\equiv} D_2}{\smile} (T, j_{2b})$ $(Ks, K_3)_{D_1 \overset{f}{\equiv} D_2}$, locally f changes orientation on D_1 and on $D_1 \cap \text{Fix } K_3$
Möbius strip with a hole, H - Fix h is not connected	$(H_{non}, \widetilde{h}_{11})$	$(Ks, K_3)_{D_1 \overset{f}{\equiv} D_2}$, locally f preserves orientation on D_1 and changes it on $D_1 \cap \text{Fix } K_3$
$D^2 \sqcup S^1 \times D^1$	(H_{or}, h_{11}) $(H_{non}, \widetilde{h}_{12})$	$(D^3, i_3) \underset{D_1 \overset{f}{\equiv} D_2}{\smile} (T, j_{2b})$ $(D^3, i_3) \underset{D_1 \overset{f}{\equiv} D_2}{\smile} (Ks, K_2)$
$D^2 \sqcup Mb$	(H_{or}, h_{12}) $(H_{non}, \widetilde{h}_{13})$	$(D^3, i_3) \underset{D_1 \overset{f}{\equiv} D_2}{\smile} (T_A, j_{2e})$ $(D^3, i_3) \underset{D_1 \overset{f}{\equiv} D_2}{\smile} (Ks, K_3)$
$D^2 \sqcup D^2 \sqcup D^2$	(H_{or}, h_{13})	$(D^3, i_3) \underset{D_1 \overset{f}{\equiv} D_2}{\smile} (T, j_{2c})$
point $\sqcup S^1$ U(Fix h) is orientable	$(H_{non}, \widetilde{h}_{14})$	$(D^3, i_1) \underset{D_1 \overset{f}{\equiv} D_2}{\smile} (T, j_{1a})$
point $\sqcup S^1$ U(Fix h) is nonorientable	$(H_{non}, \widetilde{h}_{15})$	$(D^3, i_1) \underset{D_1 \overset{f}{\equiv} D_2}{\smile} (Ks, K_1)$
$D^2 \sqcup S^1$ H - $(D^2 \cup S^1)$ is orientable	$(H_{non}, \widetilde{h}_{16})$	$(D^3, i_3) \underset{D_1 \overset{f}{\equiv} D_2}{\smile} (T, j_{1a})$

$D^2 \sqcup S^1$ $H - (D^2 \sqcup S^1)$ is nonorientable	$(H_{non}, \widetilde{h}_{17})$	$(D^3, i_3) \underset{D_1 \overset{f}{\equiv} D_2}{\smile} (Ks, K_1)$
point $\sqcup S^1 \times D^1$	(H_{or}, h_{14}) $(H_{non}, \widetilde{h}_{18})$	$(D^3, i_1) \underset{D_1 \overset{f}{\equiv} D_2}{\smile} (T, j_{2b})$ $(D^3, i_1) \underset{D_1 \overset{f}{\equiv} D_2}{\smile} (Ks, K_2)$
point $\sqcup$ Mb	(H_{or}, h_{15}) $(H_{non}, \widetilde{h}_{19})$	$(D^3, i_1) \underset{D_1 \overset{f}{\equiv} D_2}{\smile} (T_A, j_{2e})$ $(D^3, i_1) \underset{D_1 \overset{f}{\equiv} D_2}{\smile} (Ks, K_3)$
$D^1 \sqcup S^1 \times D^1$ U(Fix h) is orientable	$(H_{non}, \widetilde{h}_{20})$	$(D^3, i_2) \underset{D_1 \overset{f}{\equiv} D_2}{\smile} (T, j_{2b})$
$D^1 \sqcup S^1 \times D^1$ U(Fix h) is nonorientable	$(H_{non}, \widetilde{h}_{21})$	$(D^3, i_2) \underset{D_1 \overset{f}{\equiv} D_2}{\smile} (Ks, K_2)$
$D^1 \sqcup$ Mb U(Fix h) is orientable	$(H_{non}, \widetilde{h}_{22})$	$(D^3, i_2) \underset{D_1 \overset{f}{\equiv} D_2}{\smile} (T_A, j_{2e})$
$D^1 \sqcup$ Mb U(Fix h) is nonorientable	$(H_{non}, \widetilde{h}_{23})$	$(D^3, i_2) \underset{D_1 \overset{f}{\equiv} D_2}{\smile} (Ks, K_3)$
point $\sqcup D^1 \sqcup D^2$	$(H_{non}, \widetilde{h}_{24})$	$(D^3, i_3) \underset{D_1 \overset{f}{\equiv} D_2}{\smile} (Ks, K_4)$
point $\sqcup D^2 \sqcup D^2$	(H_{or}, h_{16})	$(D^3, i_1) \underset{D_1 \overset{f}{\equiv} D_2}{\smile} (T, j_{2c})$

point $\sqcup$ point $\sqcup$ D^2	(H_{or},h_{17})	$(D^3,i_3)\underset{D_1 \overset{f}{\equiv} D_2}{\smile}(T,j_{2d})$
point $\sqcup$ $D^1 \sqcup D^1$	$(H_{non},\widetilde{h}_{25})$	$(D^3,i_2)\underset{D_1 \overset{f}{\equiv} D_2}{\smile}(Ks,K_4)$
point $\sqcup$ point $\sqcup$ D^1	$(H_{non},\widetilde{h}_{26})$	$(D^3,i_1)\underset{D_1 \overset{f}{\equiv} D_2}{\smile}(Ks,K_4)$
$D^1 \sqcup D^1 \sqcup D^2$	$(H_{non},\widetilde{h}_{27})$	$(D^3,i_2)\underset{D_1 \overset{f}{\equiv} D_2}{\smile}(Ks,K_5)$
$D^1 \sqcup D^2 \sqcup D^2$	$(H_{non},\widetilde{h}_{28})$	$(D^3,i_3)\underset{D_1 \overset{f}{\equiv} D_2}{\smile}(Ks,K_5)$

References

[Fl] E.E. Floyd, Periodic Maps via Smith Theory, in: A. Borel, Seminar on Transformation Groups, Annals of Math. Studies 46, Princeton, New Jersey 1960.

[G-L] C. McGordon, R.A. Litherland, Incompressible Surfaces in Branched Coverings, preprint

[K-T] P.K. Kim, J.L. Tollefson, Splitting the PL-involutions of Nonprime 3-manifolds, Michigan Math. J. 27 (1980)

[Li] C.R. Livesay, Involutions with Two Fixed Points on the Three-sphere, Annals of Math., vol 78, N° 3 (1963)

[M] R. Myers, Free Involutions on Lens Spaces, Topology, vol 20, 1981

[N-1] R.B. Nelson, Some Fiber Preserving Involutions of Orientable 3-dimensional Handlebodies, preprint

[N-2] R.B. Nelson, A Unique Decomposition of Involutions of Handlebodies, preprint

[P-1] J.H. Przytycki, Z_n-actions on Some 2-and 3-manifolds, Proc. of the Inter. Conf. on Geometric Topology, PWN, Warszawa 1980.

[P-2] J.H. Przytycki, Actions of Z_n on Some Surface-bundles over S^1, Colloquium Mathematicum vol. XLVII, Fasc. 2, 1982

[Wa] F. Waldhausen, Über Involutionen der 3-sphare, Topology 8, 1969.

NORMAL COMBINATORICS OF G-ACTIONS ON MANIFOLDS

Gabriel Katz
Department of Mathematics, Ben Gurion University, Beer-Sheva 84105, Israel

This paper is the first in a series of papers developing a certain approach to the following general problem. What are the relations between the combinatorics of smooth G-actions on (closed) manifolds, in particular between the normal representations to fixed point sets, and global invariants (one can think about multisignatures as a model example) of different strata in the stratification of a manifold by the sets of points of different slice-types?

We have a pretty complete understanding of this problem for the special case $G = \mathbb{Z}_p$, p an odd prime. The answer is in terms of nontrivial numerical invariants, in particular, it depends essentially on the first factor h_1 of the class number of the cyclotomic field $\mathbb{Q}(e^{2\pi i/p})$. In this way one gets, for example, interesting conditions on the normal representations which can arise from exotic actions on $\mathbb{Z}_{(p)}$-homology complex projective spaces.

Our point is that to answer the question stated in the beginning it is very useful to organize all compact smooth G-manifolds into a ring, identifying G-manifolds having "similar" (bordant) combinatorial data with the "similar" lists of global invariants [4]. This can be viewed as an analogue of the classical relationship between the Burnside ring $\Omega(G)$ (which is a result of a Grothendick's construction, applied to finite G-sets) and the set of equivalence classes of G-CW-complexes [2]. The last equivalence deals with the Euler characteristics of different strata in the stratification of the CW-complex by different orbit-types. So, roughly speaking, the idea is to replace in the classical context the orbit-type stratification by slice-type one, and the Euler characteristic of strata by the corresponding Witt or multisignature invariants of different slice-types. For these purposes, one has to create the "discrete" objects, playing the same role with respect to the new context as finite G-sets do with respect to the classical one. We call these objects normal G-portraits (see Definitions A and B). Similar, but different, notations were considered by Dovermann-Petrie in the framing of their G-surgery program [3]. Our definitions are more accurately adjusted to the category of smooth G-actions.

The present paper is the foundation of the program described above.

In fact, any smooth G-action on a compact manifold M produces a normal G-portrait π_M. Roughly speaking, π_M is a collection of the following data (which satisfy certain relations): 1) the list of subgroups G_x of G, i.e. the stationary groups of the points x in M; 2) the list of G_x-representations ψ_x

(ψ_x is determined by the G-action on the fiber of the normal bundle $\nu(M^{G_x}, M)$ over x); 3) the list of groups which leave components of $M^{G_x}(x \in M)$ invariant and which are maximal with respect to this property; 4) the partial ordering on the set of components of M^{G_x}, $x \in M$, induced by inclusion.

It is known that there is a significant difference between the possibility of realizing data 1), 3), 4) in the category of G-manifolds and in the category of G-CW-complexes.

For example, the partially ordered set of subgroups of $\mathbb{Z}_{pqr}$ (p,q,r are distinct primes), represented on Figure 1, is realizable as the set of stationary groups on some connected $\mathbb{Z}_{pqr}$-CW-complex. But from the representation theory and data 2) it follows that this picture is not realizable on connected $\mathbb{Z}_{pqr}$-manifolds [3]. In contrast to this, the partially ordered set on Figure 2 can be realized on a G-manifold. We assume that the inclusions of various stationary groups on these diagrams correspond with the inclusions of the closures of the appropriate orbit-types of the action.

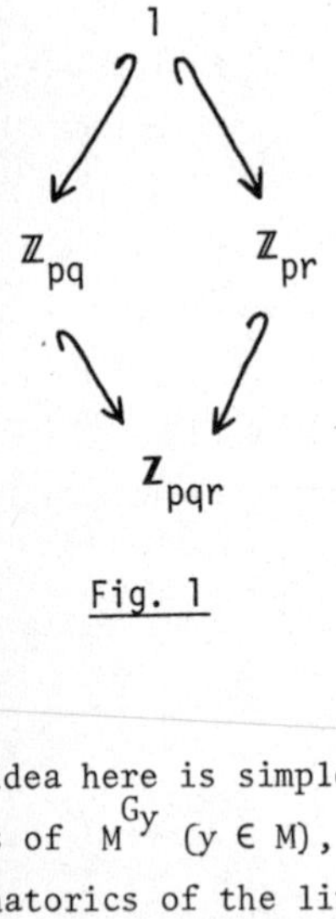

Fig. 1

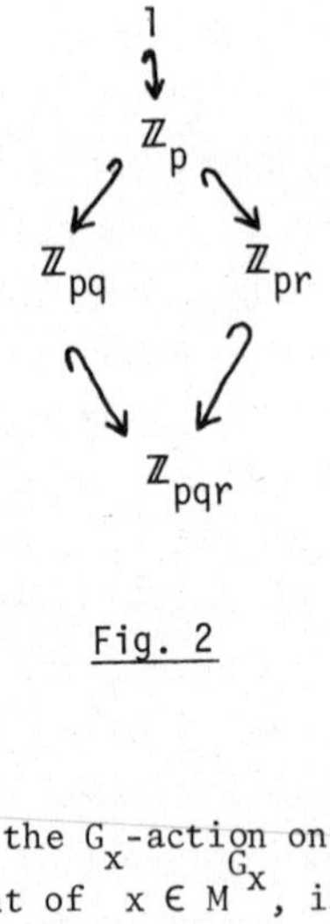

Fig. 2

The idea here is simple: the combinatorics of the G_x-action on the set of components of M^{G_y} ($y \in M$), containing the component of $x \in M^{G_x}$, is the same as the combinatorics of the linear G_x-action on the underlying space of the representation ψ_x (defined above). Basically, this observation is formalized in the notation of a normal G-portrait (see Definition B). General normal combinatorics are the result of gluing combinatorics of linear representations together.

It turns out that the notation of normal G-portraits is adequate to describe the combinatorics of G-actions. Namely, any compact smooth G-manifold determines a normal G-portrait (Lemma 2).

Our main result (Theorem) states that any normal G-portrait π can be realized by a smooth G-action on a compact manifold M_π.

Moreover, one can construct this manifold M_π with homology concentrated only in dimension 2, and the closure of each set of a given slice-type also has a similar homological structure.

Our construction allows us to "minimize" the fundamental groups of different components of the slice-type stratification of M_π. This is important if one wishes to use M_π as a basis for an equivariant version of Wall's construction [9] with the purpose of realizing geometrically equivariant surgery obstructions.

If we make no restrictions on the dimension of M, then there is no difference between the realization of a given normal combinatorial structure on a closed G-manifold or on a compact G-manifold (with ∂M realizing the same normal G-portrait as M does).

Under certain weak orientability assumptions (all the representations ψ_x taken to be SO-representations) one can prove (see Corollary) that such normal G-portraits are realizable on G-manifolds of the homotopy type of a bouquet of 2-dimensional spheres. Moreover, if all ψ_x are complex, then any fixed point set will also be of the homotopy type of $\vee_k (S^2_k)$.

These general results should be compared with more precise results obtained by other authors in important special cases (of G-actions on disks). We would like to mention two results of this sort. T. Petrie proved that any list of complex representations (up to some stabilization*), satisfying some necessary Oliver type conditions and Smith theory restrictions, are realizable as normal representations to the G-fixed points on some G-disk for G-abelian [9] (c.f. Pawałowski [8] and Tsai [10]). The geometrical construction that we use to prove our Theorem also requires some weak (+5-dimensional) stabilization not "in the normal direction to fixed point sets" as in [9], but in the "tangential one". In our approach we are flexible with dimensions of fixed point sets, but rigid with codimensions and normal representations.

The second result is due to K. Pawałowski [8]. For finite G, the following conditions are equivalent: (i) for any smooth G-action on a disk D, the tangential representations at any two G-fixed points are isomorphic, (ii) for any smooth G-action on a disk D, all the components of D^G have the same dimension, (iii) all the elements of G have prime power order. This theorem shows that for G with all the elements of prime power order, the normal portraits of G-actions on disks are the result from gluing a few copies of the G-portrait of a linear G-representation together.

Thus, it is well understood that normal G-portraits which are realizable on contractible G-manifolds (on G-disks) satisfy quite strong restrictions (see, e.g.

---

*Unfortunately this stabilization destroys the original combinatorics of these representations.

[7]). In contrast to this, as we mentioned above, any normal oriented G-portrait can be realized on a 1-connected G-manifold with non-trivial homology only in dimension 2. This 2-dimensional homology group, as a $\mathbb{Z}[G]$-module, is not projective in general (so, our construction does not associate a projective obstruction with a given normal G-portrait as one might expect).

In [4] using the results of this paper we will show that any normal G-portrait π together with an arbitrary list of multisignatures (or Witt invariants), parametrized by π, is realizable on smooth G-manifolds with boundary. For G-manifolds with boundary this completes the algebraization of the general problem stated in the very beginning. The analysis of closed G-manifolds is more complicated and leads to different integrality theorems.

I am grateful to J. Shaneson for stimulating discussions and to K.H. Dovermann and J. Shaneson for their help in making this text more readable.

Let M be a compact smooth manifold with a smooth right action of a finite group G on it. We will describe a stratification of M, defined by the G-action.

Let H be a subgroup of G. Denote by ${}^{\circ}M^H$ the set $\{x \in M | G_x = H\}$, where G_x is the stationary group of the point x. Let ${}^{\bullet}M^H$ be the closure of ${}^{\circ}M^H$ in M. It is a closed and open subset in $M^H = \{x \in M | G_x \supseteq H\}$ and a compact manifold. In fact, ${}^{\bullet}M^H$ consists of those connected components of M^H, which have a dense subset with the stationary subgroup H.

Consider the set π_M, which by definition is the connected component set $\pi_0(\coprod_{H \subseteq G} {}^{\circ}M^H)$. If $\operatorname{codim}({}^{\bullet}M^H, {}^{\bullet}M^K) > 1$ for any ${}^{\bullet}M^H \underset{\neq}{\subset} {}^{\bullet}M^K$, then π_M coincides with $\pi_0(\coprod_{H \subseteq G} {}^{\bullet}M^H)$.

Example. Let $G = \mathbb{Z}_{12}$ and $M = \mathbb{C}P^4$. Consider the G-action, which in homogeneous coordinates $(z_0:z_1:z_2:z_3:z_4)$ is given by the formula:

$(z_0:z_1:z_2:z_3:z_4)g = (\lambda z_0:\lambda^2 z_1:\lambda^4 z_2:\lambda^3 z_3:\lambda^9 z_4)$. Here g is a generator of $\mathbb{Z}_{12}$ and $\lambda = \exp(\pi i/6)$. The components of the set M^H, where $H \subseteq \mathbb{Z}_{12}$ is a subgroup, are in one-one correspondence with the nontrivial eigenspaces in $\mathbb{C}^5$ of a generator of H. Considering $\{{}^{\bullet}M^H\}$-stratification we are selecting only components of M^H which have H as a stationary group of a generic point.

Figure 3 describes $\pi = \pi_0(\coprod_k {}^{\bullet}M^{g^k})$, $k = 0,1,2,3,4,6$. The elements of π are denoted by vertices of the graph and the inclusion of components one into another by arrows (the directions of arrows are opposite to the inclusion). The right side of the picture describes the partially-ordered set $S(\mathbb{Z}_{12})$ of all subgroups in $\mathbb{Z}_{12}$. The horizontal arrows, pointing from π to $S(\mathbb{Z}_{12})$, associate with each component the stationary group of its generic point.

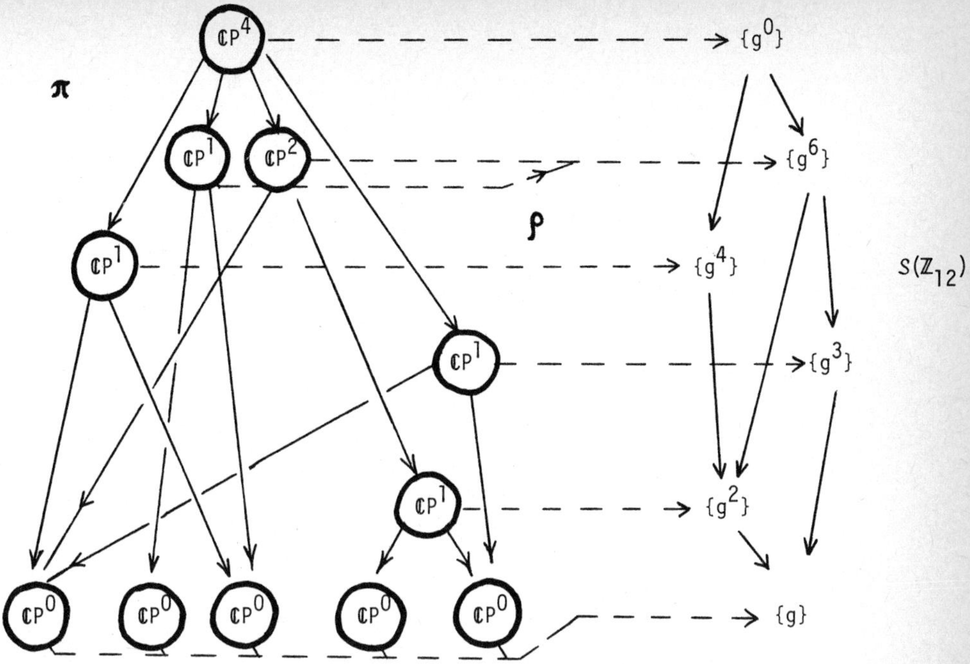

Fig. 3

Now we are going to axiomatize the properties illustrated by this example very much in the style of [3], [7]. We do this by introducing further structure on the set π.

Let G be finite and let $S(G)$ denote the set of all subgroups of G. The group G acts on $S(G)$ by the conjugation: $Ad_g : H \to g^{-1}Hg$ for any $H \in S(G)$.

Let π be a finite partially-ordered right G-set with a G-map $\rho: \pi \to S(G)$. The map ρ is consistent with the partial order $\geq$ in $S(G)$ in the following sense: for any two elements $\alpha > \beta$ of π, the group $\rho(\alpha)$ is a proper subgroup of $\rho(\beta)$. As usual, $>$ means $\geq$ and $\neq$.

Denote by G_α the stationary subgroup of $\alpha \in \pi$ (with respect to the G-action on π). We assume that $\rho(\alpha) \subseteq G_\alpha$.

Moreover, an isomorphism class of an orthogonal representation $\psi_\alpha : \rho(\alpha) \to O(V_\alpha)$ is associated with every element $\alpha \in \pi$, and we assume that the following two properties hold.

I. The representations $\{\psi_\alpha\}$ are consistent with the G-action on π

in the following sense: for any $\alpha \in \pi$ and $g \in G$, the representation $\rho(\alpha g) \xrightarrow{Ad_{g^{-1}}} \rho(\alpha) \xrightarrow{\psi_\alpha} O(V_\alpha)$ is isomorphic to the representation $\psi_{\alpha g}$. In particular, ψ_α and $\psi_\alpha \circ Ad_{g^{-1}}$ are isomorphic for any $g \in G_\alpha$.

II. The representations $\{\psi_\alpha\}$ are consistent with the partial order in π in the following sense. For any two elements $\alpha \leq \beta$, in the canonical decomposition of the $\rho(\beta)$-representation $Res_{\rho(\beta)}(\psi_\alpha)$ into the direct sum of the trivial summund and its orthogonal complement, the latter is isomorphic to ψ_β. By the definition, for any maximal $\alpha \in \pi$, the space V_α is 0-dimensional.

Remark. In fact, property I describes a nontrivial relationship between G_α, $\rho(\alpha)$ and ψ_α. Let $N_{\psi_\alpha}[\rho(\alpha)]$ denote the subgroup of elements in the normalizer $N(\rho(\alpha))$ which preserve the character of ψ_α under the action by conjugation. By I, G_α has to be a subgroup of $N_{\psi_\alpha}[\rho(\alpha)]$.

Property II implies $V_\alpha^{\rho(\alpha)} = \{0\}$ for any element α in π.

Definition A. A partially ordered right G-set π with a G-map ρ (as above) and with a list of representations $\{\psi_\alpha\}$ satisfying Properties I and II, we will call a *discrete portrait of a* G-*action,* or more briefly a G-*portrait*. One can find this notion (with minor changes) in [3], [7] under the name of POG-set.

One can replace orthogonal groups $O(V_\alpha)$ in the previous definition by the classical groups $SO(V_\alpha)$, $U(V_\alpha)$ (or any other classical Lie groups). The corresponding discrete portraits of G-action will be called (correspondingly) G-*portraits with an oriented orthogonal or complex structure.*

The following definition plays the central role in our considerations.

Definition B. A (discrete) G-portrait π is called *normal* if for every $\alpha \in \pi$, the G-map ρ maps the partially-ordered G_α-set $\pi_{\geq\alpha} = \{\beta \in \pi | \beta \geq \alpha\}$ isomorphically onto $|\psi_\alpha|$. Hereafter, $|\psi_\alpha|$ denotes the partially-ordered G_α-set of subgroups of G which are stationary groups of vectors $v \in V_\alpha$ with respect to the $\psi_\alpha(\rho(\alpha))$-action ($G_\alpha$ acts on $|\psi_\alpha|$ by the conjugation).

In the following lemma we are underlying a few properties of normal G-portraits.

Lemma 1. *For any normal* G-*portrait* π *the following holds:*

1) *for any three elements* $\alpha,\beta,\gamma \in \pi$ *such that* $\alpha > \gamma$, $\beta > \gamma$, *there exists a unique element* $\delta \in \pi$ *with the properties* $\delta \geq \alpha$, $\delta \geq \beta$ *and* $\rho(\delta) = \rho(\alpha) \cap \rho(\beta)$.

2) *as an immediate consequence of* 1), *for any* $\alpha \in \pi$, *there is a unique maximal element in the set* $\pi_{\geq\alpha}$.

3) *for any* $\alpha \in \pi$, *there is not more than one element* $\beta \in \pi_{\geq\alpha}$ *with a given value* $\rho(\beta) = H \in S(G)$.

4) *for any two elements* $\beta \geq \alpha$, *the group* $G_\beta \cap G_\alpha$ *is the normalizer*

$N_{G_\alpha}(\rho(\beta))$ *of* $\rho(\beta)$ *in* G_α.

To prove point 1) of the lemma consider the set $\pi_{\geq\gamma}$. By the definition of normal G-portrait it is isomorphic to $|\psi_\gamma|$ and the isomorphism $\rho:\pi_{\geq\gamma} \to |\psi_\gamma|$ is G_α-isovariant and order-preserving. So, it is enough to show that if $\rho(\alpha)$, $\rho(\beta)$ are stationary groups of some vectors in V_γ with respect to the $\rho(\gamma)$-action, then $\rho(\alpha) \cap \rho(\beta)$ is also a stationary group of this action. Consider subspaces $V_\gamma^{\rho(\alpha)}$, $V_\gamma^{\rho(\beta)}$ and $V_\gamma^{\rho(\alpha)\cap\rho(\beta)}$ in V_γ. It is easy to see that the stationary group of a generic point in $V_\gamma^{\rho(\alpha)\cap\rho(\beta)}$ is precisely $\rho(\alpha) \cap \rho(\beta)$.

Point 3) of the lemma just reflects the fact that the restriction of the map at $\pi_{\geq\alpha}$ is a one-one map onto $|\psi_\alpha| \subset S(G)$.

Point 4) is also quite simple. If $\rho(\beta)$ is a stationary group of some vector in V_α and $g \in N_{G_\alpha}(\rho(\beta))$, then $\rho(\beta g) = g^{-1}\rho(\beta)g = \rho(\beta)$. Because βg also belongs to $\pi_{\geq\alpha}$ for $g \in G_\alpha$, and because $\rho(\beta g) = \rho(\beta)$, by point 3) one concludes that $\beta g = \beta$, which means that $g \in G_\beta$. Now if $g \in G_\beta \cap G_\alpha$, then $\rho(\beta g) = \rho(\beta)$. So, $g^{-1}\rho(\beta)g = \rho(\beta)$ and $g \in N_G(\rho(\beta)) \cap G_\alpha \equiv N_{G_\alpha}(\rho(\beta))$. Lemma 1 is proved.

In particular, Lemma 1 shows that G-portrait in Figure 1 in the introduction is not normal. More precisely, it is impossible to introduce any normal structure in the partially ordered set of subgroups of $\mathbb{Z}_{pqr}$, described on Figure 1.

Lemma 2. *Every compact smooth manifold* M *with a smooth* G*-action* (G *is finite*) *determines a normal discrete* G*-portrait* π_M. *If the normal bundles* $\nu({}^\bullet M^H, M)$ *are oriented* (*or have a complex structure*) *for all* $H \in S(G)$ *and if* G *acts on them orientation-preserving* (*or preserving the complex structure*), *then* π_M *will have oriented orthogonal* (*or complex*) *structure.*

Proof. The manifold M determines the set $\pi = \pi_0(\coprod_{H\in S(G)} {}^\circ M^H)$ as it was described above. An element $\alpha \in \pi$ is associated with any connected component ${}^\circ M^H_\alpha$ in ${}^\circ M^H$.

By the definition, $\rho(\alpha) = H$. The group G acts on π by permuting the components $\{{}^\circ M^H_\alpha\}$. So, the stationary group G_α of an element $\alpha \in \pi$ is, in fact, the maximal subgroup in G keeping ${}^\circ M^H_\alpha$ invariant. One can check that the map $\rho:\pi \to S(G)$ is a G-equivariant map. The partial order in π is induced by the inclusion of components $\{{}^\bullet M^H_\alpha\}$ one into another. It is clear that this order is consistent with the G-action on π, and by the map ρ it is also consistent with the natural partial order in $S(G)$.

For $x \in {}^\circ M^H_\alpha$, the G-action on M defines a representation $\psi_x : H \to O(V_x)$ (or an H-representation into $SO(V_x)$, $U(V_x)$) in the fiber V_x of the normal bundle $\nu({}^\bullet M^H_\alpha, M)$ over x. Because ${}^\circ M^H_\alpha$ is connected, the isormorphism class of ψ_x does not depend on $x \in {}^\circ M^H_\alpha$ (and even on $x \in {}^\bullet M^H_\alpha$). We put $\psi_\alpha = \psi_x$ for some $x \in {}^\circ M^H_\alpha$. It follows easily from the Slice Theorem that Properties I and II in Definition A both hold, as well as that the portrait π is normal (see Definition B).

Let π be a normal G-portrait. Denote by cd_α the real dimension of ψ_α. Let $cd(\pi)$ be $\max_{\alpha\in\pi} cd_\alpha$.

Now we are able to formulate the main result.

Theorem. a). *Any normal G-portrait π is realizable on a compact smooth G-manifold W of a G-homotopy type of a 2-dimensional CW-complex. The boundary M of W realizes the same G-portrait.*

b). *If π is oriented (has a complex structure) one can construct W and all ${}^{\bullet}W^H$ $(H \in S(G))$ to be oriented manifolds (correspondingly all $\nu({}^{\bullet}W^H, W)$ have equivariant complex structure).*

c). *Assume π is oriented. Let $Z\psi_\alpha$ denote the centralizer of the group $\psi_\alpha(\rho(\alpha))$ in $O(V_\alpha)$ (correspondingly in $SO(V_\alpha)$ or in $U(V_\alpha)$), and $Z_0\psi_\alpha$ denote the connected component of the unit in $Z\psi_\alpha$. Then one can assume the fundamental groups $\pi_1({}^{\circ}M_\alpha^{\rho(\alpha)}) \cong \pi_1({}^{\circ}W_\alpha^{\rho(\alpha)})$ to be isomorphic to $Z\psi_\alpha/Z_0\psi_\alpha$. In particular, if all ψ_α are complex representations, then one can realize π on a manifold of a G-homotopy type of a 2-dimensional CW-complex with one-connected components ${}^{\circ}W^{\rho(\alpha)}$, ${}^{\circ}M_\alpha^{\rho(\alpha)}$ for each $\alpha \in \pi$.*

d). *The dimension of W, satisfying a), b), c), can be any natural $n \geq cd(\pi) + 5$. If the condition $cd_\alpha = cd(\pi)$ implies $G_\alpha = \rho(\alpha)$ and the condition $cd_\alpha < cd(\pi)$ implies $cd_\alpha \leq cd(\pi) - 5$, then one can construct W of any dimension $n \geq cd(\pi)$. In this case the G-portrait of M will be $\pi\smallsetminus\theta$, where $\theta = \{\alpha \in \pi \mid cd_\alpha = cd(\pi)\}$, and $\pi_1({}^{\circ}M^{\rho(\alpha)}) \approx Z\psi_\alpha/Z_0\psi_\alpha$ for any $\alpha \in \pi\smallsetminus\theta$. The set ${}^{\bullet}W_\alpha^{\rho(\alpha)}$ is a point for any $\alpha \in \theta$.*

Corollary. a). *Let π will be a normal oriented G-portrait. Let $cd_\alpha > 2$ for any $\alpha \in \pi$ which is not a maximal element. Then π is realizable on a compact oriented G-manifold of a homotopy type of a bouquet of 2-dimensional spheres.*

b). *If, in addition, for any $\alpha \in \pi$, $\pi_0(Z\psi_\alpha) = 1$ and for any two elements $\beta > \alpha$, $cd_\alpha - cd_\beta > 2$, then one can realize π on a manifold W of a G-homotopy type of a 2-dimensional CW-complex and each component ${}^{\bullet}W_\alpha^{\rho(\alpha)}$ will be of homotopy type of a bouquet of 2-dimensional spheres.*

Before we will prove the theorem we need to describe a classifying space for certain type of G-vector bundles. More precisely, let H be normal in G and $\xi: E(\xi) \to X$ be a G-vector bundle, satisfying the properties:

1) ${}^{\bullet}E(\xi)^H = X$,

2) G/H acts freely on X,

3) The H-representations in the fibers ξ_x are isomorphic to a given representation $\psi: H \to O(V)$.

Depending on context, 3) can be replaced by:

3a) ξ is an oriented vector bundle, G-action on $E(\xi)$ preserves the orientation of fibers ξ_x, and ψ is an H-representation in $SO(V)$.

3b) ξ is a complex vector bundle, G-action on $E(\xi)$ preserves the complex structure of fibers, and ψ is a unitary H-representation in $U(V)$.

The natural problem to classify G-bundles satisfying 1)-3) was first studied by Conner and Floyd [1]. An explicit classification of general equivariant bundles of this type is given in [6]. See also [5] for abelian G. Actually, in the case of G-vector bundles one may use an idea due to tom Dieck: the associated principal A-bundle P over X, $A = O(V)$, $SO(V)$, $U(V)$, etc. ..., with the induced G-action on it may be viewed as a $A \times G$ - space with a single orbit type $H_\psi = \{(a,g) \in A \times G \mid g \in H \text{ and } a = \psi(g)\}$. Thus using the slice theorem, P is the associated bundle over X/G of the principal $N_{A\times G}(H_\psi)/H_\psi$-bundle P^{H_ψ} with fiber $A \times G/H_\psi$. In fact one has the following exact sequence $0 \to Z\psi \to N_{A\times G}(H_\psi)/H_\psi \to G/H \to 0$, where $Z\psi$ is the centralizer of ψ in A. Note that $N_{A\times G}(H_\psi)/H_\psi$ is the group of $A\times G$-equivalences of $A \times G/H_\psi$. Hence $N_{A\times G}(H_\psi)/H_\psi$ can be viewed as the centralizer of a representation Ψ: $G \to$ {group of A-equivalences of $A \times G/H_\psi$} $\equiv A \times_H G$. The last group is in the same time the group of A-equivalences of $V \times_H G$ and, in fact, is isomorphic to the Wreath product $A \wr S_n$ of A with the symmetric group S_n, $n = |G/H|$. Let us denote $N_{A\times G}(H_\psi)/H_\psi$ by $Z\Psi$. Since $X = P/A$, using the exact sequence above, one has the following lemma.

<u>Lemma 3[6]</u>. *Isomorphism classes of* G-*vector bundles over a* G-*space* X *with a single orbit type* (H), *satisfying the properties* 1)-3) *are in one-to-one correspondence with the homotopy classes of lifts of a classifying map* f: $X/G \to B(G/H)$ *to* $BZ\Psi = B[N_{A\times G}(H_\psi)/H_\psi]$.

Now we will prove the following lemma.

<u>Lemma 4</u>. *There exists a connected 2-dimensional* (G/H)-CW-*complex* X^2 *with a* G-*vector bundle* ξ *over it, satisfying the properties* 1)-3) (or 3a),3b)). *The fundamental group* $\pi_1(X^2)$ *is isomorphic to* $Z\psi/Z_0\psi$, *where* $Z_0\psi$ *denotes the connected component of* 1 *in the centralizer* $Z\psi$.

<u>Remark 5</u>. In the case, when the short exact sequence

$$0 \to Z\psi/Z_0\psi \to Z\Psi/Z_0\psi \to G/H \to 0$$

splits, one can construct X^2 to be one-connected. In particular, if $H^2(G/H; Z\psi/Z_0\psi) = 0$, this can be done. Recall that in the case $A = O(V)$, $Z\psi/Z_0\psi = \oplus \mathbb{Z}_2$, so that if, for example, $|G/H|$ is odd, X^2 can be taken simply-connected.

Proof. The right G-action on $V\times_H G$ will produce some representation $\Psi: G \to \mathrm{Iso}_A(V\times_H G)$.

Consider the fibration $\theta: BZ\Psi \to K(G/H,1)$ with fiber $BZ\psi$ induced by the extension $0 \to Z\psi \to Z\Psi \to G/H \to 0$.

Let T denote the fundamental group $\pi_1(BZ\Psi)$. Choose some finite presentation of T. Let Y^2 be a 2-dimensional connected CW-complex, realizing this presentation. Let us take a map $s: Y^2 \to BZ\Psi$ inducing an isomorphism of the fundamental groups. By Lemma 3, s induces a G-vector bundle ξ over some space X^2 over Y^2. This covering $X^2 \to Y^2$ (with fiber G/H), induced by the map $Y^2 \xrightarrow{\theta o s} K(G/H.1)$, corresponds to the subgroup π' in $\pi_1(Y^2)$ which is the kernel of the map $(\theta o s)_*: \pi_1(Y^2) \to \pi_1(K(G/H,1))$. So, $\pi_1(X^2)$ is isomorphic to the fundamental group of the fiber $BZ\psi$. The last group is isomorphic to the group $Z\psi/Z_0\psi$. Lemma 4 is proved.

Now we are able to prove the main theorem. The proof goes by induction.

Let π be a given normal G-portrait and θ a closed G-invariant subset in it (by "closed" we mean that if $\beta < \alpha$ then $\beta \in \theta$ for any $\alpha \in \theta$).

Suppose there exist a compact smooth G-manifold ${}_\theta W$ and its boundary ${}_\theta M$ both realizing the same normal G-portrait ${}_\theta\pi$, and the following list of properties is satisfied.

1. There is a G-map $\hat{}: {}_\theta\pi \to \pi$, such that:

   a) $\hat{}$ is onto and $\rho(\alpha) = \rho(\hat\alpha)$ for any $\alpha \in {}_\theta\pi$;

   b) the partial order in ${}_\theta\pi$ is the "pull-back image" of the partial order in π: for any $\alpha,\beta \in {}_\theta\pi$, $\alpha > \beta$ if $\hat\alpha > \hat\beta$;*

   c) the map $\hat{}$ is a G-isomorphism of G-sets $\hat{}^{\,-1}(\theta)$ and θ.

2. For any $\alpha \in {}_\theta\pi$ the representations ψ_α and $\psi_{\hat\alpha}$ are isomorphic

3. ${}_\theta W$ has a G-homotopy type of a 2-dimensional CW-complex and if π is oriented, $\pi_1({}_\theta{}^\circ W_\alpha^{\rho(\alpha)}) \cong \pi_1({}_\theta{}^\circ M_\alpha^{\rho(\alpha)}) \cong Z\psi_\alpha/Z_0\psi_\alpha$ for every $\alpha \in \hat{}^{\,-1}(\theta)$.

4. The dimension of ${}_\theta W$ can be any natural $n \geq cd(\pi)+5$. In the case when the condition $cd_{\hat\alpha} = cd(\pi)$ implies $G_{\hat\alpha} = \rho(\hat\alpha)$ $(\hat\alpha \in \pi)$ and the condition $cd_{\hat\alpha} < cd(\pi)$ implies $cd_{\hat\alpha} \leq cd(\pi)-5$, one can construct ${}_\theta W$ of any dimension not less than $cd(\pi)$.

5. If all $\psi_{\hat\alpha}$ are oriented orthogonal (or unitary) representations [in other words, π is oriented (has complex structure)], then the G-action on ${}_\theta W$ is orientation-preserving, moreover, all normal bundles $\nu({}_\theta{}^\bullet W_{\hat\alpha}^{\rho(\hat\alpha)}, {}_\theta W)$ are oriented (have a complex structure), and the G-action preserves this preferred orientation (complex structure).

---

*but $\hat{}$ is not a one-to-one map, $\hat\alpha = \hat\beta$ does not imply $\alpha = \beta$.

Let $\hat{\beta} \in \pi \setminus \theta$ be an element, such that any element $\hat{\alpha} < \hat{\beta}$ belongs to θ.

The inductive step will be to construct a new G-manifold $_{\theta'}W$ with the G-portrait $_{\theta'}\pi$ also satisfying all the properties 1.-5. for θ replaced by $\theta' = \theta \cup \hat{\beta}G \subset \pi$ ($\hat{\beta}G$ denotes G-orbit of $\hat{\beta} \in \pi$).

Using Lemma 4, we can construct a connected 2-dimensional $G_{\hat{\beta}}$-CW-complex $X^2_{\hat{\beta}}$ with $G_{\hat{\beta}}$-vector bundle $\xi_{\hat{\beta}}$ over it, satisfying the following properties: 1) ${}^{\bullet}E(\xi_{\hat{\beta}})^{\rho(\hat{\beta})} = X^2_{\hat{\beta}}$; 2) $G_{\hat{\beta}}/\rho(\hat{\beta})$ acts freely on $X^2_{\hat{\beta}}$; 3) the $\rho(\hat{\beta})$-representation in the fibers of $\xi_{\hat{\beta}}$ is isomorphic to $\psi_{\hat{\beta}}$. Moreover, $\pi_1(X^2_{\hat{\beta}}) \cong Z\psi_{\hat{\beta}}/Z_0\psi_{\hat{\beta}}$.

Let us take an imbedding of $X^2_{\hat{\beta}}/G_{\hat{\beta}}$ into the euclidean space of the dimension $n\text{-dim}\,\psi_{\hat{\beta}} = n\text{-cd}_{\hat{\beta}}$. According to our assumption about n, $n\text{-cd}_{\hat{\beta}} \geq 5$. Denote by $Z_{\hat{\beta}}$ a regular neighborhood of $X^2_{\hat{\beta}}/G_{\hat{\beta}}$ in $\mathbb{R}^{n\text{-cd}_{\hat{\beta}}}$.

One can extend the classifying map $X^2_{\hat{\beta}}/G_{\hat{\beta}} \to BZ\Psi_{\hat{\beta}}$ to a map $Z_{\hat{\beta}} \to BZ\Psi_{\hat{\beta}}$, and in this way to extend the bundle $\xi_{\hat{\beta}}$ from $X^2_{\hat{\beta}}$ to the corresponding $G_{\hat{\beta}}/\rho(\hat{\beta})$-covering space $U_{\hat{\beta}}$ over $Z_{\hat{\beta}}$. Denote by $\tilde{\xi}_{\hat{\beta}}$ this extension. It is obvious that $\tilde{\xi}_{\hat{\beta}}$ satisfies the same properties 1)-3) as $\xi_{\hat{\beta}}$ does, and the base of $\tilde{\xi}_{\hat{\beta}}$ is an orientable $G_{\hat{\beta}}$-manifold $U_{\hat{\beta}}$. Note that if π is oriented (has a complex structure), then $\xi_{\hat{\beta}}$ will be oriented (will be complex) too. Moreover, for $\hat{\gamma} > \hat{\beta}$, ${}^{\bullet}E(\tilde{\xi}_{\hat{\beta}})^{\rho(\hat{\gamma})}$ will also be oriented (complex) according to the definition of an oriented (complex) G-portrait.

Let $B \subset {}_{\theta}\pi$ denotes the preimage of $\hat{\beta}$ by the map $\hat{}$. By the property 1,a) and 2) of the induction assumption, $\rho(\hat{\beta}) = \rho(\beta)$, $\psi_{\hat{\beta}} \cong \psi_{\beta}$ for any $\beta \in B$.

Now consider the set ${}^{\bullet}M_B^{\rho(\hat{\beta})} = \cup_{\beta \in B} ({}_{\theta}{}^{\bullet}M_{\beta}^{\rho(\beta)})$ in $_{\theta}M$. It is $G_{\hat{\beta}}$-invariant.

Let $D(\ )$ stand for the corresponding disk bundle. It is possible to form an equivariant connected sum of $D(\tilde{\xi}_{\hat{\beta}}) \times_{G_{\hat{\beta}}} G$ and $_{\theta}W$ by attaching equivariantly 1-handles one boundary component to $\partial U_{\hat{\beta}} \times_{G_{\hat{\beta}}} G \subset \partial D(\xi_{\beta}) \times_{G_{\hat{\beta}}} G$, and the other to

$({}_{\theta}{}^{\bullet}M^{\rho(\hat{\beta})}) \times_{G_{\hat{\beta}}} G \subset {}_{\theta}M$ (see Figure 4).

To make this construction let us consider the decomposition of the set B into different $G_{\hat{\beta}}$-orbits. For each $G_{\hat{\beta}}$-orbit we are picking up a representative β and a point x_{β} in ${}_{\theta}^{\circ}M^{\rho(\hat{\beta})}$. Let $x_{\beta}G_{\hat{\beta}}$ denote the $G_{\hat{\beta}}$-orbit of x_{β} in ${}_{\theta}^{\bullet}M^{\rho(\hat{\beta})}$ (this orbit is $G_{\hat{\beta}}$-isomorphic to $\rho(\beta)\setminus G_{\hat{\beta}}$). Let x'_{β} be some point in $\partial U_{\hat{\beta}} \subset \partial D(\tilde{\xi}_{\hat{\beta}})$. Then $x'_{\beta}G_{\hat{\beta}}$ is also isomorphic to $\rho(\beta)\setminus G_{\hat{\beta}}$. Moreover, if $D_{x_{\beta}}$ will be some $\rho(\hat{\beta})$-invariant neighborhood of x'_{β} in $\partial D(\tilde{\xi}_{\hat{\beta}})$, then the two G-sets $D_{x'_{\beta}}G \subset \partial D(\tilde{\xi}_{\hat{\beta}}) \times_{G_{\hat{\beta}}} G$ and $D_{x_{\beta}}G \subset {}_{\theta}M$ are equivariantly diffeomorphic. We are using, of course, that fact that

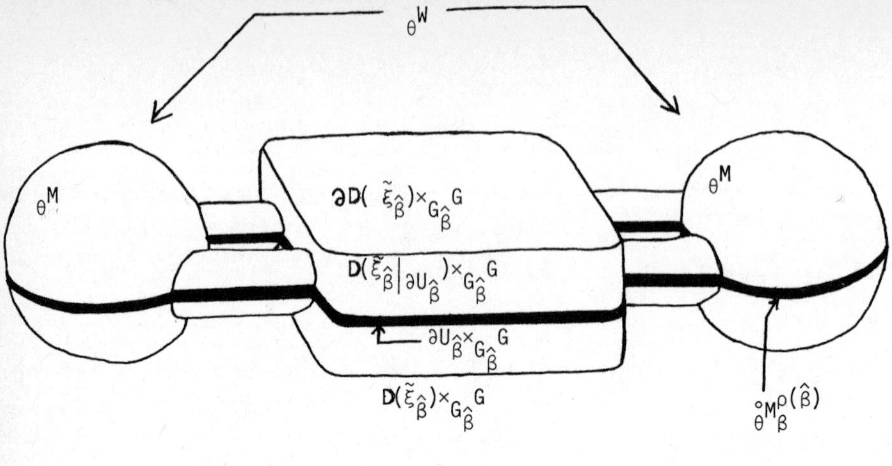

Fig. 4

by the inductive assumption $\psi_\beta \cong \psi_{x_\beta}$ and $\psi_{\hat\beta} \cong \psi_{x'_\beta}$ are isomorphic $\rho(\hat\beta)$-representations.

In the case, when all $\psi_{\hat\alpha}$ are oriented orthogonal (or unitary) representations of $\rho(\hat\alpha)$, this diffeomorphism is orientation-reversing.

So, we can realize a 1-dimensional G-surgery on the 0-dimensional sphere $x_\beta \amalg x'_\beta$. Let us repeat this procedure for each $G_{\hat\beta}$-orbit in B. Denote by $_{\theta'}W'$ the result of these surgeries.

We claim that $_{\theta'}W'$ satisfies all the properties of the induction assumption, except for the property 3). In fact, the G-portraits $_\theta\pi$ and $_{\theta'}\pi$ of $_\theta W$ and $_{\theta'}W'$ differ only by the "collapse of the set B to the element $\hat\beta$" and by gluing together $(_\theta\pi)_{\geq\beta g}$, $(_\theta\pi)_{\geq\beta' g}$ for any two elements $\beta,\beta' \in B$ and for any $g \in G$.

So, this 1-dimensional G-surgery induces a map $A_B: {_\theta\pi} \to {_{\theta'}\pi}$, identifying the elements of the $G_{\hat\beta g}$-set $\{\gamma \in {_\theta\pi} | \gamma \geq Bg\}$ $(g \in G)$ with the corresponding elements of the $G_{\hat\beta g}$-set $\{\gamma' \in {_{\theta'}\pi} | \gamma' \geq A_B(Bg)\}$. The last set is isomorphic to $|\psi_{\hat\beta g}| \cong \{\hat\gamma \in \pi | \hat\gamma \geq \hat\beta g\}$.

One can show that A_B is order-preserving and $\rho(\gamma) = \rho(A_B(\gamma))$ for $\gamma \in {_\theta\pi}$. Therefore, the original map $\hat{}: {_\theta\pi} \to \pi$ factors through A_B, and one can define a canonical map $\hat{}': {_{\theta'}\pi} \to \pi$ such that $\hat{} = \hat{}' \circ A_B$.

The new map $\hat{}'$ satisfies the same properties 1, 2, 4, 5 as $\hat{}$ does, but for the new closed subset $\theta' = \theta \cup \hat\beta G$. It is still onto and, obviously, $\rho(\alpha) = \rho(\hat\alpha')$ for any $\alpha \in {_{\theta'}\pi}$. Because A_B identifies only incomparable elements in ${}_{\theta}\pi$, one can see that $\alpha > \beta$ if and only if $\hat{\alpha}' > \hat{\beta}'$ for any $\alpha,\beta \in {}_{\theta'}\pi$. It is clear that $\hat{}'$ is an isomorphism of the G-sets $(\hat{}')^{-1}(\theta')$ and θ'.

It follows from the geometry of the previous construction that $\psi_{\alpha} \cong \psi_{\hat{\alpha}'}$ for any $\alpha \in {}_{\theta'}\pi$ (recall that we are connecting the components in ${}_{\theta}W \coprod D(\tilde{\xi}_{\hat{\beta}}) \times_{G_{\hat{\beta}}} G$ with isomorphic representations of the corresponding stationary groups).

The dimensional assumptions (property 4 of the induction assumptions) cannot be destroyed by surgery on the boundary.

An important remark has to be made. Namely, we claim that the G-portraits of ${}_{\theta'}W'$ and its boundary ${}_{\theta'}M'$ are the same. In fact, by connecting ${}_{\theta}W$ and $D(\tilde{\xi}_{\hat{\beta}}) \times_{G_{\hat{\beta}}} G$ we did not change the set $\hat{}^{-1}(\theta) \subset {}_{\theta}\pi$. Recall that, by the construction, $D(\tilde{\xi}_{\hat{\beta}})$ is a bundle over the manifold $U_{\hat{\beta}}$ with the boundary $\partial U_{\hat{\beta}}$. If $\dim U_{\hat{\beta}} \geq 2$, then $\partial U_{\hat{\beta}}$ is nonempty and connected. Therefore ${}^{\bullet}[\partial D(\tilde{\xi}_{\hat{\beta}})]^{\rho(\beta)}_{\hat{\beta}} = \partial U_{\hat{\beta}}$ consists only of one component as does ${}^{\bullet}D(\xi_{\hat{\beta}})^{\rho(\hat{\beta})}$.

The group $\rho(\hat{\gamma})$ is a stationary group of $G_{\hat{\beta}}$-action on $\partial D(\xi_{\hat{\beta}})$ if and only if it is a stationary group of $\rho(\hat{\beta})$-action on the space of the representation $\psi_{\hat{\beta}}$. On the other hand, ${}^{\bullet}\partial[D(\tilde{\xi}_{\hat{\beta}})]^{\rho(\hat{\gamma})} = {}^{\bullet}[D(\tilde{\xi}_{\hat{\beta}}|\partial U_{\hat{\beta}}) \cup \partial D(\tilde{\xi}_{\hat{\beta}})]^{\rho(\hat{\gamma})}$ is connected. So, $G_{\hat{\beta}}$-portraits of $\partial[D(\tilde{\xi}_{\hat{\beta}})]$ and $D(\tilde{\xi}_{\hat{\beta}})$ are isomorphic to $|\psi_{\hat{\beta}}|$.

By 1-dimensional G-surgeries we have connected all the components ${}^{\bullet}({}_{\theta}M)^{\rho(\beta g)}_{\rho g}$, $(\beta \in B,\ g \in G)$, with the component ${}^{\bullet}\partial[D(\tilde{\xi}_{\hat{\beta}}) \times_{G_{\hat{\beta}}} G]^{\rho(\hat{\beta}g)}$. Therefore every component ${}^{\bullet}[{}_{\theta'}W']^{\rho(\hat{\gamma})}_{\hat{\gamma}}$, being the space of a vector bundle over $U_{\hat{\beta}}$, has nonempty and connected intersection with the boundary ${}^{\bullet}[{}_{\theta'}M']^{\rho(\hat{\gamma})}_{\hat{\gamma}}$. Hence, the G-portraits of ${}_{\theta'}M'$ and ${}_{\theta'}W'$ are isomorphic.

Now we would like to have some control on the fundamental groups of the sets ${}^{\circ}({}_{\theta'}W')^{\rho(\beta)}_{\beta}$, ${}^{\circ}({}_{\theta'}M')^{\rho(\beta)}_{\beta}$, where $\beta \in {}_{\theta'}\pi$ has its image $\hat{\beta}' = \hat{\beta} \in \pi$.

The manifold ${}_{\theta'}W'$ has the G-homotopy type of a 2-dimensional G-CW-complex Y. Therefore there exists an equivariant retraction $r_t : {}_{\theta'}W' \to {}_{\theta'}W'$, $0 \leq t \leq 1$, such that $r_0 = \mathrm{id}$, $r_1 : {}_{\theta'}W' \to Y$. Moreover, r_t is an isovariant G-map, inducing the identity map of ${}_{\theta'}\pi$ into itself for all t, except $t = 1$. So, there is an isovariant, combinatoric preserving map $r_t^{-1} : {}_{\theta'}W' \smallsetminus Y \to {}_{\theta'}W' \smallsetminus Y$, $0 \leq t < 1$, such that $r_t \circ r_t^{-1} = \mathrm{id}$.

If $\dim({}_{\theta}W')^{\rho(\beta)} \geq 5$, any loop and homotopy of it in ${}^{\circ}({}_{\theta'}W')^{\rho(\beta)}_{\beta}/G_{\beta}$ can be removed away from $[Y \cap {}^{\circ}({}_{\theta'}W')^{\rho(\beta)}_{\beta}]/G_{\beta}$. By the map r_t^{-1}/G, t is close to 1, this loop or any homotopy of it are mapped into a regular neighborhood of ${}^{\circ}({}_{\theta'}M')^{\rho(\beta)}_{\beta}/G_{\beta}$.

So, $\pi_1[{}^\circ({}_{\theta},M')^{\rho(\beta)}_{\beta}/G_\beta]$ is isomorphic to $\pi_1[{}^\circ({}_{\theta},W')^{\rho(\beta)}_{\beta}/G_\beta]$.

The normal G_β-bundle of ${}^\circ({}_{\theta},W')^{\rho(\beta)}_{\beta}$ in ${}_{\theta},W'$ determines a homotopy class of the map ${}^\circ({}_{\theta},W')^{\rho(\beta)}_{\beta}/G_\beta \to BZ\Psi_\beta$. Consider the kernel K of the induced map $\pi_1[{}^\circ({}_{\theta},W')^{\rho(\beta)}_{\beta}/G_\beta] \to \pi_1[BZ\Psi_\beta]$ of the corresponding fundamental groups.

If ${}^\circ({}_{\theta},M')^{\rho(\beta)}_{\beta}/G_\beta$ is orientable (see property 5 of the induction's assumptions), the normal bundle of any loop $i:S^1 \to [{}^\circ({}_{\theta},M')^{\rho(\beta)}_{\beta}]/G_\beta$ is trivial, and one can do surgery on the immersion class of $i(S^1)$. If $i(S^1)$ belongs to the kernel K, one can extend the map $[{}^\circ({}_{\theta},M')^{\rho(\beta)}_{\beta}]/G_\beta \to BZ\Psi_\beta$ to the 2-handle $D^2{\times}D^{d(\beta)-2}$ attached by the map i ($d(\beta)$ is the dimension of ${}^\bullet({}_{\theta},W')^{\rho(\beta)}_{\beta}$, and we use here the fact that $d(\beta) > 4$). This extension produces an extension of the normal G_β-bundle of ${}^\circ({}_{\theta},M')^{\rho(\beta)}_{\beta}$ in ${}_{\theta},M'$ to a G_β-bundle ν_β over

$$[(D^2{\times}D^{d(\beta)-2}){\times}_{\rho(\beta)}G_\beta)] \cup_{\tilde{i}\times id} [{}^\circ({}_{\theta},M')^{\rho(\beta)}_{\beta}] .$$

The map $\tilde{i}$ is a lifting on the ${}^\circ({}_{\theta},M')^{\rho(\beta)}$ of the imbedding i. This lifting is possible because $i_*(S^1) \in K$ and the covering ${}^\circ({}_{\theta},M')^{\rho(\beta)}_{\beta} \to {}^\circ({}_{\theta},M')^{\rho(\beta)}_{\beta}/G_\beta$ is induced by the map into $K(G_\beta/\rho(\beta),1)$, which factors through the map $({}_{\theta},M')^{\rho(\beta)}_{\beta}/G_\beta \to BZ\Psi_\beta$.

The attaching imbedding $\tilde{i}{\times}id:(S^1{\times}D^{d(\beta)-2}){\times}_{\rho(\beta)}G_\beta \to {}^\circ({}_{\theta},M')^{\rho(\beta)}_{\beta}$ can be extended G-equivariantly to a G-imbedding of $(S^1{\times}D^{d(\beta)-2}]{\times}_{\rho(\beta)}G$ into $\cup_{g\in G}[{}^\circ({}_{\theta},M')^{\rho(\beta g)}_{\beta g}] \subset {}_{\theta},M'$. In this way one can extend the bundle-system $\nu_\beta{\times}_{G_\beta}G$ over 2-handles $(D^2{\times}D^{d(\beta)-2}){\times}_{\rho(\beta)}G$ and form a new G-manifold ${}_{\theta},W'' = {}_{\theta},W' \cup_\phi [D(\nu_\beta){\times}_{G_\beta}G]$. Here ϕ denotes a G-imbedding of $D\left(\nu_\beta\big|_{(S^1{\times}D^{d(\beta)-2}){\times}_{\rho(\beta)}G}\right)$ into ${}_{\theta},M'$.

Let us repeat this procedure, killing step by step all elements of the kernel K. Let ${}_{\theta},W$ denote the resulting G-manifolds.

It is obvious that a 2-surgery on the boundary does not affect the combinatorics of a G-manifold (if the dimension of the surgered component is > 2). Therefore the G-portraits of ${}_{\theta},W$ and ${}_{\theta},M = \partial({}_{\theta},W)$ are still ${}_{\theta},\pi$. Moreover, ${}_{\theta},W$ is G-orientable (the normal bundles system has a complex G-structure) if ${}_{\theta},W'$ is (we used oriented orthogonal (or unitary) bundles in the process of G-surgery).

But now $\pi_1[{}^\circ({}_{\theta},W)^{\rho(\beta)}_{\beta}] \cong \pi_1[{}^\circ({}_{\theta},M)^{\rho(\beta)}]$ are subgroups of $\pi_1(\widetilde{BZ\Psi_\beta})$, where $\widetilde{BZ\Psi_\beta} \to BZ\Psi_\beta$ is the $G_\beta/\rho(\beta)$-covering induced from the universal $G_\beta/\rho(\beta)$-covering over $K(G_\beta/\rho(\beta),1)$ by the canonical map $BZ\Psi_\beta \to K(G_\beta/\rho(\beta),1)$. By Lemma 3, $\widetilde{BZ\Psi_\beta}$ is homotopy equivalent to $BZ\psi_\beta$, and therefore $\pi_1(\widetilde{BZ\Psi_\beta}) \cong \pi_0(Z\psi_\beta) \cong Z\psi_\beta/Z_0\psi_\beta$. In fact, by the construction of $U_\beta \subset D(\tilde{\xi}_\beta)$, the fundamental groups of $°({}_{\theta'}W)^{\rho(\beta)}_\beta$ and $°({}_{\theta'}M)^{\rho(\beta)}_\beta$ are isomorphic to $Z\psi_\beta / Z_0\psi_\beta$.

Since we did equivariant 2-surgeries on the *boundary*, the resulting manifold ${}_{\theta'}W$ still will be of the G-homotopy type of a 2-dimensional G-CW-complex.

The induction step $\theta \to \theta' = \hat{\beta}G \cup \theta$ of the theorem is proved.

Now we have to prove the basic statement of the induction.

Let θ be the set of all minimal elements in π. Let ${}_\theta\pi$ be the G-set $\coprod_{\alpha\in\theta} \pi_{\geq\alpha}$. There is an obvious onto-map $\hat{}: \coprod_{\alpha\in\theta} \pi_{\geq\alpha} \to \pi$. Define $\rho(\alpha) = \rho(\hat{\alpha})$ for any $\alpha \in {}_\theta\pi$.

The partial order in ${}_\theta\pi$ is induced by the partial order in π: $\alpha > \beta$ if and only if $\hat{\alpha} > \hat{\beta}$. The G-action on π also induces a G-action on ${}_\theta\pi$. By the definition, βg is $[\hat{}^{-1}(\hat{\beta}g)] \cap \pi_{\geq\alpha g}$ for $\beta \in \pi_{\geq\alpha}$ and $g \in G$. This makes sense because $\hat{}: \pi_{\geq\alpha} \to \pi$ is a one-one map for any $\alpha \in \theta$.

Let ψ_β be $\psi_{\hat{\beta}}$ for any $\beta \in {}_\theta\pi$. It is clear that under these definitions, ${}_\theta\pi$ also becomes a normal G-portrait.

The map $\hat{}: {}_\theta\pi \to \pi$ induces an equivariant isomorphism of the sets of minimal elements in ${}_\theta\pi$ and π.

Consider the compact G-manifold ${}_\theta W \cong \coprod_\alpha D(\tilde{\xi}_\alpha) \times_{G_\alpha} G$ where α is a chosen representative in each G-orbit in θ. By the construction, the portrait of the G-action on ${}_\theta W$ is ${}_\theta\pi$. As we mentioned before, the property 4 of the unduction assumption implies that $\partial({}_\theta W)$ has the same G-portrait as ${}_\theta W$ does. The only exception could be if we want to realize an element $\alpha \in \theta$ with the maximal $\dim \psi_\alpha$ by 0-dimensional (but not by $\geq$ 5-dimensional) components in ${}_\theta W$. In this case the G-portrait of the boundary $\partial({}_\theta W)$ will differ from the portrait of ${}_\theta W$ by the elements $\{\alpha \in \theta\}$ with the maximal $\dim \psi_\alpha$. The Theorem is proved.

The proof of the Corollary now follows easily. If π is realizable on a G-orientable manifold W of a G-homotopy type of a 2-dimensional CW-complex, then one can equivariantly attach 2-handles to the "free part" of the top strate of W (or even of ∂W) to kill the fundamental group of the set $°W$ of generic points in W.

If $cd_\alpha > 2$ for every nonmaximal $\alpha \in \pi$, then $\operatorname{codim}(W \setminus °W)$ in W is greater than 2, and W will be 1-connected. So, one can construct W of the homotopy type of a bouquet of 2-spheres.

If $\pi_0(Z\psi_\alpha) = 1$ for any $\alpha \in \pi$, then each component $°(W)^{\rho(\alpha)}_\alpha$ is one-connected by the Theorem, and if, in addition, $\dim \psi_\alpha - \dim \psi_\beta > 2$ for any $\beta > \alpha$, then $^\bullet W^{\rho(\alpha)}_\alpha$ is of the homotopy type of a bouquet of 2-dimensional spheres. This ends the Corollary's proof.

References

[1] Conner P.E., Floyd E.E., Maps of Odd Period, Ann. of Math. 84, 132-156 (1966).

[2] tom Dieck T., Transformation Groups and Representation Theory, Lecture Notes, in Math., 766 Springer-Verlag (1979).

[3] Dovermann K.H., Petrie T., G -Surgery II. Memoirs of A.M.S., Vol. 37, N. 260 (1982).

[4] Katz G., Witt Analogs of the Burnside Ring and Integrality Theorems I & II, to appear in Amer. J. of Math.

[5] Kosniowski C., Actions of Finite Abelian Groups. Research Notes in Math. Pitman, 1978.

[6] Lashof R., Equivariant Bundles over a Single Orbit Type, III. J. Math. 28, 34-42 (1984).

[7] Oliver R., Petrie T., G-CW-Surgery and $K_0(ZG)$. Mathematische Zeit. 179, 11-42 (1982).

[8] Pawałowski K., Group Actions with Inequivalent Representations of Fixed Points, Math. Z., 187, 29-47 (1984).

[9] Petrie T. Isotropy Representations of Actions on Disks. Preprint, (1982).

[10] Tsai Y.D., Isotropy Representations of Nonabelian Finite Group Actions, Proc. of the Conference on Group Actions on Manifolds (Boulder, Colorado, 1983), Contemp. Math. 36, 269-298 (1985).

Topological invariance of equivariant rational Pontrjagin classes

Dedicated to the memory of Andrzej Jankowski and Wojtek Pulikowski

K. Kawakubo
Department of Mathematics
Osaka University
Toyonaka Osaka 560/Japan

1. Introduction.

In [7], Milnor showed that the integral Pontrjagin classes of an open manifold are not topological invariants. Afterward Novikov showed topological invariance of the rational Pontrjagin classes [9].

In [3], we defined equivariant Pontrjagin classes and equivariant Gysin homomorphisms. Concerning these concepts, we studied equivariant Riemann-Roch type theorems and localization theorems in general.

The purpose of the present paper is to show topological invariance of the equivariant rational Pontrjagin classes and to give some applications connected with the equivariant Gysin homomorphisms.

Let G be a compact Lie group. Given a right G-space A and a left G-space B, G acts on $A \times B$ by

$$g \circ (a , b) = (ag^{-1} , gb) \qquad g \in G , a \in A , b \in B .$$

The quotient space of the action on $A \times B$ is denoted by

$$A \times_G B .$$

Denote by

$$G \longrightarrow EG \longrightarrow BG$$

the universal principal G-bundle. For a G-vector bundle $\xi \longrightarrow X$ over a G-space X , we associate a vector bundle:

$$EG \times_G \xi \longrightarrow EG \times_G X .$$

Then we define our equivariant rational total Pontrjagin class $P_G(\xi)$ by

$$P_G(\xi) = P(EG \times_G \xi) \in H^*(EG \times_G X ; \mathbb{Q})$$

Research supported in part by Grant-in-Aid for Scientific Research.

where $\mathbb{Q}$ is the field of rational numbers and $P(EG \times_G \xi)$ is the classical rational total Pontrjagin class of the bundle $EG \times_G \xi \longrightarrow EG \times_G X$.

Similarly we define our equivariant total Stiefel-Whitney class $W_G(\xi)$ by

$$W_G(\xi) = W(EG \times_G \xi) \in H^*(EG \times_G X ; \mathbb{Z}_2)$$

where $\mathbb{Z}_2$ is the field $\mathbb{Z}/2\mathbb{Z}$ of order 2 and $W(EG \times_G \xi)$ is the classical total Stiefel-Whitney class of the bundle $EG \times_G \xi \longrightarrow EG \times_G X$.

For G-spaces X, Y and for a G-map $f : X \longrightarrow Y$, we denote by f_G the map

$$f_G = id \times_G f : EG \times_G X \longrightarrow EG \times_G Y .$$

For a G-manifold M, we denote by $T(M)$ the tangent G-vector bundle of M.

Then our main theorem of the present paper is the following.

Theorem 1. Let M_1, M_2 be compact smooth G-manifolds and $f : M_1 \longrightarrow M_2$ a G-homeomorphism. Then we have

$$P_G(T(M_1)) = f_G^* P_G(T(M_2))$$

where f_G^* denotes the induced homomorphism

$$f_G^* : H^*(EG \times_G M_2 , \mathbb{Q}) \longrightarrow H^*(EG \times_G M_1 ; \mathbb{Q}) .$$

The author wishes to thank Professor Z. Yosimura for enlightening him on cohomology of infinite CW-complexes.

2. Approximation by manifolds

Let G be an arbitrary compact Lie group. By the classical result [2], G is isomorphic to a closed subgroup of an orthogonal group $O(k)$ for k sufficiently large. We can suppose that $G \subset O(k)$. For any non negative integer n, we regard $O(k)$ (resp. $O(n)'$) as the closed subgroup

$$\left\{ \begin{pmatrix} A & 0 \\ 0 & I_n \end{pmatrix} \;\middle|\; A \in O(k) \right\}$$

$$\left(\text{resp. } \left\{ \begin{pmatrix} I_k & 0 \\ 0 & B \end{pmatrix} \;\middle|\; B \in O(n) \right\}\right)$$

of $O(k + n)$, where I_s denotes the unit matrix of degree s. Then the sugroups $O(k)$ and $O(n)'$ of $O(k + n)$ commute; and one may identify their direct product $O(k) \times O(n)'$ with the subgroup

$$\left\{ \begin{pmatrix} A & 0 \\ 0 & B \end{pmatrix} \middle| A \in O(k) \ , \ B \in O(n) \right\}$$

of $O(k+n)$. Since $G \subset O(k)$, the same is true of $G \times O(n)'$.
Let

$$EG^n = O(k+n)/O(n)'$$

$$BG^n = O(k+n)/G \times O(n)'$$

be left coset spaces. As is well-known, EG^n and BG^n inherit unique smooth structures such that the projections $O(k+n) \longrightarrow EG^n$, $O(k+n) \longrightarrow BG^n$ are smooth maps and that they have smooth local sections. Moreover by the inclusions

$$G \subset O(k) \subset O(k+n) \ ,$$

G acts on EG^n freely and smoothly so that the ordinary smooth structure on the orbit space EG^n/G coincides with that of BG^n and that the projection $p : EG^n \longrightarrow BG^n$ gives a principal G-bundle.
According to [10], we have

$$\pi_i(EG^n) = 0 \qquad \text{for} \qquad 0 \leqq i \leqq n-1 \ .$$

Namely the bundle above is n-universal in the sense of [10].

The correspondence

$$A \longmapsto \begin{pmatrix} A & 0 \\ 0 & 1 \end{pmatrix}$$

gives rise to an inclusion map

$$O(k+n) \longrightarrow O(k+n+1) \ .$$

Clearly this inclusion map induces the following inclusion maps

$$\overline{j_n} : EG^n \longrightarrow EG^{n+1} \ , \ j_n : BG^n \longrightarrow BG^{n+1}$$

and the following diagram

$$\begin{array}{ccc} EG^n & \xrightarrow{\overline{j_n}} & EG^{n+1} \\ \downarrow & & \downarrow \\ BG^n & \xrightarrow{j_n} & BG^{n+1} \end{array}$$

is commutative. Then $\overline{j_n}$ is a bundle map of the principal bundles.

Let EG (resp. BG) denote the direct limit (or union) of the sequence

$$EG^1 \subset EG^2 \subset EG^3 \subset \cdots \ ,$$

$$(\text{resp.} \quad BG^1 \subset BG^2 \subset BG^3 \subset \cdots \ ) \ .$$

Then the induced projection map $p : EG \longrightarrow BG$ gives a universal principal G-bundle.

Let M be a smooth G-manifold. Since G acts freely and smoothly on EG^n, the quotient space

$$EG^n \times_G M$$

inherits the smooth structure. Then observe that the following is a smooth fiber bundle

$$M \longrightarrow EG^n \times_G M \xrightarrow{\pi} BG^n$$

where π is induced from the projection map $EG^n \times M \longrightarrow EG^n$.

Since G acts on the tangent bundle $T(M)$ as a group of bundle automorphisms, we get the bundle along the fibers [1]

$$EG^n \times_G T(M) \longrightarrow EG^n \times_G M$$

of the above fibration.
Then the following lemma is well-known [1].

Lemma 2.

$$T(EG^n \times_G M) \cong EG^n \times_G T(M) \oplus \pi^! T(BG^n)$$

Here $\cong$ stands for a bundle isomorphism and $\pi^! T(BG^n)$ denotes the induced bundle of $T(BG^n)$ via the map π.

3. Topological invariance of equivariant rational Pontrjagin classes.

Let M_1, M_2 be G-manifolds and $f : M_1 \longrightarrow M_2$ a G-homeomorphism. In §2, we showed that $EG^n \times_G M_1$ and $EG^n \times_G M_2$ are smooth G-manifolds for any non negative integer n. It is clear that f induces a homeomorphism

$$f_G^n = id \times_G f : EG^n \times_G M_1 \longrightarrow EG^n \times_G M_2 .$$

Then we first show the following lemma on which Theorem 1 is based.

Lemma 3. $\quad P(EG^n \times_G T(M_1)) = f_G^{n*} P(EG^n \times_G T(M_2))$.

Proof. Notice first that the rational total Pontrjagin class satisfies the product formula:

$$P(\xi \oplus \eta) = P(\xi) \cdot P(\eta)$$

for vector bundles ξ, η over X in general.

Consider the following commutative diagram:

$$\begin{array}{ccc} EG^n \times_G M_1 & \xrightarrow{f_G^n} & EG^n \times_G M_2 \\ \downarrow \pi_1 & & \downarrow \pi_2 \\ BG^n & \xrightarrow{id} & BG^n \end{array} .$$

Then we have

$$f_G^{n*}P(\pi_2^! T(BG^n)) = P(f_G^{n!}\pi_2^! T(BG^n)) = P(\pi_1^! T(BG^n)) .$$

It follows from Lemma 2 that

$$\begin{aligned} & f_G^{n*}P(T(EG^n \times_G M_2)) \\ &= f_G^{n*}P(EG^n \times_G T(M_2) \oplus \pi_2^! T(BG^n)) \\ &= f_G^{n*}\{P(EG^n \times_G T(M_2)) \cdot P(\pi_2^! T(BG^n))\} \\ &= f_G^{n*}P(EG^n \times_G T(M_2)) \cdot f_G^{n*}P(\pi_2^! T(BG^n)) \\ &= f_G^{n*}P(EG^n \times_G T(M_2)) \cdot P(\pi_1^! T(BG^n)) . \end{aligned}$$

On the other hand, we have

$$\begin{aligned} & P(T(EG^n \times_G M_1)) \\ &= P(EG^n \times_G T(M_1) \oplus \pi_1^! T(BG^n)) \\ &= P(EG^n \times_G T(M_1)) \cdot P(\pi_1^! T(BG^n)) . \end{aligned}$$

According to [9], there holds

$$P(T(EG^n \times_G M_1)) = f_G^{n*}P(T(EG^n \times_G M_2)) .$$

Combining the above results, we have

$$P(EG^n \times_G T(M_1)) \cdot P(\pi_1^! T(BG^n)) = f_G^{n*}(P(EG^n \times_G TM_2)) \cdot P(\pi_1^! T(BG^n)) .$$

Since $P(\pi_1^! T(BG^n))$ is invertible, we have

$$P(EG^n \times_G TM_1) = f_G^{n*}P(EG^n \times_G TM_2) .$$

This makes the proof of Lemma 3 complete.

Remark. Milnor's example means that f_G^n does not induce a bundle map

$$T(EG^n \times_G M_1) \longrightarrow T(EG^n \times_G M_2)$$

in general.

Lemma 4. For a compact G-manifold M, the natural map

$$\phi : \varinjlim (EG^n \times_G M) \longrightarrow (\varinjlim EG^n) \times_G M = EG \times_G M$$

is a homeomorphism.

Proof. Consider the following commutative diagram:

$$\begin{array}{ccc} \varinjlim (EG^n \times M) & \xrightarrow{\bar{\phi}} & EG \times M \\ \Big\downarrow \varinjlim p_n & & \Big\downarrow p \\ \varinjlim (EG^n \times_G M) & \xrightarrow{\phi} & EG \times_G M \end{array}$$

where $\bar{\phi}$ is also the natural map, $\varinjlim p_n$ is induced from the projection maps $p_n : EG^n \times M \longrightarrow EG^n \times_G M$ and p is also the projection map. Clearly both ϕ and $\bar{\phi}$ are bijective maps.

In the following, we employ the terminology of Steenrod [11]. Since EG^n is a closed subset of EG^{n+1} for each n, the sequence

$$EG^1 \subset EG^2 \subset EG^3 \subset \cdots ,$$

is an expanding sequence of spaces $\{EG^n\}$. The union $EG = \varinjlim EG^n$ is given the weak topology. Namely a subset A of EG is closed if $A \cap EG^n$ is closed in EG^n for every n.

As is well-known EG has a CW-complex structure such that each EG^n is a finite CW-subcomplex. It turns out that EG is a compactly generated space. Hence EG is a filtered space as well.

Since M is a finite CW-complex, M is also a filtered space by setting $M_i = M$ $(n = 1,2,3,\cdots)$.

We now get the product $EG \times M$ filtered by

$$(EG \times M)_n = \bigcup_{i=0}^{n} EG^i \times M_{n-i} = EG^n \times M .$$

It follows from Theorem 10.3 of [11] that the product space $EG \times M$ of filtered spaces has the topology of the union

$$\varinjlim (EG \times M)_n = \varinjlim (EG^n \times M) .$$

Remark that the topology on $EG \times M$ is given by the associated compactly generated space $k(EG \times_c M)$ where $\times_c$ denotes the product with the usual cartesian topology. However the topology $EG \times_c M$ coincides with $k(EG \times_c M)$, since $EG \times_c M$ is a CW-complex.

It follows that $EG \times M$ coincides with the usual cartesian topology.

Thus we have shown that

$$\bar{\phi} : \varinjlim (EG^n \times M) \longrightarrow EG \times M$$

is a homeomorphism.

In order to prove Lemma 4, it suffices to show that the topology $\varinjlim (EG^n \times_G M)$ coincides with the quotient topology <u>via</u> the surjective map

$$\varinjlim p_n : \varinjlim (EG^n \times M) \longrightarrow \varinjlim (EG^n \times_G M) .$$

Let C be a subset of $\varinjlim (EG^n \times_G M)$. By definition, $(\varinjlim p_n)^{-1}(C)$ is closed if and only if

$$(\varinjlim p_n)^{-1}(C) \cap (EG^n \times M)$$

is closed in $EG^n \times M$ for every n . Clearly there holds

$$(\varinjlim p_n)^{-1}(C) \cap (EG^n \times M) = p_n^{-1}(C \cap (EG^n \times_G M)) .$$

Hence we have that $(\varinjlim p_n)^{-1}(C)$ is closed if and only if $p_n^{-1}(C \cap (EG^n \times_G M))$ is closed in $EG^n \times M$ for every n . Since $EG^n \times_G M$ has the quotient topology <u>via</u> the projection map $p_n : EG^n \times M \longrightarrow EG^n \times_G M$, $p_n^{-1}(C \cap (EG^n \times_G M))$ is closed in $EG^n \times M$ if and only if $C \cap (EG^n \times_G M)$ is closed in $EG^n \times_G M$. Furthermore $C \cap (EG^n \times_G M)$ is closed in $EG^n \times_G M$ for every n if and only if C is closed in $\varinjlim (EG^n \times_G M)$ by definition.

Putting all this together, we have that $(\varinjlim p_n)^{-1}(C)$ is closed if and only if C is closed in $\varinjlim (EG^n \times_G M)$. Namely $\varinjlim (EG^n \times_G M)$ has the quotient topology <u>via</u> the map $\varinjlim p_n$.

This makes the proof of Lemma 4 complete.

We are now in a position to prove Theorem 1. Consider the following commutative diagram:

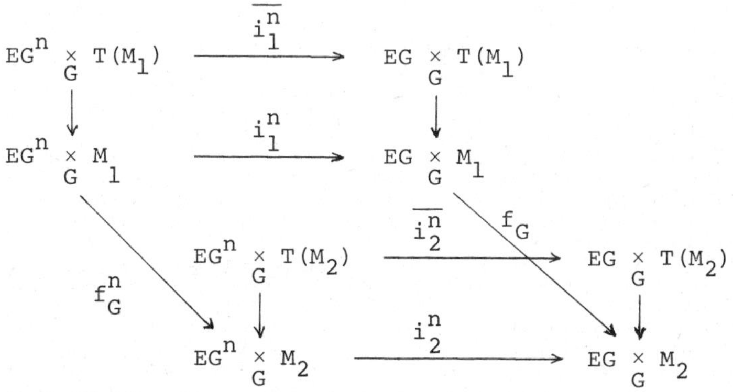

where the horizontal arrows are induced from the inclusion map $EG^n \longrightarrow EG$ and give bundle maps. Note that there are no bundle maps

$$EG^n \times_G T(M_1) \longrightarrow EG^n \times_G T(M_2) \quad ,$$

$$EG \times_G T(M_1) \longrightarrow EG \times_G T(M_2)$$

in general.

It follows from the above diagram that

$$\begin{aligned} & i_1^{n*} \cdot f_G^* P_G(T(M_2)) \\ &= f_G^{n*} \cdot i_2^{n*} P_G(T(M_2)) \\ &= f_G^{n*} P(EG^n \times_G T(M_2)) \\ &= P(EG^n \times_G T(M_1)) \qquad \text{(Lemma 3)} \\ &= i_1^{n*} P_G(T(M_1)) \ . \end{aligned}$$

According to Proposition 4 of [13], the following homomorphism

$$\psi : H^*(\varinjlim (EG^n \times_G M_1) ; \mathbb{Q}) \longrightarrow \varprojlim H^*(EG^n \times_G M_1 ; \mathbb{Q})$$

is an isomorphism.

By virtue of Lemma 4, we have an isomorphism

$$\phi^* : H^*(EG \times_G M_1 ; \mathbb{Q}) \longrightarrow H^*(\varinjlim (EG^n \times_G M_1) ; \mathbb{Q}) \ .$$

It turns out that the composition

$$\psi \cdot \phi^* : \quad H^*(EG \times_G M_1 ; \mathbb{Q}) \longrightarrow \varprojlim H^*(EG^n \times_G M_1 ; \mathbb{Q})$$

is an isomorphism.

Since there holds

$$i_1^{n*}(f_G^* P_G(T(M_2)) - P_G(T(M_1))) = 0$$

for any n, we may assert that

$$\psi \cdot \phi^*(f_G^* P_G(T(M_2)) - P_G(T(M_1))) = 0 \ .$$

Consequently we have

$$f_G^* P_G(T(M_2)) - P_G(T(M_1)) = 0 \ .$$

This makes the proof of Theorem 1 complete.

4. G-homotopy type invariance of equivariant Stiefel-Whitney classes.

In [3] and [5], we showed G-homotopy type invariance of equivariant Stiefel-Whitney classes in different ways. In this section, we shall give the third proof of it. Namely we show the following theorem.

Theorem 5. Let M_1 , M_2 be closed G-manifolds and $f : M_1 \longrightarrow M_2$ a G-homotopy equivalence. Then we have

$$W_G(T(M_1)) = f_G^* W_G(T(M_2)) .$$

where $f_G^* : H^*(EG \underset{G}{\times} M_2 ; \mathbb{Z}_2) \longrightarrow H^*(EG \underset{G}{\times} M_1 ; \mathbb{Z}_2)$ denotes the homomorphism induced from $f_G : EG \underset{G}{\times} M_1 \longrightarrow EG \underset{G}{\times} M_2$.

Proof. It is clear that f induces a homotopy equivalence

$$f_G^n = \mathrm{id} \underset{G}{\times} f : EG^n \underset{G}{\times} M_1 \longrightarrow EG^n \underset{G}{\times} M_2$$

for any n . Then the same technique as the proof of Lemma 3 applies to prove the following lemma.

Lemma 6. $W(EG^n \underset{G}{\times} T(M_1)) = f_G^{n*} W(EG^n \underset{G}{\times} T(M_2))$.

By making use of Lemmas 2 and 6, we can show the following equality

$$i_1^{n*} f_G^* W_G(T(M_2)) = i_1^{n*} W_G(T(M_1))$$

as in the proof of Theorem 1 where i_1^{n*} denotes the induced homomorphism

$$i_1^{n*} : H^*(EG \underset{G}{\times} M_1 ; \mathbb{Z}_2) \longrightarrow H^*(EG^n \underset{G}{\times} M_1 ; \mathbb{Z}_2) .$$

As is well-known, the following homomorphism

$$\psi : H^*(\varinjlim (EG^n \underset{G}{\times} M_1) ; \mathbb{Z}_2) \longrightarrow \varprojlim H^*(EG^n \underset{G}{\times} M_1 ; \mathbb{Z}_2)$$

is an isomorphism as well (see for example [12]).

Furthermore by virtue of Lemma 4, we have an isomorphism

$$\phi^* : H^*(EG \underset{G}{\times} M_1 ; \mathbb{Z}_2) \longrightarrow H^*(\varinjlim (EG^n \underset{G}{\times} M_1) ; \mathbb{Z}_2) .$$

Hence the rest of the proof is the same as that of Theorem 1.

5. Topological invariance of equivariant genera.

Let G be a compact Lie group and $h_G(\ )$ an equivariant multiplicative cohomology theory. Let M and N be closed h_G-oriented G-manifolds. Then for a G-map $f : M \longrightarrow N$ we defined an equivariant Gysin homomorphism

$$f_! : h_G(M) \longrightarrow h_G(N)$$

in general [3]. Concerning the equivariant Gysin homomorphism $f_!$, we got a localization theorem and an equivariant Riemann-Roch theorem and so on.

We now make use of the equivariant cohomology theory $H^*(EG \underset{G}{\times} M ; \mathbb{Q})$ as $h_G(M)$. When N is a point with trivial G-action, our equivariant Gysin homomorphism

$$f_! : H^*(EG \underset{G}{\times} M ; \mathbb{Q}) \longrightarrow H^*(BG ; \mathbb{Q})$$

is called an <u>index homomorphism</u> and is denoted by Ind. Using the index homomorphism, we define equivariant Pontrjagin numbers as follows. Let m be a positive integer and $I = i_1 \cdots i_k$ a partition of m. Then for a vector bundle $\xi \longrightarrow X$, we set

$$P_I(\xi) = P_{i_1}(\xi) \cdots P_{i_k}(\xi)$$

where $P_{i_j}(\xi)$ are the ordinary rational Pontrjagin classes. Let M be a closed oriented G-manifold such that the G-action is orientation preserving. Then M is $H^*(EG \underset{G}{\times} \cdot ; \mathbb{Q})$ oriented and we have

$$\mathrm{Ind} : H^*(EG \underset{G}{\times} M ; \mathbb{Q}) \longrightarrow H^*(BG ; \mathbb{Q}) .$$

We now define our equivariant Pontrjagin number $P_{GI}(M)$ by

$$P_{GI}(M) = \mathrm{Ind}\, P_I(EG \underset{G}{\times} T(M)) \in H^*(BG ; \mathbb{Q}) .$$

Note that even if m is larger than $\dim M/4$, $P_{GI}(M)$ makes sense and gives us important informations in general.

In this section, we will show that equivariant Pontrjagin numbers are topological invariants under some conditions. Accordingly equivariant genera defined by equivariant Pontrjagin numbers are also topological invariants.

We now prepare some lemmas whose proofs are easy excercises.

<u>Lemma 7</u>. Let M_1 and M_2 be closed oriented manifolds and $f : M_1 \longrightarrow M_2$ a degree 1 map. Then we have

$$f_! \cdot f^* = \mathrm{id}$$

where $f_! : H^*(M_1) \longrightarrow H^*(M_2)$ denotes the ordinary Gysin homomorphism. Namely $f_!$ is defined by the following commutative diagram

$$\begin{array}{ccc} H^*(M_1) & \xrightarrow{f_!} & H^*(M_2) \\ \big\downarrow D & & \big\downarrow D \\ H_*(M) & \xrightarrow{f_*} & H_*(M_2) \end{array}$$

where D denote the Poincaré duality isomorphisms and f_* denotes the induced homomorphism of homology groups.

<u>Lemma 8</u>. Suppose that EG^n is an oriented manifold and that G acts on EG^n preserving the orientation for every n. Let M_1 and M_2 be closed oriented G-manifolds such that the G-actions on M_1 and M_2 are orientation preserving. Let $f : M_1 \longrightarrow M_2$ be an orientation

preserving G-homeomorphism. Then $EG^n \times_G M_1$ and $EG^n \times_G M_2$ inherit the orientations so that

$$f_G^n = id \times_G f : EG^n \times_G M_1 \longrightarrow EG^n \times_G M_2$$

is an orientation preserving homeomorphism.

By combining Lemmas 3, 7 and 8, we shall show the following lemma.

<u>Lemma 9</u>. Under the conditions of Lemma 8, we have

$$f_{G!}^n P_I(EG^n \times_G T(M_1)) = P_I(EG^n \times_G T(M_2))$$

where $f_{G!}^n$ denotes the ordinary Gysin homomorphism of $f_G^n : EG^n \times_G M_1 \longrightarrow EG^n \times_G M_2$.

Proof. It follows from Lemmas 7 and 8 that

$$f_{G!}^n \cdot f_G^{n*} P_I(EG^n \times_G T(M_2)) = P_I(EG^n \times_G T(M_2)) .$$

On the other hand, by virtue of Lemma 3, we have

$$f_G^{n*} P_I(EG^n \times_G T(M_2)) = P_I(EG^n \times_G T(M_1)) .$$

Hence we obtain the reguired equality.

<u>Theorem 10</u>. Under the conditions of Lemma 8, we have

$$f_! P_I(EG \times_G T(M_1)) = P_I(EG \times_G T(M_2)) .$$

Proof. As in the proof of Lemma 4.1 in [4], one verifies the commutativity of the following diagram:

$$\begin{array}{ccc} H^*(EG \times_G M_1 ; \mathbb{Q}) & \xrightarrow{f_!} & H^*(EG \times_G M_2 ; \mathbb{Q}) \\ \downarrow i_1^{n*} & & \downarrow i_2^{n*} \\ H^*(EG^n \times_G M_1 ; \mathbb{Q}) & \xrightarrow{f_{G!}^n} & H^*(EG^n \times_G M_2 ; \mathbb{Q}) \end{array}$$

where i_j^{n*} are induced from the inclusion maps $(j = 1,2)$.
From this, we have

$$\begin{aligned} & i_2^{n*} \cdot f_! P_I(EG \times_G T(M_1)) \\ = & f_{G!}^n i_1^{n*} P_I(EG \times_G T(M_1)) \\ = & f_{G!}^n P_I(EG^n \times_G T(M_1)) \end{aligned}$$

Hence by virtue of Lemma 9, we have

$$i_2^{n*}(f_! P_I(EG \times_G T(M_1)) - P_I(EG \times_G T(M_2)))$$

$$= f^n_{G!} P_I(EG^n \times_G T(M_1)) - P_I(EG^n \times_G T(M_2))$$

$$= 0 \ .$$

Since $H^*(EG \times_G M_2 \ ; \mathbb{Q}) \cong \varprojlim H^*(EG^n \times_G M_2 \ ; \mathbb{Q})$, we may assert that

$$f_! P_I(EG \times_G T(M_1)) = P_I(EG \times_G T(M_2)) \ .$$

Theorem 11. Under the conditions of Lemma 8, we have

$$P_{GI}(M_1) = P_{GI}(M_2) \ ,$$

for any partition I .

Proof. Since our equivariant Gysin homomorphism has the functional property ((iii) of Lemma 2.2 in [3]), we have the following commutative diagram:

$$\begin{array}{ccc} H^*(EG \times_G M_1 \ ; \mathbb{Q}) & \xrightarrow{\text{Ind}} & \\ \downarrow f_! & & H^*(BG \ ; \mathbb{Q}) \\ H^*(EG \times_G M_2 \ ; \mathbb{Q}) & \xrightarrow{\text{Ind}} & \end{array}$$

Hence by Theorem 10, we have

$$P_{GI}(M_1) = \text{Ind}\ P_I(EG \times_G T(M_1))$$

$$= \text{Ind}\ f_! P_I(EG \times_G T(M_1)) = \text{Ind}\ P_I(EG \times_G T(M_2))$$

$$= P_{GI}(M_2) \ .$$

This completes the proof of Theorem 11.

It follows from Theorem 11 that any equivariant genera defined by equivariant Pontrjagin classes are topological invariants. In the following, we pick up one of them.

Let β be a multiplicative sequence in the sense of [8]. Then as an application of Theorem 11, we have the following corollary.

Corollary 12. Under the conditions of Lemma 8, we have

$$\beta_G(M_1) = \beta_G(M_2)$$

where $\beta_G(M_i)$ are defined by $\text{Ind}\ \beta(EG \times_G T(M_i))$ $(i = 1,2)$.

Concerning the localization theorem and the equivariant Riemann-Roch type theorem in [3], we have similar formulae.

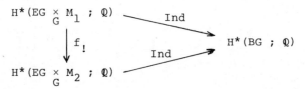

We conclude the present paper giving the following conjecture which seems to be an application of Theorem 11.

<u>Conjecture</u>. S^1-homeomorphic S^1-manifolds are S^1-bordant.

References.

1. A. Borel and F. Hirzebruch, Characteristic classes and homogeneous spaces I, Amer. J. Math., 80, 458-538 (1958).
2. C. Chevalley, Theory of Lie groups, Princeton Univ. Press, 1946.
3. K. Kawakubo, Equivariant Riemann-Roch theorems, localization and formal group law, Osaka J. Math., 17, 531-571 (1980).
4. K. Kawakubo, Global and local equivariant characteristic numbers of G-manifolds, J. Math. Soc. Japan, 32, 301-323 (1980).
5. K. Kawakubo, Compact Lie group actions and fiber homotopy type, J. Math. Soc. Japan, 33, 295-321 (1981).
6. J. Milnor, On axiomatic homology theory, Pacific J. Math., 12, 337-341 (1962).
7. J. Milnor, Microbundles: I, Topology 3 (Suppl. 1), 53-80 (1964).
8. J. Milnor and J. Stasheff, Characteristic classes, Ann. of Math. Stud. Princeton Univ. Press, 1974.
9. S. P. Novikov, Topological invariance of rational Pontrjagin classes, Doklady Tom 163, 921-923 (1965).
10. N. Steenrod, Topology of fiber bundles, Princeton Univ. Press, 1951.
11. N. Steenrod, A convenient category of topological spaces, Michigan Math. J., 14, 133-152 (1967).
12. Z. Yosimura, On cohomology theories of infinite CW-complexes, I, Publ. RIMS, Kyoto Univ., 8, 295-310 (1972/73).
13. Z. Yosimura, On cohomology theories of infinite CW-complexes, III, Publ. RIMS, Kyoto Univ., 9, 683-706 (1974).

ON THE EXISTENCE OF ACYCLIC Γ COMPLEXES OF THE LOWEST POSSIBLE DIMENSION

by

Tadeusz Kozniewski

Department of Mathematics, University of Warsaw, PKiN IXp, 00-901 Warszawa, Poland

Introduction.

Let Γ be a discrete group which contains a torsion-free subgroup of finite index. By a Γ complex we will understand a proper Γ CW complex (i.e. a Γ CW complex which has all isotropy groups finite). In the present paper we investigate connections between the existence of $\mathbb{Z}_p$ acyclic (or contractible), finite dimensional Γ complexes and the following algebraic properties of the group Γ. We say that Γ has cohomological dimesion n ($\mathrm{cd}\,\Gamma = n$) if $\mathrm{pd}_{\mathbb{Z}\Gamma}\mathbb{Z} = n$, where $\mathbb{Z}$ has trivial $\mathbb{Z}\Gamma$ module structure and for any ring Λ and any Λ module M $\mathrm{pd}_\Lambda M$ denotes the projective dimesion of M, i.e. the length of the shortest Λ projective resolution of M. The group Γ has virtual comological dimesion n ($\mathrm{vcd}\,\Gamma = n$) if there exists a subgroup Δ of finite index in Γ such that $\mathrm{cd}\,\Delta = n$. $\mathrm{vcd}\,\Gamma$ is well defined (it does not depend on Δ, see [11]). For every prime p one defines $\mathrm{cd}_p\Gamma = \mathrm{pd}_{\mathbb{Z}\Gamma}\mathbb{Z}_p$ and $\mathrm{vcd}_p\Gamma = \mathrm{cd}_p\Delta$ for any torsion-free subgroup of finite index in Γ.

Observe that if X is a $\mathbb{Z}_p$ acyclic, finite dimensional Γ complex then it follows from Smith theory that for every finite p subgroup P in Γ and every torsion-free subgroup Δ in N(P)/P (where N(P) denotes the normalizer of P in Γ) the cellular chains $C_*(X^P)\otimes\mathbb{Z}_p$ form a $\mathbb{Z}_p\Delta$-free resolution of $\mathbb{Z}_p$. Therefore $\mathrm{vcd}_p\, N(P)/P \leqslant \dim X^P$, in particular $\mathrm{vcd}_p\,\Gamma \leqslant \dim X$. The first results in the opposite direction, i.e. results showing that if $\mathrm{cd}\,\Gamma = n$ (resp. $\mathrm{vcd}_p\,\Gamma = n$) then there exists a contractible (resp. $\mathbb{Z}_p$ acyclic) Γ complex of dimension n were proved (for $n \geqslant 3$) by Eilenberg and Ganea (see [6]) and by Quillen (see [9]). Our goal is to generalize these results to the case $n = \mathrm{vcd}_p\,\Gamma$ or $n = \mathrm{vcd}\,\Gamma$.

For a given prime p we will say that a Γ complex is of type p if all its isotropy groups are p groups. Also we will say that a Γ complex is of prime power type if the order of its every isotropy group is a power of a prime (prime may vary from one isotropy group to another). To phrase our theorems we will use the posets:

$$\mathcal{F}_H(\Gamma) = \{K \mid K \text{ is a finite subgroup of } \Gamma \text{ and } H \subsetneq K\},$$

$$\mathcal{F}_{H,p}(\Gamma) = \{K \mid K \text{ is a finite p subgroup of } \Gamma \text{ and } H \subsetneq K\}.$$

By homology of a poset we mean the homology of its geometric realization.

We also use the notion of reduced equivariant cohomology $\tilde{H}^i{}_\Gamma(X;B)$ of a Γ complex. For any Γ complex X and any $\mathbb{Z}\Gamma$ module B $\tilde{H}^i{}_\Gamma(X;B)$ is defined as $H^{i+1}(\mathrm{Hom}_{\mathbb{Z}\Gamma}(C_*(p_X),B))$ where P_X denotes the canonical projection map $E\Gamma\times X \longrightarrow E\Gamma\times pt$, $C_*(p_X)$ is the algebraic mapping cone of $(p_X)_* : C_*(E\Gamma\times X) \longrightarrow C_*(E\Gamma\times pt)$ and $E\Gamma$ is the universal cover of a CW complex of type $K(\Gamma,1)$. Then we have

COROLLARY 3.1 Let $\mathrm{vcd}_p\,\Gamma = k \geqslant 2$. Then the conditions (1) and (2) below are equivalent:

(1) There exists a k dimensional $\mathbb{Z}_p$ acyclic Γ complex of type p

(2) For every finite p subgruop H in Γ we have:

(a) $H^k(\mathcal{F}_{H,p}(\Gamma);\mathbb{Z}) = 0$,

(b) $\tilde{H}^k{}_\Delta(\mathcal{F}_{H,p}(\Gamma);B) = 0$ for some subgroup Δ of finite index in $N(H)/H$

and every $\mathbb{Z}_p\Delta$ module B.

We also get

COROLLARY 3.2 If there exists a contractible k dimensional Γ complex of prime power type then the conditions (2) of 3.1 are satisfied.

A partial converse to Corollary 3.2 is given by

PROPOSITION 3.3 Assume that $\mathrm{vcd}\,\Gamma = k \geqslant 2$ and that for every prime p conditions (2) of 3.1 are satisfied. Then there exists a contractible k+1 dimensional Γ complex of prime power type.

The paper is organized as follows. In § 1 we give conditions for the projectivity of modules over group rings. In § 2 we construct Γ complexes with the property that their fixed poit sets are $\mathbb{Z}_p$ acyclic and have dimensions prescribed by a given function k from a set of finite subgroups in Γ to integers $\geqslant 2$. In § 3 we apply these constructions to the question of the existence of $\mathbb{Z}_p$ acyclic (resp. contractible) Γ complexes of dimension equal to $vcd_p \Gamma$ (resp. vcd Γ).

The paper is a revised version of a part of the author's doctoral dissertation which was written under direction of Professor Frank Connolly and submitted to the University of Notre Dame in 1985. The author would like to express deep gratitude to Professor Connolly for his help and encouragement.

§ 1. Projective modules over group rings.

We start with algebraic lemmas which give conditions for projectivity of modules over group rings.

1.1 LEMMA. If Γ is any group and Δ is a subgroup of finite index in Γ then Δ contains a subgroup Δ' which is normal in Γ and has finite index in Γ.

Proof: Let Δ be a subgroup of finite index in Γ. Define $\Delta' = \bigcap_{g\in(\Gamma/\Delta)} g\Delta g^{-1}$. Then Δ' is normal in Γ and has finite index in Γ.

□

1.2 LEMMA. Generalized projective criterion.
Let Γ be a group, let Δ be a subgroup of finite index in Γ and let R be a commutative ring with unit element $1\neq 0$. Let M be an $R\Gamma$ module. Then the conditions (1) and (2) below are equivalent:
(1) M is $R\Gamma$ projective,
(2) M is $R\Delta$ projective and $pd_{R\Gamma}M < \infty$.

Proof: (1) ⇒ (2) is clear.
(2) ⇒ (1). We start with the following two claims:
Claim 1. For any $R\Gamma$ module A

$$\mathrm{Ext}^i_{R\Gamma}(R,A) \cong H^i(\Gamma;A).$$

Proof of Claim 1: Let F_* be any $\mathbb{Z}\Gamma$ projective resolution of $\mathbb{Z}$. Then $R \otimes_{\mathbb{Z}} F_*$ is an $R\Gamma$ projective resolution of R and $\mathrm{Ext}^i_{R\Gamma}(R,A) \cong H^i(\mathrm{Hom}_{R\Gamma}(R \otimes_{\mathbb{Z}} F_*, A)) \cong H^i(\mathrm{Hom}_{\mathbb{Z}\Gamma}(F_*,A)) = H^i(\Gamma;A)$ which proves Claim 1.

Claim 2. For any two $R\Gamma$ modules N and L such that N is R projective we have:

$$\mathrm{Ext}^i_{R\Gamma}(N,L) \cong H^i(\Gamma; \mathrm{Hom}_R(N,L)).$$

Proof of Claim 2: Let F_* be any $R\Gamma$ projective resolution of R. Then for each F_i we have $\mathrm{Hom}_{R\Gamma}(F_i \otimes_R N, L) \cong \mathrm{Hom}_{R\Gamma}(F_i, \mathrm{Hom}_R(N,L))$. N is R projective so the functor $\mathrm{Hom}_R(N, \ )$ is exact and consequently the functor $\mathrm{Hom}_{R\Gamma}(F_i, \mathrm{Hom}_R(N, \ ))$ is exact. Therefore the functor $\mathrm{Hom}_{R\Gamma}(F_i \otimes_R N, \ )$ is exact, so $F_i \otimes_R N$ is projective. This shows that $F_* \otimes_R N$ is an $R\Gamma$ projective resolution of N and we have:

$\mathrm{Ext}^i_{R\Gamma}(N,L) = H^i(\mathrm{Hom}_{R\Gamma}(F_* \otimes N, L)) \cong H^i(\mathrm{Hom}_{R\Gamma}(F_*, \mathrm{Hom}_R(N,L)) = \mathrm{Ext}^i_{R\Gamma}(R, \mathrm{Hom}_R(N,L)) \cong H^i(\Gamma; \mathrm{Hom}_R(N,L))$.

The last isomorphism follows from Claim 1 and ends the proof of Claim 2.

Now observe that by Lemma 1.1 we may assume here that Δ is normal in Γ (if not replace Δ by a smaller subgroup which is normal and has finite index in Γ). Denote the quotient group $G = \Gamma/\Delta$ and let $\pi : \Gamma \longrightarrow G$ be the canonical epimorphism. For every subgroup H of G denote $\Gamma(H) = \pi^{-1}(H)$.

Claim 3. If M is an $R\Gamma$ module which is $R\Delta$ projective then for every $R\Gamma$ module N

$$\mathrm{Ext}^i_{R\Gamma}(M,N) \cong H^i(G; \mathrm{Hom}_{R\Delta}(M,N))$$

and more generally

$$\mathrm{Ext}^i_{R\Gamma(H)}(M,N) \cong H^i(H; \mathrm{Hom}_{R\Gamma(H)}(M,N))$$

Proof of Claim 3: Consider the Lyndon - Hochschild - Serre spectral sequence for $\Delta < \Gamma$ and the $R\Gamma$ module $\mathrm{Hom}_R(M,N)$.

$$E^{pq}_2 = H^p(G; H^q(\Delta; \mathrm{Hom}_R(M,N))) \cong H^p(G; \mathrm{Ext}^q_{R\Delta}(M,N)) =$$

$$\begin{cases} H^p(G; \mathrm{Hom}_{R\Delta}(M,N)) & \text{if } q = 0 \\ 0 & \text{if } q > 0. \end{cases}$$

The first isomorphism follows from Claim 2. The fact that all lines except $q = 0$ are 0 follows from $R\Delta$ projectivity of M. Therefore we get:

$H^p(\Gamma; \mathrm{Hom}_R(M,N)) \cong H^p(G; \mathrm{Hom}_{R\Delta}(M,N))$. This combined with Claim 2 prove the first isomorphism of Claim 3. The proof of the second isomorphism is analogous.

Now observe that $\mathrm{pd}_{R\Gamma}M < \infty$ implies $\mathrm{pd}_{R\Gamma(H)}M < \infty$. Therefore by the second isomorphism of Claim 3 we get that for each subgroup H in Γ $H^i(H; \mathrm{Hom}_{R\Delta}(M,N)) = 0$ for big i. It follows ([10], Theorem 4.12) that the $\mathbb{Z}G$ module $\mathrm{Hom}_{R\Delta}(M,N)$ is cohomologically trivial. In particular $\mathrm{Ext}^1{}_R(M,N) = H^1(G; \mathrm{Hom}_{R\Delta}(M,N)) = 0$ which proves $R\Gamma$ projectivity of M because N is arbitrary.

□

1.3 LEMMA. Assume that $\mathrm{vcd}_p\, \Gamma < \infty$. Let M be a $\mathbb{Z}_p\Gamma$ module. If for every finite subgroup H in Γ M is $\mathbb{Z}_pH$ projective, then $H^i(\Gamma;M) = 0$ for big i.

Proof: Let K be a finite dimensional, $\mathbb{Z}_p$ acyclic Γ CW complex (see [9]). Then there is a Leray type spectral sequence

$$E^{pq}_2 = H^p(K/\Gamma\ ; \{H^q(\Gamma_\sigma\ ; M)\}) \Rightarrow H^*{}_\Gamma(K;M) \cong H^*(\Gamma;M)$$

where Γ_σ denotes the isotropy group of a cell σ in K (see e.g. [3]). Because all isotropy groups are finite we get that $E^{pq}_2 = 0$ if $p > \dim K$ or $q > 0$. Therefore $H^i(\Gamma;M) = 0$ for $i > \dim K$.

□

1.4 LEMMA. Assume that $\mathrm{vcd}_p\, \Gamma < \infty$. Let Δ be a subgroup of finite index in Γ. Let M be a $\mathbb{Z}_p\Gamma$ module which is $\mathbb{Z}_p\Delta$ projective and $\mathbb{Z}_pP$ projective for every finite p group P in Γ. Then M is $\mathbb{Z}_p\Gamma$ projective.

Proof: By Lemma 1.1 we may as well assume that Δ is normal in Γ. Denote $G = \Gamma/\Delta$ and for every finite subgroup H in G denote $\Gamma(H) = \pi^{-1}(H)$ where $\pi : \Gamma \longrightarrow G$ is the natural projection. Let N be a $\mathbb{Z}_p\Gamma$ module. M is $\mathbb{Z}_p\Delta$ projective, so by Lemma 1.2, Claim 3:

$$\mathrm{Ext}^i{}_{\mathbb{Z}_p\Gamma}(M,N) \cong H^i(G; \mathrm{Hom}_{\mathbb{Z}_p\Delta}(M,N)).$$

It is therefore enough to show that $\mathrm{Hom}_{\mathbb{Z}_p\Delta}(M,N)$ is G cohomologically trivial. By [10], Theorem 4.12, it is then enough to show that $\mathrm{Hom}_{\mathbb{Z}_p\Delta}(M,N)$ is H cohomologically trivial for every q group H in G, where H ranges over all primes.

If $q \neq p$ this is clear since $\mathrm{Hom}_{\mathbb{Z}_p\Delta}(M,N)$ is torsion prime to q. If $q = p$ consider the subgroup $\Gamma(H)$. $\Gamma(H)$ does not contain torsion other than p-torsion. But if P is a finite p group in $\Gamma(H)$ then for $i > 0$ $H^i(P; \mathrm{Hom}_{\mathbb{Z}_p}(M,N)) \cong \mathrm{Ext}^i{}_{\mathbb{Z}_pP}(M,N) = 0$ because M is $\mathbb{Z}_pP$ projective. So we can apply Lemma 1.3 to $\Gamma(H)$ and $\mathrm{Hom}_{\mathbb{Z}_p}(M,N)$ and we get that

$H^i(\Gamma(H); Hom_{\mathbb{Z}_p}(M,N))= 0$ for big i. Lemma 1.2, Claims 2 and 3 says:

$H^i(\Gamma(H); Hom_{\mathbb{Z}_p}(M,N)) \cong H^i(H; Hom_{\mathbb{Z}_p\Delta}(M,N))$. So for every p group H in G

$H^i(H; Hom_{\mathbb{Z}_p\Delta}(M,N)) = 0$ for big i and therefore $Hom_{\mathbb{Z}_p\Delta}(M,N)$ is G cohomologically trivial.

□

To construct contractible Γ complexes we will need the following fact:

1.5 PROPOSITION. Let X be n dimensional, n-1 connected Γ complex, where $n \geq vcd\,\Gamma - 1$. Assume that for each prime p and each finite, nontrivial p group P in Γ X^P is $\mathbb{Z}_p$ acyclic. Then $H_n(X)$ is a projective $\mathbb{Z}\Gamma$ module.

Proof: A $\mathbb{Z}\Gamma$ module is projective if it is projective over some subgroup of finite index and over all finite p subgroups, for all primes p, ([5], Corollary 4.1.b).

Let Δ be a torsion-free subgroup of finite index in Γ. Then Δ acts freely on X and $C_*(X)$ - the cellular chain complex of X is a complex of free $\mathbb{Z}\Delta$ modules. X is n dimensional, n-1 connected, so

$$0 \longrightarrow H_n(X) \longrightarrow C_n(X) \longrightarrow C_{n-1}(X) \longrightarrow \ldots \longrightarrow C_0(X) \longrightarrow \mathbb{Z} \longrightarrow 0$$

is a resolution of $\mathbb{Z}$ in which all $C_i(X)$ $i = 0, \ldots, n$ are $\mathbb{Z}\Delta$ free. $vcd\,\Gamma \leq n+1$ implies $cd\,\Delta \leq n+1$ and therefore $H_n(X)$ is $\mathbb{Z}\Delta$ projectve by the generalized Schanuel's lemma (e.g. [4], Chapter VIII, Lemma 4.4).

Now let p be a prime and let P be a finite p group in Γ. Let $\mathcal{S}$ be the singular set of the P complex X. $\mathcal{S}$ is $\mathbb{Z}_p$ acyclic (by Mayer - Vietoris sequence and induction). Therefore for every i $\tilde{H}_i(X;\mathbb{Z}_p) \cong H_i(X,\mathcal{S};\mathbb{Z}_p)$ and we get that $H_n(X)\otimes\mathbb{Z}_p \cong H_n(X;\mathbb{Z}_p) \cong H_n(X,\mathcal{S};\mathbb{Z}_p)$ is the only nonzero homology group of a free, n dimensional $\mathbb{Z}_pP$ chain complex $C_*(X,\mathcal{S})\otimes\mathbb{Z}_p$. It follows ([13], Lemma 2.3) that $H_n(X;\mathbb{Z}_p)$ is $\mathbb{Z}_pP$ projective. But $H_n(X)$ is also $\mathbb{Z}$ free, so $H_n(X)$ is $\mathbb{Z}P$ projective.

□

§ 2. Γ complexes with fixed point sets having prescribed dimensions.

2.1 LEMMA. Let X be a Γ complex which has dimension $\leq n$ and is n -2 connected, $n \geq 2$.

Then the conditions (1) and (2) below are equivalent:

(1) There exists a $\mathbb{Z}_p$ acyclic, n dimensional Γ complex Z containing X as a subcomplex and such that Z - X is free

(2) $H_n(X;\mathbb{Z}_p) = 0$ and $H_{n-1}(X;\mathbb{Z}_p)$ is $\mathbb{Z}_p\Gamma$ projective.

Proof: (1)$\Rightarrow$(2). For every i $H_i(Z,X;\mathbb{Z}_p) \cong \tilde{H}_{i-1}(X;\mathbb{Z}_p)$. It follows that $H_n(X;\mathbb{Z}_p) = 0$ (because Z is n dimensional) and it follows that $H_n(Z,X;\mathbb{Z}_p)$ is the only nonzero homology group of a free, n dimensional $\mathbb{Z}_p\Gamma$ chain complex $C_*(Z,X)\otimes\mathbb{Z}_p$. Therefore $H_n(Z,X;\mathbb{Z}_p)$ is $\mathbb{Z}_p\Gamma$ projective ([13] Lemma 2.3).

(2)$\Rightarrow$(1). $H_{n-1}(X;\mathbb{Z}_p)$ is $\mathbb{Z}_p\Gamma$ projective. Therefore by "Eilenberg trick" (see e.g. [4] Chapter VIII, Lemma 2.7) there exists a free $\mathbb{Z}_p\Gamma$ module F such that $H_{n-1}(X;\mathbb{Z}_p)\oplus F$ is $\mathbb{Z}_p\Gamma$ free. Attach trivially free Γ cells of dimension n - 1 to X, one for each basis element of F. We obtain a new n dimensional Γ complex, X', which is n - 2 connected, has $H_{n-1}(X';\mathbb{Z}_p) \cong H_{n-1}(X;\mathbb{Z}_p)\oplus F$ and $H_n(X';\mathbb{Z}_p) = 0$. Use the epimorphism

$$\pi_{n-1}(X') \cong H_{n-1}(X') \longrightarrow H_{n-1}(X')\otimes\mathbb{Z}_p \cong H_{n-1}(X';\mathbb{Z}_p)$$

to represent basis elements of the free $\mathbb{Z}_p\Gamma$ module $H_{n-1}(X';\mathbb{Z}_p)$ by continuous maps $S^{n-1} \longrightarrow X'$ and use these maps to attach free Γ cells of dimension n to X'. The new Γ complex, Z, obtained this way still is n - 2 connected. Moreover $\delta : H_n(Z,X';\mathbb{Z}_p) \longrightarrow H_{n-1}(X';\mathbb{Z}_p)$ is an isomorphism which implies that: $H_{n-1}(Z;\mathbb{Z}_p) = 0 = H_n(Z;\mathbb{Z}_p)$, so Z is $\mathbb{Z}_p$ acyclic.

□

This lemma has an obvious analogue when $\mathbb{Z}_p$ is replaced by $\mathbb{Z}$ (see [7], Lemma 1.3).

Now let X be a Γ complex and let $\sigma(X)$ be the singular set of X. It was proved in [5] that there exists a Γ map $f : \sigma(X) \longrightarrow |\mathcal{F}_{\{1\}}(\Gamma)|$ such that for every finite subgroup H in Γ f restricts to N(H)/H map $f_H : \sigma_H(X) \longrightarrow |\mathcal{F}_H(\Gamma)|$, where $\sigma_H(X) = \{x\in X \mid \Gamma_x \supsetneq H\}$. It is specially easy to construct the map f in the case when X is a Γ simplicial complex. Namely: let X' denotes the barycentric subdivision of X. If σ is a vertex in X' (i.e. a simplex in X) define $f(\sigma) = \Gamma_\sigma$ = the isotropy group of σ. If $\sigma_1 < \sigma_2 < \ldots < \sigma_k$ is a simplex in X' then $\Gamma_{\sigma_1} \supset \Gamma_{\sigma_2} \supset \ldots \supset \Gamma_{\sigma_k}$. Therefore f is a simplicial map. Also for every $g \in \Gamma$ $f(g\sigma) = \Gamma_{g\sigma} = (\Gamma_\sigma)^g = (f(\sigma))^g$ so f is a Γ map. For the general construction see [5], Lemma 2.4. Another way of identifying $\sigma(X)$ and $|\mathfrak{F}_{\{1\}}(\Gamma)|$ is given in [4], Chapter IX, Lemma11.2.

Observe that if there is a prime p such that all isotropy groups in X are p groups then $f : \sigma_H(X) \longrightarrow |\mathfrak{F}_{H,p}(\Gamma)|$. Also, note that if for every $K \in \mathfrak{F}_H(\Gamma)$ X^K is acyclic (resp. $\mathbb{Z}_p$ acyclic) then by Mayer-Vietoris sequence and induction we get that f is a homology equivalence (resp. a $\mathbb{Z}_p$ homology equivalence).

Let's fix a prime p. To formulate our next resoult we need the following notation. Let k be a function from the set of all finite subgroups of Γ to integers ≥ 2. Assume that k satisfies:

(A) For each H $k(H) \geq \mathrm{vcd}_p\, N(H)/H$,

(B) If $H < K$ then $k(H) \geq k(K)$,

(C) For each H and each $g \in \Gamma$ $k(H) = k(H^g)$.

The following theorem gives the necessary and sufficient conditions for the existence of a Γ complex which has all fixed point sets $\mathbb{Z}_p$ acyclic and of dimensions prescribed by the function k.

2.2 THEOREM. Let $\mathrm{vcd}_p\, \Gamma < \infty$. Then the conditions (1) and (2) below are equivalent:

(1) There exists a Γ complex X such that for every finite subgroup H in Γ X^H is $\mathbb{Z}_p$ acyclic and $\dim X^H = k(H)$,

(2) For every finite subgruop H in Γ

(a) $H_{k(H)}(\mathfrak{F}_H(\Gamma);\mathbb{Z}_p) = 0$,

(b) There exists a subgroup Δ of finite index in N(H)/H such that $\tilde{H}^{k(H)}{}_{\Delta}(\mathfrak{F}_H(\Gamma);B) = 0$ for every $\mathbb{Z}_p\Delta$ module B.

Proof: (1) $\Rightarrow$ (2). Let H be a finite subgroup of Γ and let $k = k(H)$. In the exact sequence:

$$H_{k+1}(X^H,\sigma_H(X);\mathbb{Z}_p) \longrightarrow H_k(\sigma_H(X);\mathbb{Z}_p) \longrightarrow H_k(X^H;\mathbb{Z}_p)$$

the first group is 0 because $\dim X^H = k$ and the last group is 0 because X^H is $\mathbb{Z}_p$ acyclic.

So $H_k(\mathcal{F}_H(\Gamma); \mathbb{Z}_p) \cong H_k(\sigma_H(X); \mathbb{Z}_p) = 0$ which proves 2 (a). Now let $W = N(H)/H$ and let B be any $\mathbb{Z}_pW$ module. Then in the exact sequence

$$H^k{}_W(X^H; B) \longrightarrow H^k{}_W(\sigma_H(X); B) \longrightarrow H^{k+1}{}_W(X^H, \sigma_H(X); B)$$

the first group is $H^k(W;B)$, the second is $H^k{}_W(\mathcal{F}_H(\Gamma); B)$ and because $(X^H, \sigma_H(X))$ is a free W complex the third group is isomorphic to $H^{k+1}(X^H/W, \sigma_H(X)/W; B)$ which is 0 because $\dim X^H = k$. So we get that $H^k(W; B) \longrightarrow H^k(\mathcal{F}_H(\Gamma); B)$ is an epimorphism. This proves 2 (b).

(2) ⇒ (1). Let n be an integer bigger or equal to the order of every finite subgroup of Γ. We will construct a sequence $X_n \subset X_{n-1} \subset \ldots \subset X_1$ of Γ complexes which satisfies:

(i) $X_i{}^H$ is $\mathbb{Z}_p$ acyclic for all finite subgroups H of Γ such that $|H| \geqslant i$,

(ii) If H is a subgroup of Γ such that $|H| = i$ then all open Γ cells of type H lie in $X_i - X_{i+1}$ and have dimensions $\leqslant k(H)$.

In particular $X = X_1$ will satisfy condition (1) of 2.2.

To prove the existence of the sequence we will procede by induction. Let $|H| = i$, $k = k(H)$. First observe that $X_{i+1}{}^H = \sigma_H(X_{i+1}) = \cup_{K \supsetneq H} X^K$ and therefore $\dim X_{i+1}{}^H \leqslant k$ by the inductive assumption. Also by the inductive assumption $X_{i+1}{}^K$ is $\mathbb{Z}_p$ acyclic for each $K \supsetneq H$ and therefore $H_*(\sigma_H(X_{i+1}); \mathbb{Z}_p) \cong H_*(\mathcal{F}_H(\Gamma); \mathbb{Z}_p)$. Now attach cells of type H and of dimension $\leqslant k-1$ to X_{i+1} to get that X_{i+1} is k-2 connected. We still have $H_k(X_{i+1}{}^H; \mathbb{Z}_p) \cong H_k(\mathcal{F}_H(\Gamma); \mathbb{Z}_p)$ and the last group is 0 by 2 (a). So to end the construction of X_i such that $H_*(X_i{}^H; \mathbb{Z}_p) = 0$ it is enough to prove that $H_{k-1}(X_{i+1}{}^H; \mathbb{Z}_p)$ is $\mathbb{Z}_p(N(H)/H)$ projective (Lemma 2.1).

By Lemma 1.2 to prove that $H_{k-1}(X_{i+1}{}^H; \mathbb{Z}_p)$ is $\mathbb{Z}_p(N(H)/H)$ projective it is enough to show that $H_{k-1}(X_{i+1}{}^H; \mathbb{Z}_p)$ is:

1^0 $\mathbb{Z}_p\Delta$ projective (where Δ is some subgroup of finite index in N(H)/H),

2^0 $\mathrm{pd}_{\mathbb{Z}_p(N(H)/H)} H_{k-1}(X_{i+1}{}^H; \mathbb{Z}_p) < \infty$.

Proof of 1^0: Assume that Δ is torsion free. Let $f : X_{i+1}{}^H \longrightarrow E\Delta$ be a classifying map.

Then $H_{k-1}(X_{i+1}{}^H; \mathbb{Z}_p) \cong H_k(f; \mathbb{Z}_p)$ which is projective provided $H^{k+1}(f; B) = 0$ for all $\mathbb{Z}_p\Delta$ modules B ([13], Lemma 2.3). Consider the exact sequence:

$$H^k(\Delta;B) \longrightarrow H^k(X_{i+1}{}^H; B) \longrightarrow H^{k+1}(f; B) \longrightarrow H^{k+1}(\Delta;B).$$

The last group is 0 (because $cd_p\ \Delta \leqslant k$) and $H^k(X_{i+1}{}^H; B) = H^k{}_\Delta(X_{i+1}{}^H; B) \cong H^k{}_\Delta(\mathfrak{F}_H(\Gamma); B)$ so the condition 2 (b) gives $H^{k+1}(f; B) = 0$.

Proof of 2^0: Let $N \geqslant \max(k, \dim Z)$ for some Γ complex Z which has the property that all its fixed point sets are $\mathbb{Z}_p$ acyclic and highly connected (for the existence of Z see e.g. [9]). Let $X^{(k-1)} = X_{i+1}$ and for $j \geqslant k$ let $X^{(j)}$ be a j-1 connected N(H)/H complex obtained from $X^{(j-1)}$ by attaching free N(H)/H cells of dimension j.

Then $H_N(X^{(N)},X^{(N-1)};\mathbb{Z}_p) \xrightarrow{\delta_N} \ldots \longrightarrow H_k(X^{(k)},X^{(k-1)};\mathbb{Z}_p) \longrightarrow H_{k-1}(X^{(k-1)};\mathbb{Z}_p) \longrightarrow 0$ is an N-k stage free $\mathbb{Z}_p(N(H)/H)$ resolution of $H_{k-1}(X_{i+1}{}^H; \mathbb{Z}_p)$ with $\ker \delta_N \cong H_N(X^{(N)}; \mathbb{Z}_p)$. To end the proof we will show that $H_N(X^{(N)}; \mathbb{Z}_p)$ is a projective $\mathbb{Z}_p(N(H)/H)$ module. Let $f : X^{(N)} \longrightarrow Z^H$ be a classifying map. $\sigma(f)_* : H_*(\sigma(X^{(N)}); \mathbb{Z}_p) \longrightarrow H_*(\sigma(Z^H); \mathbb{Z}_p)$ is an isomorphism, so $H_N(X^{(N)}; \mathbb{Z}_p) \cong H_{N+1}(f; \mathbb{Z}_p) \cong H_{N+1}(f,\sigma(f); \mathbb{Z}_p)$. This is the first nonzero homology group of a free $\mathbb{Z}_p(N(H)/H)$ complex $C_*(f,\sigma(f))\otimes\mathbb{Z}_p$ and for any $\mathbb{Z}_p(N(H)/H)$ module B $H^{N+2}(f,\sigma(f); B) = 0$ because $\dim(f,\sigma(f)) = N+1$. Therefore $H_N(X^{(N)}; \mathbb{Z}_p)$ is $\mathbb{Z}_p(N(H)/H)$ projective by [13], Lemma 2.3.

If we perform the above construction on X_{i+1} for all subgroups H such that $|H| = i$ we will get X_i in our sequence. This ends the proof of the existence of the sequence and the proof of the theorem.

□

2.3 DEFINITION.

(a) Let p be a prime. A Γ complex X is of type p if all isotropy groups of X are p groups.

(b) A Γ complex is of prime power type if every isotropy group of X has prime power order.

Fix a prime p. Our goal is to give the necessary and sufficient conditions for the existence of a $\mathbb{Z}_p$ acyclic Γ complex of type p such that $\dim X = vcd_p\ \Gamma$. To do this we will first consider the analog of Theorem 2.2 which will take into account only finite p subgroups of Γ. Let k be a function from the set of all finite p subgroups of Γ (trivial subgroup included) to the set of integers ≥ 2 and assume that k satisfies conditions (A), (B), (C) above. Then we have:

2.4 THEOREM. Let $\mathrm{vcd}_p \Gamma < \infty$. The conditions (1) and (2) below are equivalent:

(1) There exists a $\mathbb{Z}_p$ acyclic Γ complex X of type p such that for every finite p group H in Γ $\dim X^H = k(H)$,

(2) For evey finite p group H in Γ

(a) $H_{k(H)}(\mathfrak{F}_{H,p}(\Gamma); \mathbb{Z}_p) = 0$,

(b) There exists a subgroup Δ of finite index in N(H)/H such that $\tilde{H}^{k(H)}{}_{\Delta}(\mathfrak{F}_{H,p}(\Gamma); B) = 0$ for every $\mathbb{Z}_p$ module B.

Proof: The methods of the proof are similar to Theorem 2.2, so we will only point out the differences.

(1) $\Rightarrow$ (2) is like in the proof of Theorem 2.2.

(2) $\Rightarrow$ (1). As before we will construct a sequence of Γ complexes $X_n \subset X_{n-1} \subset \ldots \subset X_1$ but now we require:

(i) $X_i{}^H$ is $\mathbb{Z}_p$ acyclic for all finite p subgroups H of Γ such that $|H| \geq i$,

(ii) $X_i - X_{i+1}$ consists of open Γ cells of type H ond of dimensions $\leq k(H)$,

where H runs over all p subgroups H in Γ which have $|H| = i$.

In particular note that the only subgroups of Γ which have nonempty fixed point sets are finite p subgroups.

The proof of the inductive step in the construction of the sequence is based, as before, on $\mathbb{Z}_p(N(H)/H)$ projectivity of $H_{k-1}(X_{i+1}{}^H; \mathbb{Z}_p)$. To prove that this module is $\mathbb{Z}_p(N(H)/H)$ projective we use again Lemma 1.2. The proof that $H_{k-1}(X_{i+1}{}^H; \mathbb{Z}_p)$ is $\mathbb{Z}_p\Delta$ projective remains the same. The proof that $\mathrm{pd}_{\mathbb{Z}_p(N(H)/H)} H_{k-1}(X_{i+1}{}^H; \mathbb{Z}_p) < \infty$ requires a new argument. As before we want to show that $H_N(X^{(N)}; \mathbb{Z}_p)$ is $\mathbb{Z}_p(N(H)/H)$ projective. First note that the generalized Schanuel's lemma implies that $H_N(X^{(N)}; \mathbb{Z}_p)$ is $\mathbb{Z}_p\Delta$

projective because $X^{(N)}$ is N dimensional, N-1 connected and $cd_p \Delta \leq N$. Moreover for every finite p group P in N(H)/H $H_N(X^{(N)})$ is $\mathbb{Z}P$ projective by Proposition 1.5. So $H_N(X^{(N)};\mathbb{Z}_p) \cong H_N(X^{(N)})\otimes\mathbb{Z}_p$ is $\mathbb{Z}_pP$ projective. Now we can use Lemma 1.4 to conclude that $H_N(X^{(N)};\mathbb{Z}_p)$ is $\mathbb{Z}_p(N(H)/H)$ projective and therefore

$$pd_{\mathbb{Z}_p(N(H)/H)} H_{k-1}(X_{i+1}{}^H; \mathbb{Z}_p) < \infty.$$

So we get that $H_{k-1}(X_{i+1}{}^H; \mathbb{Z}_p)$ is $\mathbb{Z}_p(N(H)/H)$ projective and the rest of the proof proceeds as in 2.2.

□

§ 3. $\mathbb{Z}_p$ acyclic Γ complexes and contractible Γ complexes.

Let $k = vcd_p \Gamma$ (resp. $k = vcd\, \Gamma$). We will apply the resoults of the previous section to examine the question of the existence of a $\mathbb{Z}_p$ acyclic (resp. contractible) Γ complex of dimension k. First note that the following is an immediate consequence of Theorem 2.4.

3.1 COROLLARY. Let $vcd_p \Gamma = k \geq 2$. The conditions (1) and (2) below are equivalent:

(1) There exists a k dimensional, $\mathbb{Z}_p$ acyclic Γ complex of type p,

(2) For every finite p group H in Γ we have:

(a) $H_k(\mathcal{F}_{H,p}(\Gamma); \mathbb{Z}_p) = 0$

(b) $\tilde{H}^k{}_\Delta(\mathcal{F}_{H,p}(\Gamma); B) = 0$ for some subgroup Δ of finite index in N(H)/H and every $\mathbb{Z}_p\Delta$ module B.

Note that it follows from Theorem 2.4 that if $vcd_p \Gamma < \infty$ then there exists a finite dimensional, $\mathbb{Z}_p$ acyclic Γ complex of type p.

We also get

3.2 COROLLARY. If there exists a contractible k dimensional Γ complex of prime power type then for every prime p the conditions (2) of 3.1 are satisfied.

Proof: If X is a contractible Γ complex of prime power type then for every prime p such that Γ has nontrivial p torsion $X_p = \{x \in X \mid \Gamma_x$ is a nontrivial p group$\}$ is a Γ subcomplex of type p. The corollary now follows from Theorem 2.4.

□

We can use Proposition 1.5 to give a partial converse to 3.2

3.3 PROPOSITION. Assume that vcd $\Gamma = k \geqslant 2$ and that for every prime p conditions (2) of 3.1 are satisfied. Then there exists a contractible, k+1 dimensional Γ complex of prime power type.

Proof: For every prime p such that Γ has nontrivial p torsion use Corollary 3.1 to construct a k dimensional $\mathbb{Z}_p$ acyclic Γ complex X_p of type p (N.B. there are only finitely many such primes p). Attach free Γ cells to $\coprod X_p$ to get a k dimensional, k-1 connected Γ complex X'. It follows from Proposition 1.5 that $H_k(X')$ is $\mathbb{Z}\Gamma$ projective. The existence of a contractible Γ complex of dimension k+1 follows now from the arguments which are analogous to the proof of Lemma 2.1.

□

References.

1. R. Bieri, Homological dimension of discrete groups, Queen Mary College Mathematical Notes, London, 1976.
2. G. Bredon, Introduction to compact transformation groups, Academic Press, New York, 1972.
3. K. S. Brown, Groups of virtually finite dimension, Homological group theory (C. T. C. Wall, ed.), London Math. Soc. Lecture Notes 36, Cambridge University Press,Cambridge, 1979, 27-70.
4. K. S. Brown, Cohomology of groups, Springer-Verlag, New York, 1982.
5. F. Connolly and T. Koźniewski, Finiteness properties of classifying spaces of proper Γ actions, to appear in: J. Pure Appl. Algebra.
6. S. Eilenberg and T. Ganea, On the Lusternik-Schnirelmann category of abstract groups, Ann. of Math. 65, 1957, 517-518.

7. T. Koźniewski, Proper group actions on acyclic complexes, Ph. D. dissertation, University of Notre Dame (1985)

8. R. Oliver, Fixed-point sets of group actions on finite acyclic complexes, Comment. Math. Helv. 50, 1875, 155-177.

9. D. Quillen, The spectrum of an equivariant cohomology ring, I, II, Ann. of Math. 94, 1971, 549-572 and 573-602.

10. D. Rim. Modules over finite groups, Ann. of Math. 69, 1959, 700-712.

11. J-P. Serre, Cohomologie des groupes discretes, Ann. of Math. Studies 70, 1971, 77-169.

12. C. T. C. Wall, Finitness conditions for CW complexes II, Proc. Royal Soc. A275, 1966, 129-139.

13. C. T. C. Wall, Surgery on compact manifolds, Academic Press, New York, 1970.

Unstable homotopy theory of homotopy representations

by Erkki Laitinen

Introduction

Let G be a finite group. A homotopy representation X of G is a G - CW -complex such that for each subgroup H of G the fixed point set X^H is a finite-dimensional CW-complex homotopy equivalent to a sphere of the same dimension. The stable theory of homotopy representations has been well explored by tom Dieck and Petrie. We shall initiate here an unstable theory.

We first describe the main problems, which all compare two homotopy representations X and Y of G. A starting point for the whole paper was the cancellation problem

A. If X and Y are stably G-homotopy equivalent, are they G-homotopy equivalent?

More generally, we should give invariants which decide the classification problem

B. When are X and Y (stably) G-homotopy equivalent?

An obvious invariant of the G-homotopy type of a homotopy representation X is the dimension function Dim X, which assigns to each subgroup H the dimension of X^H. Let X and Y be homotopy representations with the same dimension function. Then a G-map $f: X \to Y$ induces maps $f^H: X^H \to Y^H$ between spheres of the same dimensions. After a choice of orientations we may attach to f the degree function $d(f)$ whose value at the subgroup H is the degree of f^H. It turns out that the degree function determines the stable homotopy class of a G-map. Hence the analogue of the cancellation problem for maps is

C. Are G-maps f and $g: X \to Y$ with the same degree function G-homotopic?

Finally a G-map $f: X \to Y$ is a G-homotopy equivalence if and only if $\deg f^H = \pm 1$ for each subgroup H of G, so the classification problem B is a special case of problem

D. What are the possible degree functions of a G-map $f: X \to Y$?

We shall answer the problems A - D in reverse order.

A fundamental example of a homotopy representation is a linear G-sphere, the unit sphere of an orthogonal representation of G on a vector space. It is elementary to see that a linear G-sphere admits a triangulation as a finite simplicial G-complex, and can therefore be considered as a finite homotopy representation (i.e.

it is finite as a CW-complex).

If X is a linear G-sphere, the celebrated theorem of Segal tells that G-maps $f: X \to X$ are stably classified by the degree function $d(f)$ which may take arbitrary values in the Burnside ring $A(G)$. There have been two approaches to Segal's theorem: transversality and equivariant K-theory. They rely on smooth (and even analytic) techniques. We propose in section 1 an alternative based on a new type of equivariant Lefschetz class $[\Lambda(f)]$ defined for equivariant self-maps $f: X \to X$ of finite G-CW-complexes. The class $[\Lambda(f)]$ lies in the Burnside ring $A(G)$ and its characters are the Lefschetz numbers $\Lambda(f^H)$. The existence of this class immediately shows that the degrees $\deg f^H$ satisfy the usual Burnside ring congruences, when X is a finite homotopy representation. In fact it suffices that each fixed point set X^H has the R-homology of some sphere with any ring R of coefficients. This kind of a finite G - CW-complex is called a finite R-homology representation.

<u>Theorem 1</u>. Let G be a finite group and let R be any commutative ring. If X is a finite R-homology representation of G then

$$\deg f^H \equiv - \sum \phi(|K/H|)\deg f^K \mod |WH|R, \quad H \leq G$$

for all G-maps $f: X \to X$.

(The summation is over those subgroups K of G which correspond to non-trivial cyclic subgroups K/H of the Weyl group $WH = NH/H$ of H, and ϕ denotes the Euler function.)

When X and Y are different homotopy representations with the same dimension function, some care is needed in choosing the orientations. Section 2 is devoted to this question. In general it seems to be impossible to orient all fixed point sets coherently. Instead we orient the spheres X^H and Y^H only for a sufficiently small collection of subgroups H. A subgroup H is called an <u>essential isotropy group</u> of X if $X^H \neq \emptyset$ and $\dim X^K < \dim X^H$ for each subgroup K strictly larger than H. If X and Y have the same dimension function we orient them by first choosing a set of representatives of the conjugacy classes of essential isotropy groups and by then fixing orientations of X^H and Y^H for this set of subgroups H. Then a G-map $f: X \to Y$ has well-defined degrees $\deg f^H$ for the chosen subgroups H, and we show that they can be uniquely extended to all subgroups H by requiring that $\deg f^H = \deg f^K$ either when H and K are conjugate or when $H \leq K$ and $\dim X^H = \dim X^K$.

In section 3 we show following tom Dieck-Petrie [9] that there always exist G-maps $g: Y \to X$ with <u>invertible degrees</u>, i.e. $\deg g^H$ is prime to $|G|$ for each subgroup H of G. Composing an arbitrary G-map $f: X \to Y$ with such a g we are back in the situation of Theorem 1 and get congruences for the degrees $\deg f^H$.

Conversely constructing G-maps with preassigned degrees by equivariant obstruction theory as in [8] we prove

<u>Theorem 2</u>. Let X and Y be finite homotopy representations of a finite group G with the same dimension function n. There exist integers $n_{H,K}$ such that the congruences

$$\deg f^H \equiv - \sum n_{H,K} \deg f^K \bmod |WH|, \quad H \leq G$$

hold for all G-maps $f: X \to Y$. Conversely. given a collection of integers $d = (d^H)$ satisfying these congruences there exists a G-map $f: X \to Y$ with $\deg f^H = d^H$ for each $H \leq G$ if and only if d fulfils the unstability conditions

i) $d^H = 1$ when $n(H) = 1$

ii) $d^H = 1$, 0 or -1 when $n(H) = 0$

iii) $d^H = d^K$ when $n(H) = n(K)$ and $H \leq K$.

(Here d^H is assumed to be constant on conjugacy classes, and $n(H) = -1$ means $X^H = Y^H = \emptyset$.) Tom Dieck and Petrie [8] prove a stable version of Theorem 2 for complex linear G-spheres, and Tornehave [20] proves it for real linear G-spheres. Both papers determine the numbers $n_{H,K}$ explicitly. We can only say that $n_{H,K} = \phi(|K/H|)\deg g^K/\deg g^H \bmod |WH|$ where $g: Y \to X$ is any fixed G-map with invertible degrees.

Two G-maps $f,g: X \to Y$ with the same degree function need not be G-homotopic even when X and Y equal the same linear G-sphere. However, for nilpotent groups the situation is satisfying:

<u>Theorem 3</u>. Let G be a finite nilpotent group and let X and Y be finite homotopy representations of G with the same dimension function n. Two G-maps $f,g: X \to Y$ are G-homotopic if and only if

i) $\deg f^H = \deg g^H$ for each $H \leq G$

ii) $f^H = g^H$ when $n(H) = 0$

(note that $X^H = Y^H = S^0$ when $n(H) = 0$.)

Tornehave [20] proves Theorem 3 for linear G-spheres. Our proof proceeds by comparison with the linear case. A crucial fact is tom Dieck's theorem that the dimension function of any homotopy representation of a 2-group is linear [6].

If G is not nilpotent we must impose stability conditions on the dimension function to guarantee a conclusion of the type of Theorem 3. In particular we get the following generalization of Segal's theorem: the stable mapping group $\omega_G(X,X)$ is isomorphic to the Burnside ring A(G) for any finite homotopy representation X of a finite group G.

In section 4 we study the problem of G-homotopy equivalence of two homotopy representations X and Y with the same dimension function. Choose a G-map

$g: Y \to X$ with invertible degrees. By our orientation conventions the degree function $d(g)$ lies in $C(G)$, the group of integral-valued functions on the conjugacy classes of subgroups of G. Since g has invertible degrees, $d(g)$ defines an element in $\bar{C}(G)^{\times}$, the group of units of the quotient ring $\bar{C}(G) = C(G)/|G|C(G)$. The invariant which distinguishes the G-homotopy type will be the value of $d(g)$ in some quotient group of $\bar{C}(G)^{\times}$, depending on the situation.

Consider first the oriented case. We assume that X and Y have fixed orientations and require that G-homotopy equivalences are oriented, i.e. have degree 1 on each fixed point set. For two choices of $g: Y \to X$ the degree functions differ by the reduction of some invertible element in the Burnside ring. Hence the value of $d(g)$ in the <u>oriented Picard group</u>

$$\text{Inv}\,(G) = \bar{C}(G)^{\times}/\bar{A}(G)^{\times}$$

of tom Dieck and Petrie is well-defined and it may be denoted by $D^{or}(X,Y)$.

<u>Theorem 4</u>. Let X and Y be finite homotopy representations of a finite group G with the same dimension function. Choose orientations for X and Y. The following conditions are equivalent:

i) X and Y are oriented G-homotopy equivalent

ii) X and Y are stably oriented G-homotopy equivalent

iii) $D^{or}(X,Y) = 1$ in $\text{Inv}\,(G)$.

Theorem 4 was known earlier for complex linear G-spheres: tom Dieck [3] proves the equivalence of i) and ii) and tom Dieck and Petrie [8] the equivalence of ii) and iii).

The reason why the oriented cancellation law holds is easily explained. Indeed, the congruences of Theorem 2 are stable and they can be determined from a stable G-map $g: Y \to X$ with invertible degrees. If X and Y are stably oriented G-homotopy equivalent, we can find a stable map g with all degrees equal to 1, and then the congruences become the Burnside ring congruences. The constant function 1 satisfies them and is clearly unstable, so there exists an oriented G-homotopy equivalence $f: X \to Y$.

The oriented case is more complicated. A change of orientations of X and Y multiplies the value of $d(g)$ by a unit $\varepsilon = (\varepsilon^H)$ where $\varepsilon^H = \pm 1$ for each subgroup H of G. If the group of such units is denoted by $C^{\times}$, the value of $d(g)$ in the <u>Picard group</u>

$$\text{Pic}\,(G) = \bar{C}(G)^{\times}/\bar{A}(G)^{\times}C^{\times}$$

is well-defined and denoted by $D(X,Y)$. This invariant unfortunately detects only stable G-homotopy type. The <u>unstable Picard group</u> $\text{Pic}_n(G)$ is obtained by replacing all three groups $\bar{C}(G)^x, \bar{A}(G)^x$ and C^x in the definition of Pic (G) by the subgroups determined by the unstability conditions of Theorem 2. Let $D_n(X,Y)$ be the value of d(g) in Pic_n (G).

<u>Theorem 5</u>. Let X and Y be finite homotopy representations of a finite group G with the same dimension funcion n. Then

i) X and Y are stably G-homotopy equivalent if and only if D(X,Y) = 1 in Pic (G)

ii) X and Y are G-homotopy equivalent if and only if $D_n(X,Y) = 1$ in Pic_n (G).

The stable part i) is a result of tom Dieck and Petrie [9]. There is a canonical map $\mathrm{Pic}_n (G) \to \mathrm{Pic}(G)$ which is neither injective nor surjective in general. Doing a little computation in the Burnside ring, we show that nilpotent groups are again singled out:

<u>Theorem 6</u>. Let G be a finite nilpotent group with an abelian Sylow 2-subgroup and let X and Y be finite homotopy representations of G. If X and Y are stably G-homotopy equivalent they are G-homotopy equivalent.

Rothenberg [16] proves Theorem 6 for linear G-spheres and abelian groups G. At present one knows that it holds in the linear case for a wide variety of groups, e.g. all groups of odd order [21]. In fact, no counterexamples is known to the cancellation law of linear G-spheres. We give an example which shows that the cancellation law fails as soon as we step outside the linear category. If p and q are distinct odd primes and G is a metacyclic group of order pq, we show that there exist two free smooth actions of G on a sphere which are stably but not unstably G-homotopy equivalent (Example 4.11). This contradicts some results in Rothenberg [17].

We have tried to keep the paper rather self-contained at the expence of length. Except for some examples, only basic knowledge of G - CW-complexes and obstruction theory is needed. A special feature is the absence of linear representation theory which is replaced by permutation representation theory, that is, the Burnside ring. The assumption of the finiteness of the homotopy representations is due to the elementary treatment of the equivariant Lefschetz class. It can be removed from all theorems.

1. The equivariant Lefschetz class

Let G be a finite group. We define in this section an equivariant Lefschetz class $[\Lambda(f)]$ for equivariant self-maps $f\colon X \to X$ of finite G - CW-complexes X. It lies in the Burnside ring $A(G)$ and its characters coincide with the ordinary Lefschetz numbers $\Lambda(f^K)$ of the mappings $f^K\colon X^K \to X^K$. Hence the well-known congruences between the characters of an element of the Burnside ring also hold for the Lefschetz numbers $\Lambda(f^K)$. In the case of a homology representation X this gives congruences for the mapping degrees $\deg f^K$.

Let us first recall the definition and some basic properties of the Burnside ring (see [5, Ch. 1]). The isomorphism classes of finite G-sets form a semiring $A^+(G)$ under disjoint union and Cartesian product, and the Burnside ring $A(G)$ is the universal ring associated to $A^+(G)$. Let H be a subgroup of G. The function which assigns to each G-set S the number of H-fixed points $\chi_H(S) = |S^H|$ induces a ring homomorphism $\chi_H\colon A(G) \to Z$, called a <u>character</u>. Clearly $\chi_H = \chi_{H'}$, when H and H' are conjugate. Let $\phi(G)$ denote the set of conjugacy classes of subgroups of G. The maps χ_H combine to give an injective ring homomorphism

$$\chi = (\chi_H)\colon A(G) \to \prod_{\phi(G)} Z$$

whose image is characterized by the congruences

$$\chi_H(X) \equiv -\sum_{\substack{H \Delta K \leq G \\ K/H \text{ cyclic} \neq 1}} \phi(|K/H|)\chi_K(x) \qquad \text{mod } |WH|, \quad H \leq G \tag{1.1}$$

where ϕ denotes the Euler function and $WH = N_G(H)/H$.

As the congruences (1.1) are central for this paper, we shall indicate a proof. It is enough to consider the case where $H = 1$, $WH = G$ and x is a G-set S. Assume first that S is a transitive G-set $S = G/L$. Counting the number of elements of the set

$$X = \{(g,s) \mid gs = s\} \subset G \times S$$

in two ways, first according to g and then according to s, gives

$$\sum_{g \in G} |S^g| = \sum_{s \in S} |G_s| = |G/L|\,|L| = |G|.$$

More generally, decomposing an arbitrary G-set S into G-orbits we get the formula

$$\sum_{g \in G} |S^g| = |S/G|\,|G| \tag{1.2}$$

known already to Burnside [2, Th. VII p. 191]. This implies (1.1) and it shows also that the nature of the congruences is purely combinatorial, although they are usually derived from the theory of linear representations.

Let X be a finite G - CW-complex, briefly a G-complex. Then the set S_n of (ordinary) n-cells of X is a finite G-set. Following tom Dieck, we define

__Definition 1.3.__ __The equivariant Euler characteristic__ of a finite G-complex X is the element

$$[X] = \sum_m (-1)^m S_m$$

of the Burnside ring $A(G)$.

It follows at once from the definitions that

$$\chi_H[X] = \chi(X^H) \quad \text{for} \quad H \leq G. \tag{1.4}$$

Indeed, the subcomplex X^H of the CW-complex X has a cell decomposition (S_m^H), so the ordinary Euler characteristic of X^H is

$$\chi(X^H) = \Sigma_m (-1)^m |S_m^H| = \chi_H[X].$$

Two elements x and y in $A(G)$ are equal precisely when $\chi_H(x) = \chi_H(y)$ for all $H \leq G$. By (1.4) it follows for any pair of finite G-complexes X and Y that

$$[X] = [Y] \quad \text{if and only if} \quad \chi(X^H) = \chi(Y^H) \quad \text{for each} \quad H \leq G. \tag{1.5}$$

This shows that $[X]$ is an invariant of the G-homotopy type of X. In particular it does not depend on the G - CW-structure (S_m). One could also define $A(G)$ as the set of equivalence classes of finite G-complexes under the relation (1.5), and this definition of the Burnside ring works also for compact Lie groups, see [5, Ch. 5.5].

__Corollary 1.6.__ Let G be a finite group. If X is a finite G-complex then

$$\sum_{g \in G} \chi(X^g) = \chi(X/G)|G|.$$

__Proof.__ Combine (1.4) to (1.2). □

Let X be a finite G-complex with cell structure (S_m). If $f: X \to X$ is a G-map, we may assume up to G-homotopy that it is cellular. Then f induces maps $f_m: C_m(X) \to C_m(X)$ between the integral cellular chain groups

$$C_m(X) = H_m(X^m, X^{m-1}; Z) \cong Z[S_m].$$

If $c \in S_m$ is an m-cell of X let $n_f(x) \in Z$ be the coefficient of c in $f_m(c) \in C_m(X)$ with respect to the basis S_m. By equivariance $n_f(gc) = n_f(c)$ for each cell gc in the orbit of c and we may denote unambiguously $n_f(Gc) = n_f(c)$. Decompose that G-set S_m into orbits $S_m = \bigcup_{S_m/G} G_c$ and define the equivariant trace of $f_m: C_m(X) \to C_m(X)$ by

$$Tr_G(f_m) = \sum_{S_m/G} n_f(Gc)Gc \in A(G).$$

<u>Definition 1.7</u>. Let X be a finite G-complex. <u>The equivariant Lefschetz class</u> of a G-map $f: X \to X$ is the element

$$[\Lambda(f)] = \sum (-1)^m Tr_G(f_m)$$

of the Burnside ring $A(G)$.

<u>Proposition 1.8</u>. The class $[\Lambda(f)]$ depends only on the G-homotopy class of f. It satisfies $\chi_H[\Lambda(f)] = \Lambda(f^H)$ for all subgroups $H \leq G$.

<u>Proof</u>. The character $\chi_e[\Lambda(f)]$ is

$$\chi_e[\Lambda(f)] = \sum_m (-1)^m \sum_{S_m/G} n_f(Gc)|Gc| = \sum_m (-1)^m Tr(f_m: C_m(X) \to C_m(X)) = \Lambda(f),$$

the ordinary Lefschetz number of $f: X \to X$ computed from cellular chains. More generally $\chi_H[\Lambda(f)] = \Lambda(f^H)$ for all $H \leq G$ since $C_*(X^H) = Z[S_*^H]$. If f is G-homotopic to g, then f^H and g^H are homotopic and $\Lambda(f^H) = \Lambda(g^H)$ for all $H \leq G$. This implies that $[\Lambda(f)] = [\Lambda(g)]$. □

<u>Corollary 1.9.</u> Let G be a finite group and let X be a finite G-complex. Then

$$\Lambda(f^H) \equiv - \sum_{\substack{H \triangleleft K \leq G \\ K/H \text{ cyclic } \neq 1}} \phi(|K/H|)\Lambda(f^K) \qquad \text{mod } |WH|,\ H \leq G,$$

for any G-map $f: X \to X$. □

It is easy to compute the Lefschetz number of the action $g: X \to X$ of group element $g \in G$. The result is the Lefschetz fixed point formula

$$\Lambda(g) = \chi(X^g), \quad g \in G. \tag{1.10}$$

Indeed, g induces a permutation representation on $C_m(X) = Z[S_m]$. Hence $Tr(g_m: C_m(X) \to C_m(X)) = |S_m^g|$ and summing up gives

$$\Lambda(g) = \Sigma\ (-1)^m |S_m^g| = \chi(X^g).$$

Remarks. 1. W. Marzantowicz informed the author in the Poznań conference of his unpublished thesis [14] where he had developed an equivariant Lefschetz class for finite G-complexes along somewhat different lines. See also [15, Th. 1.2].

2. The equivariant Euler and Lefschetz classes can be defined for finite groups G and finite-dimensional G-complexes X provided each fixed point set X^H has finitely generated integral homology. The formulae (1.4), (1.6), (1.8), (1.9) and (1.10) remain valid. The idea is to approximate the cellular chain complex by finitely generated projective complexes over the orbit category O_G. For this and a generalization to arbitrary G-spaces, see [13].

3. S. Illman pointed out to the author that the construction of the equivariant Lefschetz class can be modified to cover finite G-complexes where G is a compact Lie group. Let X be a finite G-complex with equivariant m-cells $S_m = (d_j)$, d_j corresponding to $G/H_j \times D^m$, and let X^m denote the equivariant m-skeleton. Then

$$H_m(X^m, X^{m-1};Z) \cong \bigoplus_{d_j \in S_m} H_m(D^m \times G/H_j, S^{m-1} \times G/H_j;Z) \cong \bigoplus_{d_j \in S_m} H_o(G/H_j;Z).$$

Hence the group $C_m(X) = H_m(X^m, X^{m-1};Z)$ is free abelian with a basis consisting of the path components of G/H_j, $d_j \in S_m$. (The homology of the chain complex $C_*(X)$ is $H_*(X/G_o;Z)$ where $G_o \leq G$ is the identity component.)

Define now the equivariant trace of a cellular G-map $f: X \to X$ by

$$Tr_G f_m = \sum_{d_j \in S_m} n_f(d_j)[G/H_j] \in A(G)$$

where $n_f(d_j)$ is the coefficient of any component c_j of G/H_j in $f_*(c_j)$. Then the equivariant Lefschetz class $[\Lambda(f)] = \Sigma\ (-1)^m Tr_G f_m \in A(G)$ satisfies $\chi_H[\Lambda(f)] = \Lambda(f^H)$ for every closed subgroup $H \leq G$ such that WH is finite. A crucial point is that X^H is always a finite WH-complex.

Let R be a commutative ring (with 1). An R-<u>homology</u> <u>representation</u> of G is a finite-dimensional G-complex X such that for each subgroup H of G the fixed point set X^H has the R-homology of a sphere $S^{n(H)}$, i.e. $H_*(X^H;R) \cong H_*(S^{n(H)};R)$. We set $n(H) = -1$ when $X^H = \emptyset$. It is not required that X^H is $n(H)$ dimensional as a CW-complex. An R-homology representation X is <u>finite</u> if X is a finite G-complex.

Every self-map $f: X \to X$ of an R-homology representation has degrees $\deg f^H$ in R defined for all subgroups $H \leq G$ by

$$f_*^H(x) = \deg f^H \cdot x, \quad x \in \tilde{H}_{n(H)}(X^H;R) \cong R$$

for $n(H) \geq 0$ and by the convention $\deg f^H = 1$ when $X^H = \emptyset$.

<u>Theorem 1</u>. Let G be a finite group and let R be a commutative ring. If X is a finite R-homology representation of G then

$$\deg f^H \equiv - \sum_{\substack{H \triangleleft K \leq G \\ K/H \text{ cyclic} \neq 1}} \phi(|K/H|)\deg f^K \qquad \text{mod } |WH|R$$

for all G-maps $f: X \to X$.

<u>Proof</u>. Assume first that $R = Z$. Since $H_*(X^H) \cong H_*(S^{n(H)})$ the equivariant Euler class of X satisfies by (1.4)

$$\chi_H[X] = 1 + (-1)^{n(H)}, \quad H \leq G.$$

Similarly from (1.8) we get the formula

$$\chi_H[\Lambda(f)] = 1 + (-1)^{n(H)}\deg f^H, \quad H \leq G$$

for the equivariant Lefschetz class. Hence the product

$$\{f\} = ([X] - 1)([\Lambda(f)] - 1) \in A(G)$$

has characters $\chi_H\{f\} = \deg f^H$, and the claim follows from the Burnside ring congruences (1.1).

If R is arbitrary, $[X]$, $[\Lambda(f)]$ and $\{f\}$ are still defined as elements of $A(G)$ and we must show that the R-degree $\deg f^H$ is the image of the integer $\chi_H\{f\}$ under the canonical homomorphism $Z \to R$. Denote by $C_n = C_n(X,R)$ the cellular chain of group of X with coefficients in R in dimension n. As S is a finite CW-complex each C_n is a finitely generated free R-module. Let Z_n and B_n denote the cycle and boundary subgroups of C_n. The fact that $H_n = H_n(X,R)$ is free in all dimensions n implies by induction starting from dimension 0 that the sequences

$$0 \to Z_{n+1} \to C_{n+1} \to B_n \to 0, \quad 0 \to B_n \to Z_n \to H_n \to 0$$

split. Hence B_n and Z_n are finitely generated projective R-modules for all n.

The Bourbaki trace is therefore defined for the R-modules C_n, Z_n, B_n and H_n. The usual proof of the Hopf trace formula then works and we can compute the Euler and Lefschetz classes either from homology or from chains. On the chain level they correspond to the elements $[X]$ and $[\Lambda(f)]$ and on homology we have the same situation as in the case $R = Z$. □

In order to get congruences for G-maps $f: X \to Y$ between two different representations we have to construct suitable maps $g: Y \to X$. This is a problem in obstruction theory. so we must restrict the dimensions and connectivity of the fixed point sets. We shall work with the homotopy representations of tom Dieck and Petrie [9].

Definition 1.11. Let G be a finite group. A homotopy representation X of G is a finite-dimensional G-complex such that for each subgroup H of G the fixed point set X^H is an $n(H)$-dimensional CW-complex homotopy equivalent to $S^{n(H)}$. We set $n(H) = -1$ when X^H is empty. The homotopy representation X is finite if X is a finite CW-complex.

A homotopy representation is a Z-homology representation. Hence Theorem 1 implies

Corollary 1.12. If X is a finite homotopy representation then any G-map $f: X \to X$ satisfies the congruences of Theorem 1 with $R = Z$. □

As remarked above, the finiteness assumption on X can be removed from Theorem 1 and Corollary 1.12.

As an illustration of Theorem 1 we give simple proofs of two classical results, one of which is used in the sequel. We need the following fact from obstruction theory: If (X,A) is a relative n-dimensional G-complex such that G acts freely on $X \setminus A$ and if Y is an $(n-1)$-connected G-space, then any G-map $A \to G$ can be extended equivariantly over X. This is easily proved by induction over cells (no obstruction groups are necessary).

Proposition 1.13. (Smith) Let p be a prime and let X and Y be finite homotopy representations of Z_p such that $\dim X = \dim Y$ and $\dim X^{Z_p} = \dim Y^{Z_p}$. If $f: X \to Y$ is equivariant, then

$$\deg f \not\equiv 0 \pmod p \text{ if and only if } \deg f^{Z_p} \not\equiv 0 \pmod p.$$

Proof. Since both X^{Z_p} and Y^{Z_p} are homotopy equivalent to the same sphere $S^{n(Z_p)}$ (or empty) we can choose $g_1: Y^{Z_p} \to X^{Z_p}$ with $\deg g_1 = 1$. An equivariant extension $g: Y \to X$ exists since (Y, Y^{Z_p}) is a relatively free Z_p-complex and $\pi_i(X) = 0$ for $i < \dim Y$. By Theorem 1 the composite $h = g \circ f: X \to X$ satisfies $\deg h \equiv \deg h^{Z_p} \pmod p$ or

$$\deg g \deg f \equiv \deg g_1 \deg f^{Z_p} = \deg f^{Z_p} \pmod p.$$

If $\deg f^{Z_p} \not\equiv 0 \pmod p$ then clearly $\deg f \not\equiv 0 \pmod p$. Applying this to the map g we see that $\deg g \not\equiv 0 \pmod p$, whence the other implication. □

Proposition 1.14 (Borsuk-Ulam) Let G be a non-trivial finite group and let X and Y be free finite homotopy representations of G. If $f: X \to Y$ is equivariant then $\dim X \leq \dim Y$.

Proof. It is enough to asume that G has prime order p. If $\dim X > \dim Y$, we can find a G-map $g: Y \to X$ as above. Theorem 1 implies

$$\deg h \equiv \deg h^{Z_p} \pmod p$$

for the composite $h = g \circ f$. But $\deg h^{Z_p} = 1$ since the actions are free and $\deg h = 0$ since up to homotopy h factors as $S^n \to S^m \to S^n$ with $m < n$. The contradiction shows that $\dim X \leq \dim Y$. □

The idea of this proof of the Borsuk-Ulam theorem is due to Dold [10]. However we don't need any manifold structure in computing the degrees.

2. The isotropy group structure and orientation

This preliminary section is concerned with the isotropy group structure of homotopy representations. We introduce the concept of an essential isotropy group. It is an isotropy group H such that the fixed point sets X^K have strictly smaller dimension than X^H when $K > H$. The essential isotropy groups are needed in order to define orientations for arbitrary homotopy representations. Finally we discuss certain manifold-like conditions on the isotropy groups used by tom Dieck in his work on dimension functions, and we show that for nilpotent groups they are always satisfied.

Let G be a finite group and let X be a homotopy representation of G. Hence each fixed point set X^H is an $n(H)$-dimensional CW-complex homotopy equivalent to $S^{n(H)}$. If $f\colon X \to X$ is equivariant then the degrees $\deg f^H$ satisfy the congruences of Theorem 1. The next lemma implies further relations.

Lemma 2.1. Let X be a homotopy representation of a finite group G. If $n(H) = n(K)$ for subgroups $H \geq K$ of G then the inclusion $X^K \subset X^H$ is a homotopy equivalence. Each subgroup H of G is contained in a unique maximal subgroup $\overline{H}$ with $n(H) = n(\overline{H})$.

Proof. Let $H \leq K$ be subgroups with $n(H) = n(K) = n$. If $n = -1$, then $X^H = X^K$ is empty and if $n = 0$ then $X^H = X^K$ consists of two points. Assume $n > 0$. As X^K and X^H are CW-complexes of the homotopy type of S^n, it suffices to show that the inclusion $i\colon X^K \subset X^H$ induces an isomorphism on integral homology. The exact sequence of the pair (X^H, X^K) contains the portion

$$0 \to H_n(X^K) \xrightarrow{i_*} H_n(X^H) \to H_n(X^H, X^K) \to 0$$

where the first two groups are Z. The third group is torsion free since (X^H, X^K) is an n-dimensional relative CW-complex. It follows that i_* is an isomorphism.

If K_1 and K_2 contain H and $n(H) = n(K_1) = n(K_2) = n$, let $K = \langle K_1, K_2\rangle$ be the subgroup generated by K_1 and K_2. Then $X^K = X^{K_1} \cap X^{K_2}$ and the second claim follows if we can show that $n(K) = n$ for all possible choices of K_1 and K_2. Assume $n > 0$ and let $i\colon X^{K_1} \subset X^{K_1} \cup X^{K_2}$ and $j\colon X^{K_1} \cup X^{K_2} \subset X^H$ denote the inclusions. The composite

$$H_n(X^{K_1}) \xrightarrow{i_*} H_n(X^{K_1} \cup X^{K_2}) \xrightarrow{j_*} H_n(X^H)$$

is an isomorphism by the first part of the proof. Hence j_* is surjective. But j_* is injective as well since the CW-pair $(X^H, X^{K_1} \cup X^{K_2})$ is n-dimensional. Therefore j_* and i_* are isomorphisms. The Mayer-Vietoris sequence

$$0 \to H_n(X^K) \to H_n(X^{K_1}) \oplus H_n(X^{K_2}) \to H_n(X^{K_1} \cup X^{K_2}) \to \dots$$

now shows that $H(X^K) = Z$ so that $n(K) = n$. □

If $X^{\bar{H}}$ is not empty then $\bar{H}$ is an isotropy group such that the singular set

$$X^{>\bar{H}} = \{x \in X | G_x > \bar{H}\}$$

of $X^{\bar{H}}$ is a subcomplex of smaller dimension than $X^{\bar{H}}$. These isotropy groups resemble the principal isotropy groups in smooth G-manifolds.

Definition 2.2. Let X be a homotopy representation of a finite group. A subgroup H of G is called an essential isotropy group of X if $H = \bar{H}$ and $X^H \neq \emptyset$. The set of essential isotropy groups is denoted by EssIso (X).

For linear representation spheres or more generally for locally smooth G-manifolds all isotropy groups are essential, since all fixed point sets X^H are connected manifolds so that $X^K \subset X^H$ and $\dim X^K = \dim X^H$ imply $X^K = X^H$. The set EssIso (X) depends only on the dimension function Dim X: $\phi(G) \to Z$,

$$\operatorname{Dim} X(H) = \dim X^H, \quad H \leq G,$$

and it will also be denoted Iso (Dim X) as in [6]. It follows that EssIso (X) = EssIso (Y) when X and Y are G-homotopy equivalent.

It is now clear that a G-map $f: X \to X$ satisfies

The unstability conditions 2.3

i) $\deg f^H = 1$ when $n(H) = -1$

ii) $\deg f^H = 1, 0$ or -1, when $n(H) = 0$

iii) $\deg f^H = \deg f^{\bar{H}}$.

The congruences of Theorem 1 and the unstability conditions 2.3 turn out to be necessary and sufficient conditions for the existence of a G-map $f: X \to X$ with given degrees $\deg f^H$. We shall prove a more general version for maps $f: X \to Y$ between two homotopy representations X and Y such that Dim X = Dim Y, i.e.

$$\dim X^H = \dim Y^H \quad \text{for each } H \leq G. \tag{2.4}$$

But here it is no longer clear how to orient the fixed point spaces X^H and Y^H coherently.

If $X^H \neq \emptyset$ then the Weyl grop $WH = NH/H$ acts on X^H and therefore on $\tilde{H}^{n(H)}(X^H;Z) = Z$. Let $e_H^X(g) = 1$ (resp. -1) if $g \in NH$ preserves (resp. changes) a generator of $\tilde{H}^{n(H)}(X^H;Z)$. This defines the orientation homomorphism

$$e_H^X \colon WH \to \{\pm 1\}$$

when $X^H \neq \emptyset$ and we agree that $e_H^x \equiv 1$ when $X^H = \emptyset$, cf. [9, 1.7]. If $\mathrm{Dim}\, X = \mathrm{Dim}\, Y$ we have the following important observation of tom Dieck and Petrie, left unproved in [9, p. 135].

Lemma 2.5. If X and Y are finite homotopy representations of G with $\mathrm{Dim}\, X = \mathrm{Dim}\, Y$ then $e_H^X = e_H^Y$ for each subgroup H of G.

Proof. We may assume that $H = 1$ and $WH = G$. Denote $n(g) = \dim X^g$ for $g \in G$ and let $n = \dim X$. The Lefschetz number of g is $\Lambda(g) = 1 + (-1)^n e_1^X(g)$. By the Lefschetz fixed point formula (1.10) it equals $\chi(X^g) = 1 + (-1)^{n(g)}$. Hence

$$e_1^X(g) = (-1)^{n-n(g)}, \quad g \in G,$$

is determined by he (co)dimension function of X. □

Let X and Y be homotopy representations of G with $\mathrm{Dim}\, X = \mathrm{Dim}\, Y$ and let $f\colon X \to Y$ be a G-map. To define $\deg f^H$ we must orient X^H and Y^H. If there exists $g \in NH$ with $e_H^X(g) = -1$ then $X^H = X^{gHg^{-1}}$ but left translation by g changes the orientation. Hence it is difficult to orient the subspaces X^H coherently unless all orientation homomorphisms e_H^X are trivial. Instead of requiring this we choose one subgroup H from each conjugacy class of EssIso (X) = Iso (n), where $n = \mathrm{Dim}\, X$ is the dimension function. Let $\phi_n(G)$ be the set of the chosen representatives.

Definition 2.6. An orientation of a homotopy representation X of G is a choice of generator of $\tilde{H}^{n(H)}(X^H;Z)$ for each H in $\phi_n(G)$, $n = \mathrm{Dim}\, X$.

If X and Y are oriented by using the same set $\phi_n(G)$ then a G-map $f\colon X \to Y$ has degrees $\deg f^K$ for subgroups $K \in \phi_n(G)$. We define $\deg f^H$ for all subgroups H as follows. The group $\overline{H}$ is conjugate to a unique group K in $\phi_n(G)$, say $\overline{H} = gKg^{-1}$. The left translation by g induces a homeomorphism $1_g\colon X^K \to X^{\overline{H}}$ and the inclusion $X^{\overline{H}} \subset X^H$ is a homotopy equivalence by lemma 2.1. We transport the orientation of X^K to X^H along the composite homotopy equivalence

$$X^K \xrightarrow{1_g} X^{\overline{H}} \subset X^H.$$

If Y^H is oriented by using the same translation and inclusion, we get $\deg f^H = \deg f^K$. Another choice of g may result to different orientations of X^H and Y^H, but since $e_K^X = e_K^Y$ (2.5) they are either both preserved or both reversed, and $\deg f^H$ remains unchanged.

With these conventions we conclude

Proposition 2.7. Let X and Y be finite homotopy representations of G with $\operatorname{Dim} X = \operatorname{Dim} Y = n$. Orient X and Y using the same set of representatives of conjugacy classes in Iso (n). Then any G-map $f: X \to Y$ has well-defined degrees $\deg f^H$ for each subgroup H of G. The degrees $\deg f^H$ depend only on the conjugacy class of H and they satisfy the unstability conditions 2.3. □

One of the principal motives for studying homotopy representations is the construction of complexes which can be used as first approximation to smooth or PL actions on spheres. Therefore it is desirable that the isotropy group structure of a homotopy representation X would resemble that of an action on a genuine sphere. Consider the following two conditions:

(A) EssIso (X) = Iso (X), i.e. for any isotropy group H of X $L > H$ implies $n(L) < n(H)$.

(B) Iso (X) is closed under intersections.

They are both satisfied if X is a locally smooth G-manifold, since in that case the fixed point sets are connected submanifolds, and therefore $X^H = X^{\overline{H}}$ for any $H \leq G$.

Remark. Our terminology follows tom Dieck and Petrie [9]. Tom Dieck uses later a more restrictive notion of homotopy representation where conditions (A) and (B) are required as a part of definition [6, (1.4),(1.5) p. 231].

We first note that condition (A) always implies condition (B).

Proposition 2.8. If X is a homotopy representation and EssIso (X) = Iso (X) then Iso (X). is closed under intersections.

Proof. Assume that EssIso (X) = Iso (X). In order to prove that Iso (X) is closed under intersections it suffices to show that each subgroup $H \leq G$ is contained in a unique minimal isotropy group, viz. $\overline{H}$. If $H \in$ Iso (X) the claim is obvious. If $H \notin$ Iso (X) then $X^H = \cup X^L$ where the union is over all isotropy groups $L > H$. Let $n = n(H)$. Then $\dim X^{\overline{H}} = n$ and $\dim X^L < n$ for other isotropy groups $L > H$. We claim that they all contain $\overline{H}$.

Otherwise, let $K \in$ Iso (X) be minimal with respect to $H < K$, $\overline{H} \nleq K$. Then $m = n(K) < n$ and $\dim X^K \cap X^{\overline{H}} < m$ by condition (A). If $L \in$ Iso (X), $L > K$ and $L \neq K$, then $\dim X^K \cap X^L < m$. Indeed, if $\overline{H} \leq L$ then $X^L \subset X^{\overline{H}}$ and $\dim X^K \cap X^L \leq \dim X^K \cap X^{\overline{H}} < m$. On the other hand if $\overline{H} \nleq L$ then $KL > K$ by the minimality of K and $\dim X^K \cap X^L = \dim X^{KL} < m$ by condition (A). Hence if we denote $Y = \cup X^L$, union over isotropy groups $L > H$ different from K, we have $X^H = X^K \cup Y$ and $\dim (X^K \cap Y) < m$. The Mayer-Vietoris sequence

$$0 = H_m(X^K \cap Y) \to H_m(X^K) \oplus H_m(Y) \to H_m(X^H) = 0$$

now leads to contradiction since $H_m(X^K) \cong Z$. □

If $\mathrm{EssIso}(X) = \mathrm{Iso}(X)$, we can thus characterize $\overline{H}$ either as the minimal isotropy group containing H or as the maximal subgroup of G such that $X^H = X^{\overline{H}}$.

It is easy to give an example which demonstrates that condition (B) does not imply condition (A). Let X be the union of a circle S^1 with trivial G-action and $|G|$ copies of the unit interval freely permuted by G, glued together at one end to a common base point $x_o \in S^1$. Then $\mathrm{Iso}(X) = \{1,G\}$ is closed under intersection but the isotropy group 1 is not essential since $X^1 = X$ and $X^G = S^1$ have dimension 1. Moreover we see that adding suitable whiskers any subgroup $H \leq G$ may appear as an isotropy group. However, X is G-homotopy equivalent to the trivial G-space S^1 which satisfies both (A) and (B).

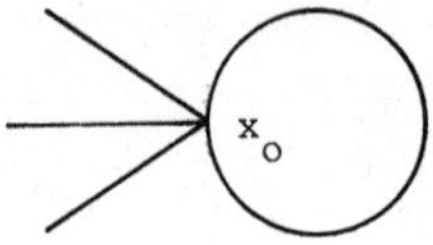

We shall show that if condition (B) is modified to the homotopy invariant form "$\mathrm{EssIso}(X)$ is closed under intersection" then condition (B) implies condition (A) up to homotopy. This is done by collapsing away the inessential orbits of X.

Lemma 2.9. The following conditions on a homotopy representation X are equivalent:

i) $\mathrm{EssIso}(X)$ is closed under intersection

ii) If $H \leq K$ then $\overline{H} \leq \overline{K}$.

Proof. Let $\mathrm{EssIso}(X)$ be closed under intersection and let $H \leq K$. If $X^K = \emptyset$ then $\overline{K} = G$ and clearly $\overline{H} \leq \overline{K}$. Otherwise both $\overline{H}$ and $\overline{K}$ are essential isotropy groups and therefore $\overline{H} \cap \overline{K}$ is an essential isotropy group, too. Since $H \leq \overline{H} \cap \overline{K} \leq \overline{H}$ must have $\overline{H} \cap \overline{K} = \overline{H}$ i.e. $\overline{H} \leq \overline{K}$.

Conversely, assume that ii) holds. Let H and K be essential isotropy groups. Then $H \cap K \leq H$ implies $\overline{H \cap K} \leq \overline{H} = H$ and similarly $\overline{H \cap K} \leq \overline{K} = K$. It follows that $\overline{H \cap K} \leq H \cap K$. As the other inclusion holds trivially, we get $\overline{H \cap K} = H \cap K$. Moreover $X^{H \cap K} \supset X^H \neq \emptyset$. Hence $H \cap K$ is an essential isotropy group. □

Proposition 2.10. Let X be a homotopy representation of a finite group G. If $\mathrm{EssIso}(X)$ is closed under intersection then X is G-homotopy equivalent to a homotopy representation Y which satisfies $\mathrm{Iso}(Y) = \mathrm{EssIso}(X)$ and

(A) If $H \in \mathrm{Iso}(Y)$ and $L > H$ then $n(L) < n(H)$

(B) $\mathrm{Iso}(Y)$ is closed under intersection.

If X is finite then Y can be chosen finite.

Proof. A family F of subgroups of G is called closed if F is closed under conjugation and each subgroup containing a member of F belongs to F. If F is a closed family, let $X(F) = \cup_{H \in F} X^H$ be the set of points of X whose isotropy groups belong to F. For each closed family F we shall construct a G-complex $Y(F)$ and a G-map $f_F\colon X(F) \to Y(F)$ such that

a) f_F is a cellular G-homotopy equivalence

b) Iso $Y(F) = \text{EssIso}(X) \cap F$.

If $F = \{G\}$, we put $X(F) = Y(F) = X^G$ and let f_F be the identity. Assume by induction that $f_F\colon X(F) \to Y(F)$ satisfying a and b is already constructed. Choose a maximal subgroup H not in F and let $F' = F \cup (H)$. Then $X(F') = X(F) \cup X^{(H)}$ contains $X(F)$ as a G-subcomplex. If $H \in \text{EssIso}(X)$ we define $Y(F')$ as the adjunction space $X(F') \cup_{f_F} Y(F)$. It is a G-complex since f_F is a cellular G-map. As f_F is a homotopy equivalence and $(X(F'),X(F))$ is a CW-pair it is a standard fact that the canonical map $f_{F'}\colon X(F') \to Y(F')$ is a homotopy equivalence [23, I 5.12]. This applies also to $f_{F'}^K$ for each $K \leq G$. Hence $f_{F'}$ is a G-homotopy equivalence which satisfies a and b.

In the case $H \notin \text{EssIso}(X)$ we let $Y(F') = Y(F)$. In order to construct an extension of $f_F\colon X(F) \to Y(F)$ to $X(F')$ it is enough to find a cellular G-deformation retraction $X(F') \to X(F)$, or equivalently a cellular WH-deformation retraction $X^H \to X^{>H}$. Hence it suffices to show that the inclusion $i\colon X^{>H} \subset X^H$ is a WH-homotopy equivalence.

If $K/H \leq WH$ is non-trivial then $K > H$ and $(X^{>H})^K = (X^H)^K = X^K$ so that i^K is the identity. Thus we are left with proving that $i\colon X^{>H} \subset X^H$ is an ordinary homotopy equivalence. We have assumed that $\bar{H} > H$. Consider the inclusions $X^{\bar{H}} \subset X^{>H} \subset X^H$. The middle term is a finite union $X^{>H} = \cup_{K>H} X^K$ of subcomplexes closed under intersection and $X^{\bar{H}} \subset X^H$ is a homotopy equivalence. If we prove that $X^{\bar{H}} \cap X^K \subset X^K$ is a homotopy equivalence for each $K > H$, an easy induction shows that $X^{>H} \subset X^H$ is a homotopy equivalence. But $X^{\bar{H}} \cap X^K = X^{\bar{H}K}$ and $H < K$ implies $\bar{H} \leq \bar{K}$ since EssIso (X) is closed under intersections (2.9). Hence $K \leq \bar{H}K \leq \bar{K}K = \bar{K}$ which implies $n(\bar{H}K) = n(K)$. By lemma 2.1 the inclusion $X^{\bar{H}K} \subset X^K$ is a homotopy equivalence.

Finally we see that the G-cells of Y consist precisely of those G-cells of X which have type G/H with $H \in \text{EssIso}(X)$. Thus Y has fewer cells than X and obviously Y is finite whenever X is finite. □

Remark. Although the map $X \to Y$ is a G-homotopy equivalence, it may be geometrically complicated. Here is an example where it is not a simple G-homotopy equivalence. One can realize the generator of $\text{Wh}(Z_5) = Z$ as the equivariant Whitehead torsion of a pair (W,x) where W is a 3-dimensional finite Z_5-complex and Z_5 acts freely outside the fixed point x [12, Ex. 1.13]. Let Z_5 act trivially on S^3 and form the wedge $X = S^3 \vee W$ along x. Since W is contractible, $X \simeq S^3$ and $X^{Z_5} = S^3$ so that X is a 3-dimensional homotopy representation of Z_5. The inclusion $Y = S^3 \subset X$ is a Z_5-homotopy equivalence which is not simple.

If some restrictions must be put on the isotropy group structure of a homotopy representation, we propose the condition "EssIso (X) is closed under intersection" since it is G-homotopy invariant and it implies conditions (A) and (B) up to G-homotopy. However, in this paper we need no additional assumptions. One reason is that we obtain the sharpest results in the case of nilpotent groups and for them the condition is automatically fulfilled as we shall see below in Proposition 2.12.

Let X be a homotopy representation of G. Then $H = \overline{1}$ is the union of all subgroups $K \leq G$ such that $n(K) = n(1)$. Since this set is closed under conjugation, H is a normal subgroup of G. We call H the <u>homotopy kernel</u> of X. If G is solvable then X is G-homotopy equivalent to the representation X^H where the action of G has kernel H in the usual sense:

<u>Proposition 2.11</u>. Let G be a finite solvable group. If X is a homotopy representation of G with homotopy kernel H, then the inclusion $X^H \subset X$ is a G-homotopy equivalence.

<u>Proof</u>. It suffices to show that the inclusion $X^{HK} \subset X^K$ is an ordinary homotopy equivalence for each subgroup $K \leq G$. Since K is solvable we can find a tower

$$1 = K_o < K_1 < \ldots < K_n = K$$

such that $K_i \triangleleft K_{i+1}$ and K_{i+1}/K_i has prime order for $i = 0,\ldots,n-1$. We shall show by induction that $X^{HK_i} \subset X^{K_i}$ is a homotopy equivalence for all i. When $i = 0$ the inclusion $X^H \subset X^1$ is a homotopy equivalence by lemma 2.1 since $n(H) = n(1)$. Assume the claim holds for the value i. Then $K_i \triangleleft K_{i+1}$ and $H \triangleleft G$ imply that $K_{i+1} \leq N(HK_i)$. Hence K_{i+1}/K_i acts on the pair (X^{K_i}, X^{HK_i}). Now $K_{i+1}/K_i \cong Z_p$ for some prime p and the induction assumption implies $H_*(X^{K_i}, X^{HK_i}; Z_p) = 0$. By Smith theory the fixed point pair $(X^{K_{i+1}}, X^{HK_{i+1}})$ has also trivial Z_p-homology [1, III 4.1]. Hence $n(K_{i+1}) = n(HK_{i+1})$ and $X^{HK_{i+1}} \subset X^{K_{i+1}}$ is a homotopy equivalence by lemma 2.1. □

<u>Proposition 2.12</u>. Let X be a homotopy representation of a finite nilpotent group G. Then EssIso (X) is closed under intersection.

<u>Proof</u>. We must show that $H \leq K$ implies $\overline{H} \leq \overline{K}$ (2.9). It is enough to consider the case $K = \overline{K}$. We fix $K = \overline{K}$ and prove the claim by downwards induction on H. If $H = K$ or $H = \overline{H}$ the claim holds trivially. Let then $H < K$ be such that $H < \overline{H}$ and assume we have already proved that $H < L \leq K$ implies $\overline{L} \leq K$. Since K is nilpotent and $H < K$ we have $K_1 = N_K(H) > H$ [11, Th. 3.4 p. 22]. Hence the inductive assumption applies to K_1 and $\overline{K}_1 \leq K$.

Consider X^H as a homotopy representation of $NH = N_G(H)$. It has kernel $H_1 = NH \cap \overline{H} = N_{\overline{H}}(H)$, and $H_1 > H$ as above. The inclusion $X^{H_1} \subset X^H$ is an NH-homotopy equivalence by Proposition 2.11. Since K_1 is a subgroup of NH, $X^{H_1K_1} \subset X^{K_1}$ is an ordinary homotopy equivalence so that $H_1K_1 \leq \overline{K}_1$ or $H_1 \leq \overline{K}_1$. Then $\overline{K}_1 \leq K$ gives $H_1 \leq K$. Hence the inductive assumption applies to H_1, too, and $\overline{H}_1 \leq K$. But $\overline{H}_1 = \overline{H}$ since $H \leq H_1 \leq \overline{H}$. Hence $\overline{H} \leq K$. □

The main result is now an immediate corollary of Propositions 2.10 and 2.12:

Proposition 2.13. Let G be a finite nilpotent group. Every homotopy representation of G is G-homotopy equivalent to a homotopy representation X such that

(A) If $H \in \mathrm{Iso}(X)$ and $L > H$ then $n(L) < n(H)$

(B) Iso (X) is closed under intersection.

X can be chosen finite if the original homotopy representation is finite. □

Remarks. 1. If G is abelian, the proof of Proposition 2.13 simplifies considerably. The only geometric imput is the fact that the fixed point set of Z_p acting on a finite-dimensional contractible complex is mod p acyclic.

2. In the case of a p-group G Proposition 2.12 can be deduced from the work of tom Dieck. He shows that each homotopy representation of a p-group has the same dimension function as some linear representation sphere [6, Satz 2.6]. Since the dimension dunction determines the essential isotropy groups and EssIso (X) = Iso (X) is closed under intersection when X is linear, Proposition 2.12 follows for p-groups. This argument does not apply to general nilpotent groups or even to abelian groups since their dimension functions are only stably linear.

We close this section by an example of a homotopy representation X of G with homotopy kernel H such that $X^H \subset X$ is not a G-homotopy equivalence and EssIso (X) is not closed under intersection. It shows that some restrictions on the group G are necessary in Propositions 2.11, 2.12 and 2.13.

Example 2.14. The binary icosahedral group I* acts on the unit quaternions S^3 by left and right multiplication. The space of the right cosets $\Sigma = S^3/I^*$ is the Poincaré homology 3-sphere, an it inherits a smooth left action of the icosahedral group $A_5 = I^*/Z_2$ with precisely one fixed point $\Sigma^{A_5} = \{eI^*\}$ (for more details, see [1, I.8 (A)]). Choose a small open slice U around the fixed point. It is A_5-homeomorphic to a 3-dimensional linear representation space V of A_5. Clearly V cannot be the trivial representation. As the degrees of the non-trivial irreducible real representations of A_5 are 3, 4 and 5 [18, 18.6], V must be irreducible, hence conjugate to the icosahedral representation. It follows that $\dim V^H = 1$ for cyclic subgroups $H \neq 1$ of A_5 and $\dim V^H = 0$ for other subgroups $H \neq 1$. By Smith theory Σ^H is a mod p homology sphere when H is one of the cyclic subgroups Z_p, p = 2, 3 or 5 [1, III 5.1]. The only possibility is then that

$$\Sigma^{Z_p} \cong S^1, \quad p = 2, 3 \text{ or } 5.$$

The normalizer of Z_p in A_5 is the dihedral group $D2p$ and

$$\Sigma^{D2p} = (\Sigma^{Z_p})^{Z_2} \cong (S^1)^{Z_2} = S^o, \quad p = 2, 3 \text{ or } 5.$$

Finally $D4 = Z_2 \oplus Z_2$ has normalizer A_4 and

$$\Sigma^{A_4} = (\Sigma^{D4})^{Z_3} \cong (S^o)^{Z_3} = S^o.$$

This describes the fixed point sets of all non-trivial subgroups $H \leq A_5$.

The complement $Y = \Sigma \setminus U$ is an acyclic 3-manifold with boundary $\partial Y \cong S(V)$ and it can be given the structure of a finite A_5-complex. Then $Z = Y \cup_{\partial Y} Y$ is a homology 3-sphere and $Z^H \cong \Sigma^H$ for $1 < H < A_5$. However $Z^{A_5} = \emptyset$ because the fixed point Σ^{A_5} lies in $U = \Sigma \setminus Y$. The join $Z*Z$ is simply-connected and therefore a 7-dimensional homotopy representation of A_5 (it is in fact homeomorphic to S^7 by the double suspension theorem, but this in inessential). From the adjunction space

$$X = (Z*Z) \underset{Y}{\cup} SY$$

where Y lies inside one copy of Z in $Z*Z$ and in the middle of the suspension $SY = S^o*Y$. X is an A_5-complex in the obvious fashion. It is simply-connected since $Z*Z$ and SY are simply connected, and $H_*(X) = H_*(S^7)$ since Y and SY are acyclic. Hence $X \simeq S^7$. We claim that X is also a 7-dimensional homotopy representation of A_5. Indeed, X^{Z_p} is S^3 with D^2 attached along a diameter so $X^{Z_p} \simeq S^3$. Similarly X^{D2p} or X^{A_4} is S^1 with D^1 attached along the middle point, so $X^{D2p} \simeq X^{A_4} \simeq S^1$. Finally $X^{A_5} = S^o$ consists of the two cone points of SY.

Consider X as a homotopy representation of $G = A_5 \times Z_2$ where A_5 acts as earlier but Z_2 switches the two cones in SY and leaves $Z*Z$ invariant. The fixed point sets of the G-action on X are those of the A_5-actions on $Z*Z$ and X. Hence the homotopy kernel of the G-action is Z_2 with $X^{Z_2} = Z*Z$. X cannot be G-homotopy equivalent to X^{Z_2} since

$$X^{A_5} = X^o, \quad (X^{Z_2})^{A_5} = (Z*Z)^{A_5} = \emptyset.$$

We also see that A_5 and Z_2 are essential isotropy groups but their intersection 1 is a non-essential one.

3. Classification of G-maps

In this section we characterize the set of fixed point degrees of a G-map $f\colon X \to Y$ between two finite homotopy representations with the same dimension function. This computes the stable mapping groups $\omega_G(X,Y)$ since two G-maps with the same fixed point degrees are stably G-homotopic. In particular we prove that $\omega_G(X,X)$ is canonically isomorphic to the Burnside ring $A(G)$ for any finite homotopy representation X. Finally we show that for nilpotent groups G the degrees $\deg f^H$ already determine the G-homotopy class of f.

We shall need an unstable version of the equivariant Hopf theorem [5, Th. 8.4.1]. We start by recalling some equivariant obstruction theory. Let G be a finite group. Assume that (X,A) is a relatively free G-complex i.e. (X,A) is a relative G - CW-complex such that G acts freely on $X \setminus A$. If $\dim (X \setminus A) = n \geq 1$ and Y is an $(n-1)$-connected and n-simple G-space then every G-map $f\colon A \to Y$ extends to a G-map $F\colon A \to Y$ and the G-homotopy classes of extensions relative to A are classified by the equivariant cohomology group $H^n_G(X,A;\pi_n Y)$.

The group $H^n_G(X,A;\pi_n Y)$ is defined as follows. Let X^k be the k-skeleton of X relative to A and denote by

$$C_k = C_k(X,A) = H_k(X^k,X^{k-1};Z)$$

the cellular chain groups. Then C_* is a chain complex of free ZG-modules. The equivariant cohomology groups $H^*_G(X,A;\pi)$ with coefficients in a ZG-module π are the homology groups of the complex $\operatorname{Hom}_{ZG}(C_*,\pi)$ of equivariant cochains.

For any ZG-module M let

$$M^G = \{m \in M \mid gm = m, g \in G\}, \quad M_G = M/\langle m - gm \mid g \in G\rangle$$

denote the modules of invariants and coinvariants. The norm $N(m) = \Sigma_{g\in G}\, gm$ induces a canonical map $N\colon M_G \to M^G$ whose kernel and cokernel are by definition the Tate groups

$$\hat{H}_o(G,M) = \operatorname{Ker}(N\colon M_G \to M^G), \quad \hat{H}^o(G,M) = \operatorname{Coker}(N\colon M_G \to M^G).$$

The unequivariant chains $\operatorname{Hom}_Z(C_k,\pi)$ can be considered as a G-module by defining the translate of $f\colon C_k \to \pi$ by $g \in G$ to be the function $gf\colon x \to gf(g^{-1}x)$. Then the equivariant chains are the invariants $\operatorname{Hom}_{ZG}(C_k,\pi) = \operatorname{Hom}_Z(C_k,\pi)^G$. But it is easy to see that the norm homeomorphism

$$N\colon \operatorname{Hom}_Z(C_k,\pi)_G \to \operatorname{Hom}_Z(C_k,\pi)^G$$

is an isomorphism because C_k is ZG-free (in fact the ZG-module $\operatorname{Hom}_Z(C^k,\pi)$ is coinduced, hence cohomologically trivial). It follows that $H^*_G(X,A;\pi)$ is also the homology of the complex of coinvariants

$$H^*_G(X,A;\pi) \cong H_*(\mathrm{Hom}_Z(C_*,\pi)_G).$$

As $\dim(X \setminus A) = n$ we have an exact sequence

$$\mathrm{Hom}_Z(C_{n-1},\pi) \to \mathrm{Hom}_Z(C_n,\pi) \to H^n(X,A;\pi) \to 0.$$

Applying the right exact functor $M \to M_G$ gives the exact sequence

$$\mathrm{Hom}_Z(C_{n-1},\pi)_G \to \mathrm{Hom}_Z(C_n,\pi)_G \to H^n(X,A;\pi)_G \to 0.$$

We just saw that the cokernel of the first map is $H^n_G(X,A;\pi)$. Hence we get the amusing formula

$$H^n_G(X,A;\pi) = H^n(X,A;\pi)_G \tag{3.1}$$

which holds for any n-dimensional relatively free G-complex (X,A) and any ZG-module π. Note that the coinvariants $H^n(X,A;\pi)_G$ cannot be replaced by the invariants since the functor $M \to M^G$ is left but not right exact. Note also that G acts on $H^n(X,A;\pi)$ by acting both on the chains and on the module π. If $H_*(X,A;Z)$ or π has finite type over Z so that we can use the universal coefficient formula

$$H^n(X,A;\pi) \cong H^n(X,A);Z) \otimes_Z \pi,$$

then G acts diagonally on the tensor product.

For any G-module M there are natural homomorphisms

$$t\colon M \to M_G, \quad p\colon M_G \to M$$

where t is the quotient map and p is induced by the norm. In the situation of (3.1) they induce homomorphisms

$$t\colon H^n(X,A;\pi) \to H^n_G(X,A;\pi), \quad p\colon H^n_G(X,A;\pi) \to H^n(X,A;\pi).$$

If $A = \emptyset$ then $H^n_G(X;\pi) = H^n(X/G;\pi)$ is the cohomology of X/G with twisted coefficients, t is the cohomology transfer and p is induced by the covering projection $X \to X/G$.

<u>Lemma 3.2</u>. Let (X,A) be a relatively free G - CW-pair of dimension n. Assume that $H^n(X;Z) \cong Z$ and that π is isomorphic to $H^n(X;Z)$ as a ZG-module. If $\dim A \leq n - 1$ then the composite homomorphism

$$H_G^n(X,A;\pi) \xrightarrow{p} H^n(X,A;\pi) \to H^n(X;\pi) = Z$$

has image $|G|Z$. If moreover $\dim A \leq n-2$ then $H_G^n(X,A;\pi) \cong Z$.

Proof. If $\dim A \leq n-1$ the homomorphism $H^n(X,A;\pi) \to H^n(X,\pi)$ is a surjection and it induces an epimorphism $H^n(X,A;\pi)_G \to H^n(X;\pi)_G$. The tensor product $H^n(X;\pi) \cong H^n(X;Z) \otimes_Z \pi$ is a trivial ZG-module since G acts diagonally through the same homomorphism $\varepsilon\colon G \to \{\pm 1\} = \mathrm{Aut}\, Z$ on both factors. Hence $H^n(X;\pi)_G = H^n(X;Z) = Z$ and the norm $p\colon H^n(X:\pi)_G \to H^n(X;\pi)$ is multiplication by $|G|$. The first claim follows from the diagram

$$\begin{array}{ccccc}
H^n(X,A;\pi)_G & \longrightarrow & H^n(X;\pi)_G & \longrightarrow & 0 \\
\big\downarrow p & & \big\downarrow p & & \\
H^n(X,A;\pi) & \longrightarrow & H^n(X;\pi) & \longrightarrow & 0
\end{array}$$

with exact rows.

If $\dim A \leq n-2$ then $H^n(X,A;\pi) \cong H^n(X;\pi) \cong Z$ is a trivial ZG-module so that $H_G^n(X,A;\pi) \cong Z_G = Z$. $\square$

We apply now these remarks to the case of homotopy representations.

Proposition 3.3. Let X and Y be finite homotopy representations of a finite group G with the same dimension function. Then

i) there exist G-maps $f\colon X \to Y$.

ii) If $f\colon X \to Y$ is a G-map, $H \in \mathrm{EssIso}\,(X)$ and $\dim X^H \geq 1$ then for each integer k there is a G-map $g\colon X \to Y$ such that $\deg g^H = \deg f^H + k|WH|$ and g coincides with f on $X^{>H}$.

iii) If $\dim X^H \geq \dim X^{>H} + 2$ for each $H \in \mathrm{Iso}\,(H)$ then G-maps $f,g\colon X \to Y$ with $\deg f^H = \deg g^H$ for all $H \leq G$ are G-homotopic.

Proof. We construct G-maps by induction over the orbit types. In the inductive step we must extend a G-map $GX^{>H} \to Y$ to GX^H or equivalently a WH-map $X^{>H} \to Y^H$ to X^H. It can always be done in some way since $(X^H, X^{>H})$ is a relatively free WH-complex and $\pi_i Y^H = 0$ for $i < \dim X^H$. This proves claim i).

Let a G-map $f\colon X \to Y$ be given. To prove claim ii) we must change $f^H\colon X^H \to Y^H$ outside $X^{>H}$. If we can find a WH-extension $g^H\colon X^H \to Y^H$ of $f^{>H}\colon X^{>H} \to Y^H$ with degree $\deg f^H + k|WH|$, it can be further extended to a G-map $g\colon X \to Y$ as above. The extensions rel $X^{>H}$ are classified by $H^n(X^H, X^{>H}; \pi_n Y^H)$ where $n = n(H)$, and the obstruction to finding a homotopy between g^H and f^H is precisely the difference $\deg g^H - \deg f^H$. The assumptions of Lemma 3.2 are satisfied. Indeed, the ZWH-modules $\pi_n Y^H \cong H_n Y^H$ $(n \geq 1)$ and $H^n(X^H)$ are isomorphic by Lemma 2.5 and $\dim X^{>H} \leq n - 1$ since $H \in \text{EssIso}(X)$. Hence we are free to change the degree of f^H by any multiple of $|WH|$, and ii) follows.

If $\dim X^H \geq \dim X^{>H} + 2$ for each $H \in \text{Iso}(X)$ then in particular $\dim X^H \geq 1$ by the convention $\dim \emptyset = -1$. Since $\dim X^{>H} \leq n - 2$, Lemma 3.2 shows that the only obstructions to constructing a G-homotopy between f and g are the differences $\deg f^H - \deg g^H$ and iii) follows. □

The following result is crucial in deriving the mapping degree congruences. Its proof is a direct modification of [9, Th. 3.8].

<u>Proposition 3.4</u>. Let X and Y be finite homotopy representations of a finite group G with the same dimension function. Then there exists a G-map $f: X \to Y$ such that $\deg f^H$ is prime to $|G|$ for all $H \leq G$.

(We say that f as <u>invertible</u> <u>degrees</u>.)

<u>Proof</u>. By Prop. 3.3 i) there exists at least one G-map $f: X \to Y$. We try to correct its degrees. If $X^H = \emptyset$ then $\deg f^H = 1$ is already prime to $|G|$. Since the 0-dimensional fixed point set X^H consists of two points, two such sets are either disjoint or coincide, and we may choose f in such a way that $\deg f^H = 1$ also when $X^H \cong S^o$.

Assume then that $\dim X^H \geq 1$ and that $\deg f^K$ is prime to $|G|$ for all $K > H$. If H is not an essential isotropy group then $\overline{H} > H$ and $\deg f^H = \deg f^{\overline{H}}$ is already prime to $|G|$. Otherwise $\dim X^H > \dim X^{>H}$ and we claim that at least $\deg f^H$ is prime to $|WH|$. Indeed, if p is a prime divisor of $|WH|$ then there exists $K \leq G$ such that $H \triangleleft K$ and $K/H \cong Z_p$. The K/H-map $f^H: X^H \to Y^H$ has fixed point degree

$$\deg (f^H)^{K/H} = \deg f^K \not\equiv 0 \bmod p.$$

Hence $\deg f^H \not\equiv 0 \bmod p$, too, by Proposition 1.13. For some $k \in Z$ the integer $\deg f^H + k|WH|$ is then prime to $|G|$. By Proposition 3.3 ii) we may modify f outside $X^{>(H)}$ so that $\deg f^K$ is prime to $|G|$ for all $K \geq H$. □

We are now ready for the classification of the degrees $\deg f^H$ for G-maps $f: X \to Y$ between two homotopy representations with the same dimension function. Let $C = C(G)$ be the product of integers over the set of conjugacy classes of subgroups of G. If X and Y are oriented as in Proposition 2.7, every G-map $f: X \to Y$ has a well-defined <u>degree</u> <u>function</u> $d(f) \in C$,

$$d(f)(H) = \deg f^H, \quad H \leq G.$$

<u>Theorem 2</u>. Let X and Y be finite homotopy representations of a finite group G with the same dimension function n. There exists integers $n_{H,K}$ such that the congruences

$$\deg f^H \equiv - \sum_{\substack{H \triangleleft K \leq G \\ K/H \text{ cyclic} \neq 1}} n_{H,K} \deg f^K \mod |WH|, \quad H \leq G$$

hold for all G-maps $f: X \to Y$. Conversely, given a collection of integers $d = (d^H) \in C$ satisfying these congruences, there is a G-map $f: X \to Y$ with $\deg h^H = d^H$ for each $H \leq G$ if and only if d fulfils the unstability conditions

i) $d^H = 1$ when $n(H) = -1$
ii) $d^H = 1$, 0 or -1 when $n(H) = 0$
iii) $d^H = d^K$ when $n(H) = n(K)$ and $H \leq K$.

Proof. According to Proposition 3.4 there exist G-maps $g: Y \to X$ with invertible degrees. We choose one of them and fix it. If $f: X \to Y$ is any G-map then the composite $g \circ f: X \to X$ satisfies the congruences

$$\deg g^H \deg f^H \equiv - \Sigma\, \phi(\, K/H\,) \deg g^K \deg f^K \mod |WH|$$

for all $H \leq G$ by theorem 1. The element $\deg g^H$ is invertible in the ring $Z_{|G|}$, hence also in the quotient ring $Z_{|WH|}$, and we can find integers $n_{H,K}$ with redidue class

$$n_{H,K} = \phi(|K/H|) \deg g^K / \deg g^H \in Z_{|WH|}.$$

Using the integers $n_{H,K}$ the congruences follow.

Assume then that $d \in C$ satisfies the congruences. If there is a G-map $f: X \to Y$ with $d(f) = d$ then d fulfils the unstability conditions by Proposition 2.7 (note that iii) is equivalent to $d^H = d^{\overline{H}}$). Conversely, if the unstability conditions hold for d, we shall construct a G-map $f: X \to Y$ with $d(f) = d$ by induction over orbit types. We start with conjugacy classes (H) such that $X^H \cong S^o$. Since two 0-dimensional fixed point sets either coincide or are disjoint we are free to choose the degrees 1, 0 and -1 on them arbitrarily. Extending over X we get a G-map $f: X \to Y$ with $\deg f^H = d^H$ when $n(H) \leq 0$.

Suppose we have obtained a G-map $f: X \to Y$ such that $\deg f^K = d^K$ for $(K) > (H)$. In the induction step we modify it to a G-map $\overline{f}: X \to Y$ such that $\deg \overline{f}^K = d^K$ for $(K) \geq (H)$. If $H \notin \text{EssIso}\,(X)$, then $\overline{H} > H$ and

$$\deg f^H = \deg f^{\overline{H}} = d^{\overline{H}} = d^H$$

by Proposition 2.7 and condition iii). Hence f qualifies as $\overline{f}$ in this case. On the other hand if $H \in \text{EssIso}\,(X)$ then the congruences

$$\deg f^H \equiv - \Sigma\, n_{H,K} \deg f^K = - \Sigma\, n_{H,K} d^K \equiv d^H \mod |WH|$$

hold for $\deg f^H$ by the first part of the proof and for d^H by assumption. Using Proposition 3.3 ii) we can modify f as desired. □

Remark. A stable version of Theorem 2 was proved for unit spheres of complex linear representations by Petrie and tom Dieck [8, Th. 3]. It was generalized to the unstable situation and real representations by Tornehave [20, Th. A]. These proofs are based on the Thom isomorphism in equivariant K-theory and they yield precise information on the numbers $n_{H,K}$. There is an alternative method using transversability which works more generally in the smooth case. Our elementary approach to the congruences seems appropriate if one only needs the existence, not the actual values of $n_{H,K}$. In the construction of G-maps with given degrees we have followed tom Dieck and Petrie.

Regard the Burnside ring $A(G)$ as a subring of C as in section 1. The subgroup of C satisfying the congruences of Theorem 2 can be compactly described as

$$C(X,Y) = \{d \in C \mid d(g)d \in A(G)\} \tag{3.5}$$

where $g: Y \to X$ is a fixed G-map with invertible degrees. Let $[X,Y]_G$ denote the set of G-homotopy classes of equivariant maps $f: X \to Y$ (no base-points are considered). Theorem 2 describes the image of the degree function

$$d: [X,Y]_G \to C(X,Y).$$

As a direct corollary we get

Corollary 3.6. Let X and Y be homotopy representations of a finite group G with the same dimension function n. Then $d: [X,Y]_G \to C(X,Y)$ is

i) surjective if and only if $\dim X^G > 0$ and

$$\dim X^H \geq \dim X^{>H} + 1 \quad \text{for each } H \leq G$$

ii) injective if $\mathrm{EssIso}(X)$ is closed under intersection and

$$\dim X^H \geq \dim X^{>H} + 2 \quad \text{for each } H \in \mathrm{EssIso}(X).$$

Proof. It is clear that the unstability conditions vanish precisely when condition i) holds. Assume that the conditions ii) hold. By Proposition 2.10 we may replace X with a G-homotopy equivalent homotopy representation Y with $\mathrm{Iso}(Y) = \mathrm{EssIso}(X)$. Then the injectiveness of d follows from Proposition 3.3 iii). □

Remark. The formulation chosen in 3.6 ii) may seem complicated. Clearly d is injective under the single condition

$$\dim X^H \geq \dim X^{>H} + 2 \quad \text{for each } H \in \text{Iso}(X). \qquad (*)$$

If (*) holds then EssIso (X) = Iso (X) and it follows from Proposition 2.8 that EssIso (X) is closed under intersection. Hence the conditions in 3.6 ii) are weaker than (*), although by no means necessary.

The join $X*Z$ of two homotopy representations is again a homotopy representation. If $f: X \to Y$ is equivariant then $f*id_Z: X*Z \to Y*Z$ has the same degree function as f when product orientations with a fixed orientation of Z are used on $X*Z$ and $Y*Z$. The stable G-homotopy sets $\omega_G(X,Y)$ are defined as

$$\omega_G(X,Y) = \varinjlim_V [X*S(V),Y*S(V)]_G$$

where the limit is taken over all linear representations V. The degree function defines a map $d: \omega_G(X,Y) \to C(X,Y)$. The set $\omega_G(X,Y)$ admits a group structure by using a trivial representation as the suspension coordinate, and d is a group homomorphism. Let CG be the complex regular representation and $S = S(CG)$ its unit sphere. Then $X*S$ satisfies all conditions in 3.6 and $d: [X*S,Y*S] \to C(X,Y)$ is an isomorphism. Hence $[X*S(V),Y*S(V)]_G$ is isomorphic to the stable group $\omega_G(X,Y)$ for every V containing CG. We have arrived to the following form of Segal's theorem:

Corollary 3.7. The degree function $d: \omega_G(X,Y) \to C(X,Y)$ is an isomorphism for all finite homotopy representations X and Y such that $\dim X^H = \dim Y^H$ for each $H \leq G$. □

The stable group $\omega_G(X,Y)$ is an invariant of X and Y but the isomorphism d and the subgroup $C(X,Y)$ of C depend on the choice of orientation for X and Y. If $X = Y$ this does not matter when we use the same orientations for the source and the target. Hence $\omega_G(X,X)$ is canonically isomorphic to $C(X,X) = A(G)$. The Burnside ring of G, for any finite homotopy representation X of G. We denote the group $\omega_G(X,X)$ by $\omega = \omega_G$ and identify it with $A(G)$.

If X and Y are unit spheres of complex linear representations then $d: [X,Y]_G \to C(X,Y)$ is always injective by Proposition 3.6. Thornehave shows in [20, Prop. 3.1] that this holds for real representations, too, when the group G is nilpotent. The problem is to show that $H^n_{WH}(X^H,X^{>H};\pi) \cong Z$ for each isotropy group H with $n = \dim X^H > 0$. In the linear case $X_H = X^H \setminus X^{>H}$ is an open n-manifold where WH acts freely, and $H^n(X^H,X^{>H};\pi)$ can be identified with $H_o(X_H;Z)$ by duality. Hence one is reduced to study the permutation action of WH on the components of X_H, when X is linear or more generally a locally smooth G-manifold.

On arbitrary homotopy representations no kind of duality can be expected. For example, consider the A_5-space X of example 2.14. The fixed point set X^H of $H = A_4$ is a wedge of a circle S^1 with an interval I^1 with the middle point as the wedge point. The singular set $X^{>H} = X^{A_5}$ consists of the two free end-points so that $H^1(X^H,X^{>H};Z) = H^1(S^1 \vee S^1;Z) = Z \oplus Z$ but $X^{>H}$ does not disconnect X^H.

However, homotopy representations of nilpotent groups are sufficiently close to linear representations to admit a generalization of Tornehave's result:

Theorem 3. Let G be a finite nilpotent group and let X and Y be finite homotopy representations of G with the same dimension function n. Two G-maps $f,g: X \to Y$ are G-homotopic if and only if

i) $\deg f^H = \deg g^H$ for each $H \leq G$

ii) $f^H = g^H$ when $n(H) = 0$.

Remark. The 0-dimensional condition has sometimes been overlooked.

The following example should make it obvious.

Let $X = Y$ be the unit circle S^1 where $G = Z_2$ acts by complex conjugation. Then the constant maps $f = 1$ and $g = -1$ are not G-homotopic although all degrees are 0, since f^G and $g^G: S^o \to S^o$ cannot be homotopic. In fact, it is esy to see that $[S^1,S^1]_{Z_2} = \{\pm f_n | f_n(z) = z^n,\ n \in Z\}$.

Proof. If $f,g: X \to Y$ are G-homotopic then f^H and g^H are homotopic and have the same degree for each $H \leq G$. If $\dim X^H = 0$ then X^H and Y^H consist of two points, and the homotopic maps $f^H,g^H: S^o \to S^o$ must coincide.

Conversely, let $f^H = g^H$ for each $H \leq G$ with $\dim X^H = 0$. Then f and g agree on the union of 0-dimensional fixed point sets and they can be connected by the constant homotopy. The further obstructions to constructing a G-homotopy between f and g are the groups $H^n_{WH}(X^H,X^{>H};\pi_n Y^H)$ where H is an isotropy group with $n = n(H) > 0$. Since G is nilpotent we may assume that EssIso (X) = Iso (X) by Proposition 2.13. Hence $\dim X^{>H} \leq \dim X^H - 1$.

As a first reduction we note that $K > H$ implies that $K_1 = K \cap NH > H$ since G is nilpotent. Hence

$$X^{>H} = \bigcup_{K>H} X^K = \bigcup_{NH \geq K_1 > H} X^{K_1}$$

is the singular set of X^H considered as a WH-space. Therefore it suffices to consider the case where $H = 1$ is the homotopy kernel of X. If $\dim X = n$ and we denote $A = X^{>1}$ then $\dim A \leq n - 1$ and $H^n_G(X,A;\pi)$ has rank at least 1 by lemma 3.2. For each subgroup $K \leq G$ there are epimorphisms

$$H^n(X,A;\pi) \to H^n_K(X,A;\pi) \to H^n_G(X,A;\pi)$$

by (3.1). Hence it is enough to show that $H^n_K(X,A;\pi) \cong Z$ for some subgroup $K \leq G$.

Let $K_1,\ldots,K_m \leq G$ be the isotropy groups with $\dim X^{K_i} = n - 1$ and let $H_1,\ldots,H_l \leq G$ be the isotropy groups such that $X^{H_j} \not\subset X^{K_i}$ for any $i = 1,\ldots,m$.

Then $A = A_1 \cup A_2$, where $A_1 = \cup_{i=1}^m X^{K_i}$ has dimension $n - 1$ (or is empty) and $A_2 = \cup_{j=1}^l X^{H_j}$ has dimension at most $n - 2$. Since all isotropy groups are essential, $A_o = A_1 \cap A_2$ has dimension at most $n - 3$. The cohomology group $H^n(X,A;\pi)$ is an extension

$$0 \to H^{n-1}(A;\pi) \to H^n(X,A;\pi) \to H^n(X;\pi) \to 0.$$

The Mayer-Vietoris sequence of $A = A_1 \cup A_2$ shows that $H^{n-1}(A) \xrightarrow{\sim} H^{n-1}(A_1)$ and consequently $H^n(X,A;\pi) \cong H^n(X,A_1;\pi)$.

Let $K_i \leq G$ be an isotropy group of X such that $\dim X^{K_i} = n - 1$. Then $\dim X^L = n - 1$ for any $L \leq K_i$ with $L \neq 1$, since 1 is the homotopy kernel of $X \simeq S^n$. Let $x = [X] - \chi(S^{n-1})1 \in A(K_i)$. Then

$$\chi_e(x) = \chi(S^n) - \chi(S^{n-1}) = \pm 2, \quad \chi_L(x) = \chi(S^{n-1}) - \chi(S^{n-1}) = 0 \quad \text{for} \quad 1 < L \leq K_i.$$

The Burnside ring relations in $A(K_i)$ imply that $\pm 2 \equiv 0 \bmod |K_i|$ so that $K_i \cong Z_2$. Let K be the subgroup of G generated by K_i, $i = 1,\dots,m$. The Sylow subgroup G_2 of G is normal since G is nilpotent. Hence $K \leq G_2$ is a 2-group. We shall show that $H_K^n(X,A;\pi) \cong Z$ for a homotopy representation $X \simeq S^n$ of a 2-group K such that $A = X^{>1} = \cup_{i=1}^m X^{K_i}$ is a union of subcomplexes $X^{K_i} \simeq S^{n-1}$.

By tom Dieck [6, Satz 2.6], X has the same dimension function as some linear representation of K. In particular, if L is contained in the subgroups L_1 and L_2 and $\dim X^{L_i} = \dim X^L - 1$ for $i = 1,2$, then $X^{L_1} \cap X^{L_2}$ has dimension $\dim X^L - 2$.

Now a double induction on $n = \dim X$ and on the number m of the components in $A = \cup_{i=1}^m X^{K_i}$ shows that

$$H^k(X,A;Z) = \begin{cases} \text{free}, & k = n \\ 0, & k \neq n. \end{cases}$$

The induction starts in dimension $n = 1$ where X/A is a connected CW-complex of dimension 1, hence homotopy equivalent to a wedge of circles. The induction on m is based on a Mayer-Vietoris argument: if $B = X^{K_{m+1}}$ is not contained in A then $B \cap A = \cup_{i=1}^m X^{L_i}$ where $L_i = K_{m+1}K_i$, $i = 1,\dots,m$ and each X^{L_i} has codimension 1 in B. Hence the induction hypothesis applies to the pair $(B,\cup_{i=1}^m X^{L_i})$ and one may apply the relative Mayer-Vietoris sequence of (X,A) and (X,B).

In particular, $H^n(X,A;Z)$ is torsion free. Since the composite

$$H_K^n(X,A;\pi) \xrightarrow{p} H^n(X,A;\pi) \xrightarrow{t} H_K^n(X,A;\pi)$$

is multiplication by $|K|$, a power of 2, all torsion in $H^n_K(X,A;\pi)$ is 2-torsion. On the other hand, if $S(V)$ is a linear representation sphere of K with $\mathrm{Dim}\, S(V) = \mathrm{Dim}\, X$, there exists by Proposition 3.4 a K-map $f\colon X \to S(V)$ such that all degrees $\deg f^H$ are odd, $H \leq K$. Comparing the Mayer-Vietoris sequences used to compute $H^n(X,A;Z)$ and $H^n(S(V),S(V)^{>1};Z)$ we get an exact sequence

$$0 \to H^n(S(V),S(V)^{>1};Z) \xrightarrow{f^*} H^n(X,A;Z) \to C \to 0 \qquad (*)$$

where C is a torsion group of odd order. Recall that $H_0(K;M) = M_K$ for any ZK-module M. The exact sequence of homology of the extension (*) of K-modules now gives

$$H_1(K,C) \to H^n_K(S(V),S(V)^{>1};\pi) \xrightarrow{f^*} H^n_K(X,A;\pi) \to C_K \to 0.$$

The group $H_1(K,C) = 0$ since K is a 2-group and C is an odd torsion module. In the linear case it is known that $H^n_K(S(V),S(V)^{>1},\pi) \cong Z$ [20, Proof of Prop. 3.1]. The resulting extension

$$0 \to Z \xrightarrow{f^*} H^n_K(X,A;\pi) \to C_K \to 0$$

where C_K is an odd torsion group shows that $H^n_K(X,A;\pi) \cong Z$ since it only has 2-torsion. □

<u>Remarks</u>. 1. The 2-group $K \leq G$ which appears in the proof is a finite group of reflexions and we may be much more specific. From the classification of Coxeter groups it follows that K is a direct product of an elementary abelian group $(Z_2)^k$ and dihedral groups $D_1,\ldots,D_1$. The components of $S(V) \setminus S(V)^{>1}$ are Weyl chambers, open n-simplices which are permuted freely and transitively by K. This implies that $S(V)/S(V)^{>1} \cong \bigvee_{k\in K} S^n_k$ so that $H^n(S(V),S(V)^{>1};Z)$ is isomorphic to ZK as a K-module. Hence the group of coinvariants in $H^n_K \cong Z$. (see Bourbaki, Groupes et algebres de Lie, Ch. 4-5).

2. The abelian case presents some short-cuts, again. Then $K \cong (Z_2)^k$ is elementary abelian and Borel's dimension formula implies

$$\dim X^H = \dim X - r, \quad H \cong (Z_2)^r.$$

Hence the representation V with $\mathrm{Dim}\, S(V) = \mathrm{Dim}\, X$ is found directly without appeal to [6] and $S(V) \setminus S(V)^{>1}$ is easy to analyze. Of course, Borel's dimension formula is an essential ingredient of tom Dieck's theorem.

4. Homotopy equivalence of homotopy representations

Let G be a finite group and let X and Y be finite homotopy representations of G. A G-homotopy equivalence $f: X \to Y$ is oriented, if $\deg f^H = 1$ for each subgroup $H \leq G$. We shall show that X and Y are stably oriented G-homotopy equivalent if and only if they are oriented G-equivalent. A similar destabilization result holds for ordinary G-homotopy equivalence if G is nilpotent and has abelian Sylow 2-subgroup but not in general. We give an example of smooth free actions of a metacyclic group on a sphere which are not G-homotopy equivalent but become such after adding a linear representation sphere of the same dimension.

Let X and Y be finite homotopy representations with $\mathrm{Dim}\, X = \mathrm{Dim}\, Y$. This is clearly a necessary condition for X and Y to be stably G-homotopy equivalent. Choose a set of representatives for the essential isotropy groups of X and orient X and Y using this set (2.7). By theorem 2 the degree functions of G-maps $f: X \to Y$ belong to a subgroup $C(X,Y)$ of C and clearly $|G|C$ is contained in $C(X,Y)$. Especially $C(X,X) = A(G)$ contains $|G|C$ and we may define

$$\bar{A}(G) = A(G)/cC, \quad \bar{C} = C/cC,$$

where c is any multiple of $|G|$. Then $\bar{A}(G)$ is a subring of $\bar{C}$. Let $\bar{A}(G)^\times$ and $\bar{C}^\times$ be the groups of units of the rings $\bar{A}(G)$ and $\bar{C}$. Note that

$$\bar{A}(G) \cap \bar{C}^\times = \bar{A}(G)^\times$$

since $\bar{C}^\times = \Pi\, Z_c^\times$ is a finite group. If $g: Y \to X$ has invertible degrees, we regard $d(g)$ as an element of $\bar{C}^\times$.

<u>Lemma 4.2</u>. If G-maps $g: Y \to X$ and $g': Y \to X$ have invertible degrees then

$$d(g')/d(g) \in \bar{A}(G)^\times.$$

<u>Proof</u>. Since $d(g)$ is an element of the finite group $\bar{C}^\times$, we can find a positive integer k such that $d(g)^k = 1$ in $\bar{C}^\times$. Then the function $d = d(g)^{k-1}$ in C belongs to $C(X,Y)$ since

$$d(g)d \in 1 + |G|C \subset A(G)$$

(see 3.5), and it also fulfils the unstability conditions since $d(g)$ does. By theorem 2 there exists a G-map $f: X \to Y$ with $d(f) = d$. In the group $\bar{C}^\times$ we have

$$d(g')/d(g) = d(g')d = d(g')d(f) = d(g'f) \in \bar{C}(X,X) = \bar{A}(G)$$

and 4.1 implies the claim. $\square$

Following tom Dieck and Petrie [8] we define the <u>oriented Picard group</u> of G as

$$\mathrm{Inv}\,(G) = \bar{C}^{\times} \,/\, \bar{A}(G)^{\times}. \tag{4.3}$$

It is a finite group which depends only on G, not on the multiple c of $|G|$ used. Let X and Y be finite homotopy representations with $\mathrm{Dim}\,X = \mathrm{Dim}\,Y$. Then we can attach to the pair (X,Y) the invariant

$$D^{or}(X,Y) = d(g) \in \mathrm{Inv}\,(G) \tag{4.4}$$

where $g\colon Y \to X$ is any map with invertible degrees. By lemma 4.2 this does not depend of the choice of g. In fact, $D^{or}(X,Y)$ is the class of the invertible module $C(X,Y) \cong \omega_G(X,Y)$ over the Burnside ring $A(G)$, but this will not be needed in the sequel. However, $D^{or}(X,Y)$ depends on the choice of orientations for X and Y. By a stable oriented G-homotopy equivalence between X and Y we mean an oriented G-homotopy equivalence $f\colon X*Z \to Y*Z$ where Z is any finite homotopy representation.

<u>Theorem 4</u>. Let X and Y be finite homotopy representations of a finite group G with the same dimension function. Choose orientations for X and Y. The following conditions are equivalent:

i) X and Y are oriented G-homotopy equivalent

ii) X and Y are stably oriented G-homotopy equivalent

iii) $D^{or}(X,Y) = 1$ in $\mathrm{Inv}\,(G)$.

<u>Proof</u>. Clearly i) implies ii). Let Z be a finite homotopy representation of G and let $f\colon Y*Z \to X*Z$ be an oriented G-homotopy equivalence. Choose a G-map $g\colon Y \to X$ with invertible degrees. Since both $g*\mathrm{id}_Z$ and f have invertible degrees, $d(g*\mathrm{id}_Z)/d(f) = d(g)$ belongs to $\bar{A}(G)^{\times}$ by lemma 4.2. Hence $D^{or}(X,Y) = 1$.

If $D^{or}(X,Y) = 1$, we have $d(g) \in A(G)$ for any $g\colon Y \to X$ with invertible degrees. Then the constant degree function 1 belongs to

$$C(X,Y) = \{d \mid d(g)d \in A(G)\}$$

(3.5). Since 1 obviously satisfies the unstability conditions, theorem 2 shows that there exists a G-map $f\colon X \to Y$ with $\deg f^H = 1$ for each $H \leq G$. The map f is the required oriented G-homotopy equivalence between X and Y. $\square$

<u>Remark</u>. Tom Dieck and Petrie proved theorem 4 for unit spheres of complex linear representations, see [3, Th. 5] and [8, Th. 2].

Theorem 4 is useful when X and Y can be oriented in a canonical way, e.g. when they are unit spheres of complex representations. Usually this is impossible. However, the product orientation on $X*X$ and $Y*Y$ is canonical and we can state as a corollary

Corollary 4.5. If X and Y are stably G-homotopy equivalent finite homotopy representations, then $X*X$ and $Y*Y$ are oriented G-homotopy equivalent.

Proof. If $f: X*Z \to Y*Z$ is a G-homotopy equivalence, then $f*f$ is an oriented stable G-homotopy equivalence between $X*X$ and $Y*Y$. By theorem 4 $X*X$ and $Y*Y$ are oriented G-homotopy equivalent. □

In general we must study the effect of a change in orientations of X and Y Let $g: Y \to X$ be a G-map with invertible degrees. If new orientations are used for X and Y, the degrees $\deg g^H$ are changed by signs $\varepsilon^H = \pm 1$, and so the degree function $d(g)$ is multiplied by a unit ε in $C^\times = \Pi\{\pm 1\}$. Hence the class of $d(g)$ in the Picard group

$$\text{Pic}\ (G) = \text{Inv}\ (G)/C^\times = \bar{C}^\times \ /\ \bar{A}(G)^\times C^\times \tag{4.6}$$

only depends on the pair (X,Y), not on the orientations. The resulting invariant

$$D(X,Y) = d(g) \in \text{Pic}\ (G) \tag{4.7}$$

detects unfortunately only stable G-homotopy equivalence. For G-homotopy equivalence we must take into account the unstability conditions.

Let X and Y be finite homotopy representations with the same dimension function n, i.e. $n(H) = \dim X^H = \dim Y^H$ for $H \leq G$. Note that the essential isotropy subgroups of X and Y can be recovered from n. If the G-map $g: Y \to X$ has invertible degrees, then $d(g) = d$ satisfies by (2.7)

$$\begin{aligned} &\text{i)} \quad d^H = 1 \quad \text{when} \quad n(H) = -1 \\ &\text{ii)} \quad d^H = \underline{+}1 \quad \text{when} \quad n(H) = 0 \\ &\text{iii)} \quad d^H = d^{\bar{H}}. \end{aligned} \tag{4.8}$$

Hence $d(g)$ belongs to the subgroup $\bar{C}_n^\times$ of $\bar{C}^\times$ defined by the conditions (4.8). Let $\bar{A}(G)_n^\times$ be the corresponding subgroup $\bar{A}(G)^\times$. We see that $D^{or}(X,Y)$ lies actually in the subgroup $\text{Inv}\ (G)_n = \bar{C}_n^\times \ /\ \bar{A}(G)_n^\times$. A change of orientations of X and Y multiples $d(g)$ by a unit ε which satisfies (4.8). Denote by $C_n^\times$ the group of such units in C. We thus get an unstable invariant

$$D_n(X,Y) = d(g) \in \text{Pic}_n\ (G) = \bar{C}_n^\times \ /\ \bar{A}(G)_n^\times C_n^\times \tag{4.9}$$

and the proof of theorem 4 gives immediately

Theorem 5. Let X and Y be finite homotopy representations of a finite group G with the same dimension function n. Then

i) X and Y are stably G-homotopy equivalent if and only if $D(X,Y) = 1$ in $\text{Pic}\ (G)$

ii) X and Y are G-homotopy equivalent if and only if $D_n(X,Y) = 1$ in $\text{Pic}_n\ (G)$. □

The difference of theorems 4 and 5 is that the map $\mathrm{Inv}_n(G) \to \mathrm{Inv}(G)$ is always injective, whereas $\mathrm{Pic}_n(G) \to \mathrm{Pic}(G)$ usually has nontrivial kernel. This may be explained as follows: If $g: Y \to X$ is a G-map with invertible degrees and $D(X,Y) = 1$ in $\mathrm{Pic}(G)$, there exists a unit $\varepsilon \in C^x$ such that $x = \varepsilon d(g)$ belongs to $A(G)$. Although the product $d(g) = \varepsilon x$ satisfies the unstability conditions (4.8), the factors ε and x need not satisfy them. However, if ε and x can be replaced by unstable ε' and x', then $D_n(X,Y) = 1$ and X and Y are G-homotopy equivalent. In this case $\varepsilon\varepsilon'$ is a unit of $A(G)$. Hence the difference between stable and unstable G-homology equivalence is connected with the units of the Burnside ring.

We shall now prove that stable G-homotopy equivalence implies ordinary G-homotopy equivalence for homotopy representations of certain nilpotent groups.

Theorem 6. Let G be a finite nilpotent group with an abelian Sylow 2-subgroup and let X and Y be finite homotopy representations of G. If X and Y are stably equivalent, they are G-homotopy equivalent.

Proof. Since X and Y are stably G-homotopy equivalent, $\mathrm{Dim}\, X = \mathrm{Dim}\, Y$. Choose a map $g: Y \to X$ with invertible degrees. Then there exists a unit ε in C^x such that $\varepsilon d(g)$ lies in $A(G)$, and ε can be realized as the degree function of a stable G-homotopy equivalence $h: Y*Z \to X*Z$. We construct a G-homotopy equivalence $f: Y \to X$ by induction over the orbit types.

Start with a map $\bar{f}: Y^G \to X^G$ with degree one. Assume that we have already found $\bar{f}: GY^{>H} \to X$ with $\deg \bar{f}^K = \pm 1$ for all $(K) > (H)$. Extend $\bar{f}^H: Y^{>H} \to X^H$ to a WH-map $f: Y^H \to X^H$. If $\deg f \equiv \pm 1 \bmod |WH|$, it can be modified to a map with degree ± 1. Now both $f*\mathrm{id}$ and h^H have invertible degree functions as WH-maps. Hence $d(f)/d(h^H)$ belongs to $\bar{A}(WH)^x$ by Lemma 4.2. Since $\deg f^K/\deg h^K = \pm 1$ for each $K > H$, it suffices to prove the following algebraic lemma

Lemma 4.10. Let G be a finite nilpotent group with abelian Sylow 2-subgroup. If the element x of $A(G)$ has $\chi_K(x) = \pm 1$ for each $K \neq 1$, then $\chi_e(x) = \pm 1 \bmod |G|$.

Proof. Assume first that G is abelian. We may clearly multiply x by a unit ε of $A(G)$ without changing the assertion. If $\chi_G(x) = -1$, we first multiply x with -1. Let H be a subgroup of G of index 2. Then the unit $\varepsilon_H = 1 - G/H$ in $A(G)$ has $\chi_H(\varepsilon_H) = -1$ and $\chi_K(\varepsilon_H) = 1$ for all K not contained in H. Hence by using units ε_H we may assume that $\chi_H(x) = 1$ for each $H \leq G$ of index at most 2. It follows that $\chi_H(x) = 1$ for each $H \neq 1$. Indeed, if this is always proved for $K > H$, the congruences (1.1) imply that $\chi_H(x) \equiv 1 \bmod |G/H|$. But $|G/H|$ is at least 3 and $\chi_H(x) = \pm 1$, so we have an equality $\chi_H(x) = 1$. Finally the congruences (1.1) once again show that $\chi_e(x) \equiv 1 \bmod |G|$.

Let now G be a general nilpotent group. The square x^2 in $A(G)$ satisfies $\chi_K(x^2) = 1$ for each $K \neq 1$ so that $\chi_e(x)^2 = \chi_e(x^2) \equiv 1 \bmod |G|$. Especially

$\chi_e(x)^2 \equiv 1 \mod p^n$ if the Sylow p-subgroup G_p has order p^n. When p is odd, this implies that $\chi_e(x) \equiv \pm 1 \mod p^n$ since $(Z/P^n)^\times$ is cyclic. Hence, for odd p there is a sign $\varepsilon_p = \pm 1$ such that

$$\chi_e(x) \equiv \varepsilon_p \mod |G_p|. \qquad (*)$$

By the abelian case, this holds also for $p = 2$. We are ready if we can show that $\varepsilon_p = \varepsilon_q$ for all p and q.

Choose a central subgroup H_p of order p in G_p for each odd prime divisor p of $|G|$. Since G is a direct product of its Sylow subgroups G_p, H_p is also central in G and the subgroup $H = G_2 \prod_p H_p$ of G is abelian. Then we know that

$$\chi_e(x) \equiv \varepsilon \mod |H| \qquad (**)$$

where $\varepsilon = \pm 1$. Comparing (*) and (**) we see that $\varepsilon_p = \varepsilon$ when p is odd and $\varepsilon_2 = \varepsilon$ when the order of G_2 is at least 4. If $G_2 = Z_2$, ε_2 can be arbitrary since $Z^\times_{|G|} = Z^\times_{|G|/2}$. □

Remark. The lemma fails for dihedral and semidihedral 2-groups G. Both groups contain a noncentral subgroup H of order 2 with $|WH| = 2$ and then $x = 1 - G/H \in A(G)$ has characters $\chi_e(x) = 1 - |G|/2$, $\chi_H(x) = -1$ and $\chi_K(x) = 1$ for other subgroups $K \leq G$. Conversely, using the multiplicative congruences of tom Dieck [7] one can extend lemma 4.10 to all nilpotent groups G such that the Sylow 2-subgroup is not dihedral or semidihedral. From the point of view of theorem 6 such a generalization is useless since one must be able to apply the lemma to all quotient groups of subgroups of G.

We conclude with an example which shows that stable G-homotopy equivalence does not imply G-homotopy equivalence in general even for smooth free actions on spheres.

Example 4.11. Let G be a metacyclic group of order pq where p and q are odd primes, i.e. G is an extension of a cyclic group H of order q by a cyclic group K of order p such that K embeds into $\mathrm{Aut}(H) = Z^\times_q$. The cohomology of G is periodic with period $2p$ and it follows from Swan [19] that there exists a free G-complex X of dimension $2p - 1$ homotopy equivalent to S^{2p-1}. The oriented G-homotopy type of X is determined by the k-invariant $e(X)$, a generator of $H^{2p}(G;Z) = Z_{|G|}$. All generators occur as k-invariants and X can be chosen finite if and only if the image of $e(X)$ under the Swan homomorphism $\theta\colon Z^X_{|G|} \to \tilde{K}_o(ZG)$ vanishes. In this case the kernel of θ consists of d which are p'th powers mod q. By using surgery Madsen, Thomas and Wall show that each finite X is G-homotopy equivalent to a free smooth action of G on S^{2p-1} [22, Th. 1, Th. 3].

Let X and Y be smooth free G-spheres diffeomorphic to S^{2p-1} and let

$g\colon Y \to X$ be a G-map. The conjugacy classes of subgroups of G are $\{1,H,K,G\}$. If $L \leq G$ is nontrivial then Y^L is empty and $\deg g^L = 1$. By theorem 2 the degree $d = \deg g$ is determined mod pq. If we define the k-invariants $e(X)$ and $e(Y)$ as the classes of the angmented cellular chain complexes

$$0 \to Z \to C_{2p-1} \to \dots \to C_o \to Z \to 0$$

in $\mathrm{Ext}^{2p}_{ZG}(Z,Z) = H^{2p}(G;Z)$ then it is immediate that $de(Y) = e(X)$. We may choose X and Y so that $d \equiv 1 \bmod q$ and $d \equiv -1 \bmod p$, since $\theta(d) = 0$. Then X and Y are not G-homotopy equivalent. However, the Burnside ring $A(G)$ consists of x such that

$$\chi_H(x) \equiv x_G(x) \bmod p,\quad \chi_1(x) \equiv \chi_H(x) \bmod q,\quad \chi_1(x) \equiv \chi_K(x) \bmod p.$$

If $\varepsilon \in C^x$ is the unit with $\varepsilon^K = -1$ and $\varepsilon^K = 1$ otherwise, then $x = \varepsilon d(g)$ belongs to $A(G)$ and X and Y are stably G-homotopy equivalent by theorem 5.

We can realize this geometrically as follows. Induce a faithful representation of H on C up to a representation V of G. Then V is irreducible and has complex degree p. The subgroup H acts freely on V but $\dim_V V^K = 1$. Hence $S(V)$ has isotropy groups 1 and K. By theorem 2 we can find a G-homotopy equivalence $f\colon X*S(V) \to Y*S(V)$ with degree function ε. Note that X,Y and $S(V)$ have dimension $2p - 1$. The lowest dimension 5 occurs for metacyclic groups of order 21.

Remark. Theorem 6 was proved in the special case of unit spheres of linear representations of abelian groups by Rothenberg [16, Cor. 4.10]. In [17] he considers finite homotopy representations X such that X^H is a PL-homeomorphic to $S^{n(H)}$ and WH acts trivially on $H_*(X^H)$ for each subgroup H of G. Example 4.11 contradicts the destabilization theorems 1.8 and 5.7 of [17], which claim that stable G-homotopy equivalence implies G-homotopy equivalence if the Sylow 2-subgroup G_2 is "very nice". Indeed, the groups of 4.11 have odd order and a trivial group should certainly be very nice. Note that lemma 4.10 is an algebraic version of the basic result [17, Prop. 2.2], which fails for metacyclic groups of odd order and dihedral groups as pointed out by Oliver [MR 81c: 57044].

References

[1] G. Bredon, Introduction to compact transformation groups, Academic Press, New York and London, 1972.

[2] W. Burnside, Theory of groups of finite order, 2nd Edition 1911, Reprinted by Dover Publications, New York, 1955.

[3] T. tom Dieck, Homotopy-equivalent group representations, J. Reine Angew. Math. 298 (1978), 182-195.

[4] T. tom Dieck, Homotopy-equivalent group representations and Picard groups of the Burnside ring and the character ring, Manuscripta math. 26 (1978), 179-200.

[5] T. tom Dieck, Transformation groups and representation theory, Lecture Notes in Mathematics 766, Springer-Verlag, Berlin Heidelberg New York, 1979.

[6] T. tom Dieck, Homotopiedarstellungen endlicher Gruppen: Dimensionsfunktionen, Invent. math. 67 (1982), 231-252.

[7] T. tom Dieck, Die Picard-Gruppe des Burnside-Ringes, pp. 573-586 in Algebraic Topology, Aarhus 1982, Lecture Notes in Mathematics 1051, Springer-Verlag, Berlin Heidelberg New York, 1984.

[8] T. tom Dieck and T. Petrie, Geometric modules over the Burnside ring, Invent. math. 47 (1978), 273-287.

[9] T. tom Dieck and T. Petrie, Homotopy representations of finite groups, Publ. Math. IHES 56 (1982), 337-377.

[10] A. Dold, Simple proofs of some Borsuk-Ulam results, pp. 65-69 in Proceedings of the Northwestern homotopy theory conference, AMS, Providence, 1983.

[11] D. Gorenstein, Finite groups, Harper & Row, New York, Evanston and London, 1968.

[12] S. Illman, Whitehead torsion and group actions, Ann. Acad. Sci. Fennicae A I 588, 1974.

[13] E. Laitinen, The equivariant Euler and Lefschetz classes, to appear.

[14] W. Marzantowicz, Liczby Lefschetza odwzorowań przemiennych z działaniem grupy, Ph.D. Thesis, Warsaw 1977.

[15] W. Marzantowicz, On the nonlinear elliptic equations with symmetry, J. Math. Anal. Appl. 81 (1981), 156-181.

[16] M. Rothenberg, Torsion invariants and finite transformation groups, pp. 267-311 in Algebraic and Geometric Topology, Proc. Symp. Pure Math. 32 Part 1, AMS, Providence, 1978.

[17] M. Rothenberg, Homotopy type of G spheres, pp. 573-590 in Algebraic Topology, Aarhus 1978, Lecture Notes in Mathematics 763, Springer-Verlag, Berlin Heidelberg New York, 1979.

[18] J.-P. Serre, Représentations linéaires des groupes finis, 2. éd., Hermann, Paris, 1971.

[19] R.C. Swan, Periodic resolutions for finite groups, Ann. of Math. 72 (1960), 267-291.

[20] J. Tornehave, Equivariant maps of spheres with conjugate orthogonal actions, pp. 275-301 in Current Trends in Algebraic Topology, CMS Conference Proceedings Vol. 2 Part 2, AMS, Providence, 1982.

[21] P. Traczyk, Cancellation law for homotopy equivalent representations of groups of odd order, Manuscripta Math. 40 (1982), 135-154.

[22] C.T.C. Wall, Free actions of finite groups on spheres, pp. 115-124 in Algebraic and Geometric Topology, Proc. Symp. Pure Math. 32 Part 1, AMS, Providence, 1978.

[23] G.W. Whitehead, Elements of homotopy theory, Graduate Texts in Mathematics 61, Springer-Verlag, New York Heidelberg Berlin, 1978.

Department of Mathematics
University of Helsinki
Hallituskatu 15
00100 Helsinki, Finland

DUALITY IN ORBIT SPACES

Arunas Liulevicius* and Murad Özaydın

Our aim in this paper is to present a new technique for studying symmetric products of G-sets. The motivation for this work originally came from the study of exterior powers in the Burnside ring of a finite group motivated by the work of Dold [2] which presents a new model for the universal λ-ring [1], [3] on one generator. Some of the results mentioned here will only have sketch proofs - for more detail the reader can consult [5].

Let G be a group and X a finite G-set. The k-fold symmetric product $S_k X$ is defined as follows. The symmetric group S_k operates on the k-fold Cartesian product X^k by permutation of coordinates, and we define $S_k X = X^k/S_k$. The diagonal action of G on X^k commutes with the action of S_k, so this means that $S_k X$ inherits an action of G.

The key idea in our approach to the study of $S_k X$ is that it is convenient to study all of them at the same time. We define the graded set $S_* X = \{S_k X\}_{k \in N}$, where N is the set of natural numbers.

PROPOSITION 1. Suppose X is a finite set. Then $S_* X = \mathrm{Map}(X,N)$.

Proof. If $z \in S_* X$, let $\langle z, \ \rangle : X \longrightarrow N$ be the counting function determined by z, that is $\langle z,x \rangle$ is the number of times the element x occurs in z. Even more precisely, if $z = (z_1, \dots, z_k).S_k$, then $\langle z,x \rangle$ is the number of i such that $x = z_i$. Notice that k is recaptured from the counting function for z by the identity

$$k = \sum_{x \in X} \langle z,x \rangle .$$

---

* Research partially supported by NSF grant DMS 8303251.

Conversely, given a function $c : X \longrightarrow N$, there exists a unique element $z \in S_k X$ such that $c = \langle z, \rangle$. Indeed, here k is given by

$$k = \sum_{x \in X} c(x) .$$

COROLLARY 2. If X and Y are finite sets and $X \sqcup Y$ denotes their disjoint sum, then $S_*(X \sqcup Y) = S_*X \times S_*Y$.

Proof. A function $C : X \sqcup Y \longrightarrow N$ is completely determined by the restrictions $C_1 : X \longrightarrow N$ and $C_2 : Y \longrightarrow N$.

Notice that if X is a G-set with G acting on the right, then under the correspondence $S_*X = \mathrm{Map}(X,N)$ the action of G on S_*X inherited from the diagonal action on X^k corresponds to the right G-action on Map(X,N) defined by $(c.g)(x) = c(x.g^{-1})$. This allows us to prove

COROLLARY 3. If X is a finite right G-set and $H \subset G$ is a subgroup, then $(S_*X)^H = S_*(X/H)$.

Proof. To say that a counting function $c : X \longrightarrow N$ is in the fixed point set $(S_*X)^H$ is the same as saying that c is constant on the orbits of H in X, that is, it corresponds to a function $\underline{c} : X/H \longrightarrow N$.

For both the statement and the proof of the statement above it is essential to use S_*X. Without it the statement becomes more complicated, since the orbits of H need not have the same number of elements.

COROLLARY 4 (Duality). If X is a finite G-set and G is a finite group, then $S_*X = \mathrm{Map}_G(X,S_*G)$. Here the usual right action on $S_*G = \mathrm{Map}(G,N)$ is used in defining the set $\mathrm{Map}_G(X,S_*G)$. There is a second (commuting) right action of G on $S_*G = \mathrm{Map}(G,N)$ defined by $(c_*g)(y) = c(gy)$, and this action corresponds to the standard action of G on S_*X.

Proof. It is enough to check this on orbits G/H. We have just seen that $S_*(G/H) = (S_*G)^H = Map_G(G/H, S_*G)$.

Our second duality result involves the infinite group G = (Z,+), the additive group of the integers. We wish to determine the structure of the finite Z-sets S_*X, and according to Corollary 2 it is enough to do this for the cycles Z/(n).

PROPOSITION 5. The multiplicity of the cycle Z/(r) in S_k(Z/(rs)) is zero if s does not divide k. If k = ms , then the multiplicity of Z/(r) in S_k(Z/(rs)) is the same as the multiplicity A(m,r) of Z/(r) in S_m(Z/(r)).

Proof. This is a consequence of Corollary 3. See [4] for an alternative argument and [5] for a more detailed discussion.

COROLLARY 6. Let A(k,n) be the multiplicity of Z/(n) in S_k(Z/(n)). Then if μ is the Möbius function, we have

$$A(k,n) = \sum_{(n,k)\subset(s)} (\mu(s)/s).(n/s+k/s-1)!(n/s)!(k/s)! .$$

Proof. Use the Möbius inversion formula to solve the recursion relations for A(k,n) coming from Proposition 5.

COROLLARY 7 (Reciprocity Law). If A(k,n) is the multiplicity of the cycle Z/(n) in S_k(Z/(n)) , then A(k,n) = A(n,k) for all k,n.

Proof. Notice that the formula for A(k,n) in Corollary 6 is symmetric in k and n.

This is not entirely satisfactory, since the reciprocity seems to be an accidental result of a complicated number-theoretical formula. The key which explains the reciprocity law is the following duality map of orbit spaces:

THEOREM 8 (Duality map). There exists a one-to-one isotropy preserving correspondence

$$D : (S_k(Z/(n))/ Z/(n) \longrightarrow (S_n(Z/(k))/ Z/(k) .$$

That is, for each $(k,n) \subset (s)$ the multiplicity of the cycle $Z(n/s)$ in $S_k(Z/(n))$ is the same as the multiplicity of the cycle $Z/(k/s)$ in $S_n(Z/(k))$.

Proof. The key point in the proof [5] is to identify the orbit space $(S_k(Z/(n)))/ Z/(n)$ as the set of all circular Lazy Susans having n walls and k balls distributed in the n chambers. The duality map interchanges the roles of the walls with that of the balls.

REFERENCES

[1] M.F.Atiyah and D.O.Tall, Group representations, λ-rings and the J-homomorphism, Topology 8 (1969), 253-297.

[2] A.Dold, Fixed point indices of iterated maps. Preprint, Forschungsinstitut für Mathematik ETH Zurich, February 1983.

[3] D.Knutson, λ-rings and the Representation Theory of the Symmetric Group, Springer LNM 308 (1973).

[4] A.Liulevicius, Symmetric products of cycles, Max Planck Institut für Mathematik, Bonn, 1983.

[5] A.Liulevicius and M.Özaydın, Duality in Symmetric Products of Cycles, Preprint, University of Chicago, June 1985.

Department of Mathematics
The University of Chicago
Eckhart Hall
5734 University Avenue
Chicago, IL 60637 U S A

and

Department of Mathematics
University of Wisconsin
Van Vleck Hall
480 Lincoln Drive
Madison, WI 53706 U S A

CYCLIC HOMOLOGY AND IDEMPOTENTS IN GROUP RINGS

Zbigniew Marciniak
Warsaw, Poland

We present here an algebraic approach to the Burghelea Theorem on cyclic homology of group rings. The original proof involves arguments from the theory of bundles with S^1-action and it is not easily accesible to algebraists. As an application we offer a new criterion for non-existence of idempotents in a group ring. In particular, we give a completely different proof of Formanek's Theorem on polycyclic-by-finite groups.

Cyclic homology

Let k be a commutative ring with 1. For an associative k-algebra Λ with 1 one can consider the Hochschild homology of Λ which is, by definition, the homology of the chain complex:

$$\mathcal{H} : 0 \longleftarrow \Lambda \xleftarrow{b_1} \Lambda^2 \xleftarrow{b_2} \Lambda^3 \xleftarrow{b_3} \cdots ,$$

where $\Lambda^n = \Lambda \otimes_k \cdots \otimes_k \Lambda$ (n times) and $b_n : \Lambda^{n+1} \to \Lambda^n$ is given by

$$b_n(a_o \otimes \ldots \otimes a_n) = \sum_{i=o}^{n-1} (-1)^i a_o \otimes \ldots \otimes a_i a_{i+1} \otimes \ldots \otimes a_n + (-1)^n a_n a_o \otimes a_1 \otimes \ldots \otimes a_{n-1} .$$

A. Connes observed that the above construction, when suitably modified, leads to interesting applications. The resulting homology is called "cyclic homology" and the most useful definition seems to be the following [6].

In addition to the chain complex $\mathcal{H}$ we consider its modified version

$$\mathcal{H}' : 0 \longleftarrow \Lambda \xleftarrow{b'_1} \Lambda^2 \xleftarrow{b'_2} \Lambda^3 \xleftarrow{b'_3} \cdots$$

with

$$b'_n(a_o \otimes \ldots \otimes a_n) = \sum_{i=o}^{n-1} (-1)^i a_o \otimes \ldots \otimes a_i a_{i+1} \otimes \ldots \otimes a_n .$$

This complex can be contracted via $s : \Lambda^n \longrightarrow \Lambda^{n+1}$, $s(a_o \otimes \ldots \otimes a_{n-1}) = 1 \otimes a_o \otimes \ldots \otimes a_{n-1}$. We put the complexes $\mathcal{H}$ and $\mathcal{H}'$ together to form a double complex

$$D(\Lambda) : \mathcal{H} \xleftarrow{1-T} \mathcal{H}' \xleftarrow{N} \mathcal{H} \xleftarrow{1-T} \mathcal{H}' \xleftarrow{N} \cdots$$

where T and N are chain maps defined as follows:

$T_n : \Lambda^{n+1} \to \Lambda^{n+1}$ is the cyclic permutation of coordinates:

$$T_n(a_o \otimes \ldots \otimes a_n) = (-1)^n a_n \otimes a_o \otimes \ldots \otimes a_{n-1} ;$$

$$N_n = T_n^{(o)} + T_n^{(1)} + \ldots + T_n^{(n)} , \quad T_n^{(k)} = T_n \circ \ldots \circ T_n \quad (k \text{ times}) .$$

The cyclic homology of Λ is just the homology of the total complex of $D(\Lambda)$:

$$HC_*(\Lambda) = H_*(\mathrm{Tot}\ D(\Lambda)) .$$

We notice for further reference that the double complex $D(\Lambda)$ has a shift map $S : D(\Lambda) \to D(\Lambda)$ which sends the first two columns $\mathcal{H}$, $\mathcal{H}'$ of $D(\Lambda)$ to zero and shifts the other columns two places to the left. Consequently, we obtain shift maps

$$S : HC_n(\Lambda) \to HC_{n-2}(\Lambda) \quad \text{for all } n \geq 2 .$$

Group rings

Among the algebras which are of interest for topologists we have group algebras kG, defined for any group G. D. Burghelea skilfully used in [2] the theory of circle bundles over the classifying space of G to determine the groups $HC_n(kG)$. To present his result we need some notation.

For a group G let TG denote the set of its conjugacy classes. Let $T_\infty G$ be the subset of those classes, which consist of elements of infinite order. Let $c \in TG$ and $z \in c$. We denote by G_c the quotient group $C_G(z)/\langle z \rangle$ where $C_G(z)$ is the centralizer of z in G. We need the following weak form of Burghelea's result.

Burghelea Theorem

Let G be a group and let k be any commutative ring with unity. Then

$$HC_*(kG) \approx \bigoplus_{c \in T_\infty G} H_*(G_c) \oplus T_* .$$

Here $H_*(G_c)$ stands for the homology of groups with trivial coefficients k. The summand T_* can be completely described in terms of homology of some nice fibrations associated with G. However, for our purposes it is not necessary to go deeper into the structure of T_*.

We gave a purely algebraic proof of the precise formulation of the

Burghelea Theorem in the case when k is a field of char 0 in [7]. In this paper we offer an application.

Idempotents

One way of studing a k-algebra Λ is to investigate its idempotents: $e = e^2 \in \Lambda$. If $e \neq 0,1$ then it splits Λ into a direct sum $\Lambda = \Lambda e \oplus \Lambda(1-e)$ of left Λ-modules.

Any idempotent $e \in \Lambda$ generates a sequence of special elements $e_n \in HC_{2n}(\Lambda)$ for all $n \geq 0$ (see [4, Prop. 14, Ch. II]). They can be defined in the following way.

Let $e^{(i)} = e \otimes \ldots \otimes e$ (i times) belong to Λ^i. Set $\alpha_1 = 1$ and for $i \geq 1$

$$\alpha_{2i} = \frac{(-1)^{i-1}}{2} \cdot \frac{(2i)!}{i!}, \quad \alpha_{2i+1} = (-1)^i \cdot \frac{(2i)!}{i!}.$$

All these numbers are integers. Consider $e_n = \sum_{i=1}^{2n+1} \alpha_i e^{(i)} \in \operatorname{Tot} D(\Lambda)_{2n}$.

A straightforward calculation shows that e_n are cycles of the chain complex $\operatorname{Tot} D(\Lambda)$. It is also clear from the definition of the shift S that we have $S(e_{n+1}) = e_n$ for all $n \geq 0$.

From now on we assume that k is a field of characteristic 0. Let Λ be a group algebra kG. It is easy to produce an idempotent $e \in kG$ once you have an element $g \in G$ of finite order n : we set $e = 1/n(1+g+\ldots+g^{n-1}) \in kG$. Another method of producing idempotents is described in [5] but it still requires the existence of torsion in G. Moreover, we have the following long-standing.

The Idempotent Conjecture

If a group G is torsion free then its group algebra kG has no idempotents different from 0 and 1. We will prove the following result.

Main Theorem

Let G be a torsion free group and let k be a field of characteristic zero. If for every conjugacy class $c \in TG\backslash\{1\}$ there exists a number $n_c > 0$ such that $H_{2n_c}(G_c;k) = 0$, then the group algebra kG contains only two idempotents: 0 and 1.

The basic tool in the work with idempotents in kG is the trace function $tr : kG \longrightarrow k$ given by $tr(\sum e(g)g) = e(1)$. It is very efficient because of the following result.

Kaplansky Theorem [8, Thm. 2.1.8]

Let $e = e^2 \in kG$. Then $tr(e) = 0$ implies $e = 0$ and $tr(e) = 1$ implies $e = 1$. ∎

We have also other trace functions on kG. For any $c \in TG$ we have a function $t_c : kG \longrightarrow k$ defined as $t_c(e) = \sum\{e(g) \mid g \in c\}$. In particular $t_{\{1\}} = tr$. These functions are substitutes for characters from finite group theory and they indeed share some of their properties.

As the augmentation homomorphism $\epsilon : kG \longrightarrow k$ is a ring homomorphism, we have

$$\sum_{c \in TG} t_c(e) = \epsilon(e) = 0 \text{ or } 1 .$$

Thus, by the Kaplansky Theorem, the Idempotent Conjecture is equivalent to saying that if G is torsion free and $e = e^2 \in kG$ then $t_c(e) = 0$ for all $c \in TG\setminus\{1\}$.

Proof of the Main Theorem:

Let G be a torsion free group and let e be an idempotent in kG. As remarked earlier, e generates a sequence $\{e_n\}$ of elements lying in $HC_{2n}(kG)$ for $n = 0,1,\ldots,$ such that $S(e_{n+1}) = e_n$.

By the Burghelea Theorem we have

$$HC_o(kG) \approx \bigoplus_{c \in T_\infty G} H_o(G_c) \oplus T_o .$$

From the explicit description of the above isomorphism given in [7] it is easy to see that the element $e_o \in HC_o(kG)$ corresponds to the vector of its traces $t_c(e)$. Further, from the proof of the Burghelea Theorem presented there it is clear that the shift S respects the direct sum decomposition

$$HC_*(kG) \approx \bigoplus_{c \in T_\infty G} H_*(G_c) \oplus T_* .$$

Thus, for any $c \in TG\setminus\{1\} = T_\infty G$ and for any $n \geq 1$ we have a homomorphism $S_c : H_{2n}(G_c) \longrightarrow H_{2n-2}(G_c)$.

Fix now a conjugacy class $c \in TG\setminus\{1\}$ - For any $n \geq 0$ let $x_n \in H_{2n}(G_c)$ be the coordinate of e_n corresponding to c. Then we have $S_c(x_{n+1}) = x_n$ and $x_o = t_c(e)$.

Suppose there is an integer $n_c > 0$ such that $H_{2n_c}(G_c) = 0$. Then $x_{n_c} = 0$ and hence $t_c(e) = 0$. If the same holds for all $c \in TG\setminus\{1\}$ then all traces $t_c(e)$ vanish and e must be 0 or 1 ∎

<u>Corollary:</u> (Compare with Thm. 2.3.10 in [8])

If G is a torsion free polycyclic-by-finite group and k is a field of char 0 then kG has no idempotents different from 0 and 1.

<u>Proof:</u>

Let h be the Hirsch number of G. It is well known that the cohomological dimension of G is equal to h [1]. Now, for any $c \in TG$ the group G_c is also polycyclic-by-finite and its Hirsch number does not exceed h. Consequently, for $2n > h$ we have $H_{2n}(G_c) = 0$ (we have coefficients from a field of characteristic zero!) and so the Main Theorem can be applied. ∎

<u>Remark:</u>

Whatever we have said about idempotents holds as well for finitely generated projective modules, as cyclic homology is Morita invariant. The obvious generalization of the Main Theorem is left to the reader.

<u>References</u>

[1] K. Brown: Cohomology of Groups, Springer 1982, New York

[2] D. Burghelea: The cyclic homology of the group rings, Comm. Math. Helv. 60 (1985), 354-365

[3] H. Cartan, S. Eilenberg: Homological Algebra, Princeton 1956

[4] A. Connes: Non Commutative Differential Geometry, Publ. Math. IHES 62 (1986), 257-360

[5] D. Farkas, Z. Marciniak: Idempotents in group rings - a surprise, J. Algebra 81, No. 1 (1983), 266-267

[6] J.-L. Loday: Cyclic homology, a survey, to appear in Banach Center Publications

[7] Z. Marciniak: Cyclic homology of group rings, to appear in Banach Center Publications

[8] D.S. Passman: The Algebraic Structure of Group Rings, Wiley 1977

$\mathbb{Z}_2$ surgery theory and smooth involutions on homotopy complex projective spaces

Mikiya Masuda

Department of Mathematics, Osaka City University, Osaka 558, Japan

§0. Introduction

Let a group act smoothly on a manifold M. One of the fundamental problems in transformation groups is to study relations between the global invariants of M (e.g. Pontrjagin classes) and invariants of the fixed point set. The Atiyah-Singer index theorem gives profound answers to this problem, which are necessary conditions of the action. Conversely it is interesting to ask if those are sufficient conditions. In other words, to what extent are there actions realizing such relations ? In this paper we deal with the realization problem of this kind for smooth involutions on homotopy complex projective spaces.

Let X be a $2(N-1)$-dimensional closed smooth manifold homotopy equivalent to the complex projective space $P(\mathbb{C}^N)$. We call such X a homotopy $P(\mathbb{C}^N)$ briefly. Suppose that X supports a smooth involution, that is to say, an order two group (denoted by G throughout this paper) acts on X. Then Bredon-Su's Fixed Point Theorem (see p.382 of [B]) describes the cohomological nature of the fixed point set X^G of X. It depends on the number of connected components of X^G :

<u>Type 0</u>. X^G is empty,

<u>Type I</u>. X^G is connected and has the same cohomology ring as the real projective space $P(\mathbb{R}^N)$ of dimension $N-1$ with $\mathbb{Z}_2$ coefficients,

<u>Type II</u>. X^G consists of two connected components F_1, F_2 and each F_i has the same cohomology ring as $P(\mathbb{C}^{N_i})$ with $\mathbb{Z}_2$ coefficients. Here $N_1+N_2 = N$. Moreover the restriction map from $H^*(X;\mathbb{Z}_2)$ to $H^*(F_i;\mathbb{Z}_2)$ is surjective. When the minimum of N_i-1 (= dim $F_i/2$) is ℓ, we say more specifically that the involution is of Type II_ℓ.

Type I involutions are fairly well understood due to studies of Kakutani [K], Dovermann-Masuda-Schultz [DMSc], and Stolz [S]. In a way made precise in [DMSc] we may say that almost all homotopy $P(\mathbb{C}^N)$ admit Type I involutions. As a matter of fact no homotopy $P(\mathbb{C}^N)$ has been discovered which does not admit a Type I involution.

In this paper we are concerned with Type II involutions. To illustrate our results we pose

<u>Definition</u>. Let x be a generator of $H^2(X;\mathbb{Z})$. For a fixed component F_i $(i = 1, 2)$ of dimension $2(N_i-1)$, we restrict x^{N_i-1} to F_i and evaluate it on a fundamental class of F_i. We denote the value by $D(F_i)$ and call it the <u>defect</u> of F_i. Due to choices of a genarator x and an orientation of F_i, $D(F_i)$ is defined only up to sign. The defects $D(F_i)$ are odd because the restriction map from $H^*(X;\mathbb{Z}_2)$ to $H^*(F_i;\mathbb{Z}_2)$ is surjective.

Clearly the set $\{D(F_1), D(F_2)\}$ is an invariant of the G action. It is a G homotopy invariant. For instance, if X is G homotopy equivalent to $P(\mathbb{C}^N)$ with a linear Type II involution, then $D(F_i) = \pm 1$. Therefore one may regard defects as invariants which measure the exoticness of actions. The concept of defect is relevant for general $\mathbb{Z}_m$ actions with the same definition. The reader is referred to [HS], [DM], [DMSu], [D2], [M3], [We] in this direction.

The Atiyah-Singer index theorem for Dirac operators associated with $Spin^c$ structures implies that the defects are related to the characteristic classes of X, F_i and those of the normal bundles of F_i. It gives many rather complicated integrality conditions, from which we deduce a neat congruence between the defects and the first Pontrjagin class $p_1(X)$ of X. In fact Theorem 4.3 says that if we choose suitable signs of $D(F_i)$, then the following congruence (*) holds :

$$(*) \qquad D(F_1) + D(F_2) \equiv 4k(X) \pmod 8,$$

where $k(X)$ is the integer determined by $p_1(X) = (N+24k(X))x^2$ (see Lemma 4.1). As a consequence (Corollary 4.4) one can conclude that $k(X)$ must be even if X is G homotopy equivalent to $P(\mathbb{C}^N)$ with a linear Type II involution (remember that $D(F_i) = \pm 1$ under this assumption).

We regard (*) as a guidepost for our construction of Type II involutions. One of our main results (Theorem 5.1) says that (*) is also a sufficient condition for Type $II_{N/2-1}$ involutions in case $N = 4$ or 8. The diffeomorphism types of homotopy $P(\mathbb{C}^4)$'s are classified by their first Pontrjagin classes (equivalently, the integer $k(X)$) and there are infinitely many sets $\{D(F_1), D(F_2)\}$ satisfying the congruence (*) for each $k(X)$. Hence Theorem 5.1 implies

<u>Corollary 5.3</u>. Every homotopy $P(\mathbb{C}^4)$ admits <u>infinitely many</u> Type II_1 involutions distinguished by the defects. In particular they are not G homotopy equivalent to each other.

This is an improvement of Theorem B (1) of [M1]. For a general N dvisible by 4, a rather weaker result than that of Theorem 5.1 is obtained (Theorem 5.4). For the other values of N we only see that infinitely many non-standard homotopy $P(\mathbb{C}^N)$ admit Type II involutions with non-standard fixed point sets (Theorems 5.6, 5.7).

As for the method, we apply G surgery theory developed by Petrie and Dovermann. It is a useful tool to construct G manifolds in the same homotopy (or G homotopy) type as a given G manifold Z. In fact we take $P(\mathbb{C}^N)$ with a linear Type II involution as Z. When we apply G surgery theory, we must work out two things. One is to produce a G normal map. We construct a nice G quasi-equivalence in §3, which together with the G transversality theorem produces a G normal map. The other is to analyse G surgery obstructions. In all but one case, we can compute those obstructions by using G signature and Sullivan's Characteritic Variety Formula. If $\dim Z \equiv 2 \pmod 4$, then the obstruction in an L group $L_{\dim Z}(\mathbb{Z}[G],1) \cong \mathbb{Z}_2$ is treated differently. We show the existence of a framed G manifold with the Kervaire invariant one in $L_{\dim Z}(\mathbb{Z}[G],1)$, which serves to kill the obstruction.

This paper is organized as follows. In §1 we review G surgery theory and in §2 we construct framed G manifolds with the Kervaire invariant one. A nice G quasi-equivalence due to Petrie is exhibited in §3. In §4 we apply the Atiyah-Singer index theorem to deduce congruence (*). Type II involutions are constructed in §5. In Appendix we apply the ordinary surgery theory to produce Type II involutions, where the gap hypothesis (see §1) is unnecessary but the fixed point sets are standard ones.

Throughout this paper we always work in the C^∞ category ; so the word "smooth" will be omitted.

Notations. Here are some conventions used in this paper :

G : an order two group.

$\mathbb{Z}_2$: the ring $\mathbb{Z}/2\mathbb{Z} = \{0, 1\}$.

$\mathbb{C}^{m,n}$ (resp. $\mathbb{R}^{m,n}$) : $\mathbb{C}^{m+n}$ (resp. $\mathbb{R}^{m+n}$) with the involution defined by

$$(z_1, \dots, z_{m+n}) \longrightarrow (z_1, \dots, z_m, -z_{m+1}, \dots, -z_{m+n}).$$

Such $\mathbb{C}^{m,n}$ is sometimes denoted by $\mathbb{C}_+^{m,n}$ to distinguish it from the space $\mathbb{C}^{m+n}$ with the involution defined by

$$(z_1, \dots, z_{m+n}) \longrightarrow (-z_1, \dots, -z_m, z_{m+1}, \dots, z_{m+n}).$$

The latter G space is denoted by $\mathbb{C}_-^{m,n}$.

For a complex (or real) representation V (with a metric)

$S(V)$ (resp. $D(V)$) : the unit sphere (resp. disk) of V,

$P(V)$: the space consisting of complex (or real) lines through the origin in V.

In concluding this introduction I would like to express my hearty thanks to Professor T. Petrie for suggesting this problem to me and for valuable long discussions during his visit to Japan in the summer of 1983. This paper is an outcome of discussions with him.

§1. Review of G surgery theory

G surgery theory is a tool to construct a G manifold in the same homotopy (or G homotopy) type as a given (connected) G manifold Z. For a general finite group G, we must impose complicated technical conditions on Z so that G surgery theory is applicable. But in our case G is of order two; so those conditions are simplified as follows. Let $\dim Z^G$ denote each dimension of connected components of Z^G. Then

(1.1) $\dim Z \geqq 5$

(1.2) $\dim Z^G \neq 0, 3, 4$

(1.3) (Gap hypothesis) $2\dim Z^G < \dim Z$.

For simplicity we require in addition :

(1.4) Z and each component of Z^G are simply connected,

(1.5) the action of G preserves an orientation on Z.

Throughout this section and the next section the G manifold Z will be assumed to satisfy these five conditions unless otherwise stated.

Roughly speaking G surgery theory consists of three concepts in our construction :

I. G quasi-equivalences or G fiber homotopy equivalences

II. G transversality

III. G normal maps and G surgery.

Here the meaning of these terms will be clarified below little by little. According to these concepts G surgery theory is divided into three steps. In the following the (fiber) degree of a map has a sense up to sign.

<u>First</u> we set up a G quasi-equivalence or a G fiber homopoty equivalence $\hat{\omega} : \hat{V} \longrightarrow \hat{U}$ between G vector bundles over Z. Here a G quasi-equivalence means that $\hat{\omega}$ is a proper fiber preserving G map of degree one on each fiber, and a G fiber homotopy equivalence is a G quasi-equivalence such that the restricted map $\hat{\omega}^G : \hat{V}^G \rightarrow \hat{U}^G$ to the fixed point sets is also of degree one on each fiber (note that this implies the existence of a G fiber homotopy inverse in a stable sense, see §13, Chapter I of [PR]). A G quasi-equivalence (resp. a G fiber homotpoy equivalence) is used to produce a G manifold in the same homotopy (resp. G homotopy)

type as the given G manifold Z.

Next we convert $\widehat{\omega}$ into a G map h transverse to the zero section $Z \subset \widehat{U}$ via a proper G homotopy. In a general setting we encounter obstructions to finding it at this stage. In our case, however, it is always possible because those obstructions identically vanish under the gap hypothesis (see Corollary 4.17 of [P2]). The G transverse map h produces a triple $\kappa = (W,f,b)$ where $W = h^{-1}(Z)$, $f = h|W : W \longrightarrow Z$ and $b : TW \underset{s}{\cong} f^{*}(TZ+\widehat{V}-\widehat{U})$ (the notation $\underset{s}{\cong}$ denotes that b is a stable G vector bundle isomorphism). Here we may assume $f_{*}^{G} : \pi_0(W^G) \longrightarrow \pi_0(Z^G)$ is bijective, if necessary, by doing 0-surgery. Moreover we should notice that

$$(1.6)\quad \begin{aligned} &\text{the degree of } f = \text{the fiber degree of } \widehat{\omega} = 1, \\ &\text{the degree of } f^G = \text{the fiber degree of } \widehat{\omega}^G \\ &\quad = \text{an \underline{odd} integer at each component of } Z^G \text{ (by Smith theory).} \end{aligned}$$

With these observations

Definition. A G normal map is a triple $\kappa = (W,f,b)$ such that

(i) $f : W \longrightarrow Z$ is a G map of degree one,

(ii) $f_{*}^{G} : \pi_0(W^G) \longrightarrow \pi_0(Z^G)$ is bijective,

(iii) $f^G : W^G \longrightarrow Z^G$ is of odd degree at each component of Z^G,

(iv) $b : TW \underset{s}{\cong} f^{*}(TZ+E)$ for some $E \in KO_G(Z)$.

At a final step we perform G surgery on the G normal map κ via a G normal cobordism to produce a new G normal map $\kappa' = (W',f',b')$ with $f' : W' \longrightarrow Z$ a homotopy (or a G homotopy) equivalence.

To achieve the final step we first do surgery on the G fixed point set W^G and then on the G free part $W-W^G$. Unfortunately we encounter an obstruction at each procedure. The primary one is the surgery obstruction to coverting $f^G : W^G \longrightarrow Z^G$ into a $\mathbb{Z}_2$ homology (resp. a homotopy, if f^G is of degree one) equivalence. This is denoted by $\sigma_G(f)$. Since Z^G may be disconnected, it lies in a sum of L groups :

$$\sigma_G(f) \in L_{\dim Z^G}(\mathbb{Z}_{(2)}[1]) \quad (\text{resp. } L_{\dim Z^G}(\mathbb{Z}[1]))$$

where $\mathbb{Z}_{(2)}$ denotes the localized ring of $\mathbb{Z}$ by the ideal generated by 2 and the orientation homomorphisms from $\pi_1(Z^G)$ to $\mathbb{Z}_2$ are omitted in the notation of L groups because they are trivial by (1.5). The reader should note that we must check the vanishing of $\sigma_G(f)$ for each component of Z^G.

When $\dim Z^G \equiv 2 \pmod 4$, the above L groups are isomorphic to $\mathbb{Z}_2$ componentwise. The values of $\sigma_G(f)$ via the isomorphisms are called the Kervaire invariants and denoted by $c(f^G)$. The computation of $c(f^G)$ is done in [M2] for G normal maps treated later.

When $\dim Z^G \equiv 0 \pmod 4$ and f^G is of degree one, $L_{\dim Z^G}(\mathbb{Z}[1])$ is isomorphic to $\mathbb{Z}$ componentwise. The values of $\sigma_G(f)$ via the isomorphisms are componentwise differences $\text{Sign } W^G - \text{Sign } Z^G$ of signatures of W^G and Z^G.

Suppose $\sigma_G(f)$ identically vanishes; so we may assume f^G is a $\mathbb{Z}_2$ homology (resp. a homotopy, if f^G is of degree one) equivalence. Then we do surgery on $W-W^G$ equivariantly to convert f into a homotopy (resp. a G homotopy) equivalence. We again encounter an obstuction. In fact, the vanishing of $\sigma_G(f)$ allows us to define the obstruction

$$\sigma(f) \in L_{\dim Z}(\mathbb{Z}[G]).$$

When $\dim Z \equiv 2 \pmod 4$, $L_{\dim Z}(\mathbb{Z}[G])$ is isomorphic to $\mathbb{Z}_2$ (see

§13A of [W1]). But this time there is no helpful formula to estimate $\sigma(f)$ in terms of $\kappa = (W,f,b)$ and Z. The next section is devoted to this problem.

Summing up the content of this section, we have

<u>Proposition 1.7</u>. Let Z be a connected G manifold satisfying (1.1) - (1.5) and let $\kappa = (W,f,b)$ $f : W \longrightarrow Z$ be a G normal map with $b : TW \underset{s}{\cong} f^*(TZ+E)$ for some $E \in KO_G(Z)$. Suppose

(i) $\dim Z \equiv \dim Z^G \equiv 2 \pmod 4$,

(ii) $c(f^G) = 0$ (componentwise),

(iii) $\sigma(f) = 0$ in $L_{\dim Z}(\mathbb{Z}[G]) \cong \mathbb{Z}_2$.

Then there is a G normal map $\kappa' = (W',f',b')$ $f' : W' \longrightarrow Z$ such that

(1) f' is a homotopy (a G homotopy, if f^G is of degree one) equivalence,

(2) $b' : TW' \underset{s}{\cong} f'^*(TZ+E)$.

<u>Proposition 1.8</u>. Let Z, κ and E be the same as in Proposition 1.7. Suppose

(i) $\dim Z \equiv 2 \pmod 4$ and $\dim Z^G \equiv 0 \pmod 4$,

(ii) $\operatorname{Sign} W^G - \operatorname{Sign} Z^G = 0$ (componentwise),

(iii) $\sigma(f) = 0$ in $L_{\dim Z}(\mathbb{Z}[G]) \cong \mathbb{Z}_2$.

Then the same conclusion as Proposition 1.7 holds.

§2. Framed G manifolds with the Kervaire invariant one

In this section we will show the existence of framed G manifolds with the Kervaire invariant one. This enables us to kill

$\sigma(f)$ (or $c(f^G)$) in Propositions 1.7, 1.8, if necessary, by doing equivariant conneceted sum.

A framed G manifold can be naturally regarded as a G normal map with a sphere as the target manifold; so we state our results in terms of a G normal map. We first treat low dimensional cases.

Theorem 2.1. For $m = 2$ or 4 there is a G normal map $\kappa_m = (W_m, f_m, b_m)$ $f_m : W_m \longrightarrow S(\mathbb{R}^{2m-1,2m})$ such that

(1) $W_m{}^G = S(\mathbb{R}^m) \times S(\mathbb{R}^m)$,

(2) $c(f_m{}^G) = 1$ in $L_{2m-2}(\mathbb{Z}[1]) \cong \mathbb{Z}_2$,

(3) TW_m is a trivial G vector bundle.

This theorem is obtained by making the following well known fact equivariant.

Proposition 2.2. For $m = 1, 2, 4$ there is a normal map $\kappa_m{}^0 = (W_m{}^0, f_m{}^0, b_m{}^0)$ $f_m{}^0 : W_m{}^0 \longrightarrow S(\mathbb{R}^{4m-1})$ such that

(1) $W_m{}^0 = S(\mathbb{R}^{2m}) \times S(\mathbb{R}^{2m})$; hence $TW_m{}^0$ is trivial,

(2) $c(f_m{}^0) = 1$ in $L_{4m-2}(\mathbb{Z}[1]) \cong \mathbb{Z}_2$.

We shall recall the explict construction of $\kappa_m{}^0$. The map $f_m{}^0$ is defined by collapsing the exterior of an open ball in $W_m{}^0$ to a point, and $b_m{}^0$ is the trivialization of $T(S(\mathbb{R}^{2m}) \times S(\mathbb{R}^{2m}))$ defined as follows. Remember that $\mathbb{R}^{2m}$ admits a mutiplicative structure defined by

$$(q_1, q_2)(q_1', q_2') = (q_1 q_1' - \bar{q}_2' q_2, q_2' q_1 + q_2 \bar{q}_1')$$

where (q_1, q_2) and (q_1', q_2') are ordered pairs of real numbers if $m = 1$ (complex numbers if $m = 2$ or quaternion numbers if $m = 4$) and $-$ denotes the usual conjugation. This equips $S(\mathbb{R}^{2m}) \times S(\mathbb{R}^{2m})$ with a multiplicative structure. Take a framing on

$S(\mathbb{R}^{2m})\times S(\mathbb{R}^{2m})$ at a point and transmit it to the other points using the multiplication. This defines the desired trivialization.

Proof of Theorem 2.1. Define an involution by $(q_1, q_2) \longrightarrow (q_1, -q_2)$. This preserves the multiplication and the length of (q_1, q_2). Hence $S(\mathbb{R}^{2m})$ inherits the involution and so does W_m^0 via the diagonal action. This is the required G manifold W_m. The G map f_m is defined similarly to f_m^0. But this time we need to take a G invariant open ball around a point of W_m^G. The definition of b_m is the same as b_m^0. It is immediate from our construction that $\kappa_m^G = \kappa_{m/2}^0$ for $m = 2$ or 4. This together with Proposition 2.2 proves the theorem. Q.E.D.

One can also use the normal map κ_m^0 to kill the secondary surgery obstruction $\sigma(f)$. In fact, given a G normal map $\kappa = (W,f,b)$ $f : W \longrightarrow Z$ with $\dim Z = 4m-2$ and $\sigma(f) = 1$, then we do connected sum of κ and (two copies of) κ_m^0 equivariantly away from W^G to obtain a new G normal map $\kappa = (W',f',b')$ $f' : W' \longrightarrow Z$. Here recall that the inclusion map $: 1 \longrightarrow G$ induces an isomorphism $L_{4m-2}(\mathbb{Z}[1]) \longrightarrow L_{4m-2}(\mathbb{Z}[G])$. This and the additivity of the Kervaire invariant under connected sum mean that

$$\sigma(f') = \sigma(f) + c(f_m^0) = 1 + 1 = 0.$$

Now we are in a position to prove

Theorem 2.3. Let $m = 2$ or 4. If we are given a G normal map $\kappa = (W,f,b)$ $f : W \longrightarrow Z$ such that

(i) $\dim Z = 4m-2$,

(ii) $\dim Z^G = 2m-2$ for each component of Z^G,

(iii) $b : TW \underset{s}{\cong} f^*(TZ+E)$ for some $E \in KO_G(Z)$,

then there is a G normal map $\kappa' = (W',f',b')$ $f' : W' \longrightarrow Z$ such that

(1) f' is a homotopy (or a G homotopy, if f^G is of degree one) equivalence,

(2) $b' : TW' \underset{s}{\cong} f'^*(TZ+E)$.

<u>Remark</u>. κ' is not necessarily G normally cobordant to κ.

<u>Proof</u>. Since $L_{2m-2}(\mathbb{Z}_{(2)}[1]) \cong L_{2m-2}(\mathbb{Z}[1]) \cong \mathbb{Z}_2$ and the degree of f^G is odd, the primary obstruction $\sigma_G(f) = c(f^G)$ can be killed, if necessary, by doing equivariant connected sum with κ_m at fixed points of W and Z. As for the secondary surgery obstruction, the observation preceding this theorem shows how to kill it. Finally we note that E in (iii) is unchanged through these connected sum operations because TW_m and TW_m^0 are trivial G vector bundles. Q.E.D.

Now we proceed to higher dimensional case. It is known that there is no <u>closed</u> framed manifold with the Kervaire invaiant one except dimensions 2^n-2 ; so we are obliged to weaken the results.

<u>Theorem 2.4</u>. For a positive integer $m \neq 1, 2, 4$ there is a G normal map $\kappa_m = (W_m, f_m, b_m)$ $f_m : W_m \longrightarrow S(\mathbb{R}^{2m-1,2m})$ such that

(1) W_m^G is diffeomorphic to $S(\mathbb{R}^{2m-1})$,

(2) TW_m is a stably trivial G vector bundle,

(3) f_m^G is a homotopy equivalence (hence $\sigma_G(f_m) = 0$),

(4) $\sigma(f_m) = 1$ in $L_{4m-2}(\mathbb{Z}[G]) \cong \mathbb{Z}_2$.

This time we use the following fact in place of Proposition 2.2.

<u>Proposition 2.5</u>. For $m \neq 1, 2, 4$ there is a normal map $\kappa_m^0 = (W_m^0, f_m^0, b_m^0)$ $f_m^0 : (W_m^0, \partial W_m^0) \longrightarrow (D(\mathbb{R}^{4m-2}), S(\mathbb{R}^{4m-2}))$ such that

(1) ∂f_m^0 is a homeomorphism,

(2) $c(f_m^0) = 1$ in $L_{4m-2}(\mathbb{Z}[1]) \cong \mathbb{Z}_2$,

(3) TW_m^0 is trivial.

<u>Remark</u>. A choice of b_m^0 does not effect the value of $c(f_m^0)$ provided $m \neq 1, 2, 4$.

An explict construction of κ_m^0 is as follows. Let δ be a small real number. Then W_m^0 is defined by

(2.6) $W_m^0 = \{(z_1, .. , z_{2m}) \in \mathbb{C}^{2m} \mid z_1^3 + z_2^2 + .. + z_{2m}^2 = \delta\} \cap D(\mathbb{C}^{2m})$.

Pinch the complement of a collar boundary in W_m^0 to a point. Since the boundary of W_m^0 is known to be homeomorphic to $S(\mathbb{R}^{4m-2})$, this defines the desired map f_m^0. b_m^0 is defined as a trivialization of TW_m^0.

<u>Proof of Theorem 2.4</u>. Since δ is real, the complex conjugation map : $(z_1, .. ,z_{2m}) \longrightarrow (\bar{z}_1, .. ,\bar{z}_{2m})$ preserves W_m^0 ; so this defines an involution τ on ∂W_m^0. One can easily see that τ reverses an orientation on ∂W_m^0 and $(\partial W_m^0)^\tau$ is diffeomorphic to $S(\mathbb{R}^{2m-1})$.

Now prepare a copy W_m^{0*} of W_m^0 and denote by z^* the corresponding point of W_m^{0*} to $z \in W_m^0$. We glue W_m^0 and W_m^{0*} along the boundary by identifying z^* with τz for all $z \in \partial W_m^0$. The resulting space is a closed and orientable manifold. We define an involution on it by sending z to z^* and z^* to z, which is compatible with the identification because τ is of order two. This is the required G manifold W_m. This construction is due to Lopez de Medrano [L] p.28. The action of G is orientation preserving as τ is orientation reversing, and W_m^G coincides with $(\partial W_m^0)^\tau$, which verifies (1).

The proof of (2) is as follows. Since W_m^0 is a submanifold of

$D(\mathbb{C}^{2m})$ and the involution on $\partial W_m^{\ 0}$ comes from the complex conjugatin map on $S(\mathbb{C}^{2m})$, we can regard W_m as a closed G submanifold of a G sphere $D(\mathbb{C}^{2m}) \cup_\tau D(\mathbb{C}^{2m})^* = S$. Then it is easy to see that the G normal bundle of W_m in S is isomorphic to $W_m \times \mathbb{R}^{1,1}$ and that TS is a stably trivial G vector bundle. This verifies (2).

We define the G map f_m by collapsing the exterior of an open invariant ball around a point of $W_m^{\ G}$ to a point. Then (3) is clear.

b_m is defined as a stable equivariant trivialization of TW_m. By Proposition 2.5 $c(f_m^{\ 0}) = 1$ provided $m \neq 1, 2, 4$. On the other hand, as indicated before, the inclusion map : $1 \longrightarrow G$ induces an isomorphism : $L_{4m-2}(\mathbb{Z}[1]) \longrightarrow L_{4m-2}(\mathbb{Z}[G])$. The above geometric construction exactly corresponds to this algebraic isomorphism ; so (4) follows. Q.E.D.

As a consequemce of Theorem 2.4 we have

<u>Corollary 2.7</u>. Let $m \neq 1, 2, 4$. Let $\kappa = (W,f,b)$ $f : W \longrightarrow Z$ be a G normal map such that

(i) $\dim Z = 4m-2$,

(ii) $\dim K = 2m-2$ for <u>some</u> connected component K of Z^G,

(iii) $b : TW \underset{s}{\cong} f^*(TZ+E)$ for some $E \in KO_G(Z)$.

If $\sigma(f^G) = 0$, then there is a G normal map $\kappa' = (W',f',b')$ $f' : W' \longrightarrow Z$ such that

(1) f' is a homotopy (a G homotopy, if f^G is of degree one) equivalence,

(2) $b' : TW' \underset{s}{\cong} f'^*(TZ+E)$.

§3. Construction of G quasi-equivalences

In this section we use the idea of Petrie (see §12, Chapter 3 of [PR] or §2 of [MaP]) to construct explict and nice G quasi-equivalences (or G fiber homotopy equivalences) over $P(\mathbb{C}^{m,n})$. A general method to produce G fiber homotopy equivalences by means of Adams operations is discussed in [P3].

Suppose that we are given a proper $S^1 \times G$ map $\omega : V \longrightarrow U$ of degree one between $S^1 \times G$ representations. Then we associate a proper fiber preserving G map with a principal $S^1 \times G$ bundle $S(\mathbb{C}^{m,n}_{\varepsilon}) \longrightarrow P(\mathbb{C}^{m,n})$:

$$\begin{array}{ccc} \hat{\omega}_{\varepsilon} : \hat{V}_{\varepsilon} = S(\mathbb{C}^{m,n}_{\varepsilon}) \times_{S^1} V & \longrightarrow & \hat{U}_{\varepsilon} = S(\mathbb{C}^{m,n}_{\varepsilon}) \times_{S^1} U \\ & \searrow \quad \swarrow & \\ & P(\mathbb{C}^{m,n}) & \end{array}$$

where ε denotes + or -. Since ω is of degree one, so is $\hat{\omega}_{\varepsilon}$ on each fiber, i.e. $\hat{\omega}_{\varepsilon}$ is a G quasi-equivalence. This is the desired construction.

Forgetting the G action, it is a fiber homotopy equivalence. We shall denote it by $\hat{\omega} : \hat{V} \longrightarrow \hat{U}$ by dropping the suffix ε.

Here are two interesting examples used later. We refer the reader to [MeP] for a general construction of ω.

<u>Example 3.1</u>. Let t denote the standard complex 1-dimensional representation of S^1 and t^k the k fold tensor product of t over $\mathbb{C}$. Let p and q be relatively prime integers greater than one. We set

$$U^{p,q} = t + t^{pq}, \quad V^{p,q} = t^p + t^q$$

Choosing positive integers a and b such that $-ap+bq = 1$, we define a proper $S^1 \times G$ map $\omega^{p,q} : V^{p,q} \longrightarrow U^{p,q}$ by

$$\omega^{p,q}(z_1, z_2) = (\bar{z}_1{}^a z_2{}^b,\ z_1{}^q + z_2{}^p).$$

One can check that $\omega^{p,q}$ is of degree one (see §2 of [MaP] for example).

Putting the trivial G actions on $U^{p,q}$ and $V^{p,q}$, $\omega^{p,q}$ can be regarded as an $S^1 \times G$ map. Since $\omega^{p,q}$ is of degree one, the induced map $\hat{\omega}^{p,q}_{\varepsilon}$ is a G quasi-equivalence. However it is not necessarily a G fiber homotopy equivalence. It depends on the values of p and q. Let us observe the effect for the case $\varepsilon = +$. For the case $\varepsilon = -$ the role of the components $P(\mathbb{C}^m \times 0)$ and $P(0 \times \mathbb{C}^n)$ of $P(\mathbb{C}^{m,n})^G$ is nothing but interchanged.

<u>Case 1</u>. The case where p and q are both odd. In this case one can see

$$(\hat{\omega}^{p,q}_{+})^G \quad \begin{array}{ccccc} (\hat{U}^{p,q}_{+})^G & = & \hat{U}^{p,q}_{+} | P(\mathbb{C}^m \times 0) & \cup & P(0 \times \mathbb{C}^n) \\ \uparrow & & \uparrow & & \uparrow \\ (\hat{V}^{p,q}_{+})^G & = & \hat{V}^{p,q}_{+} | P(\mathbb{C}^m \times 0) & \cup & P(0 \times \mathbb{C}^n), \end{array}$$

where the symbol $|$ denotes the restriction. We know that the fiber degree of $\hat{\omega}^{p,q}$ is one ; so this diagram shows that the fiber degree of $(\hat{\omega}^{p,q}_{+})^G$ is also one. Hence $\hat{\omega}^{p,q}_{+}$ is a G fiber homotopy equivalence.

<u>Case 2</u>. The case where p is even and q is odd. In this case we have

$$(\hat{\omega}^{p,q}_{+})^G \quad \begin{array}{ccccc} (\hat{U}^{p,q}_{+})^G & = & (\hat{U}^{p,q}_{+}) | P(\mathbb{C}^m \times 0) & \cup & S(0 \times \mathbb{C}^n) \times_{S^1} t^{pq} \\ \uparrow & & \uparrow & & \uparrow \\ (\hat{V}^{p,q}_{+})^G & = & (\hat{V}^{p,q}_{+}) | P(\mathbb{C}^m \times 0) & \cup & S(0 \times \mathbb{C}^n) \times_{S^1} t^{p}. \end{array}$$

The fiber degree of $(\hat{\omega}^{p,q}_{+})^G$ over $P(\mathbb{C}^m \times 0)$ is one as before, but that over $P(0\times\mathbb{C}^n)$ is q as is easily seen from the definition of $\omega^{p,q}$. Therefore $\hat{\omega}_+^{p,q}$ is not a G fiber homotopy equivalence in this case.

The same argument as in Case 2 works for the remaining case where p is odd and q is even.

Example 3.2. We take the double of $\omega^{p,q}$ and define an action of G by permuting them :

$$\mu^{p,q} = \omega^{p,q} \oplus \omega^{p,q} : V^{p,q} \oplus V^{p,q} \longrightarrow U^{p,q} \oplus U^{p,q}.$$

$(\hat{\mu}_\varepsilon^{p,q})^G$ over $P(\mathbb{C}^m\times 0)$ (resp. $P(0\times\mathbb{C}^n)$) is isomorphic to $\hat{\omega}^{p,q}$ over $P(\mathbb{C}^m\times 0)$ (resp. $P(0\times\mathbb{C}^n)$). In particular $\hat{\mu}_\varepsilon^{p,q}$ is necessarily a G fiber homotopy equivalence independent of values of p and q; so it has a G fiber homotopy inverse (in a stable sense). We shall denote it by $-\hat{\mu}_\varepsilon^{p,q}$. Hence Whitney sum of h copies of $\hat{\mu}_\varepsilon^{p,q}$, denoted by $h\hat{\mu}_\varepsilon^{p,q}$, has a sense for every integer h.

§4. First Pontrjagin classes and defects

In this section we apply the Atiyah-Singer index theorem for Dirac operators to a homotopy $P(\mathbb{C}^N)$ with a Type II involution and deduce some interesting congruences between the first Pontrjagin classes and the defects defined in the Introduction. This section is independent of G surgery part, so the reader may take a glance at the results (Lemma 4.1, Theorems 4.3, 4.5 and Corollaries 4.4, 4.6) and skip their proofs. The following lemma will be established in the course of the proof of Theorem 4.3.

Lemma 4.1. Let X be a homotopy $P(\mathbb{C}^N)$. Then the first

Pontrjagin class $p_1(X)$ of X is of the form

$$p_1(X) = (N + 24k(X))x^2$$

with some integer $k(X)$, where x is a generator of $H^2(X;\mathbb{Z})$.

Remark 4.2. On the dimensions, where framed closed manifolds with the Kervaire invariant one exist, the function $k(X)$ takes any integer ($N = 2, 4, 8, 16, 32$ are the cases at presnt). For the other even values of N one can see that $k(X)$ can take any even integer. Conversely the recent result of Stolz [S] (together with (4.4) of [DMSc]) implies that $k(X)$ must be even if N is an even integer except powers of 2 (note that $k(X)$ modulo 2 agrees with the μ invariant in [DMSc]). For an odd integer N the value of $k(X)$ is more restrictive and complicated.

Our main results of this section are as follows.

Theorem 4.3. Let X be a homotopy $P(\mathbb{C}^N)$ with a Type II G action. Let F_i $(i = 1, 2)$ be connected components of X^G of dimension $2(N_i-1)$. Then, choosing suitable signs of the defects $D(F_i)$, we have

$$D(F_1) + D(F_2) \equiv 4k(X) \pmod 8.$$

Corollary 4.4. If X is G homotopy equivalent to $P(\mathbb{C}^{N_1,N_2})$, then $k(X) \equiv 0 \pmod 2$.

Proof of Corollary 4.4. The assumption means $D(F_i) = \pm 1$ as remarked in the Introduction. This together with Theorem 4.3 proves the corollary. Q.E.D.

Theorem 4.5. If X is G homotopy equivalent to $P(\mathbb{C}^{N_1,N_2})$, then $2k(F_i) \equiv k(X) \pmod 4$ provided $N_i > 2$.

Corollary 4.6. Let X be the same as in Theorem 4.5. If N_i

is an even integer except powers of 2 for either i, then $k(X) \equiv 0 \pmod 4$.

Proof of Corollary 4.6. By the assumption and Remark 4.2, $k(F_i)$ is even. This and Theorem 4.5 prove the corollary. Q.E.D.

Theorems 4.3 and 4.5 are proved in a similar fashion to each other. The tools used in the proofs are based on [P1]. We shall review them briefly. See [P1] for the details.

Since $H^3(X;\mathbb{Z})$ vanishes, X admits a $\mathrm{Spin}^c(2N-2)$ structure, i.e. there is a principal $\mathrm{Spin}^c(2N-2)$ bundle over X with total space P such that

$$P \times_{\mathrm{Spin}^c(2N-2)} \mathbb{R}^{2N-2} \cong TX.$$

By [P1] the G action on X lifts to an action on P which covers the canonical G action on TX defined by the differential. Then the half $\mathrm{Spin}^c(2N-2)$ modules Δ_+ and Δ_- give G vector bundles E_+ and E_- over TX

$$E_\pm = P \times_{\mathrm{Spin}^c(2N-2)} (\mathbb{R}^{2N-2} \times \Delta_\pm)$$

and there is a G complex over TX ; $E_+ \longrightarrow E_-$ which defines an element $\delta_G \in K_G(TX)$.

Let $\mathrm{Id}_G^X : K_G(TX) \longrightarrow R(G)$ denote the Atiyah-Singer index homomorphism to the complex representation ring $R(G)$ of G. For $V \in K_G(TX)$, $\mathrm{Id}_G^X(V)(g)$ is the value of the character $\mathrm{Id}_G^X(V)$ at $g \in G$. An element $\hat{E}$ of $K_G(X)$ yields an element $\hat{E}\delta_G$ of $K_G(TX)$ through the natural $K_G(X)$ module structure on $K_G(TX)$. The following lemma is stated in the proof of Theorem 3.1 of [P1].

Lemma 4.7. Let g be the generator of G. Then the values of $\mathrm{Id}_G^X(\hat{E}\delta_G)$ at 1 and g are as follows :

(i) $\mathrm{Id}_G^X(\hat{E}\delta_G)(1) = \langle ch(E)e^{Nx/2}\hat{A}(X),\ [X]\rangle$

(ii) $\mathrm{Id}_G^X(\hat{E}\delta_G)(g) = \Sigma\ \varepsilon_i\langle ch_g(\hat{E}|F_i)e^{Nx_i/2}\hat{A}(F_i)/ch\Delta(\nu_i),\ [F_i]\rangle$

where

(a) E is the element of $K(X)$ obtained from $\hat{E}$ by forgetting the action,

(b) x_i denotes the restricted element of x to $H^2(F_i;\mathbb{Z})$,

(c) $ch_g : K_G(F_i) = R(G)\otimes K(F_i) \longrightarrow H^*(F_i;\mathbb{Q})$ is defined by $ch_g(V\otimes\alpha) = V(g)ch(\alpha)$ where $V(g)$ is the value of the character V at g,

(d) $\varepsilon_i = \pm 1$,

(e) $ch\Delta(\nu_i)$ is the unit of $H^*(F_i;\mathbb{Q})$ defined by the formal power series

$$2^{N-N_i}\prod_{j=1}^{N-N_i}\cosh(\omega_j/2)$$

where the elementary symmetric functions of the ω_j^2 give the Pontrjagin classes of the normal bundle ν_i of F_i to X,

(f) $\hat{A}(Y)$ is the $\hat{A}$ class of Y and the lower terms are expressed by $\hat{A}(Y) = 1 - p_1(Y)/24 + \ldots$.

Since $\mathrm{Id}_G^X(\hat{E}\delta_G)$ is an element of $R(G)$, the evaluated values at 1 and g are both integers and their difference must be even. This fact will give an integrality condition on the Pontrjagin classes of X and F_i if there is an element $\hat{E}$ of $K_G(X)$. The following lemma provides such an element.

<u>Lemma 4.8</u> (Corollary 1.3 of [P1]). Any complex line bundle ξ over X comes from an element of $K_G(X)$.

Lifting of the G action on X to ξ is not unique. There are exactly two kinds of liftings. The resulting two complex G

line bundles are related to each other through the tensor product by the non-trivial one dimensional complex representation t of G. Therefore a complex G line bundle, whose underlying bundle is ξ and the action on a fiber over a point of F_1 is trivial, is unique. We shall denote such a G bundle by $\hat{\xi}$. Under these preparations

Proof of Theorem 4.3. Let η be a complex line bundle over X whose first Chern class is a generator x of $H^2(X;\mathbb{Z})$. By Lemma 4.8 $\hat{E}_r = (\hat{\eta}-1)^{N_1-1}(t\hat{\eta}-1)^{N_2-2}\hat{\eta}^r$ is an element of $K_G(X)$ for any integer r. As is well known $R(G) = \mathbb{Z}[t]/(t^2-1)$; so one can express

$$Id_G^X(\hat{E}_r\delta_G) = a_r(1-t) + b_r$$

with integers a_r and b_r. This means that

$$Id_G^X(\hat{E}_r\delta_G)(1) = b_r$$

$$Id_G^X(\hat{E}_r\delta_G)(g) = 2a_r + b_r.$$

Now we shall apply Lemma 4.7 to compute these values. Remember that

$$E_r = (\eta-1)^{N-3}\eta^r$$

$$\hat{A}(X) = 1 - p_1(X)/24 + .. = 1 - (N/24+k(X))x^2 + .. .$$

Since the lowest term in $ch(E_r)$ is x^{N-3}, one can easily deduce

(4.9) $$b_r = (r+N-2)(r+N-1)/2 - k(X)$$

from (i) of Lemma 4.7. This shows the integrality of $k(X)$; so Lemma 4.1 is established.

The computation of (ii) of Lemma 4.7 is as follows. The point is that the cohomological degree of the lowest term in $ch_g(\hat{E}_r|F_1)$ (resp. $ch_g(\hat{E}_r|F_2)$) is $2(N_1-1)$ (resp. $2(N_2-2)$) and both $\hat{A}(F_i)$ and $ch\Delta(\nu_i)$ have values of cohomological degrees divisible by 4. This means that only the constant terms in $\hat{A}(F_i)$ and $ch\Delta(\nu_i)$, which are respectively 1 and 2^{N-N_i}, contribute to the computation. Thus, by an elementary calculation, (ii) reduces to

$$2a_r + b_r = \{D(F_1) + (2r+2N-3)D(F_2)\}/4 \tag{4.10}$$

(remember that $D(F_i)$ are defined up to sign).

Eliminate b_r in (4.10) using (4.9) and multiply the resulting identity by 4. Then we get

$$2(r+N-1)(r+N-2) - 4k(X) \equiv D(F_1) + (2r+2N-3)D(F_2) \pmod 8$$

because a_r is an integer. This congruence holds for every integer r ; so take $r = 2-N$ for instance. Then it turns into

$$4k(X) \equiv D(F_1) + D(F_2) \pmod 8$$

which verifies Theorem 4.3. Q.E.D.

<u>Proof of Theorem 4.5</u>. The idea is the same as in the proof of Theorem 4.3. This time we make use of $\hat{E}'_r = (\hat{\eta}-1)^{N_1-1}(t\hat{\eta}-1)^{N_2-3}\hat{\eta}$ instead of $\hat{E}_r$. Then one can deduce the desired congruence for F_2. We omit the details because the computation is similar to the before. The parallel argument works for F_1. Q.E.D.

§5. Construction of Type II involutions

In this section we apply the preceding results to construct

homotopy $P(\mathbb{C}^N)$'s with Type II involutions. The gap hypothesis then restricts our object to Type $II_{N/2-1}$ actions and $N \equiv 0 \pmod 2$. As observed in §1, the surgery obstructions which we encounter are different by the values of N modulo 4.

First we treat the case $N \equiv 0 \pmod 4$. We consider the realization problem of Theorem 4.3 and Corollaries 4.4, 4.6. The first main result is Theorem 5.1. The author believes that it is valid iff N is a power of 2 greater than 2 (cf. Remark 4.2). But it is related to the Kervaire invariant conjecture; so it would be beyond our scope.

Theorem 5.1. Let $N = 4$ or 8. Suppose we are given a triple (k, d_1, d_2) of integers satisfying these conditions :

(1) d_i are odd,

(2) $d_1 + d_2 \equiv 4k \pmod 8$ or $d_1 - d_2 \equiv 4k \pmod 8$.

Then there is a homotopy $P(\mathbb{C}^N)$ X with a Type $II_{N/2-1}$ G action such that

$$(k, |d_1|, |d_2|) = (k(X), |D(F_1)|, |D(F_2)|)$$

where F_i are connected components of X^G. In addition there is a G map $f : X \to P(\mathbb{C}^{N/2,N/2})$ giving a homotopy (or a G homotopy, if $d_i = \pm 1$) equivalence.

Proof. Since $(q^2-1)/8 \equiv 0$ or $1 \pmod 2$ according as $q \equiv \pm 1$ or $\pm 3 \pmod 8$, the assumption means that $k+(d_1^2-1)/8+(d_2^2-1)/8$ is an even integer. We denote it by $2h$ and consider a G quasi-equivalence

$$\hat{\omega} = \hat{\omega}_+^{2,d_2} \oplus \hat{\omega}_-^{2,d_1} \oplus (-h)\hat{\mu}^{2,3}$$

over $P(\mathbb{C}^{N/2,N/2})$ (see Examples 3.1 and 3.2). By Theorem 2.3 $\hat{\omega}$ yields a G normal map (X,f,b) with a homotopy equivalence f.

This is the desired one. In fact it easily follows from the definition of the above $\hat{\omega}$ that

$$D(F_1) = \text{the fiber degree of } \hat{\omega}^G|P(\mathbb{C}^{N/2}\times 0) = d_1$$

$$D(F_2) = \text{the fiber degree of } \hat{\omega}^G|P(0\times\mathbb{C}^{N/2}) = d_2$$

$$\begin{aligned} 24k(X)x^2 &= p_1(\hat{V}^{2,d_1}-\hat{U}^{2,d_1}+\hat{V}^{2,d_2}-\hat{U}^{2,d_2}-2h(\hat{V}^{2,3}-\hat{U}^{2,3})) \\ &= \{-3(d_1^2-1) - 3(d_2^2-1) + 48h\}x^2 \\ &= 24k\ x^2. \end{aligned}$$

Q.E.D.

<u>Corollary 5.2</u> (cf. Corollary 4.4). Let $N = 4$ or 8 and k be even. Then there is a G homotopy $P(\mathbb{C}^{N/2,N/2})$ X with $k(X) = k$.

<u>Proof</u>. Apply Theorem 5.1 to $(k, 1, 1)$. Q.E.D.

<u>Corollary 5.3</u>. Every homotopy $P(\mathbb{C}^4)$ admits infinitely many Type II_1 involutions distinguished by the defects. In particular they are not G homotopy equivalent to each other.

<u>Proof</u>. By [W2] the set of homotopy $P(\mathbb{C}^4)$'s bijectively corresponds to $\mathbb{Z}$ via the function $k(X)$. For a fixed integer k there are infinitely many triples (k, d_1, d_2) satisfying the conditions of Theorem 5.1. This verifies the corollary. Q.E.D.

For higher dimensional cases we use Corollary 2.7 instead of Theorem 2.3. There it must be arranged that the Kervaire invariant on the fixed point set vanishes. This forces us to put a constraint that k is even, but it is essential unless N is a power of 2 (see Remark 4.2).

<u>Theorem 5.4</u>. Let $N \equiv 0 \pmod 4$. Suppose we are given a triple (k, d_1, d_2) of integers satisfying these conditions :

(1) d_i are odd and k is even,

(ii) $d_1 + d_2 \equiv 4k \pmod{16}$ or $d_1 - d_2 \equiv 4k \pmod{16}$.

Then the same conclusion as in Theorem 5.1 holds.

Proof. The proof is similar to that of Theorem 5.1. The assumption means that $k+(d_1^2-1)/8+(d_2^2-1)/8 \equiv 0$ or $2 \pmod 4$ according as $d_i \equiv \pm 1 \pmod 8$ or $d_i \equiv \pm 3 \pmod 8$. We denote it by $2h$ and consider a G quasi-equivalence $\hat{\omega}$ defined in the proof of Theorem 5.1. Observe that

$$\hat{\omega}^G|P(\mathbb{C}^{N/2}\times 0) = (\phi_{2,2d_1} \oplus \hat{\omega}^{2,d_2} \oplus (-h)\hat{\omega}^{2,3})|P(\mathbb{C}^{N/2}\times 0)$$

$$\hat{\omega}^G|P(0\times\mathbb{C}^{N/2}) = (\hat{\omega}^{2,d_1} \oplus \phi_{2,2d_2} \oplus (-h)\hat{\omega}^{2,3})|P(0\times\mathbb{C}^{N/2})$$

where $\phi_{u,v}$ is the v times map from γ^u to γ^{uv} and where γ denotes the canonical line bundle over $P(\mathbb{C}^{N/2})$. The following assertion is proved in Lemma 3.11 and Theorem 3.1 of [M2].

Assertion. (1) $c(\phi_{u,v}) = 0$ if u is even,

(2) $c(\hat{\omega}^{p,q}) \equiv (p^2-1)(q^2-1)/24 \pmod 2$.

Since the Kervaire invariant is additive with respect to Whitney sum of odd degree fiber preserving proper maps (see [BM]), the above assertion implies $c(\hat{\omega}^G) = 0$. Therefore it follows from Corollary 2.7 that $\hat{\omega}$ yields a G normal map (X,f,b) with a homotopy equivalence f. In a similar way to the proof of Theorem 5.1 one can see that this is the desired one. Q.E.D.

As a consequence of Theorem 5.4, if we weaken the dimensional assumption $N = 4$ or 8 in Corollary 5.2 to $N \equiv 0 \pmod 4$, then we get

Corollary 5.5 (cf. Corollary 4.6). Let $N \equiv k \equiv 0 \pmod 4$. Then there is a G homotopy $P(\mathbb{C}^{N/2,N/2})$ X with $k(X) = k$.

Proof. Apply Theorem 5.4 to $(k, 1, 1)$. Q.E.D.

For the case $N \equiv 2 \pmod 4$ we again apply Corollary 2.7. This time the surgery obstructin $\sigma_G(f)$ is detected by the signature (Propsition 1.8). We shall outline the proof of Theorem 5.6 stated below.

First recall that $\hat{\omega}_\varepsilon^{p,q}$ is a G fiber homotopy equivalence over $P(\mathbb{C}^{N/2,N/2})$ if p and q are both odd (see Example 3.1). Consider an abelian group Ω generated by all such $\hat{\omega}_\varepsilon^{p,q}$. We want to find an element $\hat{\omega}$ of Ω such that the surgery obstruction of $\hat{\omega}^G$ vanishes. By Proposition 1.8 the obstruction is detected by the componentwise differences $\operatorname{Sign} W^G - \operatorname{Sign} P(\mathbb{C}^{N/2,N/2})^G$ where W is a G manifold obtained from $\hat{\omega}$. Since the fixed point set consists of two connected components, we get a map

$$\operatorname{Sign} : \Omega \longrightarrow \mathbb{Z} \oplus \mathbb{Z}$$

given by

$$\hat{\omega} \longrightarrow \operatorname{Sign} W^G - \operatorname{Sign} P(\mathbb{C}^{N/2,N/2})^G.$$

Unfortunately this is not a homomorphism. However, if we restrict it to a certain subgroup of Ω, then it turns out to be a homomorphism and hence its kernel would contain infinitely many elements provided that the rank of the subgroup is greater than two. This is the case if $N \geqq 6$. The trick to make the map Sign a homomorphism in this way is due to W.C. Hsiang [H].

Consequently we have

Theorem 5.6. Let $N \equiv 2 \pmod 4$ and $N \geqq 10$. Then there are infinitely many G homotopy $P(\mathbb{C}^{N/2,N/2})$ X such that the total Pontrjagin classes of X and F_i are not of the same form as the standard ones, where F_i are components of X^G as before.

Remark. The reason why we exclude the case $N = 6$ is to avoid 4-dimensional surgery on the fixed point set.

For the remaining case $N \equiv 1 \pmod 2$ a Type II involution on a homotopy $P(\mathbb{C}^N)$ X has a fixed point component of dimension at least $N-1$. Hence the gap hypothesis is never satisfied and hence we cannot apply the preceding G surgery theory. But, for a Type $II_{(N-1)/2}$ involution, one of the fixed point components is of dimension equal to $1/2\dim X = N-1$ and the other is of dimension less than $1/2\dim X$. The G surgery obstruction under these situations is analyzed by Dovermann. We quote it in our setting.

Proposition ([D1]). Let $\kappa = (W,f,b)$ $f : W \longrightarrow P = P(\mathbb{C}^{(N+1)/2,(N-1)/2})$ be a G normal map such that f^G is of degree one. Then if the following conditions are satisfied, then one can convert f into a G homotopy equivalence via a G normal cobordism :

(1) $c(f^G) = 0$ (componentwise)

(2) $\mathrm{Sign}\ W^G = \mathrm{Sign}\ P^G$ (componentwise)

(3) $\mathrm{Sign}(G,W) = \mathrm{Sign}(G,P)$.

As before we can produce many G normal maps from G fiber homotopy equivalences over P because the G transversality still holds ([P2]). We must carefully choose a fiber homotopy equivalence so that the associated G normal map satisfies the above (1) - (3). We may neglect (1) by virtue of additivity of the Kervaire invariant with respect to Whitney sum of fiber homotopy equivalences. We apply the Hsiang's trick to adjust (2). For (3) we again apply the Hsiang's trick. However at this last step we must evaluate $\mathrm{Sign}(G,W)$, which consists of two elements : one is the ordinary signature of W and the other is the equivariant signature of W at the generator of G. Here the later causes a problem. Namely, in order to compute it using G signature theorem, we need to know the Euler class of the normal bundle ν of the fixed point component of dimension equal to 1/2dim W. However the stable G isomorphism b does not provide us with any information for it because the Euler class is not a stable invariant. To solve this problem we consider a semi-free $\mathbb{Z}_4$ action extending the G action. Namely we consider a $\mathbb{Z}_4$ fiber homotopy equivalence. It then equips the normal bundle ν with a complex structure induced from the $\mathbb{Z}_4$ action. Since the Chern classes are stable invariants and the top Chern class agrees with the Euler class up to sign, this method enables us to evaluate the Euler class of ν through the stable $\mathbb{Z}_4$ isomorphism b.

Consequently we use the Hsiang's trick twice to obtain the following result similar to Theorem 5.6. The details are omitted.

Theorem 5.7. Let $N \equiv 1 \pmod 2$ and $N \geqq 11$. Then there are infinitely many G homotopy $P(\mathbb{C}^{(N+1)/2,(N-1)/2})$ X such that the total Pontrjagin classes of X and F_i are not of the same form as the standard ones.

Appendix

In this appendix we apply the ordinary surgery theory to exhibit infinitely many non-standard G homotopy $P(\mathbb{C}^{m,n})$. Here the gap hypothesis is unnecessary, but the fixed point sets and their equivariant tubular neighborhoods are equivariantly diffeomorphic to those of $P(\mathbb{C}^{m,n})$. The following lemma is easy.

Lemma A.1. Let P_0 be the exterior of an equivariant open tubular neighborhood of $P(\mathbb{C}^{m,n})^G$ in $P(\mathbb{C}^{m,n})$. Then P_0 is a free

G space and equivariantly diffeomorphic to the product of $(S(\mathbb{C}_{-}^{m,0})\times S(\mathbb{C}^{n}))/S^1$ and the unit interval, where the S^1 action is the diagonal one induced from the complex multiplication.

We shall denote the G orbit space of P_0 by $\overline{P}_0$. Suppose we are given a manifold $\overline{X}_0$ together with a homotopy equivalence $\overline{f}_0$: $\overline{X}_0 \to \overline{P}_0$ which restricts to a diffeomorphism on the boundary. Then we lift $\overline{f}_0$ to the double coverings and glue the equivariant tubular neighborhood of $P(\mathbb{C}^{m,n})^G$ in $P(\mathbb{C}^{m,n})$ to X_0 (the double cover of $\overline{X}_0$) and P_0 respectively along their boundaries via the lifted map. This yields a G homotopy $P(\mathbb{C}^{m,n})$ together with a G homotopy equivalence.

In order to produce such a pair $(\overline{X}_0,\overline{f}_0)$ we use the ordinary surgery theory (relative boundary). The surgery exact sequence yields

$$0 = L_{2N-1}(G) \to hS(\overline{P}_0,\partial\overline{P}_0) \longrightarrow [\overline{P}_0/\partial\overline{P}_0,F/O] \xrightarrow{\sigma} L_{2N-2}(G)$$

where $N = m+n$ and $hS(\overline{P}_0,\partial\overline{P}_0)$ denotes the set of such pairs $(\overline{X}_0,\overline{f}_0)$ identified by a natural equivalence relation (see §10 of [W1]). By Lemma A.1 $\overline{P}_0$ is diffeomorphic to the product of a closed manifold with the unit interval; so the above surgery obstruction σ turns out to be a homomorphism (see p.111 of [W1]). As is easily seen, the rank of the abelian group $[\overline{P}_0/\partial\overline{P}_0,F/O]$ is $[(N-1)/2]-[(\max(m,n)-1)/2]$ (see [DMSu] for the details). Moreover $L_{2N-2}(G) = \mathbb{Z}_2$ or $\mathbb{Z}\oplus\mathbb{Z}$ according as N is even or odd (see p.162 of [W1]). These mean that $hS(\overline{P}_0,\partial\overline{P}_0)$ contains infinitely many elements $(\overline{X}_0,\overline{f}_0)$ distinguished by the Pontrjagin classes of $\overline{X}_0$ if either

(1) $N = m+n$ is even and $\max(m,n) \leqq N-2$,

or (2) $N = m+n$ is odd and $\max(m,n) \leqq N-5$.

Thus we have established

<u>Theorem A.2</u>. Suppose m and n satisfy either of the above (1) or (2). Then there are infinitely many G homotopy $P(\mathbb{C}^{m,n})$ such that the fixed point sets and their equivariant tubular neighborhoods are equivariantly diffeomorphic to those of $P(\mathbb{C}^{m,n})$.

References

[AS] M.F. Atiyah and I.M. Singer, The index of elliptic operators III, Ann. of Math. 87 (1968), 546-604.

[B] G.E. Bredon, Introduction to Compact Transformation Groups, Academic Press, 1972.

[BM] G. Brumfiel and I. Madsen, Evaluation of the transfer and the universal surgery classes, Invent. Math. 32 (1976), 133-169.

[D1] K.H. Dovermann, $\mathbb{Z}_2$ surgery theory, Michigan Math. J. 28 (1981), 267-287.

[D2] K.H. Dovermann, Rigid cyclic group actions on cohomology complex projective spaces, preprint.

[DM] K.H. Dovermann and M. Masuda, Exotic cyclic actions on homotopy complex projective spaces, in preparation.

[DMSc] K.H. Dovermann, M. Masuda, and R. Schultz, Conjugation type involutions on homotopy complex projective spaces, Japan. J. of Math. 12 (1986), to appear.

[DMSu] K.H. Dovermann, M. Masuda, and D.Y. Suh, Rigid versus non-rigid cyclic actions, in preparation.

[H] W.-C. Hsiang, A note on free differentiable actions of S^1 and S^3 on homotopy spheres, Ann. of Math., 83 (1966), 266-272.

[HS] W.C. Hsiang and R.H. Szczarba, On embedding spheres in four manifolds, Proc. of Symp. in Pure Math. vol. XXII AMS (1971), 97-103.

[K] S. Kakutani, An application of Dovermann's $\mathbb{Z}_2$-surgery theory to 2n-dimensional complex projective spaces with the conjugate involution, Mem. Fac. Sc. Kochi Univ. (Math.) 5 (1984), 27-43.

[L] S. Lopez de Medrano, Involutions on Manifolds, Ergeb. der Math. Bd. 59, Springer, New York, 1971.

[M1] M. Masuda, Smooth involutions on homotopy $\mathbb{C}P^3$, Amer. J. Math. 106 (1984), 1487-1501.

[M2] M. Masuda, The Kervaire invariant of some fiber homotopy equivalences, Adv. Studies in Pure Math. 9, Kinokuniya North-Holland, to appear.

[M3] M. Masuda, Smooth group actions on cohomology complex projective spaces with a fixed point component of codimension 2, preprint.

[MaP] M. Masuda and T. Petrie, Lectures on transformation groups and Smith equivalences, Contemp. Math. 36 (1985), 191-242.

[MT] M. Masuda and Y.D. Tsai, Tangential representations of cyclic group actions on homotopy complex projective spaces, Osaka J. Math. 22 (1985), 907-919.

[MeP] A. Meyerhoff and T. Petrie, Quasi-equivalence of G modules, Topology 15 (1976), 69-75.

[P1] T. Petrie, Involutions on homotopy complex projective spaces and related topics, Lect. Notes in Math. 298 (1972), Springer, 234-259.

[P2] T. Petrie, Pseudoeqiovalences of G manifolds, Proc. of Symp. in Pure Math. 32 (1978), 169-210.

[P3] T. Petrie, Smith equivalence of representations, Math. Proc. Camb. Soc. 94 (1983), 61-99.

[PR] T. Petrie and J. Randall, Transformation Groups on Manifolds, Dekker Lecture Series 82, 1984.

[S] S. Stolz, A note on conjugation involutions on homotopy complex projective spaces, preprint.

[W1] C.T.C. Wall, Surgery on Compact Manifolds, Academic Press, 1970.

[W2] C.T.C. Wall, Classification problems in differential topology. V : On certain 6-manifolds, Invent. Math. 1 (1966), 355-374.

[We] S. Weinberger, Constructions of group actions : a survey of some recent developments, Contemp. Math. 36 (1985), 269-298.

PROPER SUBANALYTIC TRANSFORMATION GROUPS AND UNIQUE TRIANGULATION OF THE ORBIT SPACES

Takao Matumoto

Department of Mathematics
Faculty of Science
Hiroshima University
Hiroshima 730, Japan

Masahiro Shiota

Department of Mathematics
Faculty of General Education
Nagoya University
Nagoya 464, Japan

§ 1. Introduction

Let G be a transformation group of a topological space X. Triangulation of the orbit space X/G was treated by several people (e. g. [5], [12] and [13]) in some cases of compact differentiable transformation groups. The authors showed in [7] a unique triangulation of X/G, provided that G is a compact Lie group, X is a real analytic manifold and the action is analytic. Moreover, the uniqueness was extended to the case of differentiable G-manifolds and played an important role in defining the equivariant simple homotopy type of compact differentiable G-manifolds when G is a compact Lie group. Let us explain what the uniqueness means here. Under the above conditions we can give naturally X/G a subanalytic structure. On the other hand we know a combinatorially unique subanalytic triangulation of a locally compact subanalytic set ([3] and [11]). Hence X/G comes to admit a unique subanalytic triangulation.

Now we consider a problem under what weaker condition X/G has a natural subanalytic structure. Of course we may assume that X, G and the action are subanalytic; as a subanalytic set is Hausdorff, it is natural to assume a condition that the action is proper in the sense of [6] and [9] (see §2); moreover, in order to simplify the description we assume that X is locally compact. In this paper we shall show that these conditions are sufficient (Corollary 3.4) and hence we obtain a unique subanalytic triangulation of the orbit space of a proper subanalytic triangulation of the orbit space of a proper subanalytic transformation group of a locally compact subanalytic set (Corollary 3.5).

We shall see that a subanalytic group is homeomorphic to a Lie group. But we shall not use properties of Lie group except for the Montgomery-Zippin neighboring subgroups theorem [8].

See [7] for more references and our terminology.

§ 2. Subanalytic transformation groups

Let G be a topological group contained in a real analytic manifold M. If G is subanalytic in M then we call G a <u>subanalytic group</u> in M.

Remark 2.1. A subanalytic group in an analytic manifold is homeomorphic to a Lie group. It seems that G may be subanalytically homeomorphic to a Lie group.

Proof. As the Hilbert's fifth problem is affirmative [8] it suffices to see that G is locally Euclidean at some point of G. But this is clear by the fact that a subanalytic set admits a subanalytic stratification (see Lemma 2.2, [7]).

Let G be a subanalytic group in M_1 and X a subanalytic set in M_2. If G is a topological transformation group of X and the action $G \times X \ni (g, x) \to gx \in X$ is subanalytic (i.e. the graph is subanalytic in $M_1 \times M_2$) then we call (G, M_1) a <u>subanalyitc transformation group</u> of (X, M_2).

A transformation group G of a topological space X is called <u>proper</u> if for any $x, y \in X$, there exist neighborhoods U of x and V of y such that $\{h \in G: hU \cap V \neq \phi\}$ is relatively compact in G ([6] and [9]). This is equivalent to say that $G \times X \ni (g, x) \to (gx, x) \in X \times X$ is proper when G is locally compact and X is Hausdorff.

Remark 2.2. Let G be a locally compact proper transformation group of a completely regular space X. Then X/G is completely regular [9].

Lemma 2.3. Let (G, M_1) be a subanalytic proper transformation group of a subanalytic set (X, M_2) and $\{X_i\}$ be the decomposition of X by orbit types. Then $\{X_i\}$ is locally finite in U of X in M_2.

Proof. For each $x \in X$ let G_x denote the isotropy subgroup of G at x. Put

$$A = \bigcup_{x \in X} G_x \times x = \{(g, x) \in G \times X: gx = x\}$$

and let $\pi: M_1 \times M_2 \to M_2$ be the projection. Then A is subanalytic in $M_1 \times M_2$. Moreover, we can choose an open neighborhood U of X in M_2 so that $\pi|_{A'}: A' \to U$ is proper from the fact that a subanalytic set is σ-compact and the assumption that $\pi|_A: A \to X$ is proper, where A' is the closure of A in $G \times U$. We may consider the problem in U and an open neighborhood of G in M_1 in place of M_2 and M_1 respectively, and this U will satisfy the requirements in the lemma. Hence we can assume from the beginning that G is closed in M_1 and the map $\pi|_{\bar{A}}: \bar{A} \to M_2$ is proper where $\bar{A}$ is the closure of A in $M_1 \times M_2$. Let $\bar{X}$ also denote the closure of X in M_2. We remark $\bar{A} \cap G \times X = A$ because A is closed in $G \times X$.

Now we note that the following assertion is obtained from Hironaka's theorem [4, p.215] since $\pi|_{\bar{A}}: \bar{A} \to \bar{X}$ is a proper map.

Assertion: $\bar{A}$ and $\bar{X}$ have subanalytic stratification $\mathcal{A} = \{A_i\}$ and $\mathcal{Y} = \{Y_j\}$ respectively such that $\pi|_{\bar{A}}: \mathcal{A} \to \mathcal{Y}$ is a stratified map compatible with X: i.e.,

(i) For each stratum A_i of $\mathcal{A}$, $\pi(A_i)$ is contained in some Y_j.

(ii) For such i and j, $\pi|_{A_i}: A_i \to Y_j$ is a C^∞ submersion.

(iii) For each j, $\mathcal{A}_j = \{A_i \in \mathcal{A}: \pi(A_i) \subset Y_j\}$ is a Whitney stratification ([2] or [10]).

(iv) X is a union of some strata of $\mathcal{Y}$.

Apply the Thom's first isotopy lemma to $\pi|_{\bar{A}}: \mathcal{A} \to \mathcal{Y}$ (e.g. 5.2, Chapter II. [1]). Then for each Y_j and $x_1, x_2 \in Y_j$, $\pi^{-1}(x_1) \cap \bar{A}$ and $\pi^{-1}(x_2) \cap \bar{A}$ are homeomorphic. Here it is important that Y_j are connected. Now if $x \in X$ then

$$\pi^{-1}(x) \cap \bar{A} = \pi^{-1}(x) \cap A = G_x \times x.$$

Hence for $x_1, x_2 \in Y_j \subset Z$, G_{x_1} and G_{x_2} are homeomorphic. Furthermore, for such x_1 and x_2, G_{x_1} and G_{x_2} will be conjugate. To see this recall the Montgomery-Zippin neighboring subgroups theorem [8, p.216], which states that each compact subgroup H of G has a neighborhood O in G such that any compact subgroup of G included in O is conjugate to a subgroup of H. Hence, by the properness assumption, each $x \in X$ has a neighborhood V in X such that G_y is conjugate to a subgroup of G_x for any $y \in V$. But a proper subgroup of G_x is never homeomorphic to G_x as G_x is compact. Therefore if $y \in V$ is located in the same stratum as x then G_y is conjugate to G_x. Thus we have proved for $x_1, x_2 \in Y_j \subset X$, G_{x_1} and G_{x_2} are conjugate. Hence each of X_i in the lemma is a union of some $Y_j \subset X$. Therefore $\{X_i\}$ satisfies the requirements in the lemma, which completes the proof.

Remark 2.4. In Lemma 2.3 and Lemma 3.1 below we can replace the properness condition by a weaker condition that X is a Cartan G-space in the sense of [9], which is clear by their proofs.

In Lemma 2,3 if X is closed in M_2 we can put $U = M_2$ for the following reason (Lemma 2.1, [7]). A subset Y of an analytic manifold M is subanalytic in M if each $x \in M$ has an open neighborhood W in M such that $Y \cap W$ is subanalytic in W.

§ 3. Subanalytic structure on an orbit space and its triangulation

Let X be a topological space. A <u>subanalytic structure</u> on X is a proper continuous map $\varphi: X \to M$ to an analytic manifold such that $\varphi(X)$ is subanalytic in M and $\varphi: X \to \varphi(X)$ is a homeomorphism. Let X_1, X_2 be topological spaces with subanalytic structures (φ_1, M_1) and (φ_2, M_2) respectively. A <u>subanalytic map</u> $f: X_1 \to X_2$ is a continuous map such that the graph of $\varphi_2 \circ f \circ \varphi_1^{-1}: \varphi_1(X_1) \to \varphi_2(X_2)$ is subanalytic in $M_1 \times M_2$. Subanalytic structures (φ_1, M_1) and (φ_2, M_2) on X are <u>equivalent</u> if the identity map of X is subanalytic with respect to the structures (φ_1, M_1) on the domain and (φ_2, M_2) on the target. We shall regard equivalent subanalytic structures as the same.

If X is a locally compact subanalytic set in an analytic manifold M from the outset, then X is regarded as equipped with the subanaytic structure given by the inclusion $: X \to U$ where U is some open neighborhood of X in M such that X is closed in U. We give every polyhedron a subanalytic structure by PL embedding it in a Euclidean space so that the image is closed in the space. Then a PL map between polyhedra with such subanalytic structures is subanalytic and hence the subanalytic structure on a polyhedron is unique.

Let X be a subanalytic set or a topological space with a subanalytic structure. Then a <u>subanalytic triangulation</u> of X is a pair consisting of a simplicial complex K and a subanalytic homeomorphism

$\tau:|K| \to X$. For a family $\{X_i\}$ of subsets of X, a triangulation (K, τ) of X is <u>compatible</u> <u>with</u> $\{X_i\}$ if each X_i is a union of some $\tau(\mathrm{Int}\,\sigma)$, $\sigma \in K$.

We remark that when we consider a subanalytic structure on a topological space or a subanalytic triangulation of the space we shall treat only a locally compact space. Of course we can define a subanalytic structure and a subanalytic 'triangulation' (in this case a subanalytic 'triangulation' consists of open subanalytic simplices and may not contain the boundary of the simplices) without the locally compact assumption. But the description, e.g. the definition of equivalence relation of subanalytic structures, will be complicated, because the composition of two subanalytic maps is not necessarily subanalytic in the usual sense (but always "locally subanalytic" [11]); and to make matters worse a subanalytic finite 'triangulation' (= a decomposition into finite open subanalytic simplices) of a subanalytic set is not unique in general.

Let $q:X \to X/G$ be the natural quotient map for a transformation group G of a topological space X. The following is the key lemma to the main theorems.

Lemma 3.1. Let (G, M_1) be a subanalytic proper transformation group of a subanalytic set (X, M_2) and x_0 a point of X. Assume that X is locally compact. Then there exist a neighborhood U of x_0 in X and a G-invariant subanalytic map $f:GU \to \mathbb{R}^{2k+1}$, $k = \dim X$, such that the induced map $\bar{f}:GU/G \to f(U)$ is a homeomorphism.

Proof. By properly embedding M_2 in a Euclidean space we can assume $M_2 = \mathbb{R}^n$ and $x_0 = 0$. It is sufficient to define a G-invariant subanalytic map $f:GU \to \mathbb{R}^{2k+1}$ so that $\bar{f}:GU/G \to \mathbb{R}^{2k+1}$ is one-to-one, because GU/G is locally compact. Put

$$Z = \{(x, y) \in X \times X: q(x) = q(y)\}.$$

Then Z is the image of the projection on $X \times X$ of the graph of the action $G \times X \to X$. As the problem is local at 0 we can assume by the properness condition that the projection on $\mathbb{R}^n \times \mathbb{R}^n$ of the closure of the above graph is proper and hence by (2.6), [10] Z is subanalytic in $\mathbb{R}^n \times \mathbb{R}^n$. Let $B(\varepsilon, a)$ and $S(\varepsilon, a)$ for $\varepsilon > 0$ and $a \in \mathbb{R}^n$ or $\in \mathbb{R}^n \times \mathbb{R}^n$ denote the open ε-ball and ε-sphere with center at a respectively.

We shall construct open neighborhoods $V_0 \supset \cdots \supset V_{2k+1}$ of 0 in

X and G-invariant bounded subanalytic maps $f_i : V_i \to \mathbb{R}^i$, $i = 0, \cdots, 2k+1$, such that

$$f_{i+1} = (f_i|_{V_{i+1}}, g_{i+1}),\ V_i = X \cap B(\varepsilon_i, 0)$$

for some subanalytic function g_{i+1} and some $\varepsilon_i > 0$, and

$$Z_i = \{(x, y) \in V_i \times V_i - Z : f_i(x) = f_i(y)\}$$

is of dimension $\leq 2k - i$. If we construct these and put $U = V_{2k+1}$ and $f =$ the extension of f_{2k+1} to GU then $\bar{f} : GU/G \to \mathbb{R}^{2k+1}$ will be one-to-one, because $\dim Z_{2k+1} = -1$ means that if $x, y \in U$ belong to the distinct orbits then $f(x) \neq f(y)$.

We carry out the above construction by induction on i. For $i = 0$ we put trivially $V_0 = X \cap B(1, 0)$ and $f_0 = 0$. So assume that we have already constructed V_i and f_i. Clearly Z_i is subanalytic in $\mathbb{R}^n \times \mathbb{R}^n$. Assume that $\dim Z_i = 2k - i$, otherwise it suffices to put $V_{i+1} = V_i$ and $g_{i+1} = 0$. Let Y_{i+1} be the union of all strata of dimension $< 2k - i$ in a subanalytic stratification of Z_i. Then Y_{i+1} $(\subset Z_i)$ is a subanalytic set in $\mathbb{R}^n \times \mathbb{R}^n$, closed in $V_i \times V_i - Z$ and of dimension $\leq 2k - i - 1$ such that $Z_i - Y_{i+1}$ is an analytic manifold of dimension $2k - i$. For every large integer m we put $W_m = (Z_i - Y_{i+1}) \cap S(1/m, 0)$. Then W_m is an analytic manifold of dimension $2k - i - 1$ since $(Z_i - Y_{i+1}, 0)$ satisfies the Whiteny condition (Prof. 4.7, [8]). Choose a sequence of points $\{a_j\}_{j=1,2,\ldots}$ in $\cup W_m$ so that for any large m and $x \in W_m$, $B(\exp(-m), x)$ contains at least one a_j. Write $a_j = (a_j', a_j'')$. Then $Ga_j' \cap Ga_j'' = \phi$. Put

$$G_0 = \{g \in G : g\bar{V}_0 \cap \bar{V}_0 \neq \phi\}$$

where $\bar{V}_0$ denotes the closure of V_0. Then we have $G_0^{-1} = G_0$, G_0 is compact by the properness condition, and hence $X_0 = G_0\bar{V}_0$ is compact. Let $\{P_\alpha\}$ be the decomposition of X_0 such that x and y in X_0 are contained in the same P_α if and only if there exists a finite sequence $x = x_0, x_1, \ldots, x_\ell = y$ in X_0 with $g_i x_i = x_{i+1}$ for some g_i of G_0. Here $\ell = 3$ is sufficient for the following reason. Let $x_0, \ldots x_\ell$ be a sequence in X_0 chained by $g_0, \ldots, g_{\ell-1}$ in G_0 as above. Then by definition of X_0 there are $y_0, \ldots, y_\ell$ in $\bar{V}_0$ and $h_0, \ldots, h_\ell$ in G_0 such that $x_i = h_i y_i$. Hence we have

$$y_\ell = h_\ell^{-1} g_{\ell-1} \cdots g_1 g_0 h_0 y_0.$$

Therefore, by definition of G_0, $h_\ell^{-1} g_{\ell-1} \cdots g_1 g_0 h_0 \in G_0$. Hence the sequence x_0, y_0, y_ℓ, x_ℓ is chained by the elements h_0^{-1}, $h_\ell^{-1} g_{\ell-1} \cdots g_1 g_0 h_0$, h_ℓ of G_0, which proves that $\ell = 3$ is sufficient.

The above proof shows also that (i) for each α and $x \in P_\alpha \cap \bar{V}_0$, $P_\alpha = G_0(G_0 x \cap \bar{V}_0)$ and $P_\alpha \cap \bar{V}_0 = Gx \cap \bar{V}_0$ (i.e. $\{P_\beta \cap \bar{V}_0\}$ is the family of intersections of G-orbits with $\bar{V}_0$). From the first equality it follows that each P_α is compact and subanalytic, because $G_0 x \cap \bar{V}_0$ is compact and subanalytic. Moreover $\ell = 3$ shows the following. (ii) let α_1, α_2, ... be a sequence such that there exist $b_1 \in P_{\alpha_1}$, $b_2 \in P_{\alpha_2}$, ... converging to a point b. Then $\cap_{r=1}^{\infty} \overline{\cup_{i=r}^{\infty} P_{\alpha_i}}$ is identical with P_α which contains b.

Define a map $A: C^0(X_0) \to C^0(\bar{V}_0)$ by

$$Ah(x) = \sup\{h(y) : y \in P_\alpha \text{ for } \alpha \text{ with } x \in P_\alpha\} \text{ for } x \in \bar{V}_0.$$

Then, by (ii) and by the fact that X_0 is compact, (iii) A is well-defined (i.e. $Ah \in C^0(\bar{V}_0)$ for $h \in C^0(X_0)$) and continuous with respect to the uniform C^0 topology on $C^0(X_0)$ and $C^0(\bar{V}_0)$; (iv) by (i) Ah are G-invariant for $h \in C^0(X_0)$; and (v) if h is subanalytic then Ah is subanalytic for the following reason. Let h be subanalytic. By (i) the set

$$D = \{(x, Y) \in X_0 \times X_0 : x, y \in P_\alpha \text{ for some } \alpha\}$$

is the image under the proper projection $X_0^2 \times \bar{V}_0^2 \times G_0^3 \to X_0^2$ of the subanalytic set

$$\{(x_1, y_1, x_2, y_2, g_1, g_2, g) \in X_0^2 \times \bar{V}_0^2 \times G_0^3 : x_1 = g_1 x_2,\ y_1 = g_2 y_2,\ x_2 = g y_2\}.$$

Hence D is subanalytic. Now by definition $Ah(x) = \sup\{h(y) : (x, y) \in D\}$, and the graph of Ah is the boundary of the image by the proper projection $\bar{V}_0 \times X_0 \times \mathbb{R} \ni (x, y, t) \to (x, t) \in \bar{V}_0 \times \mathbb{R}$ of the subanalytic set

$$\{(x, y, t) \in \bar{V}_0 \times V_0 \times \mathbb{R} : (x, y) \in D,\ t \geq h(y)\}.$$

Therefore, Ah is subanalytic.

Assertion: Let $\varphi_j \in C^0(X_0)$, $j = 1, 2, \ldots$, be a sequence satisgying $A\varphi_j(a_j') \neq A\varphi_j(a_j'')$. Let also $b_j > 0$. Then there exist $c_j \geq 0$, $j = 1, 2, \ldots$, such that $c_j \leq b_j$, $\sum_j c_j \varphi_j$ uniformly converges to some $\varphi \in C^0(X_0)$ and $A\varphi(a_j') \neq A\varphi(a_j'')$ for all j.

Proof of Assertion: We define c_j inductively as follows, Put

$c_1 = b_1$. Assume we have already defined $c_1, \ldots, c_j$ so that if we put $\psi_\ell = c_1\varphi_1 + \cdots + c_\ell\varphi_\ell$ for $\ell \leq j$ then

$$(1)_\ell \qquad A\psi_\ell(a'_\ell) \neq A\psi_\ell(a''_\ell) \quad \text{and}$$

$$(2)_{\ell p} \qquad c_\ell(|A\varphi_\ell(a'_p)| + |A\varphi_\ell(a''_p)|) \leq |A\psi_p(a'_p) - A\psi_p(a''_p)|/2^{\ell-p+1} \quad \text{for } p < \ell$$

We want c_{j+1} satisfying $(1)_{j+1}$ and $(2)_{j+1p}$, $p \leq j$. If $A\psi_j(a'_{j+1}) \neq A\psi_j(a''_{j+1})$, it suffices to put $c_{j+1} = 0$. If $A\psi_j(a'_{j+1}) = A\psi_j(a''_{j+1})$, then we choose positive c_{j+1} so that $(2)_{j+1p}$, $p \leq j$, hold. In this case

$$A\psi_{j+1}(a'_{j+1}) - A\psi_{j+1}(a''_{j+1}) = c_{j+1}(A\varphi_{j+1}(a'_{j+1}) - A\varphi_{j+1}(a''_{j+1})) \neq 0,$$

hence $(1)_{j+1}$ holds. Thus we obtain a sequence $c_1, c_2, \ldots,$ with $(1)_\ell$ and $(2)_{\ell p}$ for $p < \ell$. Then for any integer $p > p' > 0$

$$(3) \qquad |A\psi_p(a'_{p'}) - A\psi_p(a''_{p'})| \geq |A\psi_{p'}(a'_{p'}) - A\psi_{p'}(a''_{p'})|/2.$$

Furthermore, diminishing c_j if necessary we can assume ψ_j uniformly converges to some φ. Then it follows from (3) that

$$A\varphi(a'_j) \neq A\varphi(a''_j) \text{ for all } j,$$

which proves Assertion.

For every a_j the polynomial approximation theorem assures the existence of a polynomial φ_j on $\mathbb{R}^n$ such that

$$A(\varphi_j|_{X_0})(a'_j) \neq A(\varphi_j|_{X_0})(a''_j).$$

Let $b_1, b_2, \ldots$ be small positive numbers such that the power series $\sum_j b_j\tilde{\varphi}_j$ is of convergence radius ∞ where $\tilde{\varphi}_j(x)$ means $\sum_\alpha |d_\alpha|x^\alpha$ when we write $\varphi_j(x) = \sum_\alpha d_\alpha x^\alpha$.

Apply Assertion to these $\varphi_j|_{\bar{X}_0}$ and b_j. Then we obtain $c_j \geq 0$ such that $\sum_{j=1}^\infty c_j\varphi_j$ converges to an analytic function φ on $\mathbb{R}^n$ and

$$A(\varphi|_{X_0})(a'_j) \neq A(\varphi|_{X_0})(a''_j) \text{ for all } j.$$

Put $g'_{i+1} = A(\varphi|_{X_0})$ on V_i. Then we have already seen that g'_{i+1} is subanalytic. Hence we only need to see that

$$Z'_{i+1} = \{(x, y) \in Z_i : g'_{i+1}(x) = g'_{i+1}(y)\}$$

is of dimension $\leq 2k-i-1$ in some small neighborhood $V_{i+1} \times V_{i+1}$ of 0. In fact $g_{i+1} = g'_{i+1}|V_{i+1}$ is what we wanted.

Assume the dimension of Z'_{i+1} at 0 is $2k-i$. Then there is a subanalytic analytic manifold $N_i (\subset Z'_{i+1} \cap (Z_i - Y_{i+1}))$ of dimension 2k-i whose closure in $\mathbb{R}^n$ contains 0. Recall the subanalytic version (Prop. 3.9, [2]) of a theorem of Bruhat-Whitney which states that there exists a real analytic map $\rho : [0, 1] \to N_i \cup \{0\}$ such that $\rho(0) = 0$ and $\rho((0, 1]) \subset N_i$. Define a continuous function χ on [0, 1] by

$$\chi(t) = \mathrm{dist}(\rho(t), Z_i - N_i).$$

Then it is easy to see that χ is subanalytic and positive outside 0 and hence that

$$\chi(t) \geq C|t|^d, \quad t \in [0, 1]$$

for some $C, d > 0$ (the Łojasiewicz' inequality). These imply

$$B(C|t|^d, \rho(t)) \cap Z_i \subset N_i$$

in other words

$$g'_{i+1}(x) = g'_{i+1}(y) \text{ for } (x, y) \in B(C|t|^d, \rho(t)) \cap Z_i.$$

On the other hand, by definition of g'_{i+1}

$$g'_{i+1}(a'_j) \neq g'_{i+1}(a''_j) \text{ for all } j.$$

Hence

(4) $$a_j \notin B(C|t|^d, \rho(t)) \text{ for all } j.$$

consider now the Łojasiewicz' inequality to the inverse function of $|\rho(t)| = \mathrm{dist}(0, \rho(t))$. Then, we have

$$|\rho(t)| \leq C''|t|^{d''} \text{ for some } C'' \text{ and } d'' > 0.$$

Hence it follows from (4) that for some C' and $d' > 0$

$$a_j \notin B(C'|\rho(t)|^{d'}, \rho(t)) \text{ for all } j.$$

But this contradicts the fact that for any large m and $x \in W_m$, $B(\exp(-m), x)$ contains at least one a_j. Hence Z'_{i+1} is of dimension $\leq 2k-i-1$ in some neighborhood of 0. Thus we have proved that $\bar{f}$ is one-to-one.

Remark 3.2 In Lemma 3.1 we can choose f to be extensible on X as a G-invariant subanalytic map by retaking $U = V_{2k+2} = X \cap B(\varepsilon_{2k+2}, 0)$ with $\varepsilon_{2k+2} < \varepsilon_{2k+1}$. Moreover, we have a G-invariant subanalytic map $F = (f, \varphi_{2k+2}) : X \to \mathbb{R}^{2k+2}$ with the properties (3.2.1) and (3.2.2) below.

Indeed let θ be a subanalytic function on X with support in V_0 such that $0 \leq \theta \leq 1$ and $\theta^{-1}(1)$ is a neighborhood of $\bar{U}$. Put

$$h(x) = \begin{cases} A(\theta|_{X_0})(y) & \text{on } G\bar{V}_0 \\ 0 & \text{on } X - G\bar{V}_0, \end{cases}$$

where $y \in V_0 \cap G_x$. Then hf is extensible on X so that the extension vanishes on $X - GU$. We denote the extension by f for simplicity.

Let $\varphi_{2k+2} : X \to \mathbb{R}$ be defined by

$$\varphi_{2k+2}(x) = \begin{cases} \inf\{|y| : (x, y) \in Z\} & \text{for } x \in GU \\ \varepsilon_{2k+2} & \text{otherwise.} \end{cases}$$

Then φ_{2k+2} is a G-invariant subanalytic function, and $F = (f, \varphi_{2k+2}) : X \to \mathbb{R}^{2k+2}$ satisfies moreover

(3.2.1) $$F(GU) \cap F(X - GU) = \phi.$$

For such F it follows from (2.6), [10] that

(3.2.2) $F(X)$ is subanalytic in $\mathbb{R}^{2k+2}$

because of $F(X) = F(X \cap B(1, 0))$ and because the closure of graph $F|_{X \cap B(1, 0)}$ is bounded and subanalytic.

Theorem 3.3. Let (G, M_1) be a subanalytic proper transformation group of a locally compact subanalytic set (X, M_2). Then there exist an open neighborhood M_2' of X in M_2 and a G-invariant subanalytic map $\varphi : X \to \mathbb{R}^{2k+1}$ with respect to subanalytic structures (inclusion, M_2') and (identity, $\mathbb{R}^{2k+1}$) such that $\varphi(X)$ is closed and subanalytic in $\mathbb{R}^{2k+1}$ and that the induced map $\bar{\varphi} : X/G \to \varphi(X)$ is a homeomorphism, where $k = \dim X$.

Proof. For each point x of X let U_x be an open neighborhood of x in M_2 such $U_x \cap X$ is contained in a neighborhood of x in X which satisfies the requirements in Lemma 3.1 and Remark 3.2. Let M_2' be the union of all U_x. By properly embedding M_2' in a Euclidean space, we can assume $M_2' = \mathbb{R}^n$ and we give always X a subanalytic structure (inclusion, $\mathbb{R}^n$).

The case where $X = G(K \cap X)$ for some compact set K in $\mathbb{R}^n$: As K is covered by a finite number (say s) of U_x, there exists a G-invariant subanalytic map $\psi : X \to \mathbb{R}^{2s(k+1)}$ by Lemma 3.1 and Remark 3.2 such that the induced map $\bar{\psi} : X/G \to \psi(X)$ is a homeomorphism. Here we use (3.2.1) for the existence of $\bar{\psi}^{-1}$, and we see that $\psi(X)$ is subanalytic in $\mathbb{R}^{2s(k+1)}$ for the same reason as in (3.2.2), because we can choose K subanalytic, e.g. $\overline{B(\varepsilon, 0)}$ for some large ε, so that $\psi(X) = \psi(K \cap X)$. We note also that $\psi(X)$ is closed in $\mathbb{R}^{2s(k+1)}$ by the compactness of $K \cap X$. Let (K, τ) be a subanalytic triangulation of $\mathbb{R}^{2s(k+1)}$ compatible with $\psi(X)$ (see Lemma 2.3, [7]), K' the family of $\sigma \in K$ whose interior is mapped by τ into $\psi(X)$ and $\pi : |K'| \to \mathbb{R}^{2k+1}$ be a PL embedding. Then $\varphi = \pi \circ \tau^{-1} \circ \tau : X \to \mathbb{R}^{2k+1}$ is what we want.

The case where there is no compact set K in $\mathbb{R}^n$ such that $X = G(K \cap X)$: Let θ be a G-invariant subanalytic function on X such that for any compact set H in $\mathbb{R}$ there exists a compact K in $\mathbb{R}^n$ with $\theta^{-1}(H) = G(K \cap X)$

$$(\text{e.g. } \theta(x) = \inf\{|gx| : g \in G\}),$$

and let α be a subanalytic function on $\mathbb{R}$ such that for each integer i

$$\alpha = \begin{cases} 1 & \text{on } [2i, 2i+1] \\ 0 & \text{on } [2i - 2/3, 2i - 1/3]. \end{cases}$$

For each i consider the G-invariant subspace

$$X_i = \theta^{-1}([2i - 1/3, 2i + 4/3])$$

of X. By the property of θ, (X_i, G) corresponds to the first case. Hence there exists a G-invariant subanalytic map $\varphi_i : X_i \to \mathbb{R}^{2k+1}$ such that $\bar{\varphi}_i : X_i/G \to \varphi_i(X_i)$ is a homeomorphism. Define $\Phi : X \to \mathbb{R}^{2k+2}$ by

$$\Phi(x) = \begin{cases} (\alpha \circ \theta(x)\varphi_i(x), \theta(x)) & \text{for } x \in X_i \\ (0, \theta(x)) & \text{for } x \notin \bigcup_{i=1}^{\infty} X_i \end{cases}$$

Then Φ is G-invariant and subanalytic, $\bar{\Phi}|_{(\bigcup_i \theta^{-1}((2i-i/3,\ 2i+4/3)))/G}$ is a homeomorphism onto the image, and for any integers $j \neq j'$

$$\operatorname{dist}(\Phi(\theta^{-1}([j+1/3, j+2/3])), \Phi(\theta^{-1}([j'+1/3.\ j'+2/3]))) > 0.$$

In the same way we obtain a G-invariant subanalytic map $\Phi' : X \to \mathbb{R}^{2k+2}$ such that $\bar{\Phi}'|_{(\bigcup_i \theta^{-1}((2i-4/3,\ 2i+1/3)))/G}$ is a homeomorphism onto the image. Hence $\psi = (\Phi,\ \Phi') : X \to \mathbb{R}^{4k+4}$ is G-invariant subanalytic map whose induced map $\bar{\psi} : X/G \to \psi(X)$ is a homeomorphism. Recalling the property of θ, we have a closed neighborhood U of x and a compact set K in $\mathbb{R}^n$ such that $\psi(K \cap X) = \psi(X) \cap U$ for any point x of $\mathbb{R}^{4k+4}$. From this it follows that $\psi(X)$ is closed and subanalytic in $\mathbb{R}^{4k+4}$, since we can choose a subanalytic K. Moreover we can diminish $4k+4$ to $2k+1$ in the same way as the first case. Therefore the theorem is proved.

Corollary 3.4. Let (G, M_1) be a subanalytic proper transformation group of a subanalytic set (X, M_2). Assume X is locally compact. Then X/G admits a unique subanalytic structure such that $q : X \to X/G$ is subanalytic.

Proof. Trivial by Theorem 3.3.

Corollary 3.5. Let (G, M_1) and (X, M_2) be as above and give X/G the above subanalytic structure. Then there exists a subanalytic triangulation of X/G compatible with the orbit type stratification and uniquely in the following sense. If there are two subanalytic triangulations (K, τ) and (K', τ'), we have subanalytic triangulation isotopies (K, τ_t) and (K', τ'_t) of X/G such that $\tau_0 = \tau$, $\tau'_0 = \tau'$ and $(\tau'_1)^{-1} \circ \tau_1 : |K| \to |K'|$ is a PL map (see [7] for the definition of subanalytic triangulation isotopy).

Proof. Follows immediately from Lemma 2.4 in [6], Corollary 3.4 and the next fact. Let $\{X_i\}$ be the decomposition of X by orbit types. Then Lemma 2.3 tells us that $\{q(X_i)\}$ is a locally finite family of subanalytic subsets of X/G.

References

[1] C. G. Gibson et al. Topological stability of smooth mappings, Lecture Notes in Math., Springer, Berlin and New York, 552 (1976).

[2] H. Hironaka, Subanalytic set, in Number theory, algebraic geometry and commutative algebra, in honor of Y. Akizuki, Kinokuniya, Tokyo (1973), 453-493.

[3] ——————, Triangulations of algebraic sets, Proc. Symp. in Pure Math., Amer. Math. Soc., 29 (1975), 165-185.

[4] ——————, Stratification and flatness, in Real and complex singularities, Oslo 1976, edited by Holm, Sijthoff & Noordhoff, Alphen aan den Rijn (1977), 199-265.

[5] S. Illman, Smooth equivariant triangulations of G-manifold for G a finite group, Math. Ann., 233 (1978), 199-220.

[6] J. L. Koszul, Lectures on groups of transformations, Tata Inst., Bombay (1965).

[7] T. Matumoto-M. Shiota, Unique triangulation of the orbit space of a differentiable transformation group and its application, (to appear in Advanced Studies in Pure Math. 9)

[8] D. Montgomery-L. Zippin, Topological transformation groups, Wiley (Interscience), New York (1955).

[9] R. S. Palais, On the existence of slices for actions of non-compact Lie groups, Ann. of Math., 73 (1961), 295-323.

[10] M. Shiota, Piecewise linearization of real analytic functions, Publ. Math. RIMS, Kyoto Univ., 20 (1984), 727-792.

[11] M. Shiota-M. Yokoi, Triangulations of subanalytic sets and locally subanalytic manifolds, Trans. Amer. Math. Soc., 286 (1984), 727-750.

[12] A. Verona, Stratified mappings-structure and triangulability, Lecture Notes in Math., Springer, Berlin-Heiderberg, 1102 (1984).

[13] C. T. Yang, The triangulability of the orbit space of a differentiable transformation group, Bull. Amer. Math. Soc., 69 (1963), 405-408.

A remark on duality and the Segal conjecture

by J. P. May

The Segal conjecture, in its nonequivariant form, provides a spectacular example of the failure of duality for infinite complexes. The purpose of this note is to point out that the Segal conjecture, in its equivariant form, implies the validity of duality for certain infinite G-complexes in theories, such as equivariant K-theory, which enjoy the same kind of invariance property that cohomotopy enjoys.

To establish context, we give a quick review of duality theory. For based spaces X, Y, and Z, there is an evident natural map

$$\nu\colon F(X,Y)\wedge Z \longrightarrow F(X,Y\wedge Z).$$

Here $F(X,Y)$ is the function space of based maps $X \to Y$ and ν is specified by $\nu(f\wedge z)(x) = f(x)\wedge z$. Any up-to-date construction of the stable category comes equipped with an analogous function spectrum functor F and an analogous natural map ν defined for spectra X, Y, and Z. If either X or Z is a finite CW-spectrum, then ν is an equivalence. The dual of X is $DX = F(X,S)$, where S denotes the sphere spectrum. Replacing Z by the representing spectrum k of some theory of interest, we obtain $\nu\colon DX\wedge k \to F(X,k)$. On passage to π_q, this gives $\nu_*\colon k_q(DX) \to k^{-q}(X)$, and ν_* is an isomorphism if X is finite. Classical Spanier-Whitehead duality amounts to an identification of the homotopy type of $D\Sigma^\infty X$ when X is a polyhedron embedded in a sphere, where Σ^∞ denotes the suspension spectrum functor. This outline applies equally well equivariantly, with spectra replaced by G-spectra for a compact Lie group G. We need only remark that a map of G-spectra is an equivalence if and only it induces an isomorphism on passage to $\pi_q^H(?) = [G/H_+\wedge S^q,?]_G$ for all integers q and all closed subgroups H of G (where the + denotes addition of a disjoint basepoint) and that homology and cohomology are specified by

$$k_q^G(X) = \pi_q^G(X\wedge k_G) \quad\text{and}\quad k_G^q(X) = \pi_{-q}^G(F(X,k_G))$$

for any G-spectra X and k_G. See [6] for details on all of this.

We restrict our discussion of the Segal conjecture to finite p-groups for a fixed prime p, and we agree once and for all to complete all spectra at p without change of notation. See [4] for a good discussion of completions of spectra. Completions of G-spectra work the same way (and have properties analogous to completions of G-spaces [7]). The nonequivariant formulation of the Segal conjecture [1,5,8] asserts that a certain map

$$\alpha \colon \vee \Sigma^\infty BWH_+ \longrightarrow DBG_+$$

is an equivalence, where B denotes the classifying space functor and the wedge runs over the conjugacy classes of subgroups H of G. Since both the mod p homology of DBG_+ and the mod p cohomology of BG_+ are concentrated in non-negative degrees, we see that the duality map $\nu_* \colon H_*(DBG_+) \to H^{-*}(BG_+)$ cannot possibly be an isomorphism. It is not much harder to see that the corresponding duality map in p-adic K-theory also fails to be an isomorphism.

As explained in [5], the map α above is obtained by passage to G-fixed point spectra from the map of G-spectra

$$\beta \colon S \cong F(S^0, S) \longrightarrow F(EG_+, S)$$

induced by the projection $EG \to pt$, where EG is a free contractible G-space. The equivariant form of the Segal conjecture asserts that β is an equivalence. More generally, the analogous map with EG_+ replaced by its smash product with any based finite G-CW complex X is an equivalence. The crux of our observation is just the following naturality diagram, where k_G is any G-spectrum.

$$\begin{array}{ccc} DX \wedge k_G & \xrightarrow{\beta \wedge 1} & D(EG_+ \wedge X) \wedge k_G \\ \nu \downarrow & & \downarrow \nu \\ F(X, k_G) & \xrightarrow{\beta} & F(EG_+ \wedge X, k_G) \end{array}$$

The left map ν is an equivalence since X is finite. The top map β, hence also $\beta \wedge 1$, is an equivalence by the Segal conjecture. If k_G^* carries G-maps which are nonequivariant homotopy equivalences to isomorphisms, then β on the bottom is an equivalence (as we see by replacing X with $G/H_+ \wedge X$ for all $H \subset G$) and we can conclude that ν on the right is an equivalence. In particular, duality holds in k_G^*-theory for the infinite G-complex $EG_+ \wedge X$; that is,

$$\nu_* \colon k_q^G(D(EG_+ \wedge X)) \longrightarrow k_G^{-q}(EG_+ \wedge X)$$

is an isomorphism. Of course, equivariant K-theory has the specified invariance property by the Atiyah-Segal completion theorem [3]. Equivariant cohomotopy with coefficients in any equivariant classifying space also has this property [5,8,9].

In the examples just mentioned, k_G and its underlying non-equivariant spectrum k (which represents ordinary K-theory or ordinary cohomotopy with coefficients in the relevant nonequivariant classifying space) are sufficiently nicely related that, for any free G-CW spectrum X,

$$k_G^*(X) \cong k^*(X/G) \quad \text{and} \quad k_*^G(X) \cong k_*(X/G).$$

(See [6,II].) With X replaced by $EG_+ \wedge X$ for a finite G-CW complex X, this may appear to be suspiciously close to a contradiction to the failure of duality in non-equivariant K-theory cited above. The point is that the dual of a free finite G-CW spectrum is equivalent to a free finite G-CW spectrum [2,8.4; 5,III.2.12], but the dual of a free infinite G-CW spectrum need not be equivalent to a free G-CW spectrum, and in fact $\Sigma^\infty EG_+ \wedge X$ provides a counterexample.

Bibliography

1. J. F. Adams. Grame Segal's Burnside ring conjecture. Bull. Amer. Math. Soc. 6(1982), 201-210.

2. J. F. Adams. Prerequisites (on equivariant theory) for Carlsson's lecture. Springer Lecture Notes in Mathematics Vol. 1051, 1986, 483-532.

3. M. F. Atiyah and G. B. Segal. Equivariant K-theory and completion. J. Diff. Geometry 3(1969), 1-18.

4. A. K. Bousfield. The localization of spectra with respect to homology. Topology 18(1979), 257-281.

5. L. G. Lewis, J. P. May, and J. E. McClure. Classifying G-spaces and the Segal conjecture. Canadian Math. Soc. Conf. Proc. Vol. 2, Part 2, 1982, 165-179.

6. L. G. Lewis, J. P. May, and Mark Steinberger (with contributions by J. E. McClure). Equivariant stable homotopy theory. Springer Lecture Notes in Mathematics. To appear.

7. J. P. May. Equivariant completion. Bull. London Math. Soc. 14(1982), 231-237.

8. J. P. May. The completion conjecture in equivariant cohomology. Springer Lecture Notes in Mathematics Vol. 1051, 1984, 620-637.

9. J. P. May. A further generalization of the Segal conjecture. To appear.

On the bounded and thin h-cobordism theorem parameterized by $\mathbb{R}^k$

by

Erik Kjaer Pedersen

0. Introduction

In this paper we consider bounded and thin h-cobordisms parameterized by $\mathbb{R}^k$. We obtain results similar to those obtained by Quinn [Q1,Q2] and Chapman [C], but in a much more restricted situation. The point of the exercise is to give a self contained proof, based on the algebra developed in [P1,P2] in the important special case, where the parameter space is euclidean space. We also get a nice explanation as to why the thin and bounded h-cobordism theorems have the same obstruction groups. Unlike the general version being developed by D.R.Anderson and H.J.Munkholm [A-M], we only consider h-cobordisms with constant (uniformly bounded) fundamental group.

In case of the bounded h-cobordism theorem, it is however clear, that the discussion we carry through will generalize to more general metric spaces than $\mathbb{R}^k$, namely to proper metric spaces (every ball compact). We mention this because in this case, we have computed K_1 of some of the relevant categories i. e. the obstruction groups, in joint work with C. Weibel.

This work was completed while the author spent a most enjoyable year at the Sonderforschungsbereich für Geometrie und Analysis at Göttingen University. The author wants to thank for support and hospitality. The author also wants to acknowledge useful conversations with D.R.Anderson and H.J.Munkholm.

1. Definitions. Statements of results.

Definition 1.1 *A manifold W parameterized by* $\mathbb{R}^k$ *consists of a manifold W together with a proper map* $W \xrightarrow{p} \mathbb{R}^k$*, which is onto.*

We use the map p to give a pseudo metric on W by which we measure size. This is distilled in the following definition.

<u>Definition</u> <u>1.2</u> *Given $K \subseteq W$, W parameterized by $\mathbb{R}^k$ by $p : W \longrightarrow \mathbb{R}^k$, we define the size of K, $S(K)$ to be*

$$S(K) = inf\{r \mid \exists\ y \in \mathbb{R}^k : p(K) \subseteq B(y, r/2)\}$$

where $B(y,r/2)$ is the closed ball in $\mathbb{R}^k$ with radius $r/2$.

$S(K)$ is thus the diameter of the smallest ball containing $p(K)$.

We shall now introduce uniformly bounded and locally constant fundamental groups. Given $t \in \mathbb{R}_+$, we shall define t-bounded fundamental group as follows:

<u>Definition</u> <u>1.3</u> *The fundamental group of W is t-bounded if the following 2 conditions hold:*

1) *For every $\{x,y\} \in W$ and for every homotopy class of paths from x to y, there is a representative $\alpha: (I,0,1) \longrightarrow (W,x,y)$ so that $S(\alpha(I)) < t + S(\{x,y\})$.*

2) *For every null homotopic map $\alpha: S^1 \longrightarrow W$, there is a null homotopy $A: D^2 \longrightarrow W$ so that $S(A(D^2)) < S(\alpha(S^1)) + t$.*

In other words, generators and relations of $\pi_1(W)$ are everywhere representable by something universally bounded. We say *the fundamental group is bounded*, if for some t it is t-bounded, and we say it is *locally constant* if it is t-bounded for all t.

We shall now consider h-cobordisms in the category of manifolds parameterized by $\mathbb{R}^k$.

<u>Definition</u> <u>1.4</u> *The triple $(W, \partial_0 W, \partial_1 W)$ parameterized by $\mathbb{R}^k$, is a bounded h-cobordism (bounded by t) if the boundary of W, ∂W, is the disjoint union of $\partial_0 W$ and $\partial_1 W$, and there are deformations $D_i: W \times I \longrightarrow W$ of W in $\partial_i W$, so that $S(D_i(w \times I)) < t$ for all $w \in W$..*

Given an h-cobordism of this kind, it is natural to ask for a product structure:

Definition 1.5 *A bounded product structure (bounded by t) on* $(W,\partial_0 W,\partial_1 W)$ *is a homeomorphism*

$$h : (\partial_0 W\times I,\partial_0 W\times 0,\partial_0 W\times 1) \longrightarrow (W,\partial_0 W,\partial_1 W)$$

which is the identity on $\partial_0 W$ *and satisfies that* $S(H(w\times I)) < t$ *for all* $w \in \partial_0 W$.

We are now able to formulate the thin and bounded h-cobordism theorems.

Bounded h-cobordism theorem. *Let* $(W,\partial_0 W,\partial_1 W)$ *be a bounded h-cobordism of dimension at least 6, parameterized by* $\mathbb{R}^k$ *with bounded fundamental group* π. *Then there is an invariant in* $\tilde{K}_{-k+1}(\mathbb{Z}\pi)$, *which vanishes if and only if W admits a bounded product structure. All such invariants are realized by bounded h-cobordisms.*

This bounded h-cobordism theorem is a formal consequence of the thin h-cobordism theorem, which we proceed to formulate. However it is much easier to prove the bounded h-cobordism theorem. In the above statement, one could replace $\mathbb{R}^k$ by any other metric space X, which is proper in the sense that every ball is compact, at the price of replacing the obstruction group by $\tilde{K}_1(C_X(\mathbb{Z}\pi))$. (see section 5 for definition and discussion of this).

We now formulate the thin h-cobordism theorem:

Thin h-cobordism theorem: *There is a function* $f : \mathbb{N}\times\mathbb{N} \longrightarrow \mathbb{R}$ *so that if* $(W,\partial_0 W,\partial_1 W)$ *is an h-cobordism of dimension n bigger that 6, parameterized by* $\mathbb{R}^k$, *bounded by t, with fundamental group bounded by t, then there is a product structure on W bounded by* $f(n,k)\cdot t$, *if and only if the obstruction to a bounded product structure, in* $\tilde{K}_{-k+1}(\mathbb{Z}\pi)$, *vanishes.*

Remark 1.6 The difference between the thin and bounded h-cobordism theorems parameterized by $\mathbb{R}^k$ thus lies in the predictability of the bound of the product structure. This of course implies that one may let t go to 0, whereas in the bounded h-cobordism theorem, that has no effect.

It is natural to relate bounded h-cobordism theorems to classical compact h-cobordism theorems. This is done in the following:

<u>Theorem</u> <u>1.7</u> *Let* $(M,\partial_0 M,\partial_1 M)$ *be a compact h-cobordism with fundamental group* $\pi\times\mathbb{Z}^k$, *and let* $M \longrightarrow T^k$ *induce the projection* $\pi\times\mathbb{Z}^k \longrightarrow \mathbb{Z}^k$ *on fundamental groups. Then the pullback over* $\mathbb{R}^k \longrightarrow T^k$ *defines a bounded h-cobordism* $(W,\partial_0 W,\partial_1 W)$ *(the* $\mathbb{Z}^k$*-covering) and the torsion invariants are related by the Bass-Heller-Swan epimorphism*

$$Wh(\pi\times\mathbb{Z}^k) \longrightarrow \tilde{\mathbb{K}}_{-k+1}(\mathbb{Z}\pi).$$

<u>Remark</u> <u>1.8</u> $\tilde{K}_{-k+1}(\mathbb{Z}\pi)$ means $Wh(\pi)$ for $k = 0$, $\tilde{K}_0(\mathbb{Z}\pi)$ for $k = 1$ and $K_{-k+1}(\mathbb{Z}\pi)$ for $k > 1$.

<u>2. Reviewing the algebra</u>.

In this section, we review some of the algebra from [P1,P2]. We also develop the algebra needed to make it possible to treat not only the bounded h-cobordism theorem, but also the thin h-cobordism theorem. This amounts to a discussion of the "size" of the "reason" for the vanishing of an invariant, which is known to vanish. A reader familiar with [P1,P2] and only interested in the bounded h-cobordism theorem, may thus skip this section.

Given a ring R we define the category $\mathcal{C}_k(R)$ to be $\mathbb{Z}^k$-graded, free, finitely generated, based R-modules and bounded homomorphisms. That means an object A is a collection of finitely generated, free, based R-modules $A(J)$, $J \in \mathbb{Z}^k$, and a morphism $\phi: A \longrightarrow B$ is a collection $\phi_J^I : A(I) \longrightarrow B(J)$ of R-module morphisms with the property that there is a $r = r(\phi)$ so that $\phi_J^I = 0$ when $\|I-J\| > r$. Here it is convenient to use the max norm on $\mathbb{Z}^k$. A morphism ϕ will be called *degree preserving* or *homogeneous* if $\phi_J^I = 0$ for I different from J.

Another way of thinking of $\mathcal{C}_k(R)$ is to think of A as $\oplus A(J)$. Then the condition on ϕ is that $\phi : A \longrightarrow B$ is a usual R-module morphism satisfying that $\phi(A(J)) \subseteq \bigoplus_{\|I-J\| \leq r} B(I)$.

The description given here differs from the one given in [P1] in that we take based R-modules. This however does not change anything and makes applications to geometry easier. In [P1] we proved that $K_1(\mathcal{C}_k(R)) \cong K_{-k+1}(R)$. The definition of $K_1(\mathcal{C}_k(R))$ is, that as generators we take $[A,\alpha]$ where A is an object and α an automorphism and as relations $[A,\alpha\beta] - [A,\alpha] - [A,\beta]$ and $A \oplus B \xrightarrow{\begin{pmatrix} 1 & \eta \\ 0 & 1 \end{pmatrix}} A \oplus B$. The reason it does not make a difference whether we consider based or unbased R-modules, is that $[A,\alpha\beta\alpha^{-1}] = [A,\beta]$. Thus a basis change will have no effect on the invariant.

Given an object A of $\mathcal{C}_{k+1}(R)$ there is an obvious object $A[t,t^{-1}]$ of $\mathcal{C}_{k+1}(R[t,t^{-1}])$. This object has a homogeneous automorphism β_t which is the identity on homogeneous elements, whose last coordinate is negative, and multiplication by t when the last coordinate is positive. If α is an automorphism of A bounded by r, then the commutator $[\alpha,\beta_t]$ is the identity on any element whose last coordinate is numerically bigger than r, since α both commutes with multiplication by t and with the identity. This means that $[\alpha,\beta_t]$ only does something interesting in a certain band. If we then restrict to that band, and forget the last coordinate in the grading (by taking direct sum), then we get a $\mathbb{Z}^k$ graded automorphism in $\mathcal{C}_k(R[t,t^{-1}])$. This is the *Bass-Heller-Swan monomorphism*

$$K_{-k}(R) = K_1(\mathcal{C}_{k+1}(R)) \longrightarrow K_1(\mathcal{C}_k(R[t,t^{-1}])) = K_{-k+1}(R[t,t^{-1}]).$$

The details are given in [P1]. Here we want to use this for some simple observations:

Let K be a fixed integral k-tuple. We may then regrade $\mathbb{Z}^k$ by vector addition of K. This will clearly induce a functor of $\mathcal{C}_k(R)$.

<u>Lemma 2.1</u> The map on $K_{-k+1}(R)$ induced by the regrading given by vector addition of K is the identity.

<u>Proof</u> The map A $\longrightarrow$ (regraded A) induced by the identity is bounded, and the map on $K_{-k+1}(R)$ is thus given by conjugation by this map.

This lemma is used to prove the more interesting

<u>Lemma 2.2</u> Let A be an object of $\mathcal{C}_k(R)$ and α and β two automorphisms of

A bounded by r. Suppose there is a $K \in \mathbb{Z}^k$ so that α and β agree on all $A(J)$ with $\|J-K\| \leq r$, i.e. on some box with sides 2r, α and β agree. Then $[A,\alpha] = [A,\beta]$ in $K_{-k+1}(R)$.

Proof Using Lemma 2.1 we may assume $K = 0$. Now conside $\gamma = \alpha\beta^{-1}$. We have $\gamma = id$ on a box with side length 2r, and after application of the Bass-Heller-Swan monomorphism this is still the case. After k applications of the B-H-S monomorphism , we thus have the identity.

The above lemma is used to show that parameterized torsion is well defined under subdivision.

Now consider the map $r : \mathbb{Z}^k \longrightarrow \mathbb{Z}^k$ multiplying by $r > 0$. This induces a functor $r_* : \mathcal{C}_k(R) \longrightarrow \mathcal{C}_k(R)$ sending A to r_*A with $r_*A(J) = A(rJ)$ and 0 otherwise, morphism induced by the identity.

Lemma 2.3 The map induced by multiplication by $r > 0$ is the identity on $K_{-k+1}(R)$.

Proof After k applications of the Bass-Heller-Swan monomorphism, we clearly have the identity.

Finally we have to do the algebra needed to get the thin h-cobordism theorem, rather than just the bounded h-cobordism theorem. At this point we need to remind the reader as to what we mean by an elementary automorphism α of A. By this we mean there is a direct sum decomposition $A = A_1 \oplus A_2$ of based submodules, so that α may be given the matrix presentation $\begin{pmatrix} 1 & \eta \\ 0 & 1 \end{pmatrix}$. We also need to remind the reader that there is an alternative description of $K_{-k+1}(R)$ as the Grothendieck construction of $\mathbb{Z}^{k-1}$-graded projections. We call a projection *geometric* when it sends any basis element either to itself or to 0.

Lemma 2.4 There is a function $f : \mathbb{N} \longrightarrow \mathbb{N}$, so that the following is true:

1) If $A \in \mathcal{C}_{k-1}(R)$ and $p : A \longrightarrow A$ is a projection bounded by 1 and so that $[A,p] = 0$ in $K_{-k+1}(R)$. Then after stabilization there is an automorphism β bounded by $f(k)\cdot r\cdot 24$ so that $\beta p \beta^{-1}$ is geometric.

2) If $A \in \mathcal{C}_k(R)$ and $\alpha : A \longrightarrow A$ is an automorphism bounded by r, so that $[A,\alpha] = 0 \in K_{-k+1}(R)$. Then stably α may be written as a product

of 24 elementary automorphisms, each of which is bounded by $f(k)\cdot r$.

Proof is by induction on k on the statements 1) and 2) for any ring. We will show, that if the ring is of the form $R = S[t,t^{-1}]$ and the given automorphism (projection) only involves finitely many t-powers, then the automorphisms produced have the same property. We shall allow ourselves to refer freely to [P1]. To facilitate the reading, we do the first two steps rather than the general step. For k = 1 statement 1 disappears, so consider statement 2. The map $p_0 : A \longrightarrow A$ is the identity in positive gradings and the 0-map in negative gradings. The map $\alpha p_0 \alpha^{-1}$ restricted to $\bigoplus_{i=-r}^{r} A(i)$ is conjugate to p_0 at least after stabilization of say A(0), so there is an automorphism β of $\bigoplus_{i=-r}^{r} A(i)$ so that $\beta\alpha p_0 \alpha^{-1}\beta^{-1} = p_0$ or $\beta\alpha p_0 = p_0\beta\alpha$. Extending β to all of A by the identity, we have an automorphism β bounded by 2r so that $\beta\alpha p_0 = p_0\beta\alpha$. We thus get $\alpha = \beta^{-1}(\beta\alpha)$ where β^{-1} and $\beta\alpha$ both are bounded by 2r. Since β is the identity away from the interval -r to r, β preserves the two halves when we split up A say at r. Denote $\beta\alpha$ or β^{-1} by γ. The trick used in [P1] is the equation

$$(\gamma\oplus 1\oplus 1\oplus\dots) = (\gamma\oplus\gamma^{-1}\oplus\gamma\dots)(1\oplus\gamma\oplus\gamma^{-1}\oplus\dots)$$

each term on the right side may be written as a product of 6 elementary isomorphisms each of which is bounded by 4r, so f(1) may be taken to be 4. If the ring R is of the form $S[t,t^{-1}]$, and the automorphism α only involves finitely many t-powers, then clearly all the elementary automorphisms produced have that same property.

For k = 2 consider a $\mathbb{Z}$-graded projection p of A as in statement 1). Then $pt + (1-p)$ is a $\mathbb{Z}$-graded automorphism of $R[t,t^{-1}]$ modules involving only finitely many t-powers and bounded by r. By what we just proved $pt + (1-p)$ may be written as a product of 24 elementary matrices, each only involving finitely many t-powers and each bounded by 4r, i. e., $pt + (1-p) = \prod_{i=1}^{24} E_i$. Turning t-powers into a grading, and conjugating the projection p_0 by this automorphism, delivers back the projection p at t-degree 0, the id in positive t-degrees and the 0-map in negative t-degrees. Considering $(pt + (1-p))p_0(pt + (1-p))^{-1}$ in a band around t-degree 0 corresponds to stabilization. Using the trick of lemma 1.10 in [P1] which turns an elementary matrix into a product of one with support in a band around t-degree 0 and one far away, we obtain β bounded by $24\cdot 4\cdot r$ so that in a broad band (of t-degrees) $\beta p \beta^{-1} = p_0$. The trick being employed is that it does not matter how high t-powers get involved, because the grading introduced by the t-powers will immediately be forgotten.

It is now clear how the induction proceeds, one essentially uses the same words.

3. Bounded simple homotopy theory parameterized by $\mathbb{R}^k$.

In this section we elaborate a little on the results of [P2], and carry these results into the manifold category. First we recall

Definition 3.1 *A finite, bounded CW complex parameterized by $\mathbb{R}^k$ consists of the following: A finite dimensional CW complex X together with a map $X \longrightarrow \mathbb{R}^k$ which is onto and proper, so that there is a $t \in \mathbb{R}^+$ so that the size, $S(C) < t$ for each cell C.*

Definition 3.2 *Let K be a space parameterized by $\mathbb{R}^k$. A simple homotopy type on K consists of*

1) a bounded, finite CW complex X parameterized by $\mathbb{R}^k$.

2) a bounded homotopy equivalence $K \longrightarrow X$

Two such are said to be equivalent if the induced bounded homotopy equivalence of finite bounded CW-complexes has 0 torsion in $K_{-k+1}(\mathbb{Z}\pi_1 K)$ (see [P2] for definitions)

Theorem 3.3 *A manifold W parameterized by $\mathbb{R}^k$ with bounded fundamental group, has a well defined simple homotopy type given by a triangulation with bounded simplices (in the PL or DIFF categories) or by a bounded handlebody structure in the TOP category.*

Proof We give the argument in the PL category. This extends to the DIFF category by smooth triangulations. The TOP category requires the usual modifications in the argument. Given $t \in \mathbb{R}_+$, we choose a triangulation with simplices of size less than t. This is a bounded finite CW complex, hence the identity defines a simple homotopy type on W. We have to compare this to another arbitrary triangulation with simplices of size less that t'. The two triangulations have a common subdivision, so as in compact topology it suffices to show that the identity is a homotopy equivalence with trivial torsion, when thought of as a map from W with some triangulation K to a subdivision $\bar{K}$. We pick out one of the coordinates in $\mathbb{R}^k$, say the last, and call this x. Rather than comparing the triangulation and its subdivision directly, we introduce an intermediate subdivision cell complex K' which is a subdivision of K and has $\bar{K}$ as a subdivision. Furthermore if a simplex of K' has barycenter with x-value bigger than 3t the simplex is also a simplex of K, whereas if the x-value is smaller than -3t, the simplex is also a simplex of $\bar{K}$. In other words the cell decomposition agrees with K for large positive values of x and with $\bar{K}$ for large negative values of x. It is not possible to have K' be a triangulation, because we have to subdivide a face of a simplex without subdividing the simplex itself. This however is no problem when we only want a cell complex. We now compare K and K'. At the level of chain complexes the identity induces a map sending a generator corresponding to a cell to the sum of the simplices it is being divided into, and the homotopy inverse sends one of these back to the generator and the rest to 0. For large positive x-values there is no subdivision, so the map is the identity. By Lemma 2.2, it suffices to know the map on a big chunk, so we are done. Comparing K' and $\bar{K}$ is treated similarly, but now using the fact that the cell decompositions agree for large negative x-values.

Note that the reason we can not simply refer to the usual compact proof is, that we may not subdivide equally much everywhere, so there may be more than finitely many steps in the subdivision procedure. We are now ready to define the obstruction and prove the theorems.

4. Proof of thin and bounded h-cobordism theorem parameterized by $\mathbb{R}^k$.

Consider an h-cobordism $(W,\partial_0 W,\partial_1 W)$ parameterized by $\mathbb{R}^k$ and bounded by t, with fundamental group π bounded by t. For the purposes of the bounded h-cobordism, these can be taken to be the same number by taking the bigger, while for the thin h-cobordism theorem it is part of the assumption. By assumption the inclusion $\partial_0 W \subseteq W$ is a bounded homotopy equivalence. Since $\partial_0 W$ as well as W have well defined simple homotopy types by theorem 3.3, this homotopy equivalence has a well defined torsion in $\tilde{K}_{-k+1}(\mathbb{Z}\pi)$. If $(W, \partial_0 W, \partial_1 W)$ is boundedly equivalent to $(\partial_0 W \times I, \partial_0 W, \partial_0 W \times 1)$ then W is obtained from $\partial_0 W$ by attaching no handles, and it is clear that this torsion must vanish. Assuming the invariant vanishes, we give W a filtration as $\partial_0 W \times I \cup$0-handles$\cup$1-handles$\cup \ldots \cup$n+1-handles$\cup \partial_1 W \times I$ in such a way that the size of each handle is bounded by t, and the size of each $w \times I$ in $\partial_0 W \times I$ or $\partial_1 W \times I$ is bounded by t. The aim now is to get rid of all the handles in between, without changing the size of the product structure lines too badly. The procedure is the usual handlebody theory, with attention paid to size, and the arguments are very similar to those applied by Quinn in [Q1], but of course with different algebra.

Cancelling 0-handles is done in standard fashion, but one has to worry that one does not get too long a sequence of 0 and 1 handles, letting the size get out of control. We have a t-bounded deformation retraction of W to $\partial_0 W$. The restriction to 0-handles defines a map

$$(0\text{-handles}) \times I \longrightarrow W$$

defining a path from the core of each 0-handle to $\partial_0 W$. Using (very small) general position, one may assume this path runs in the 1-skeleton of W, relative to $\partial_0 W$, so from the core of every 0-handle, there is a path through cores of 1 and 0-handles to $\partial_0 W$, bounded by t when measured in $\mathbb{R}^k$. If this path has any loop, we may simply discard the loop. That does not increase the size. Also if the path from one 0-handle is a part of a longer path from another 0-handle, we may forget the shorter path. In the end we would like to have an embedding

$$(\text{cores of some 0-handles}) \times I \longrightarrow W$$

which goes through all 0-handles and retaining the control of size. This is done by subdividing every 0-handle with more than 1 path going through into so many 0 and 1-handles, that they have been made disjoint. We now have a disjoint embedding of paths from $\partial_0 W$ going through 0 and 1 handles and with size being bounded by t. Cancelling these 0-handles accordingly will change the boundedness of the collar structure on the boundary to a controlled multiple of t.

The cancelling of 1-handles is now done in standard fashion by introducing 2 and 3 handles, and using the 2 handles to cancel the 1-handles. Having done this from both ends of the handlebody, we have a handlebody without any 0,1,n and n+1 handles, and the product structure on the collars of the boundary is bounded by a constant times t. All 2 and n-1 handles must be attached to the boundary by homotopically trivial maps (otherwise they would change the fundamental group), so we now have the same fundamental group π at all levels of the decomposition.

The cellular $\mathbb{Z}\pi$ chain complex of $(W,\partial_0 W)$ may be thought of as a chain complex in $\mathcal{C}_k(\mathbb{Z}\pi)$ by associating to each cell an integral lattice point in $\mathbb{R}^k$ near the points in $\mathbb{R}^k$ over which the cell sits. As elaborated in [P2], this cellular chain complex

$$0 \longrightarrow C_{n-1} \xrightarrow{\partial} C_{n-2} \xrightarrow{\partial} \cdots \xrightarrow{\partial} C_3 \xrightarrow{\partial} C_2 \longrightarrow 0$$

will be contractible in $\mathcal{C}_k(\mathbb{Z}\pi)$, with a contraction s whose bound is directly related to the bound of the deformation of W in $\partial_0 W$. We now proceed to cancel handles following the scheme indicated by the algebra: we introduce cancelling 3 and 4 handles corresponding to all the 2-handles, and sitting over the points in $\mathbb{R}^k$ where the 2-handles sit, to obtain a chain complex in $\mathcal{C}_k(\mathbb{Z}\pi)$ which in low dimensions is

$$C_2 \oplus C_4 \xrightarrow{1\oplus\partial} C_2 \oplus C_3 \xrightarrow{(0,\partial)} C_2 \longrightarrow 0.$$

At the level of 3-handles we now perform handle additions, so that to each handle x in C_2 we add s(x) in C_3. Since s is bounded, this will increase the cell size by a controllable amount. Since in dim 2 we have $\partial s = 1$, the chain complex, after having performed this handle addition, now has the form

$$C_2 \oplus C_4 \longrightarrow C_2 \oplus C_3 \xrightarrow{(1,*)} C_2 \longrightarrow 0.$$

We are now in a situation to cancel the 3-handles we introduced against the 2-handles, since we have obtained algebraic intersection 1, and after some small Whitney isotopies we will have geometric intersection 1 and can cancel handles. After the cancellation the chain complex has the form $C_2 \oplus C_4 \longrightarrow C_3 \longrightarrow 0$, and is of course still contractible in $\mathcal{C}_k(\mathbb{Z}\pi)$.

Continuing this procedure, we get into a two-index situation

$$0 \longrightarrow C_{r+1} \underset{s}{\overset{\partial}{\rightleftarrows}} C_r \longrightarrow 0$$

and the collars have bounded product structures, bounded by some predictable (even computable as a function of dim(W)) constant times t. The invariant in $\tilde{K}_{-k+1}(\mathbb{Z}\pi) = K_1\mathcal{C}_k(\mathbb{Z}\pi)$ is given by the torsion of this chain complex, which is exactly the isomorphism ∂. Of course ∂ is not an automorphism but an isomorphism. The point is that if ∂ is of the type sending a generator to a generator, then we may cancel handles. It is however easy to see that, at least stably (see e.g. [P1]) C_r and C_{r+1} are isomorphic by an isomorphism sending generators to generators, hence composing ∂ with such an isomorphism, we obtain an automorphism. At this point there is a choice involved, but for $k > 1$ the torsion of an automorphism sending generator to generator is 0. This is Lemma 1.5 of [P1]. When $k=1$ this is not true, and what Quinn calls a flux phenomen occurs. The invariant thus only becomes well defined after dividing out by automorphisms that send generators to generators, which amounts to saying the invariant lives in reduced K-groups. At this point one might mention that the choices involved in finding representing cells of the $\mathbb{Z}\pi$ modules have no effect since an automorphism multiplying generators by elements of π will have 0 torsion, because it is homogeneous.

Since we have assumed the invariant is 0 in $\tilde{K}_{-k+1}(\mathbb{Z}\pi)$, the automorphism can be written as a product of elementary automorphisms after stabilization. After stabilizing geometrically by introducing cancelling handles, we may then change ∂ to cancel one of these elementary automorphisms at a time, at the expense of letting the handles grow bigger. At this point, as in all handle addition arguments, we of course use the boundedness of the fundamental group, to be able to judge how much bigger the handles get. In the end ∂ will be equal to the isomorphism from C_{r+1} to C_r chosen, that sends generators to generators. We now cancel handles and are done.

To prove the thin h-cobordism theorem, we have to worry about how many handle additions we perform, but by lemma 2.4 this is controlled. To sum up the difference between the thin and the bounded h-cobordism theorem, to do the thin version one needs to do the following: First multiply the reference map in $\mathbb{R}^k$ by $1/\epsilon$ so the h-cobordism will be bounded by 1. Here we use lemma 2.3 to show this does not change the obstruction. To get into the 2 index situation, there is no difference between the two proofs. In the 2-index situation, we need lemma 2.4 to see that we can control how many handle additions we need to perform, and how far away the handles that have to be added can sit.

Proof of Theorem 1.7

Consider a compact h-cobordism $(M, \partial_0 M, \partial_1 M)$ with fundamental group $\pi \times \mathbb{Z}^k$. The torsion of this h-cobordism will be represented by the torsion of the based chain complex of the universal cover of $(M, \partial_0 M)$ as $\mathbb{Z}[\pi \times \mathbb{Z}^k]$ modules. This is exactly the same chain complex as that of the $\mathbb{Z}^k$-covering, but now the $\mathbb{Z}^k$ has been turned into a $\mathbb{Z}^k$-grading. On the other hand, the description of the Bass-Heller-Swan epimorphism given in [P1] is exactly that.

Realizability of obstructions

Given a manifold $\partial_0 W \longrightarrow \mathbb{R}^k$ with unifomly bounded fundamental group π and an element $\sigma \in \tilde{K}_{-k+1}(\mathbb{Z}\pi)$, we wish to construct an h-cobordism $(W, \partial_0 W, \partial_1 W)$ with obstruction σ. However σ is represented by a $\mathbb{Z}^k$-graded bounded automorphism $\alpha : C \longrightarrow C$, where C is some object of $\mathcal{C}_k(\mathbb{Z}\pi)$. We start out with $\partial_0 W \times I$. Then we attach infinitely many trivial handles of the same dimension r corresponding to the generators of C, and each placed at a point which in $\mathbb{R}^k$ is near by the integral lattice point of the generator in C. As in the standard realizability theorem we now attach r+1-handles by maps given by α above. It is easy to extend the reference map to $\mathbb{R}^k$ and we get a manifold $(W, \partial_0 W, \partial_1 W)$ with the chain complex $0 \longrightarrow C \xrightarrow{\alpha} C \longrightarrow 0$ and will thus have torsion given by the class of α which is σ. To prove it is a bounded h-cobordism, we do however need to invoke the Whitehead theorem type results of Anderson and Munkholm [A-M].

5. Parameterizing by other metric spaces.

In the proof of the bounded h-cobordism theorem (not the thin h-cobordism theorem) we have nowhere used that the metric space we parameterized by is $\mathbb{R}^k$. Any other metric space X will do, as long as X satisfies that every ball in X is compact (A proper metric space in the sense of [A-M]). The groups in which the obstructions will then take values will then be $\tilde{K}_1(\mathcal{C}_X(\mathbb{Z}\pi)$ where $\mathcal{C}_X(R)$ is an additive category described as based, finitely generated, free R-modules parameterized by X and bounded homomorphisms. That means an object A is a set of based, finitely generated, free R-modules A(x), one for each $x \in X$ with the property, that for any ball $B \subset X$, $A(x) = 0$ for all but finitely many $x \in B$. A morphism $\phi : A \longrightarrow B$ is a set of R-module morphisms $\phi_y^x : A(x) \longrightarrow B(y)$, so that there exists $k = k(\phi)$ with the property that $\phi_y^x = 0$ for $d(x,y) > k$. The study of this sort of category is the object of forthcoming joint work with C.Weibel, in which we obtain results about the K-theory of such categories. In the case of $X = \mathbb{R}^k$ we have preferred to have the modules sitting at the integral lattice points, but this is not an important difference. In general when the fundamental group is uniformally bounded with respect to the metric space X, the proof of the bounded h-cobordism theorem will go through word for word. The obstructions will be elements of $\tilde{K}_1(\mathcal{C}_X(\mathbb{Z}\pi)$, where $\sim$ stands for the reduction by automorphisms sending generators to generators. The case where the fundamental group is not necessarily being assumed to be uniformally bounded is presently being studied by D.R.Anderson and H.J.Munkholm.

References

[A-M] D.R.Anderson and H.J.Munkholm: The simple homotopy theory of controlled spaces, an announcement. Odense university preprint series no7,1984.

[C] Chapman: Controlled Simple Homotopy Theory and Applications,Springer Lecture notes 1009.

[P1] E.K.Pedersen: On the K_{-i}-functors, Journ. of Algebra,90, (1984) 461-475.

[P2] E.K.Pedersen: K_{-i}-invariants of chain complexes. Proceedings of Leningrad topological conference, Springer Lecture Notes in Mathematics 1060, 174-186.

[Q1] F.Quinn: Ends of maps I, Ann. of Math. 110 (1979) 275-331.

[Q2] F.Quinn: Ends of Maps II, Invent Math. 68 (1982) 353-424.

Sonderforschungsbereich Geometrie und Analysis
Matematisches Institut der Georg August Universität
Bunsenstraße 3-5
D-3400 Göttingen BRD

and

Matematisk Institut
Odense Universitet
DK-5230 Odense M
Danmark

ALGEBRAIC AND GEOMETRIC SPLITTINGS OF THE K- AND L-GROUPS OF POLYNOMIAL EXTENSIONS

Andrew Ranicki

Introduction

This paper is an account of assorted results concerning the algebraic and geometric splittings of the Whitehead group of a polynomial extension as a direct sum

$$Wh(\pi\times\mathbb{Z}) = Wh(\pi)\oplus\tilde{K}_0(\mathbb{Z}[\pi])\oplus\widetilde{Nil}(\mathbb{Z}[\pi])\oplus\widetilde{Nil}(\mathbb{Z}[\pi])$$

and the analogous splittings of the Wall surgery obstruction groups

$$\begin{cases} L^s_*(\pi\times\mathbb{Z}) = L^s_*(\pi)\oplus L^h_{*-1}(\pi) \\ L^h_*(\pi\times\mathbb{Z}) = L^h_*(\pi)\oplus L^p_{*-1}(\pi) \end{cases}.$$

Such a splitting of $Wh(\pi\times\mathbb{Z})$ was first obtained by Bass, Heller and Swan [2]. $\begin{cases}\text{Shaneson [29]}\\ \text{Pedersen and Ranicki [18]}\end{cases}$ obtained such a splitting of $\begin{cases}L^s_*(\pi\times\mathbb{Z})\\ L^h_*(\pi\times\mathbb{Z})\end{cases}$ geometrically. Novikov [17] and Ranicki [20] obtained such L-theory splittings algebraically.

The main object of this paper is to point out that the geometric L-theory splittings of [29] and [18] are not in fact the same as the algebraic L-theory splittings of [17] and [20] (contrary to the claims put forward in [18],[20],[23] and [24] that they coincided), and to express the difference between them in terms of algebra. The splitting maps $\begin{cases}L^s_*(\pi)\rightarrowtail L^s_*(\pi\times\mathbb{Z})\\ L^h_*(\pi)\rightarrowtail L^h_*(\pi\times\mathbb{Z})\end{cases}$, $\begin{cases}L^s_*(\pi\times\mathbb{Z})\twoheadrightarrow L^h_{*-1}(\pi)\\ L^h_*(\pi\times\mathbb{Z})\twoheadrightarrow L^p_{*-1}(\pi)\end{cases}$ are the same in algebra and geometry, the split injections being the ones induced functorially from the split injection of groups $\bar{\varepsilon}:\pi\rightarrowtail\pi\times\mathbb{Z}$. However, the splitting maps $\begin{cases}L^s_*(\pi\times\mathbb{Z})\twoheadrightarrow L^s_*(\pi)\\ L^h_*(\pi\times\mathbb{Z})\twoheadrightarrow L^h_*(\pi)\end{cases}$, $\begin{cases}L^h_{*-1}(\pi)\rightarrowtail L^s_*(\pi\times\mathbb{Z})\\ L^p_{*-1}(\pi)\rightarrowtail L^h_*(\pi\times\mathbb{Z})\end{cases}$ are in general different in algebra and geometry. In particular, the geometric split split surjections are not the algebraic split surjections induced functorially from the split surjection of groups $\varepsilon:\pi\times\mathbb{Z}\twoheadrightarrow\pi$! This may be seen by considering the composite $\varepsilon\bar{B}'$ of the geometric split injection

$$\begin{cases} \overline{B}' : L^h_{n-1}(\pi) \rightarrowtail L^s_n(\pi\times\mathbb{Z}) \ ; \\ \qquad \sigma^h_*((f,b):M \longrightarrow X) \longmapsto \sigma^s_*((f,b)\times 1 : M\times S^1 \longrightarrow X\times S^1) \\ \overline{B}' : L^p_{n-1}(\pi) \rightarrowtail L^h_n(\pi\times\mathbb{Z}) \ , \\ \qquad \sigma^p_*((f,b):M \longrightarrow X) \longmapsto \sigma^h_*((f,b)\times 1 : M\times S^1 \longrightarrow X\times S^1) \end{cases}$$

(denoted $\overline{B}'$ to distinguish from the algebraic split injection $\overline{B}$ of [20]) and the algebraic split surjection

$$\begin{cases} \varepsilon : L^s_n(\pi\times\mathbb{Z}) \twoheadrightarrow L^s_n(\pi) \ ; \\ \qquad \sigma^s_*((g,c):N \longrightarrow Y) \longmapsto \mathbb{Z}[\pi]\otimes_{\mathbb{Z}[\pi\times\mathbb{Z}]}\sigma^s_*(g,c) \\ \varepsilon : L^h_n(\pi\times\mathbb{Z}) \twoheadrightarrow L^h_n(\pi) \ ; \\ \qquad \sigma^h_*((g,c):N \longrightarrow Y) \longmapsto \mathbb{Z}[\pi]\otimes_{\mathbb{Z}[\pi\times\mathbb{Z}]}\sigma^h_*(g,c) \ . \end{cases}$$

Now $\varepsilon\overline{B}'$ need not be zero: if X is a $\begin{cases}\text{finite}\\ \text{finitely dominated}\end{cases}$ (n-1)-dimensional geometric Poincaré complex then $X\times S^1$ is a $\begin{cases}\text{simple}\\ \text{homotopy finite}\end{cases}$ n-dimensional geometric Poincaré complex, the boundary of the $\begin{cases}\text{finite}\\ \text{finitely dominated}\end{cases}$ (n+1)-dimensional geometric Poincaré pair $(X\times D^2, X\times S^1)$, but not in general the boundary of a $\begin{cases}\text{simple}\\ \text{homotopy finite}\end{cases}$ pair $(W, X\times S^1)$ with $\pi_1(W) = \pi_1(X)$, so that ε and $\overline{B}'$ do not belong to the same direct sum system.

The geometrically significant splittings of $L_*(\pi\times\mathbb{Z})$ obtained in §6 are compatible with the geometrically significant variant in §3 of the splitting of $Wh(\pi\times\mathbb{Z})$ due to Bass, Heller and Swan [2]. In both K- and L-theory the algebraic and geometric splitting maps differ in 2-torsion only, there being no difference if $Wh(\pi) = 0$.

I am grateful to Hans Munkholm for our collaboration on [16]. It is the considerations of the appendix of [16] which led to the discovery that the algebraic and geometric L-theory splittings are not the same.

This is a revised version of a paper first written in 1982 at the Institute for Advanced Study, Princeton. I should like to thank the Institute and the National Science Foundation for their support in that year. Thanks also to the Göttingen SFB for a visit in June 1985.

Detailed proofs of the results announced here will be found in Ranicki [26],[27],[28].

§1. Absolute K-theory invariants

The definitions of the Wall finiteness obstruction $[X] \in \tilde{K}_0(\mathbb{Z}[\pi_1(X)])$ of a finitely dominated CW complex X and the Whitehead torsion $\tau(f) \in Wh(\pi_1(X))$ of a homotopy equivalence $f: X \longrightarrow Y$ of finite CW complexes are too well known to bear repeating here. The reduced algebraic K-groups $\tilde{K}_0$, Wh are not as well-behaved with respect to products as the absolute K-groups K_0, K_1. Accordingly it is necessary to deal with absolute versions of the invariants. The projective class of a finitely dominated CW complex X

$$[X] = (\chi(X), [X]) \in K_0(\mathbb{Z}[\pi_1(X)]) = K_0(\mathbb{Z}) \oplus \tilde{K}_0(\mathbb{Z}[\pi_1(X)])$$

is well-known, with $\chi(X) \in K_0(\mathbb{Z}) = \mathbb{Z}$ the Euler characteristic. It is harder to come by an absolute torsion invariant.

Let A be an associative ring with 1 such that the rank of f.g. free A-modules is well-defined, e.g. a group ring $A = \mathbb{Z}[\pi]$. An A-module chain complex C is finite if it is a bounded positive complex of based f.g. free A-modules

$$C : \ldots \longrightarrow 0 \longrightarrow C_n \xrightarrow{d} C_{n-1} \longrightarrow \ldots \longrightarrow C_1 \xrightarrow{d} C_0 \longrightarrow 0 \longrightarrow \ldots,$$

in which case the Euler characteristic of C is defined in the usual manner by

$$\chi(C) = \sum_{r=0}^{n} (-)^r \mathrm{rank}_A(C_r) \in \mathbb{Z} .$$

A finite A-module chain complex C is round if

$$\chi(C) = 0 \in \mathbb{Z} .$$

The absolute torsion of a chain equivalence $f: C \longrightarrow D$ of round finite A-module chain complexes is defined in Ranicki [25] to be an element

$$\tau(f) \in K_1(A)$$

which is a chain homotopy invariant of f such that

i) if f is an isomorphism $\tau(f) = \sum_{r=0}^{\infty} (-)^r \tau(f: C_r \longrightarrow D_r)$.

ii) $\tau(gf) = \tau(f) + \tau(g)$ for $f: C \longrightarrow D$, $g: D \longrightarrow E$.

iii) The reduction of $\tau(f)$ in $\tilde{K}_1(A) = K_1(A)/\{\tau(-1: A \longrightarrow A)\}$ is the usual reduced torsion invariant of f, defined for a chain equivalence $f: C \longrightarrow D$ of finite A-module chain complexes to be the reduction of the torsion $\tau(C(f)) \in K_1(A)$ of the algebraic mapping cone C(f). Thus for $A = \mathbb{Z}[\pi]$ the reduction of $\tau(f) \in K_1(\mathbb{Z}[\pi])$ in the Whitehead group $Wh(\pi) = \tilde{K}_1(\mathbb{Z}[\pi])/\{\pi\}$ is the usual Whitehead torsion of f.

iv) $\tau(f) = \tau(D) - \tau(C) \in K_1(A)$ for contractible finite C,D.

v) In general $\tau(f) \neq \tau(C(f)) \in K_1(A)$, and $\tau(f \oplus f') \neq \tau(f) + \tau(f')$ (although the differences are at most $\tau(-1: A \longrightarrow A) \in K_1(A)$).

vi) The absolute torsion $\tau(f) \in K_1(A)$ of a self chain equivalence $f: C \longrightarrow D = C$ agrees with the absolute torsion invariant $\tau(f) \in K_1(A)$ defined by Gersten [10] for a self chain equivalence $f: C \longrightarrow C$ of a finitely dominated A-module chain complex C.

A $\begin{cases} \text{\underline{round}} \\ - \end{cases}$ finite structure on an A-module chain complex C is an equivalence class of pairs (F,ϕ) with F a $\begin{cases} \text{round} \\ - \end{cases}$ finite A-module chain complex and $\phi: F \longrightarrow C$ a chain equivalence, subject to the equivalence relation

$$(F,\phi) \sim (F',\phi') \text{ if } \tau(\phi'^{-1}\phi: F \longrightarrow C \longrightarrow F') = 0 \in \begin{cases} K_1(A) \\ \widetilde{K}_1(A) \end{cases} .$$

In the topological applications $A = \mathbb{Z}[\pi]$, and $\widetilde{K}_1(A)$ is replaced by $Wh(\pi)$.

<u>Proposition 1.1</u> A finitely dominated A-module chain complex C admits a $\begin{cases} \text{round} \\ - \end{cases}$ finite structure if and only if it has $\begin{cases} \text{absolute} \\ \text{reduced} \end{cases}$ projective class $[C] = 0 \in \begin{cases} K_0(A) \\ \widetilde{K}_0(A) \end{cases}$, in which case the set of such structures on C carries an affine $\begin{cases} K_1(A)\text{-} \\ \widetilde{K}_1(A)\text{-} \end{cases}$ structure.

[]

Let X be a (connected) CW complex with universal cover $\widetilde{X}$ and fundamental group $\pi_1(X) = \pi$. The cellular chain complex $C(\widetilde{X})$ is defined as usual, with $C(\widetilde{X})_r = H_r(\widetilde{X}^{(r)},\widetilde{X}^{(r-1)})$ $(r \geqslant 0)$ the free $\mathbb{Z}[\pi]$-module generated by the r-cells of X. The cell structure of X determines for each $C(\widetilde{X})_r$ a $\mathbb{Z}[\pi]$-module base up to the multiplication of each element by $\pm g$ $(g \in \pi)$. Thus for a finite CW complex X the cellular $\mathbb{Z}[\pi]$-module chain complex $C(\widetilde{X})$ has a canonical finite structure.

A CW complex X is <u>round finite</u> if it is finite, $\chi(X) = 0 \in \mathbb{Z}$, and there is given a choice of actual base for each $C(\widetilde{X})_r$ $(r \geqslant 0)$ in the class of bases determined by the cell structure of X.

The $\begin{cases} \text{\underline{absolute}} \\ \text{\underline{Whitehead}} \end{cases}$ <u>torsion</u> of a homotopy equivalence $f: X \longrightarrow Y$ of $\begin{cases} \text{round} \\ - \end{cases}$ finite CW complexes is defined by

$$\tau(f) = \tau(\widetilde{f}: C(\widetilde{X}) \longrightarrow C(\widetilde{Y})) \in \begin{cases} K_1(\mathbb{Z}[\pi_1(X)]) \\ Wh(\pi_1(X)) \end{cases}$$

A $\begin{cases}\underline{\text{round}} \\ - \end{cases}$ finite structure on a CW complex X is an equivalence class of pairs (F,ϕ) with F a $\begin{cases}\text{round} \\ - \end{cases}$ finite CW complex and $\phi:F\longrightarrow X$ a homotopy equivalence, subject to the equivalence relation

$$(F,\phi) \sim (F',\phi') \text{ if } \tau(\phi'^{-1}\phi:F\longrightarrow X\longrightarrow F') = 0 \in \begin{cases} K_1(\mathbb{Z}[\pi_1(X)]) \\ Wh(\pi_1(X)) \end{cases} .$$

The finiteness obstruction theory of Wall [34] gives:

Proposition 1.2 The $\begin{cases}\text{round} \\ - \end{cases}$ finite structures on a finitely dominated CW complex X are in a natural one-one correspondence with the $\begin{cases}\text{round} \\ - \end{cases}$ finite structures on the $\mathbb{Z}[\pi_1(X)]$-module chain complex $C(\tilde{X})$.

[]

The mapping torus of a self map $f:X\longrightarrow X$ is defined as usual by

$$T(f) = X\times[0,1]/\{(x,0)=(f(x),1)\,|\,x\in X\} .$$

Proposition 1.3 (Ranicki [26]) The mapping torus $T(f)$ of a self map $f:X\longrightarrow X$ of a finitely dominated CW complex X has a canonical round finite structure.

[]

The circle $S^1 = [0,1]/(0=1)$ has universal cover $\tilde{S}^1 = \mathbb{R}$ and fundamental group $\pi_1(S^1) = \mathbb{Z}$. Let $z\in\pi_1(S^1) = \mathbb{Z}$ denote the generator such that

$$z : \mathbb{R}\longrightarrow\mathbb{R} ;\ x\longmapsto x+1 .$$

The canonical round finite structure on the circle $S^1 = e^0\cup e^1 = T(id.:\{pt.\}\longrightarrow\{pt.\})$ is represented by the bases $\tilde{e}^r\in C(\tilde{S}^1)_r = \mathbb{Z}[z,z^{-1}]$ $(r=0,1)$ with

$$d = 1-z : C(\tilde{S}^1)_1 = \mathbb{Z}[z,z^{-1}]\longrightarrow C(\tilde{S}^1)_0 = \mathbb{Z}[z,z^{-1}] ;\ \tilde{e}^1\longmapsto\tilde{e}^0 - z\tilde{e}^0 ,$$

corresponding to the lifts $\tilde{e}^0 = \{0\}$, $\tilde{e}^1 = [0,1]\subset\mathbb{R}$ of e^0,e^1.

In particular, Proposition 1.3 applies to the product $X\times S^1 = T(id.:X\longrightarrow X)$, in which case the canonical round finite structure is a refinement of the finite structure defined geometrically by Mather [14] and Ferry [8], using the homotopy equivalent finite CW complex $T(fg:Y\longrightarrow Y)$ for any domination of X

$$(Y,\ f:X\longrightarrow Y,\ g:Y\longrightarrow X,\ h:gf\simeq 1:X\longrightarrow X)$$

by a finite CW complex Y.

Given a ring morphism $\alpha: A \longrightarrow B$ let

$$\alpha_! : (A\text{-modules}) \longrightarrow (B\text{-modules}) \; ; \; M \longmapsto B\otimes_A M$$

be the functor inducing morphisms in the algebraic K-groups

$$\alpha_! : K_i(A) \longrightarrow K_i(B) \quad (i=0,1) \quad ,$$

which we shall usually abbreviate to α. Given a ring automorphism $\alpha: A \longrightarrow A$ let $K_1(A,\alpha)$ be the relative K-group in the exact sequence

$$K_1(A) \xrightarrow{1-\alpha} K_1(A) \xrightarrow{j} K_1(A,\alpha) \xrightarrow{\partial} K_0(A) \xrightarrow{1-\alpha} K_0(A) \quad ,$$

as originally defined by Siebenmann [33] in connection with the splitting theorem for $K_1(A_\alpha[z,z^{-1}])$ recalled in §3 below. By definition $K_1(A,\alpha)$ is the exotic group of pairs (P,f) with P a f.g. projective A-module and $f \in \mathrm{Hom}_A(\alpha_! P, P)$ an isomorphism. The <u>mixed invariant</u> of a finitely dominated A-module chain complex C and a chain equivalence $f:\alpha_! C \longrightarrow C$ was defined in Ranicki [26] to be an element

$$[C,f] \in K_1(A,\alpha)$$

such that $\partial([C,f]) = [C] \in K_0(A)$, and such that $[C,f] = 0 \in K_1(A,\alpha)$ if and only if C admits a round finite structure $(F,\phi:F \longrightarrow C)$ with

$$\tau(\phi^{-1}f(\alpha_!\phi) : \alpha_! F \longrightarrow \alpha_! C \longrightarrow C \longrightarrow F) = 0 \in K_1(A) \quad .$$

The invariant is a mixture of projective class and torsion, and indeed for $\alpha = 1 : A \longrightarrow A$

$$[C,f] = (\tau(f),[C]) \in K_1(A,1) = K_1(A)\oplus K_0(A) \quad .$$

The absolute torsion invariant defined by Gersten [10] for a self homotopy equivalence $f:X \longrightarrow X$ of a finitely dominated CW complex X inducing $f_* = 1 : \pi_1(X) = \pi \longrightarrow \pi$

$$\tau(f) = \tau(\tilde{f}:C(\tilde{X}) \longrightarrow C(\tilde{X})) \in K_1(\mathbb{Z}[\pi])$$

was generalized in Ranicki [26]: the <u>mixed invariant</u> of a self homotopy equivalence $f:X \longrightarrow X$ of a finitely dominated CW complex X inducing any automorphism $f_* = \alpha : \pi_1(X) = \pi \longrightarrow \pi$ is defined by

$$[X,f] = [C(\tilde{X}),\tilde{f}:\alpha_! C(\tilde{X}) \longrightarrow C(\tilde{X})] \in K_1(\mathbb{Z}[\pi],\alpha) \quad .$$

This has image $\partial([X,f]) = [X] \in K_0(\mathbb{Z}[\pi])$, and is such that $[X,f] = 0$ if and only if X admits a round finite structure $(F,\phi:F \longrightarrow X)$ such that

$$\tau(\phi^{-1}f\phi : F \longrightarrow X \longrightarrow X \longrightarrow F) = 0 \in K_1(\mathbb{Z}[\pi]) \quad .$$

If X admits a round finite structure (F,ϕ) then $[X,f] = j(\tau(\phi^{-1}f\phi))$ is the image of $\tau(\phi^{-1}f\phi:F \longrightarrow F) \in K_1(\mathbb{Z}[\pi])$.

§2. Products in K-theory

For any rings A,B and automorphism $\beta:B\longrightarrow B$ there is defined a product of algebraic K-groups

$$\otimes : K_0(A)\otimes K_1(B,\beta)\longrightarrow K_1(A\otimes B,1\otimes\beta) \ ;$$

$$[P]\otimes[Q,f:\beta_!Q\longrightarrow Q]\longmapsto[P\otimes Q,1\otimes f:(1\otimes\beta)_!(P\otimes Q)=P\otimes\beta_!Q\longrightarrow P\otimes Q] \ ,$$

which in the case $\beta = 1$ is made up of the products

$$\otimes : K_0(A)\otimes K_0(B)\longrightarrow K_0(A\otimes B) \ ; \ [P]\otimes[Q]\longmapsto[P\otimes Q]$$

$$\otimes : K_0(A)\otimes K_1(B)\longrightarrow K_1(A\otimes B) \ ; \ [P]\otimes\tau(f:Q\longrightarrow Q)\longmapsto\tau(1\otimes f:P\otimes Q\longrightarrow P\otimes Q) .$$

The product of a finitely dominated A-module chain complex C and a finitely dominated B-module chain complex D is a finitely dominated $A\otimes B$-module chain complex $C\otimes D$ with projective class

$$[C\otimes D] = [C]\otimes[D] \in K_0(A\otimes B) \ ,$$

and if $f:\beta_!D\longrightarrow D$ is a chain equivalence then the product chain equivalence $1\otimes f : C\otimes\beta_!D\longrightarrow C\otimes D$ has mixed invariant

$$[C\otimes D,1\otimes f] = [C]\otimes[D,f] \in K_1(A\otimes B,1\otimes\beta)$$

The following product formula is an immediate consequence.

Proposition 2.1 Let X,F be finitely dominated CW complexes with $\pi_1(X) = \pi$, $\pi_1(F) = \rho$, and let $f : F\longrightarrow F$ be a self homotopy equivalence inducing the automorphism $f_* = \beta : \rho\longrightarrow\rho$. The mixed invariant of the product self homotopy equivalence $1\times f : X\times F\longrightarrow X\times F$ is given by

$$[X\times F,1\times f] = [X]\otimes[F,f] \in K_1(\mathbb{Z}[\pi\times\rho],1\otimes\beta) \ ,$$

identifying $\mathbb{Z}[\pi\times\rho] = \mathbb{Z}[\pi]\otimes\mathbb{Z}[\rho]$.

[]

In the case $\beta = 1 : \rho\longrightarrow\rho$ the result of Proposition 2.1 is made up of the product formula of Gersten [9] and Siebenmann [30] for the projective class

$$[X\times F] = [X]\otimes[F] \in K_0(\mathbb{Z}[\pi\times\rho])$$

and the product formula of Gersten [10] for torsion

$$\tau(1\times f:X\times F\longrightarrow X\times F) = [X]\otimes\tau(f:F\longrightarrow F) \in K_1(\mathbb{Z}[\pi\times\rho]) .$$

If also X is finite the product formula $\tau(1\times f) = [X]\otimes\tau(f)$ is an absolute version of the special case $e = 1 : X\longrightarrow X' = X$, $f_* = 1$ of the formula of Kwun and Szczarba [12] for the Whitehead torsion of the product $e\times f : X\times F\longrightarrow X'\times F'$ of homotopy equivalences $e : X\longrightarrow X'$, $f : F\longrightarrow F'$ of finite CW complexes

$$\tau(e\times f) = \chi(X)\otimes\tau(f) + \tau(e)\otimes\chi(F) \in Wh(\pi\times\rho)$$

The product $A\otimes B$-module chain complex $C\otimes D$ of a finitely dominated A-module chain complex C and a round finite B-module chain complex D was shown in Ranicki [26] to have a canonical round finite structure, with

$$\tau(e\otimes f:C\otimes D \longrightarrow C'\otimes D') = [C]\otimes\tau(f:D \longrightarrow D') \in K_1(A\otimes B)$$

for any chain equivalences $e:C \longrightarrow C', f:D \longrightarrow D'$ of such complexes. The following product structure theorem of [26] was an immediate consequence.

Proposition 2.2 The product $X\times F$ of a finitely dominated CW complex X and a round finite CW complex F has a canonical round finite structure, with

$$\tau(e\times f:X\times F \longrightarrow X'\times F') = [X]\otimes\tau(f:F \longrightarrow F') \in K_1(\mathbb{Z}[\pi_1(X)\times\pi_1(F)])$$

for any homotopy equivalences $e:X \longrightarrow X', f:F \longrightarrow F'$ of such complexes.

[]

The canonical round finite structure on $X\times S^1 = T(\mathrm{id.}:X \longrightarrow X)$ given by Proposition 1.3 coincides with the canonical round finite structure given by Proposition 2.2.

The product

$$K_0(\mathbb{Z}[\pi])\otimes K_1(\mathbb{Z}[\rho]) \longrightarrow K_1(\mathbb{Z}[\pi\times\rho])$$

has a reduced version

$$\tilde{K}_0(\mathbb{Z}[\pi])\otimes\{\pm\rho\} \longrightarrow Wh(\pi\times\rho)\ ;$$

$$[P]\otimes\tau(\pm g:\mathbb{Z}[\rho] \longrightarrow \mathbb{Z}[\rho]) \longmapsto \tau(1\otimes\pm g:P[\rho] \longrightarrow P[\rho])$$

with $\{\pm\rho\} = \{\pm 1\}\times\rho^{ab} = \ker(K_1(\mathbb{Z}[\rho]) \twoheadrightarrow Wh(\rho))$. We shall make much use of this reduced version with $\rho = \mathbb{Z}$, for which $\{\pm\mathbb{Z}\} = K_1(\mathbb{Z}[\mathbb{Z}])$.

§3. The Whitehead group of a polynomial extension

In the first instance we recall some of the details of the direct sum decomposition

$$Wh(\pi\times\mathbb{Z}) = Wh(\pi)\oplus\tilde{K}_0(\mathbb{Z}[\pi])\oplus\widetilde{Nil}(\mathbb{Z}[\pi])\oplus\widetilde{Nil}(\mathbb{Z}[\pi])$$

obtained by Bass, Heller and Swan [2] and Bass [1 ,XII] for any group We shall call this the algebraically significant splitting of $Wh(\pi\times\mathbb{Z})$. The relevant isomorphism

$$\beta_K = \begin{pmatrix}\varepsilon\\ B\\ \Delta_+\\ \Delta_-\end{pmatrix} : Wh(\pi\times\mathbb{Z}) \longrightarrow Wh(\pi)\oplus\tilde{K}_0(\mathbb{Z}[\pi])\oplus\widetilde{Nil}(\mathbb{Z}[\pi])\oplus\widetilde{Nil}(\mathbb{Z}[\pi])$$

and its inverse

$$\beta_K^{-1} = (\bar{\varepsilon}\ \bar{B}\ \bar{\Delta}_+\ \bar{\Delta}_-) : Wh(\pi)\oplus\tilde{K}_0(\mathbb{Z}[\pi])\oplus\widetilde{Nil}(\mathbb{Z}[\pi])\oplus\widetilde{Nil}(\mathbb{Z}[\pi]) \longrightarrow Wh(\pi\times\mathbb{Z})$$

involve the split $\begin{cases}\text{surjection}\\ \text{injection}\end{cases}$ of group rings

$$\begin{cases}\varepsilon : \mathbb{Z}[\pi\times\mathbb{Z}] = \mathbb{Z}[\pi][z,z^{-1}] \twoheadrightarrow \mathbb{Z}[\pi]\ ;\ \sum\limits_{j=-\infty}^{\infty} a_j z^j \longmapsto \sum\limits_{j=-\infty}^{\infty} a_j \\ \bar{\varepsilon} : \mathbb{Z}[\pi] \rightarrowtail \mathbb{Z}[\pi][z,z^{-1}]\ ;\ a \longmapsto a \qquad (a,a_j \in \mathbb{Z}[\pi])\ .\end{cases}$$

The split injection $\bar{B}:\tilde{K}_0(\mathbb{Z}[\pi]) \rightarrowtail Wh(\pi\times\mathbb{Z})$ is the evaluation of the product $\tilde{K}_0(\mathbb{Z}[\pi])\otimes K_1(\mathbb{Z}[\mathbb{Z}]) \longrightarrow Wh(\pi\times\mathbb{Z})$ (the reduction of $K_0(\mathbb{Z}[\pi])\otimes K_1(\mathbb{Z}[\mathbb{Z}]) \longrightarrow K_1(\mathbb{Z}[\pi\times\mathbb{Z}])$) on the element $\tau(z)\in K_1(\mathbb{Z}[\mathbb{Z}])$

$$\bar{B} = -\otimes\tau(z) : \tilde{K}_0(\mathbb{Z}[\pi]) \rightarrowtail Wh(\pi\times\mathbb{Z})\ ;$$
$$[P] \longmapsto \tau(z:P[z,z^{-1}] \longrightarrow P[z,z^{-1}])\ .$$

If $P = im(p)$ is the image of the projection $p = p^2 : \mathbb{Z}[\pi]^r \longrightarrow \mathbb{Z}[\pi]^r$ then

$$\bar{B}([P]) = \tau(pz+1-p:\mathbb{Z}[\pi\times\mathbb{Z}]^r \longrightarrow \mathbb{Z}[\pi\times\mathbb{Z}]^r) \in Wh(\pi\times\mathbb{Z})\ .$$

By definition, $\widetilde{Nil}(\mathbb{Z}[\pi])$ is the exotic K-group of pairs (F,ν) with F a f.g. free $\mathbb{Z}[\pi]$-module and $\nu\in Hom_{\mathbb{Z}[\pi]}(F,F)$ a nilpotent endomorphism. The split injections $\bar{\Delta}_+$, $\bar{\Delta}_-$ are defined by

$$\bar{\Delta}_{\pm} : \widetilde{Nil}(\mathbb{Z}[\pi]) \rightarrowtail Wh(\pi\times\mathbb{Z})\ ;$$
$$(F,\nu) \longmapsto \tau(1+z^{\pm1}\nu:F[z,z^{-1}] \longrightarrow F[z,z^{-1}])\ .$$

The precise definitions of the split surjections $B, \Delta_{\pm}$ need not detain us here, especially as they are the same for the algebraically and geometrically significant direct sum decompositions of $Wh(\pi \times \mathbb{Z})$.

The exact sequence

$$0 \longrightarrow Wh(\pi) \xrightarrow{\bar{\varepsilon}} Wh(\pi \times \mathbb{Z}) \xrightarrow{\begin{pmatrix} B \\ \Delta_+ \\ \Delta_- \end{pmatrix}} \tilde{K}_0(\mathbb{Z}[\pi]) \oplus \widetilde{Nil}(\mathbb{Z}[\pi]) \oplus \widetilde{Nil}(\mathbb{Z}[\pi]) \longrightarrow 0$$

was interpreted geometrically by Farrell and Hsiang [5],[7]: if X is a finite n-dimensional geometric Poincaré complex with $\pi_1(X) = \pi$ and $f : M \longrightarrow X \times S^1$ is a homotopy equivalence with M^{n+1} a compact $(n+1)$-dimensional manifold then the Whitehead torsion $\tau(f) \in Wh(\pi \times \mathbb{Z})$ is such that

$$\begin{aligned} \tau(f) &\in \operatorname{im}(\bar{\varepsilon} : Wh(\pi) \rightarrowtail Wh(\pi \times \mathbb{Z})) \\ &= \ker(\begin{pmatrix} B \\ \Delta_+ \\ \Delta_- \end{pmatrix} : Wh(\pi \times \mathbb{Z}) \twoheadrightarrow \tilde{K}_0(\mathbb{Z}[\pi]) \oplus \widetilde{Nil}(\mathbb{Z}[\pi]) \oplus \widetilde{Nil}(\mathbb{Z}[\pi])) \end{aligned}$$

if (and for $n \geqslant 5$ only if) f is homotopic to a map transverse regular at $X \times \{pt.\} \subset X \times S^1$ with the restriction

$$g = f| \; : N^n = f^{-1}(X \times \{pt.\}) \longrightarrow X$$

also a homotopy equivalence. Thus $\tau(f) \in \operatorname{coker}(\bar{\varepsilon} : Wh(\pi) \rightarrowtail Wh(\pi \times \mathbb{Z}))$ is the codimension 1 splitting obstruction of f along $X \times \{pt.\} \subset X \times S^1$. For a finitely presented group π every element of $Wh(\pi \times \mathbb{Z})$ is the Whitehead torsion $\tau(f)$ for a homotopy equivalence of pairs $(f, \partial f) : (M, \partial M) \longrightarrow (X, \partial X) \times S^1$ with $(M, \partial M)$ a compact $(n+1)$-dimensional manifold with boundary, and $(X, \partial X)$ a finite n-dimensional geometric Poincaré pair with $\pi_1(X) = \pi$, for some $n \geqslant 5$. In this case $\tau(f) \in \operatorname{coker}(\bar{\varepsilon} : Wh(\pi) \rightarrowtail Wh(\pi \times \mathbb{Z}))$ is the relative codimension 1 splitting obstruction.

The <u>geometrically significant</u> splitting

$$Wh(\pi \times \mathbb{Z}) = Wh(\pi) \oplus \tilde{K}_0(\mathbb{Z}[\pi]) \oplus \widetilde{Nil}(\mathbb{Z}[\pi]) \oplus \widetilde{Nil}(\mathbb{Z}[\pi])$$

is defined by the isomorphism

$$\beta_K' = \begin{pmatrix} \varepsilon' \\ \bar{B}' \\ \Delta_+ \\ \Delta_- \end{pmatrix} : Wh(\pi \times \mathbb{Z}) \longrightarrow Wh(\pi) \oplus \tilde{K}_0(\mathbb{Z}[\pi]) \oplus \widetilde{Nil}(\mathbb{Z}[\pi]) \oplus \widetilde{Nil}(\mathbb{Z}[\pi])$$

with inverse

$$\beta_K'^{-1} = (\bar{\epsilon}\ \bar{B}'\ \bar{\Delta}_+\ \bar{\Delta}_-) : Wh(\pi)\oplus\tilde{K}_0(\mathbb{Z}[\pi])\oplus\widetilde{Nil}(\mathbb{Z}[\pi])\oplus\widetilde{Nil}(\mathbb{Z}[\pi])\longrightarrow Wh(\pi\times\mathbb{Z})\ ,$$

where

$$\bar{B}' = -\otimes\tau(-z) : \tilde{K}_0(\mathbb{Z}[\pi])\rightarrowtail Wh(\pi\times\mathbb{Z})\ ;\ [P]\longmapsto\tau(-z:P[z,z^{-1}]\longrightarrow P[z,z^{-1}])$$

$$(=\tau(-pz+1-p)\ \text{if}\ P = \mathrm{im}(p=p^2))\ ,$$

$$\epsilon' = \epsilon(1-\bar{B}'B) : Wh(\pi\times\mathbb{Z})\twoheadrightarrow Wh(\pi)\ ;$$

$$\tau(f:P[z,z^{-1}]\longrightarrow P[z,z^{-1}])\longrightarrow\tau(\epsilon f:P\longrightarrow P)+\tau(-1:Q\longrightarrow Q)$$

with f an automorphism of the f.g. projective $\mathbb{Z}[\pi\times\mathbb{Z}]$-module $P[z,z^{-1}]$ induced from a f.g. projective $\mathbb{Z}[\pi]$-module P, and Q a f.g. projective $\mathbb{Z}[\pi]$-module such that $B(\tau(f)) = [Q]\in\tilde{K}_0(\mathbb{Z}[\pi])$.

Ferry [8] defined a geometric injection for any finitely presented group

$$\bar{B}'' : \tilde{K}_0(\mathbb{Z}[\pi])\rightarrowtail Wh(\pi\times\mathbb{Z})\ ;$$

$$[X]\longmapsto\tau(f=\phi^{-1}(1\times -1)\phi : Y\xrightarrow{\ \phi\ }X\times S^1\xrightarrow{\ 1\times -1\ }X\times S^1\xrightarrow{\ \phi^{-1}\ }Y)\ ,$$

with $[X]\in\tilde{K}_0(\mathbb{Z}[\pi])$ the Wall finiteness obstruction of a finitely dominated CW complex X with $\pi_1(X)=\pi$ and $\tau(f)\in Wh(\pi\times\mathbb{Z})$ the Whitehead torsion of the homotopy equivalence $f=\phi^{-1}(1\times -1)\phi:Y\longrightarrow Y$ defined using the map $-1:S^1\longrightarrow S^1$ reflecting the circle in a diameter and any homotopy equivalence $\phi:Y\longrightarrow X\times S^1$ from a finite CW complex Y in the finite structure on $X\times S^1$ given by the mapping torus construction of Mather [14].

<u>Proposition 3.1</u> The geometrically significant injection $\bar{B}'$ agrees with the geometric injection $\bar{B}''$

$$\bar{B}' = \bar{B}'' : \tilde{K}_0(\mathbb{Z}[\pi])\rightarrowtail Wh(\pi\times\mathbb{Z})\ .$$

<u>Proof</u>: By Proposition 2.2

$$\bar{B}''([X]) = [X]\otimes\tau(-1:S^1\longrightarrow S^1)\in Wh(\pi\times\mathbb{Z})\ ,$$

with $\tau(-1:S^1\longrightarrow S^1)\in K_1(\mathbb{Z}[z,z^{-1}])$ the absolute torsion. Now $-1:S^1\longrightarrow S^1$ induces the non-trivial automorphism $z\longmapsto z^{-1}$ of $\pi_1(S^1) = \langle z\rangle$, and the induced chain equivalence of based f.g. free $\mathbb{Z}[z,z^{-1}]$-module chain complexes is given by

$$\begin{array}{ccccc}
(-1)_!C(\tilde{S}^1) & : & \mathbb{Z}[z,z^{-1}] & \xrightarrow{\ 1-z^{-1}\ } & \mathbb{Z}[z,z^{-1}] \\
\downarrow{\scriptstyle\widetilde{(-1)}} & & \downarrow{\scriptstyle 1} & & \downarrow{\scriptstyle -z} \\
C(\tilde{S}^1) & : & \mathbb{Z}[z,z^{-1}] & \xrightarrow{\ 1-z\ } & \mathbb{Z}[z,z^{-1}]
\end{array}\ ,$$

so that

$$\tau(-1:S^1 \longrightarrow S^1) = \tau(-z:\mathbb{Z}[z,z^{-1}] \longrightarrow \mathbb{Z}[z,z^{-1}]) \in K_1(\mathbb{Z}[z,z^{-1}]) .$$

Thus

$$\bar{B}'' = -\otimes\tau(-z) = \bar{B}' : \tilde{K}_0(\mathbb{Z}[\pi]) \rightarrowtail Wh(\pi\times\mathbb{Z}) .$$

[]

Ferry [8] characterized $im(\bar{B}'') \subseteq Wh(\pi\times\mathbb{Z})$ as the subgroup of the elements $\tau \in Wh(\pi\times\mathbb{Z})$ such that $(p_n)^!(\tau) = \tau$ for some $n \geqslant 2$, with $(p_n)^! : Wh(\pi\times\mathbb{Z}) \longrightarrow Wh(\pi\times\mathbb{Z})$ the transfer map associated to the n-fold covering of the circle by itself

$$p_n : S^1 \longrightarrow S^1 ; z \longmapsto z^n .$$

See Ranicki [27] for an explicit algebraic verification that $im(\bar{B}') \subseteq Wh(\pi\times\mathbb{Z})$ is the subgroup of transfer invariant elements.

The algebraically significant decomposition of $Wh(\pi\times\mathbb{Z})$ also has a certain measure of geometric significance, in that it is related to the Bott periodicity theorem in topological K-theory - cf. Bass [1 ,XIV]. More recently, Munkholm [15] identified the infinite structure set $\mathcal{S}(X \times \mathbb{R}^2) = ker(\varepsilon:\tilde{K}_0(\mathbb{Z}[\pi\times\mathbb{Z}]) \longrightarrow \tilde{K}_0(\mathbb{Z}[\pi]))$ (X compact, $\pi_1(X) = \pi$) of Siebenmann [32] with the lower algebraic K-groups derived from the algebraically significant splitting of $Wh(\pi\times\mathbb{Z})$ by Bass [1 ,XII] - to be precise $\mathcal{S}(X \times \mathbb{R}^2) = (K_{-1}\oplus NK_0\oplus NK_0)(\mathbb{Z}[\pi])$.

Both the injections $\bar{B},\bar{B}':\tilde{K}_0(\mathbb{Z}[\pi]) \rightarrowtail Wh(\pi\times\mathbb{Z})$ can be realized geometrically for a finitely presented group π, as follows. Given a f.g. projective $\mathbb{Z}[\pi]$-module P let $p = p^2 \in Hom_{\mathbb{Z}[\pi]}(\mathbb{Z}[\pi]^r,\mathbb{Z}[\pi]^r)$ be a projection such that $P = im(p)$. Let K be a finite CW complex such that $\pi_1(K) = \pi$. For any integer $N \geqslant 2$ define the finite CW complexes

$$X = (K \times S^1 \vee \bigvee_r S^N) \cup_{pz+1-p} (\bigcup_r e^{N+1})$$

$$X' = (K \times S^1 \vee \bigvee_r S^N) \cup_{-pz+1-p} (\bigcup_r e^{N+1}) ,$$

such that the inclusions define homotopy equivalences

$$K \times S^1 \longrightarrow X , \quad K \times S^1 \longrightarrow X' .$$

<u>Proposition 3.2</u> The injections $\bar{B},\bar{B}'$ are realized geometrically by

$$\bar{B} : \tilde{K}_0(\mathbb{Z}[\pi]) \rightarrowtail Wh(\pi \times \mathbb{Z}) ; [P] \longmapsto (-)^N \tau(K \times S^1 \longrightarrow X)$$

$$\bar{B}' : \tilde{K}_0(\mathbb{Z}[\pi]) \rightarrowtail Wh(\pi \times \mathbb{Z}) ; [P] \longmapsto (-)^N \tau(K \times S^1 \longrightarrow X') .$$

[]

Nevertheless, $\bar{B}'$ is more geometrically significant than $\bar{B}$.

(Following Siebenmann [31] define a band to be a finite CW complex X equipped with a map $p:X \longrightarrow S^1$ such that the pullback infinite cyclic cover $\overline{X} = p^*(\mathbb{R})$ of X is finitely dominated. For a connected band X the infinite complex $\overline{X}$ has two ends ε^+, ε^- which are contained in finitely dominated subcomplexes $\overline{X}^+, \overline{X}^- \subset \overline{X}$ such that $\overline{X}^+ \cap \overline{X}^-$ is finite and $\overline{X}^+ \cup \overline{X}^- = \overline{X}$. The finiteness obstructions are such that

$$[\overline{X}] = [\overline{X}^+] + [\overline{X}^-] \in \widetilde{K}_0(\mathbb{Z}[\pi]) \quad (\pi = \pi_1(\overline{X})) .$$

For a manifold band X the finiteness obstructions $[\overline{X}^\pm] \in \widetilde{K}_0(\mathbb{Z}[\pi])$ are images of the end obstructions $[\varepsilon^\pm] \in \widetilde{K}_0(\mathbb{Z}[\pi_1(\varepsilon^\pm)])$ of Siebenmann [30]. For any finitely presented group π the surjection $B:Wh(\pi\times\mathbb{Z}) \twoheadrightarrow \widetilde{K}_0(\mathbb{Z}[\pi])$ is realized geometrically by

$$B(\tau(f:X \longrightarrow Y)) = [\overline{Y}^+] - [\overline{X}^+] \in \widetilde{K}_0(\mathbb{Z}[\pi]) ,$$

with $\tau(f) \in Wh(\pi\times\mathbb{Z})$ the Whitehead torsion of a homotopy equivalence of bands $f:X \longrightarrow Y$ with $\pi_1(X) = \pi\times\mathbb{Z}$, $\pi_1(\overline{X}) = \pi$. For the bands used in Proposition 3.2

$$[\overline{X}^+] = -[\overline{X}^-] = [\overline{X}'^+] = -[\overline{X}'^-] = (-)^N[P] ,$$

$$[\overline{(K\times S^1)}^+] = [\overline{(K\times S^1)}^-] = [K\times\mathbb{R}^+] = [K] = 0 \in \widetilde{K}_0(\mathbb{Z}[\pi])).$$

We shall now express the difference between the algebraically and geometrically significant splittings of $Wh(\pi\times\mathbb{Z})$ using the generator $\tau(-1:\mathbb{Z} \longrightarrow \mathbb{Z}) \in K_1(\mathbb{Z})$ $(= \mathbb{Z}_2)$ and the product map

$$\omega = -\otimes\tau(-1) : \widetilde{K}_0(\mathbb{Z}[\pi]) \longrightarrow Wh(\pi) ; [P] \longmapsto \tau(-1:P \longrightarrow P) .$$

If $P = \mathrm{im}(p)$ for a projection $p = p^2 : F \longrightarrow F$ of a f.g. free $\mathbb{Z}[\pi]$-module F then the automorphism $1-2p:F \longrightarrow F$ is such that

$$\omega([P]) = \tau(1-2p:F \longrightarrow F) \in Wh(\pi).$$

Proposition 3.3 The algebraically and geometrically significant

$$\begin{cases} \text{surjections } \varepsilon,\varepsilon' : Wh(\pi\times\mathbb{Z}) \twoheadrightarrow Wh(\pi) \\ \text{injections } \overline{B},\overline{B}' : \widetilde{K}_0(\mathbb{Z}[\pi]) \rightarrowtail Wh(\pi\times\mathbb{Z}) \end{cases} \text{ differ by}$$

$$\begin{cases} \varepsilon' - \varepsilon = \omega B : Wh(\pi\times\mathbb{Z}) \xrightarrow{B} \widetilde{K}_0(\mathbb{Z}[\pi]) \xrightarrow{\omega} Wh(\pi) \\ \overline{B}' - \overline{B} = \overline{\varepsilon}\omega : \widetilde{K}_0(\mathbb{Z}[\pi]) \xrightarrow{\omega} Wh(\pi) \xrightarrow{\overline{\varepsilon}} Wh(\pi\times\mathbb{Z}) \end{cases} .$$

[]

In particular, the difference between the algebraic and geometric splittings is 2-torsion only, since $2\omega = 0$.

It is tempting to identify the geometrically significant surjection $\varepsilon' : Wh(\pi\times\mathbb{Z}) \longrightarrow Wh(\pi)$ with the surjection induced functorially by the split surjection of rings defined by $z \longmapsto -1$

$$\eta : \mathbb{Z}[\pi\times\mathbb{Z}] = \mathbb{Z}[\pi][z,z^{-1}] \longrightarrow \mathbb{Z}[\pi] \ ; \ \sum_{j=-\infty}^{\infty} a_j z^j \longmapsto \sum_{j=-\infty}^{\infty} a_j (-1)^j \ ,$$

and indeed

$$\begin{aligned} \varepsilon'| = \eta| \ : \ & \mathrm{im}((\bar{\varepsilon}\ \bar{B}) : Wh(\pi)\oplus\widetilde{K}_0(\mathbb{Z}[\pi]) \rightarrowtail Wh(\pi\times\mathbb{Z})) \\ & = \mathrm{im}((\bar{\varepsilon}\ \bar{B}') : Wh(\pi)\oplus\widetilde{K}_0(\mathbb{Z}[\pi]) \rightarrowtail Wh(\pi\times\mathbb{Z})) \longrightarrow Wh(\pi) \ . \end{aligned}$$

However, in general

$$\begin{aligned} \varepsilon'| \neq \eta| \ : \ & \mathrm{im}((\bar{\Delta}_+\ \bar{\Delta}_-) : \widetilde{Nil}(\mathbb{Z}[\pi])\oplus\widetilde{Nil}(\mathbb{Z}[\pi]) \rightarrowtail Wh(\pi\times\mathbb{Z})) \\ & \longrightarrow Wh(\pi) \end{aligned}$$

so that $\varepsilon' \neq \eta : Wh(\pi\times\mathbb{Z}) \longrightarrow Wh(\pi)$.

For an automorphism $\alpha : \pi \longrightarrow \pi$ of a group π Farrell and Hsiang [6] and Siebenmann [33] expressed the Whitehead group of the α-twisted extension $\pi\times_\alpha\mathbb{Z}$ of π by $\mathbb{Z} = \langle z\rangle$ ($gz = z\alpha(g) \in \pi\times_\alpha\mathbb{Z}$ for $g\in\pi$) as a natural direct sum

$$Wh(\pi\times_\alpha\mathbb{Z}) = Wh(\pi,\alpha)\oplus\widetilde{Nil}(\mathbb{Z}[\pi],\alpha)\oplus\widetilde{Nil}(\mathbb{Z}[\pi],\alpha^{-1})$$

with $Wh(\pi,\alpha)$ the relative group in the exact sequence

$$Wh(\pi) \xrightarrow{1-\alpha} Wh(\pi) \xrightarrow{j} Wh(\pi,\alpha) \xrightarrow{\partial} \widetilde{K}_0(\mathbb{Z}[\pi]) \xrightarrow{1-\alpha} \widetilde{K}_0(\mathbb{Z}[\pi])$$

(the reduced version of the group $K_1(\mathbb{Z}[\pi],\alpha)$ discussed at the end of §1) and $\widetilde{Nil}(\mathbb{Z}[\pi],\alpha^{\pm 1})$ the exotic K-group of pairs (F,ν) with F a f.g. free $\mathbb{Z}[\pi]$-module and $\nu \in \mathrm{Hom}_{\mathbb{Z}[\pi]}((\alpha^{\pm 1})_! F,F)$ nilpotent. Given a f.g. projective $\mathbb{Z}[\pi]$-module P and an isomorphism $f \in \mathrm{Hom}_{\mathbb{Z}[\pi]}(\alpha_! P,P)$ there is defined a mixed invariant $[P,f] \in Wh(\pi,\alpha)$ with $\partial([P,f]) = [P] \in \widetilde{K}_0(\mathbb{Z}[\pi])$. As in the untwisted case $\alpha = 1$ there are defined an <u>algebraically significant</u> splitting of $Wh(\pi\times_\alpha\mathbb{Z})$, with inverse isomorphisms

$$Wh(\pi\times_\alpha\mathbb{Z}) \underset{(\bar{B}\ \ \bar{\Delta}_+\ \bar{\Delta}_-)}{\overset{\begin{pmatrix} B \\ \Delta_+ \\ \Delta_- \end{pmatrix}}{\rightleftarrows}} Wh(\pi,\alpha)\oplus\widetilde{Nil}(\mathbb{Z}[\pi],\alpha)\oplus\widetilde{Nil}(\mathbb{Z}[\pi],\alpha^{-1}) \ ,$$

and a <u>geometrically significant</u> splitting of $Wh(\pi\times_\alpha\mathbb{Z})$ with inverse isomorphisms

$$Wh(\pi\times_\alpha\mathbb{Z}) \underset{(\bar{B}'\ \bar{\Delta}_+\ \bar{\Delta}_-)}{\overset{\begin{pmatrix} B' \\ \Delta_+ \\ \Delta_- \end{pmatrix}}{\rightleftarrows}} Wh(\pi,\alpha)\oplus\widetilde{Nil}(\mathbb{Z}[\pi],\alpha)\oplus\widetilde{Nil}(\mathbb{Z}[\pi],\alpha^{-1}) \quad ,$$

with

$$\bar{B} : Wh(\pi,\alpha) \rightarrowtail Wh(\pi\times_\alpha\mathbb{Z}) \ ; \ [P,f] \longmapsto \tau(zf:P_\alpha[z,z^{-1}] \longrightarrow P_\alpha[z,z^{-1}])$$

$$\bar{B}' : Wh(\pi,\alpha) \rightarrowtail Wh(\pi\times_\alpha\mathbb{Z}) \ ; \ [P,f] \longmapsto \tau(-zf:P_\alpha[z,z^{-1}] \longrightarrow P_\alpha[z,z^{-1}])$$

$$\bar{\Delta}_\pm : \widetilde{Nil}(\mathbb{Z}[\pi],\alpha^{\pm 1}) \rightarrowtail Wh(\pi\times_\alpha\mathbb{Z}) \ ;$$
$$(P,\nu) \longmapsto \tau(1+z^{\pm 1}\nu:P_\alpha[z,z^{-1}] \longrightarrow P_\alpha[z,z^{-1}]) \ ,$$

identifying $\mathbb{Z}[\pi\times_\alpha\mathbb{Z}] = \mathbb{Z}[\pi]_\alpha[z,z^{-1}]$. The automorphism

$$\Omega : Wh(\pi,\alpha) \longrightarrow Wh(\pi,\alpha) \ ; \ [P,f] \longmapsto [P,-f]$$

is such that $\Omega^2 = 1$ and

$$\bar{B}' = \bar{B}\Omega : Wh(\pi,\alpha) \rightarrowtail Wh(\pi\times_\alpha\mathbb{Z})$$

$$B' = \Omega B : Wh(\pi\times_\alpha\mathbb{Z}) \twoheadrightarrow Wh(\pi,\alpha) \ .$$

In the untwisted case $\alpha = 1$ $\pi\times_\alpha\mathbb{Z}$ is just the product $\pi\times\mathbb{Z}$, and there is defined an isomorphism

$$Wh(\pi)\oplus\tilde{K}_0(\mathbb{Z}[\pi]) \longrightarrow Wh(\pi,1) \ ;$$
$$(\tau(f:P\longrightarrow P),[Q]) \longmapsto [P,f] - [P,1] + [Q,1]$$

with respect to which

$$\Omega = \begin{pmatrix} 1 & \omega \\ 0 & 1 \end{pmatrix} : Wh(\pi)\oplus\tilde{K}_0(\mathbb{Z}[\pi]) \longrightarrow Wh(\pi)\oplus\tilde{K}_0(\mathbb{Z}[\pi]) \ .$$

The algebraically (resp. geometrically) significant splitting of $Wh(\pi\times_\alpha\mathbb{Z})$ for $\alpha = 1$ corresponds under this isomorphism to the algebraically (resp. geometrically) significant splitting of $Wh(\pi\times\mathbb{Z})$ defined previously.

A self homotopy equivalence $f:X\longrightarrow X$ of a finitely dominated CW complex X has a <u>mixed invariant</u>

$$[X,f] \in Wh(\pi,\alpha)$$

with $\alpha = f_* : \pi = \pi_1(X) \longrightarrow \pi$, such that $\partial([X,f]) = [X] \in \tilde{K}_0(\mathbb{Z}[\pi])$, a reduction of the mixed invariant $[X,f] \in K_1(\mathbb{Z}[\pi],\alpha)$ described at the end of §1. Let $f^{-1}:X\longrightarrow X$ be a homotopy inverse, with homotopy $e:f^{-1}f \simeq 1:X\longrightarrow X$. The mapping tori of f and f^{-1} are related by the homotopy equivalence

$$U : T(f^{-1}) \longrightarrow T(f) \ ; \ (x,t) \longmapsto (e(x,t), 1-t)$$

inducing the isomorphism of fundamental groups

$$U_* : \pi_1(T(f^{-1})) = \pi \times_{\alpha^{-1}} \mathbb{Z} \longrightarrow \pi_1(T(f)) = \pi \times_\alpha \mathbb{Z} \ ;$$
$$g\ (\in \pi) \longmapsto g \ , \ z \longmapsto z^{-1} \ .$$

The torsion of U with respect to the canonical round finite structures given by Proposition 1.3 is

$$\tau(U) = \tau(-z\tilde{f} : C(\tilde{X})_\alpha[z,z^{-1}] \longrightarrow C(\tilde{X})_\alpha[z,z^{-1}]) \in K_1(\mathbb{Z}[\pi]_\alpha[z,z^{-1}]) \ ,$$

so that:

Proposition 3.4 The geometrically defined split injection is given geometrically by

$$\overline{B}' : Wh(\pi,\alpha) \rightarrowtail Wh(\pi \times_\alpha \mathbb{Z}) \ ; \ [X,f] \longmapsto \tau(U : T(f^{-1}) \longrightarrow T(f)) \ .$$

[]

Proposition 3.3 is just the untwisted case $\alpha = 1$ of Proposition 3.4, with $f = 1 : X \longrightarrow X$ and

$$U = 1 \times -1 : T(1:X \longrightarrow X) = X \times S^1 \longrightarrow T(1) = X \times S^1 \ ,$$
$$-1 : S^1 = \mathbb{R}/\mathbb{Z} \longrightarrow S^1 \ ; \ t \longmapsto 1-t \ .$$

The exact sequence

$$Wh(\pi) \xrightarrow{1-\alpha} Wh(\pi) \xrightarrow{\overline{\varepsilon}} Wh(\pi \times_\alpha \mathbb{Z})$$
$$\xrightarrow{\begin{pmatrix} \partial B \\ \Delta_+ \\ \Delta_- \end{pmatrix}} \tilde{K}_0(\mathbb{Z}[\pi]) \oplus \widetilde{Nil}(\mathbb{Z}[\pi],\alpha) \oplus \widetilde{Nil}(\mathbb{Z}[\pi],\alpha^{-1})$$
$$\xrightarrow{(1-\alpha \ \ 0 \ \ 0)} \tilde{K}_0(\mathbb{Z}[\pi]) \xrightarrow{\overline{\varepsilon}} \tilde{K}_0(\mathbb{Z}[\pi \times_\alpha \mathbb{Z}])$$
$$(\overline{\varepsilon} = \overline{B}j = \overline{B}'j \ , \ \partial B = \partial B')$$

has a geometric interpretation in terms of codimension 1 splitting obstructions for homotopy equivalences $f : M^n \longrightarrow X$ with $\pi_1(X) = \pi \times_\alpha \mathbb{Z}$ (Farrell and Hsiang [5],[7]), as in the untwisted case $\alpha = 1$.

The obstruction theory of Farrell [4] and Siebenmann [33] for fibering manifolds over S^1 can be used to give the injection $\overline{B}' : \tilde{K}_0(\mathbb{Z}[\pi]) \rightarrowtail Wh(\pi \times \mathbb{Z})$ a further degree of geometric significance, as follows.

Let $p:\overline{X} \longrightarrow X$ be the covering projection of a regular infinite cyclic cover of a connected space X, with $\overline{X}$ connected also. Let $\zeta:\overline{X} \longrightarrow \overline{X}$ be a generating covering translation, inducing the automorphism $\zeta_* = \alpha : \pi_1(\overline{X}) = \pi \longrightarrow \pi$. The map

$$T(\zeta) \longrightarrow X \ ; \ (x,t) \longmapsto p(x)$$

is a homotopy equivalence, inducing an isomorphism of fundamental groups $\pi_1(T(\zeta)) = \pi \times_\alpha \mathbb{Z} \longrightarrow \pi_1(X)$. If X is a finite CW complex and $\overline{X}$ is finitely dominated the canonical (round) finite structure on $T(\zeta)$ given by Proposition 1.3 can be used to define the <u>fibering obstruction</u>

$$\Phi(X) = \tau(T(\zeta) \longrightarrow X) \in Wh(\pi \times_\alpha \mathbb{Z}) \quad .$$

This is the invariant described (but not defined) by Siebenmann [31]. If X is a compact n-manifold with the finite structure determined by a handlebody decomposition then $\Phi(X) = 0$ if (and for $n \geqslant 6$ only if) X fibres over S^1 in a manner compatible with p, by the theory of Farrell [4] and Siebenmann [33].

Given a finitely dominated CW complex X with $\pi_1(X) = \pi$ let $Y \longrightarrow X \times S^1$ be a homotopy equivalence from a finite CW complex Y in the canonical finite structure. Embed $Y \subset S^N$ (N large) with closed regular neighbourhood an N-dimensional manifold with boundary $(Z,\partial Z)$, and let $(\overline{Z},\overline{\partial Z})$ be the infinite cyclic cover of $(Z,\partial Z)$ classified by the projection

$$\pi_1(Z) = \pi_1(\partial Z) = \pi_1(X \times S^1) = \pi \times \mathbb{Z} \longrightarrow\!\!\!\!\!\rightarrow \mathbb{Z} \ .$$

Thicken up the self homotopy equivalence transposing the S^1-factors

$$1 \times T : X \times S^1 \times S^1 \longrightarrow X \times S^1 \times S^1 \ ; \ (x,s,t) \longmapsto (x,t,s)$$

to a self homotopy equivalence of a pair

$$(f,\partial f) : (Z,\partial Z) \times S^1 \longrightarrow (Z,\partial Z) \times S^1$$

inducing on the fundamental group the automorphism

$$\pi \times \mathbb{Z} \times \mathbb{Z} \longrightarrow \pi \times \mathbb{Z} \times \mathbb{Z} \ ; \ (x,s,t) \longmapsto (x,t,s)$$

transposing the $\mathbb{Z}$-factors. Thus $(f,\partial f)$ lifts to a $\mathbb{Z}$-equivariant homotopy equivalence

$$(\overline{f},\overline{\partial f}) : (\overline{Z},\overline{\partial Z}) \times S^1 \longrightarrow (Z,\partial Z) \times \mathbb{R} \ .$$

In particular, this shows that ∂Z is a finite CW complex with a finitely dominated infinite cyclic cover $\overline{\partial Z}$.

<u>Proposition 3.5</u> The geometrically significant injection is such that

$$\overline{B}' : \tilde{K}_0(\mathbb{Z}[\pi]) \rightarrowtail Wh(\pi \times \mathbb{Z}) \ ; \ [X] \longmapsto \Phi(\partial Z) \ .$$

[]

§4. Absolute L-theory invariants

The duality involutions on the algebraic K-groups of a ring A with involution $\bar{}: A \longrightarrow A; a \longmapsto \bar{a}$ are defined as usual by

$$* : K_0(A) \longrightarrow K_0(A) \ ; \ [P] \longmapsto [P^*] \ , \ P^* = \mathrm{Hom}_A(P,A)$$

$$* : K_1(A) \longrightarrow K_1(A) \ ; \ \tau(f:P \longrightarrow P) \longmapsto \tau(f^*:P^* \longrightarrow P^*) \ ,$$

with reduced versions for $\tilde{K}_0(A)$, $\tilde{K}_1(A)$. We shall only be concerned with group rings $A = \mathbb{Z}[\pi]$ and the involution $\bar{g} = w(g)g^{-1}$ $(g \in \pi)$ determined by a group morphism $w : \pi \longrightarrow \mathbb{Z}_2 = \{\pm 1\}$, so that there is also defined a duality involution $*: Wh(\pi) \longrightarrow Wh(\pi)$.

The $\left\{\begin{matrix}\text{projective class}\\ \text{Whitehead torsion}\end{matrix}\right.$ of a $\left\{\begin{matrix}\text{finitely dominated}\\ \text{finite}\end{matrix}\right.$ n-dimensional geometric Poincaré complex X with $\pi_1(X) = \pi$

$$\begin{cases} [X] = [C(\tilde{X})] \in K_0(\mathbb{Z}[\pi]) \\ \tau(X) = \tau(C(\tilde{X})^{n-*} \longrightarrow C(\tilde{X})) \in Wh(\pi) \end{cases}$$

satisfies the usual duality formula

$$\begin{cases} [X]^* = (-)^n[X] \in K_0(\mathbb{Z}[\pi]) \\ \tau(X)^* = (-)^n\tau(X) \in Wh(\pi) \ . \end{cases}$$

The <u>torsion</u> of a round finite n-dimensional geometric Poincaré complex X

$$\tau(X) = \tau(C(\tilde{X})^{n-*} \longrightarrow C(\tilde{X})) \in K_1(\mathbb{Z}[\pi])$$

is such that

$$\tau(X)^* = (-)^n\tau(X) \in K_1(\mathbb{Z}[\pi]) \ .$$

The Poincaré duality chain equivalence for the universal cover $\tilde{S}^1 = \mathbb{R}$ of the circle S^1 is given by

$$\begin{array}{cccc} C(\tilde{S}^1)^{1-*} & : \mathbb{Z}[z,z^{-1}] & \xrightarrow{1-z^{-1}} & \mathbb{Z}[z,z^{-1}] \\ {\scriptstyle [S^1]\cap -}\downarrow & \downarrow{\scriptstyle 1} & & \downarrow{\scriptstyle -z} \\ C(\tilde{S}^1) & : \mathbb{Z}[z,z^{-1}] & \xrightarrow{1-z} & \mathbb{Z}[z,z^{-1}] \end{array} \ ,$$

so that S^1 has torsion

$$\begin{aligned} \tau(S^1) &= \tau([S^1]\cap - : C(\tilde{S}^1)^{1-*} \longrightarrow C(\tilde{S}^1)) \\ &= \tau(-z : \mathbb{Z}[z,z^{-1}] \longrightarrow \mathbb{Z}[z,z^{-1}]) \\ &\qquad \in K_1(\mathbb{Z}[z,z^{-1}]) \ . \end{aligned}$$

This is the special case $f = 1 : X = \{pt.\} \longrightarrow \{pt.\}$ of the following formula, which is the Poincaré complex version of Propositions 1.3,3.4.

Proposition 4.1 Let $f:X \longrightarrow X$ be a self homotopy equivalence of a finitely dominated n-dimensional geometric Poincaré complex X inducing the automorphism $f_* = \alpha : \pi_1(X) = \pi \longrightarrow \pi$ and the $\mathbb{Z}[\pi]$-module chain equivalence $\tilde{f} : \alpha_! C(\tilde{X}) \longrightarrow C(\tilde{X})$. The mapping torus T(f) is an (n+1)-dimensional geometric Poincaré complex with canonical round finite structure, with torsion

$$\tau(T(f)) = \tau(-z\tilde{f}:C(\tilde{X})_\alpha[z,z^{-1}] \longrightarrow C(\tilde{X})_\alpha[z,z^{-1}]) \in K_1(\mathbb{Z}[\pi]_\alpha[z,z^{-1}]) .$$

[]

For $f = 1 : X \longrightarrow X$ the formula of Proposition 4.1 gives

$$\begin{aligned}\tau(X \times S^1) &= \tau(-z:C(\tilde{X})[z,z^{-1}] \longrightarrow C(\tilde{X})[z,z^{-1}]) \\ &= [X]\otimes\tau(S^1) = \bar{B}'([X]) \in K_1(\mathbb{Z}[\pi][z,z^{-1}])\end{aligned}$$

with $[X] \in K_0(\mathbb{Z}[\pi])$ the projective class and $\bar{B}'$ the absolute version

$$\begin{aligned}\bar{B}' : K_0(\mathbb{Z}[\pi]) &\rightarrowtail K_1(\mathbb{Z}[\pi][z,z^{-1}]) ; \\ [P] &\longmapsto \tau(-z:P[z,z^{-1}] \longrightarrow P[z,z^{-1}])\end{aligned}$$

(also a split injection) of $\bar{B}':\tilde{K}_0(\mathbb{Z}[\pi]) \rightarrowtail Wh(\pi\times\mathbb{Z})$.

For a finitely presented group π every element $x \in \tilde{K}_0(\mathbb{Z}[\pi])$ is the finiteness obstruction $x = [X]$ of a finitely dominated CW complex X with $\pi_1(X) = \pi$, by the realization theorem of Wall [34]. We need the version for Poincaré complexes:

Proposition 4.2 (Pedersen and Ranicki [18]) For a finitely presented group π every element $x \in \tilde{K}_0(\mathbb{Z}[\pi])$ is the finiteness obstruction $x = [X]$ for a finitely dominated geometric Poincaré pair $(X,\partial X)$ with $\pi_1(X) = \pi$.

[]

The method of [18] used the obstruction theory of Siebenmann [30]. The construction of Proposition 3.5 gives a more direct method, since $(\bar{Z},\overline{\partial Z})$ is a finitely dominated (N-1)-dimensional geometric Poincaré pair with prescribed $[\bar{Z}] \in \tilde{K}_0(\mathbb{Z}[\pi])$. (Moreover, if the evident map of pairs $(e,\partial e):(Z,\partial Z) \longrightarrow S^1$ is made transverse regular at $\mathrm{pt.} \in S^1$ the inclusion

$$(M,\partial M) = (e,\partial e)^{-1}(\{\mathrm{pt.}\}) \longrightarrow (Z,\partial Z)$$

lifts to a normal map

$$(f,b) : (M,\partial M) \longrightarrow (\bar{Z},\overline{\partial Z})$$

from a compact (N-1)-dimensional manifold with boundary. This gives a more direct proof of the realization theorem of [18] for the projective surgery groups $L^p_*(\pi)$, except possibly in the low dimensions).

By the relative version of Proposition 4.1 the product of a finitely dominated n-dimensional geometric Poincaré pair $(X,\partial X)$ and the circle S^1 is an (n+1)-dimensional geometric Poincaré pair

$$(X,\partial X)\times S^1 = (X\times S^1,\partial X\times S^1)$$

with canonical round finite structure, and torsion

$$\begin{aligned}\tau(X\times S^1,\partial X\times S^1) &= \tau(-z:C(\tilde{X})[z,z^{-1}]\longrightarrow C(\tilde{X})[z,z^{-1}]) \\ &= [X]\otimes\tau(S^1) = \bar{B}'([X]) \in K_1(\mathbb{Z}[\pi][z,z^{-1}]) \ .\end{aligned}$$

Combined with Proposition 4.2 this gives:

Proposition 4.3 The geometrically significant injection is such that

$$\bar{B}' : \tilde{K}_0(\mathbb{Z}[\pi]) \rightarrowtail Wh(\pi\times\mathbb{Z}) \ ; \ [X] \longmapsto \tau(X\times S^1,\partial X\times S^1) \ ,$$

for any finitely dominated geometric Poincaré pair $(X,\partial X)$ with $\pi_1(X)=\pi$ []

In §5 this will be seen to be a special case of the product formula for the torsion of (finitely dominated) × (round finite) Poincaré complexes.

Given a $*$-invariant subgroup $S\subseteq\tilde{K}_0(\mathbb{Z}[\pi])$ (resp. $S\subseteq Wh(\pi)$) let $\begin{cases}L^n_S(\pi)\\ L_n^S(\pi)\end{cases}$ $(n\geq 0)$ be the cobordism group of finitely dominated (resp. finite) n-dimensional $\begin{cases}\text{symmetric}\\ \text{quadratic}\end{cases}$ Poincaré complexes over $\mathbb{Z}[\pi]$ $\begin{cases}(C,\phi\in Q^n(C))\\ (C,\psi\in Q_n(C))\end{cases}$ with finiteness obstruction $[C]\in S\subseteq\tilde{K}_0(\mathbb{Z}[\pi])$ (resp. Whitehead torsion $\begin{cases}\tau(C,\phi) = \tau(\phi_0:C^{n-*}\longrightarrow C)\\ \tau(C,\psi) = \tau((1+T)\psi_0:C^{n-*}\longrightarrow C)\end{cases} \in S\subseteq Wh(\pi)$).

A finitely dominated (resp. finite) n-dimensional geometric Poincaré complex X with $\pi_1(X)=\pi$ and $[X]\in S$ (resp. $\tau(X)\in S$) has a symmetric signature invariant

$$\sigma^*_S(X) = (C(\tilde{X}),\phi) \in L^n_S(\pi)$$

with $\phi_0 = [X]\cap - : C(\tilde{X})^{n-*}\longrightarrow C(\tilde{X})$, and a normal map $(f,b):M\longrightarrow X$ of such complexes has a quadratic signature invariant

$$\sigma^S_*(f,b) \in L_n^S(\pi)$$

such that $(1+T)\sigma^S_*(f,b) = \sigma^*_S(M) - \sigma^*_S(X)$. See Ranicki [22],[23] for the details. In the extreme cases $S=\{0\},\tilde{K}_0(\mathbb{Z}[\pi])$ (resp. $\{0\},Wh(\pi)$) the notation is abbreviated in the usual fashion

$$\begin{cases} L^n_{\tilde{K}_0(\mathbb{Z}[\pi])}(\pi) = L^n_p(\pi) \\ L_n^{\tilde{K}_0(\mathbb{Z}[\pi])}(\pi) = L_n^p(\pi) \end{cases} , \quad \begin{cases} L^n_{\{0\}\subseteq Wh(\pi)}(\pi) = L^n_s(\pi) \\ L_n^{\{0\}\subseteq Wh(\pi)}(\pi) = L_n^s(\pi) \end{cases}$$

$$\begin{cases} L^n_{\{0\}\subseteq\tilde{K}_0(\mathbb{Z}[\pi])}(\pi) = L^n_{Wh(\pi)}(\pi) = L^n_h(\pi) \\ L_n^{\{0\}\subseteq\tilde{K}_0(\mathbb{Z}[\pi])}(\pi) = L_n^{Wh(\pi)}(\pi) = L_n^h(\pi) \ . \end{cases}$$

In particular, the simple quadratic L-groups $L^s_*(\pi)$ are the original surgery obstruction groups of Wall [35], with $\sigma^s_*(f,b)$ the surgery obstruction.

The torsion of a round finite n-dimensional $\begin{cases}\text{symmetric}\\ \text{quadratic}\end{cases}$ Poincaré complex over $\mathbb{Z}[\pi]$ $\begin{cases}(C,\phi)\\ (C,\psi)\end{cases}$ is defined by

$$\begin{cases} \tau(C,\phi) = \tau(\phi_0 : C^{n-*} \longrightarrow C) \in K_1(\mathbb{Z}[\pi]) \\ \tau(C,\psi) = \tau((1+T)\psi_0 : C^{n-*} \longrightarrow C) \in K_1(\mathbb{Z}[\pi]) \end{cases}$$

and is such that

$$\begin{cases} \tau(C,\phi)^* = (-)^n\tau(C,\phi) \\ \tau(C,\psi)^* = (-)^n\tau(C,\psi) \end{cases} \in K_1(\mathbb{Z}[\pi]) \ .$$

Given a $*$-invariant subgroup $S\subseteq K_1(\mathbb{Z}[\pi])$ define the round $\begin{cases}\text{symmetric}\\ \text{quadratic}\end{cases}$ L-group $\begin{cases}L^n_{rS}(\pi)\\ L_n^{rS}(\pi)\end{cases}$ $(n\geqslant 0)$ to be the cobordism group of round finite n-dimensional $\begin{cases}\text{symmetric}\\ \text{quadratic}\end{cases}$ Poincaré complexes over $\mathbb{Z}[\pi]$ $\begin{cases}(C,\phi)\\ (C,\psi)\end{cases}$ with torsion $\begin{cases}\tau(C,\phi)\\ \tau(C,\psi)\end{cases} \in S\subseteq K_1(\mathbb{Z}[\pi])$. See Hambleton, Ranicki and Taylor [11] for an exposition of round L-theory. We shall only be concerned with the round symmetric L-groups L^*_{rS} here, adopting the terminology

$$L^n_{rh}(\pi) = L^n_{rK_1(\mathbb{Z}[\pi])}(\pi) \quad , \quad L^n_{rs}(\pi) = L^n_{r\{\pm\pi\}}(\pi) \ .$$

The Rothenberg exact sequence for the quadratic L-groups

$$\dots \longrightarrow L_n^s(\pi) \longrightarrow L_n^h(\pi) \longrightarrow \hat{H}^n(\mathbb{Z}_2;Wh(\pi)) \longrightarrow L_{n-1}^s(\pi) \longrightarrow \dots$$

has versions for the symmetric and round symmetric L-groups which fit together in a commutative braid of exact sequences

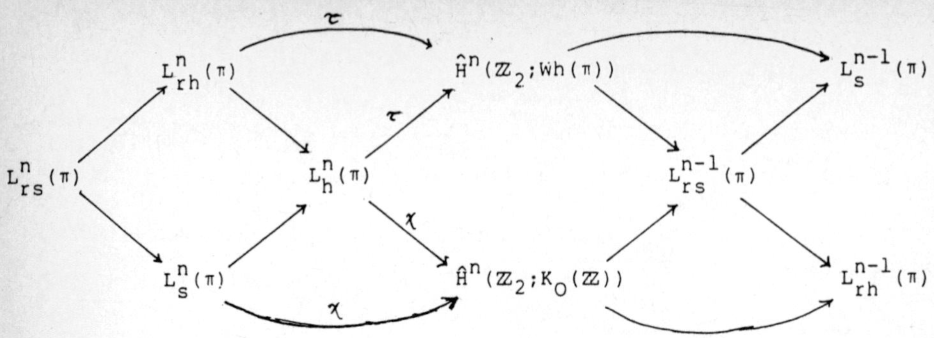

with the maps τ (resp. χ) defined by the Whitehead torsion (resp. Euler characteristic). In the case $Wh(\pi) = 0$ the L-groups $\begin{cases} L^*_{rh}(\pi) = L^*_{rs}(\pi) \\ L^*_h(\pi) = L^*_s(\pi) \end{cases}$ are abbreviated to $\begin{cases} L^*_r(\pi) \\ L^*(\pi) \end{cases}$. The L-groups of the trivial group $\pi = \{1\}$ are given by

$$L^n(\{1\}) = \begin{cases} \mathbb{Z} \\ \mathbb{Z}_2 \\ 0 \\ 0 \end{cases}, \quad L^n_r(\{1\}) = \begin{cases} \mathbb{Z} \\ \mathbb{Z}_2 \oplus \mathbb{Z}_2 \\ 0 \\ 0 \end{cases} \quad \text{if } n \equiv \begin{cases} 0 \\ 1 \\ 2 \\ 3 \end{cases} \pmod 4 ,$$

with isomorphisms

$$L^{4k}(\{1\}) \longrightarrow \mathbb{Z} \ ; \ (C,\phi) \longmapsto \text{signature}(C,\phi)$$

$$L^{4k+1}(\{1\}) \longrightarrow \mathbb{Z}_2 \ ; \ (C,\phi) \longmapsto \text{deRham}(C,\phi) = \chi_{\frac{1}{2}}(C;\mathbb{Z}_2) + \chi_{\frac{1}{2}}(C;\mathbb{Q})$$

$$L^{4k}_r(\{1\}) \longrightarrow \mathbb{Z} \ ; \ (C,\phi) \longmapsto \tfrac{1}{2}(\text{signature}(C,\phi))$$

$$L^{4k+1}_r(\{1\}) \longrightarrow \mathbb{Z}_2 \oplus \mathbb{Z}_2 \ ; \ (C,\phi) \longmapsto (\chi_{\frac{1}{2}}(C;\mathbb{Z}_2), \chi_{\frac{1}{2}}(C;\mathbb{Q})) \ .$$

(See [11] for details. The F-coefficient semicharacteristic of a (2i+1)-dimensional $\mathbb{Z}$-module chain complex C is defined by

$$\chi_{\frac{1}{2}}(C;F) = \sum_{r=0}^{i} (-)^r \text{rank}_F H_r(C) \in \mathbb{Z} ,$$

for any field F).

The torsion of a round finite n-dimensional geometric Poincaré complex X with $\pi_1(X) = \pi$ is the torsion of the associated round finite n-dimensional symmetric Poincaré complex over $\mathbb{Z}[\pi]$ $(C(\tilde{X}),\phi)$

$$\tau(X) = \tau(C(\tilde{X}),\phi) = \tau(\phi_0 = [X] \cap - : C(\tilde{X})^{n-*} \longrightarrow C(\tilde{X})) \in K_1(\mathbb{Z}[\pi]) \ .$$

If $S \subseteq K_1(\mathbb{Z}[\pi])$ is a $*$-invariant subgroup such that $\tau(X) \in S$ the <u>round</u>

<u>symmetric signature</u> of X is defined by

$$\sigma^*_{rS}(X) = (C(\tilde{X}),\phi) \in L^n_{rS}(\pi) .$$

In the case $S = K_1(\mathbb{Z}[\pi])$ (resp. $\{\pm\pi\}$) this is denoted $\sigma^*_{rh}(X) \in L^n_{rh}(\pi)$ (resp. $\sigma^*_{rs}(X) \in L^n_{rs}(\pi)$), and if also $Wh(\pi) = 0$ by $\sigma^*_r(X) \in L^n_r(\pi)$.

We shall be particularly concerned with the round symmetric signature of the circle S^1

$$\sigma^*_r(S^1) = (C(\tilde{S}^1),\phi) \in L^1_r(\mathbb{Z}) .$$

The image of the $\mathbb{Z}[z,z^{-1}]$-module chain complex

$$C(\tilde{S}^1) : \mathbb{Z}[z,z^{-1}] \xrightarrow{1-z} \mathbb{Z}[z,z^{-1}]$$

under the morphism of rings with involution

$$\begin{cases} \varepsilon : \mathbb{Z}[\mathbb{Z}] = \mathbb{Z}[z,z^{-1}] \longrightarrow \mathbb{Z} \; ; \; z \longmapsto 1 \\ \eta : \mathbb{Z}[\mathbb{Z}] = \mathbb{Z}[z,z^{-1}] \longrightarrow \mathbb{Z} \; ; \; z \longmapsto -1 \end{cases} \qquad (\bar{z} = z^{-1})$$

is the $\mathbb{Z}$-module chain complex

$$\begin{cases} \varepsilon_! C(\tilde{S}^1) : \mathbb{Z} \xrightarrow{0} \mathbb{Z} \\ \eta_! C(\tilde{S}^1) : \mathbb{Z} \xrightarrow{2} \mathbb{Z} \end{cases}$$

with mod2 and rational semicharacteristics $\begin{cases} (\chi_{1/2}(C;\mathbb{Z}_2),\chi_{1/2}(C;\mathbb{Q})) = (1,1) \\ (\chi_{1/2}(D;\mathbb{Z}_2),\chi_{1/2}(D;\mathbb{Q})) = (1,0) \end{cases}$

so that $\sigma^*_r(S^1) \in L^1_r(\mathbb{Z})$ has images

$$\begin{cases} \varepsilon_!\sigma^*_r(S^1) = (1,1) \\ \eta_!\sigma^*_r(S^1) = (1,0) \end{cases} \in L^1_r(\{1\}) = \mathbb{Z}_2 \oplus \mathbb{Z}_2 .$$

The algebraic proof of the splitting theorem for the quadratic L-groups $L^s_n(\pi\times\mathbb{Z}) = L^s_n(\pi)\oplus L^h_{n-1}(\pi)$ discussed in §6 below can be extended to prove analogous splitting theorems for the symmetric and round symmetric L-groups

$$L^n_s(\pi\times\mathbb{Z}) = L^n_s(\pi)\oplus L^{n-1}_h(\pi) \quad , \quad L^n_{rs}(\pi\times\mathbb{Z}) = L^n_{rs}(\pi)\oplus L^{n-1}_h(\pi) .$$

Thus $L^1_r(\mathbb{Z}) = L^1_r(\{1\})\oplus L^0(\{1\}) = \mathbb{Z}_2\oplus\mathbb{Z}_2\oplus\mathbb{Z}$, although we do not actually need this computation here.

§5. Products in L-theory

The product of an m-dimensional $\begin{cases}\text{symmetric}\\ \text{quadratic}\end{cases}$ Poincaré complex over A (C,ϕ) and an n-dimensional symmetric Poincaré complex over B (D,θ) is an (m+n)-dimensional $\begin{cases}\text{symmetric}\\ \text{quadratic}\end{cases}$ Poincaré complex over $A\otimes B$

$$(C,\phi)\otimes(D,\theta) = (C\otimes D,\phi\otimes\theta) \quad ,$$

allowing the definition (in Ranicki [22]) of products in L-theory of the type

$$\begin{cases} L^m(A)\otimes L^n(B) \longrightarrow L^{m+n}(A\otimes B) \\ L_m(A)\otimes L^n(B) \longrightarrow L_{m+n}(A\otimes B) \ . \end{cases}$$

We shall only be concerned with the product $L_m\otimes L^n \longrightarrow L_{m+n}$ here, with $A = \mathbb{Z}[\pi]$, $B = \mathbb{Z}[\rho]$ group rings, so that $A\otimes B = \mathbb{Z}[\pi\times\rho]$.

The product of a $\begin{cases}\text{finitely dominated}\\ \text{finite}\end{cases}$ m-dimensional symmetric (resp. quadratic) Poincaré complex over $\mathbb{Z}[\pi]$ (C,ϕ) and a $\begin{cases}\text{finitely dominated}\\ \text{finite}\end{cases}$ n-dimensional symmetric Poincaré complex over $\mathbb{Z}[\rho]$ (D,θ) is a $\begin{cases}\text{finitely dominated}\\ \text{finite}\end{cases}$ (m+n)-dimensional symmetric (resp. quadratic) Poincaré complex over $\mathbb{Z}[\pi\times\rho]$ $(C\otimes D,\phi\otimes\theta)$ with $\begin{cases}\text{projective class}\\ \text{Whitehead torsion}\end{cases}$

$$\begin{cases} [C\otimes D] = [C]\otimes[D] \in K_0(\mathbb{Z}[\pi\times\rho]) \\ \tau(C\otimes D,\phi\otimes\theta) = \tau(C,\phi)\otimes\chi(D) + \chi(C)\otimes\tau(D,\theta) \in Wh(\pi\times\rho) \ . \end{cases}$$

The following product formulae for geometric Poincaré complexes are immediate consequences.

<u>Proposition 5.1</u> The product of a $\begin{cases}\text{finitely dominated}\\ \text{finite}\end{cases}$ m-dimensional geometric Poincaré complex X with $\pi_1(X) = \pi$ and a $\begin{cases}\text{finitely dominated}\\ \text{finite}\end{cases}$ n-dimensional geometric Poincaré complex F with $\pi_1(F) = \rho$ is a $\begin{cases}\text{finitely dominated}\\ \text{finite}\end{cases}$ (m+n)-dimensional geometric Poincaré complex $X\times F$

with $\begin{cases} \text{projective class} \\ \text{Whitehead torsion} \end{cases}$

$$\begin{cases} [X\times F] = [X]\otimes[F] \in K_0(\mathbb{Z}[\pi\times\rho]) \\ \tau(X\times F) = \tau(X)\otimes\chi(F) + \chi(X)\otimes\tau(F) \in Wh(\pi\times\rho) \ . \end{cases}$$

[]

Given $*$-invariant subgroups $\begin{cases} S\subseteq \tilde{K}_0(\mathbb{Z}[\pi]) \\ S\subseteq Wh(\pi) \end{cases}$, $\begin{cases} T\subseteq \tilde{K}_0(\mathbb{Z}[\rho]) \\ T\subseteq Wh(\rho) \end{cases}$,

$\begin{cases} U\subseteq \tilde{K}_0(\mathbb{Z}[\pi\times\rho]) \\ U\subseteq Wh(\pi\times\rho) \end{cases}$ such that $\begin{cases} [P\otimes Q]\in U \\ \tau(f)\otimes 1, 1\otimes\tau(g)\in U \end{cases}$ for $\begin{cases} [P]\in S,\ [Q]\in T \\ \tau(f)\in S,\ \tau(g)\in T \end{cases}$

there is defined a product in L-theory

$$\otimes\ :\ L^S_m(\pi)\otimes L^n_T(\rho)\longrightarrow L^U_{m+n}(\pi\times\rho)\ ;\ (C,\psi)\otimes(D,\theta)\longmapsto(C\otimes D,\psi\otimes\theta)$$

with the following geometric interpretation.

<u>Proposition 5.2</u> (Ranicki [23]) If $(f,b):M\longrightarrow X$ is a normal map of $\begin{cases} \text{finitely dominated} \\ \text{finite} \end{cases}$ m-dimensional geometric Poincaré complexes with

$\pi_1(X) = \pi$ and $\begin{cases} [M]-[X]\in S\subseteq \tilde{K}_0(\mathbb{Z}[\pi]) \\ \tau(M)-\tau(X)\in S\subseteq Wh(\pi) \end{cases}$, and if F is a

$\begin{cases} \text{finitely dominated} \\ \text{finite} \end{cases}$ n-dimensional geometric Poincaré complex with

$\pi_1(F) = \rho$ and $\begin{cases} [F]\in T\subseteq \tilde{K}_0(\mathbb{Z}[\rho]) \\ \tau(F)\in T\subseteq Wh(\rho) \end{cases}$, then the quadratic signature of the

product normal map of $\begin{cases} \text{finitely dominated} \\ \text{finite} \end{cases}$ (m+n)-dimensional geometric Poincaré complexes

$$(g,c) = (f,b)\times 1\ :\ M\times F\longrightarrow X\times F$$

is given by

$$\sigma^U_*(g,c) = \sigma^S_*(f,b)\otimes\sigma^*_T(F)\in L^U_{m+n}(\pi\times\rho)\ ,$$

the product of $\sigma^S_*(f,b)\in L^S_m(\pi)$ and $\sigma^*_T(F)\in L^n_T(\rho)$.

[]

The methods of Ranicki [26] apply to the products of algebraic Poincaré complexes, giving the following analogues of Propositions 2.2, 5.2:

Proposition 5.3 i) The product of a finitely dominated m-dimensional quadratic Poincaré complex over $\mathbb{Z}[\pi]$ (C,ψ) and a round finite n-dimensional symmetric Poincaré complex over $\mathbb{Z}[\rho]$ (D,θ) is an $(m+n)$-dimensional quadratic Poincaré complex over $\mathbb{Z}[\pi\times\rho]$ $(C\otimes D,\psi\otimes\theta)$ with canonical round finite structure, and torsion

$$\tau(C\otimes D,\psi\otimes\theta) = [C]\otimes\tau(D,\theta) \in K_1(\mathbb{Z}[\pi\times\rho])$$

the product of $[C]\in K_0(\mathbb{Z}[\pi])$ and $\tau(D,\theta)\in K_1(\mathbb{Z}[\rho])$.

ii) Given *-invariant subgroups $S\subseteq\tilde{K}_0(\mathbb{Z}[\pi])$, $T\subseteq K_1(\mathbb{Z}[\rho])$, $U\subseteq Wh(\pi\times\rho)$ such that $S\otimes T\subseteq U$ there is defined a product in L-theory

$$\otimes : L^S_m(\pi)\otimes L^n_{rT}(\rho) \longrightarrow L^U_{m+n}(\pi\times\rho) \ ; \ (C,\psi)\otimes(D,\theta) \longmapsto (C\otimes D,\psi\otimes\theta) \ .$$

If $(f,b):M\longrightarrow X$ is a normal map of finitely dominated n-dimensional geometric Poincaré complexes with $\pi_1(X)=\pi$ and $[M]-[X]\in S\subseteq\tilde{K}_0(\mathbb{Z}[\pi])$, and if F is a round finite n-dimensional geometric Poincaré complex with $\pi_1(F)=\rho$ and $\tau(F)\in T\subseteq K_1(\mathbb{Z}[\rho])$ then the product map of $(m+n)$-dimensional geometric Poincaré complexes with canonical (round) finite structure

$$(g,c) = (f,b)\times 1 : M\times F \longrightarrow X\times F$$

has quadratic signature

$$\sigma^U_*(g,c) = \sigma^S_*(f,b)\otimes\sigma^*_{rT}(F) \in L^U_{m+n}(\pi\times\rho)$$

the product of $\sigma^S_*(f,b)\in L^S_m(\pi)$ and $\sigma^*_{rT}(F)\in L^n_{rT}(\rho)$.

[]

An n-dimensional geometric Poincaré complex F is <u>round simple</u> if it is round finite and

$$\tau(F) \in \{\pm\rho\}\subseteq K_1(\mathbb{Z}[\rho]) \quad (\rho = \pi_1(F)) \ ,$$

so that $\tau(F) = 0 \in Wh(\rho)$ and the round simple symmetric signature $\sigma^*_{rs}(F)\in L^n_{rs}(\rho)$ is defined.

Proposition 5.3 shows in particular that for a round $\begin{cases}\text{finite}\\\text{simple}\end{cases}$ n-dimensional geometric Poincaré complex F product with the round $\begin{cases}\text{finite}\\\text{simple}\end{cases}$ symmetric signature $\begin{cases}\sigma^*_{rh}(F)\in L^n_{rh}(\rho)\\\sigma^*_{rs}(F)\in L^n_{rs}(\rho)\end{cases}$ defines a morphism of

$$\begin{cases} -\otimes\sigma^*_{rh}(F) \ : \ L^p_m(\pi) \longrightarrow L^h_{m+n}(\pi\times\rho) \\ -\otimes\sigma^*_{rs}(F) \ : \ L^h_m(\pi) \longrightarrow L^s_{m+n}(\pi\times\rho) \ . \end{cases}$$

In the simple case these products define a map of generalized Rothenberg exact sequences

$$\begin{array}{ccccccccc} \ldots \longrightarrow & L^h_m(\pi) & \longrightarrow & L^p_m(\pi) & \longrightarrow & \hat{H}^m(\mathbb{Z}_2;\tilde{K}_0(\mathbb{Z}[\pi])) & \longrightarrow & L^h_{m-1}(\pi) & \longrightarrow \ldots \\ & \Big\downarrow {\scriptstyle -\otimes\sigma^*_{rs}(F)} & & \Big\downarrow {\scriptstyle -\otimes\sigma^*_{rh}(F)} & & \Big\downarrow {\scriptstyle -\otimes\tau(F)} & & \Big\downarrow {\scriptstyle -\otimes\sigma^*_{rs}(F)} & \\ \ldots \longrightarrow & L^s_{m+n}(\pi\times\rho) & \longrightarrow & L^h_{m+n}(\pi\times\rho) & \longrightarrow & \hat{H}^{m+n}(\mathbb{Z}_2;Wh(\pi\times\rho)) & \longrightarrow & L^s_{m+n-1}(\pi\times\rho) & \longrightarrow \ldots \end{array}$$

with $\tau(F) \in \{\pm\rho\} \subseteq K_1(\mathbb{Z}[\rho])$. The map of exact sequences in the appendix of Munkholm and Ranicki [16] is the special case $F = S^1$. Moreover, the split injection

$$\overline{B}' = -\otimes\tau(S^1) \ : \ \hat{H}^m(\mathbb{Z}_2;\tilde{K}_0(\mathbb{Z}[\pi])) \longrightarrow \hat{H}^{m+1}(\mathbb{Z}_2;Wh(\pi\times\mathbb{Z}))$$

was identified there with the connecting map δ arising from a short exact sequence of $\mathbb{Z}[\mathbb{Z}_2]$-modules

$$0 \longrightarrow Wh(\pi\times\mathbb{Z}) \longrightarrow Wh(p^!) \longrightarrow \tilde{K}_0(\mathbb{Z}[\pi]) \longrightarrow 0 \quad ,$$

with $Wh(p^!)$ the relative Whitehead group in the exact sequence of transfer maps

$$Wh(\pi) \xrightarrow{\tilde{p}^!_1 = 0} Wh(\pi\times\mathbb{Z}) \longrightarrow Wh(p^!) \longrightarrow \tilde{K}_0(\mathbb{Z}[\pi]) \xrightarrow{\tilde{p}^!_0 = 0} \tilde{K}_0(\mathbb{Z}[\pi\times\mathbb{Z}])$$

associated to the trivial S^1-bundle

$$S^1 \longrightarrow E = K(\pi,1) \times S^1 \xrightarrow{\ p\ =\ \text{projection}\ } B = K(\pi,1)$$

and $\mathbb{Z}_2$ acting by duality involutions. The relationship between transfer maps and duality in algebraic K-theory will be studied in Lück and Ranicki [13] for any fibration $F \longrightarrow E \xrightarrow{\ p\ } B$ with the fibre F a finitely dominated n-dimensional geometric Poincaré complex. In particular, there will be defined a duality involution $* : K_1(p^!) \longrightarrow K_1(p^!)$ on the relative K-group $K_1(p^!)$ in the transfer exact sequence

$$K_1(\mathbb{Z}[\pi_1(B)]) \xrightarrow{p^!_1} K_1(\mathbb{Z}[\pi_1(E)]) \longrightarrow K_1(p^!)$$
$$\longrightarrow K_0(\mathbb{Z}[\pi_1(B)]) \xrightarrow{p^!_0} K_0(\mathbb{Z}[\pi_1(E)]) \ ,$$

as well as assorted transfer maps $p^! : L_m(\pi_1(B)) \longrightarrow L_{m+n}(\pi_1(E))$ in algebraic L-theory. If F is round simple and $\pi_1(B)$ acts on F by self equivalences $F \longrightarrow F$ with $\tau = 0 \in Wh(\pi_1(E))$ (e.g. if p is a PL bundle with a round manifold fibre) then there is also defined a transfer exact sequence

$$\begin{array}{l} Wh(\pi_1(B)) \xrightarrow{\widetilde{p}_1^!} Wh(\pi_1(E)) \longrightarrow Wh(p^!) \\ \qquad\qquad \longrightarrow \widetilde{K}_0(\mathbb{Z}[\pi_1(B)]) \xrightarrow{\widetilde{p}_0^!} \widetilde{K}_0(\mathbb{Z}[\pi_1(E)]) \end{array}$$

with a duality involution $*: Wh(p^!) \longrightarrow Wh(p^!)$ on the relative Whitehead group. The connecting maps δ in Tate $\mathbb{Z}_2$-cohomology arising from the short exact sequence of $\mathbb{Z}[\mathbb{Z}_2]$-modules

$$0 \longrightarrow \mathrm{coker}(\widetilde{p}_1^!) \longrightarrow Wh(p^!) \longrightarrow \ker(\widetilde{p}_0^!) \longrightarrow 0$$

and the transfer maps in L-theory together define a morphism of exact sequences

$$\begin{array}{ccccccccc} \dots \longrightarrow & L_m^h(\pi) & \longrightarrow & L_m^{\ker(\widetilde{p}_0^!)}(\pi) & \longrightarrow & \hat{H}^m(\mathbb{Z}_2;\ker(\widetilde{p}_0^!)) & \longrightarrow & L_{m-1}^h(\pi) & \longrightarrow \dots \\ & \downarrow{\scriptstyle p^!} & & \downarrow{\scriptstyle p^!} & & \downarrow{\scriptstyle \delta} & & \downarrow{\scriptstyle p^!} & \\ \dots \longrightarrow & L_{m+n}^{\mathrm{im}(\widetilde{p}_1^!)}(\Pi) & \longrightarrow & L_{m+n}^h(\Pi) & \longrightarrow & \hat{H}^{m+n}(\mathbb{Z}_2;\mathrm{coker}(\widetilde{p}_1^!)) & \longrightarrow & L_{m+n-1}^{\mathrm{im}(\widetilde{p}_1^!)}(\Pi) & \longrightarrow \dots \end{array}$$

$$(\pi = \pi_1(B),\ \Pi = \pi_1(E)) .$$

In the case of the trivial fibration

$$F \longrightarrow E = B \times F \xrightarrow{p\ =\ \text{projection}} B$$

(with the fibre F a round simple Poincaré complex, as before) the algebraic K-theory transfer maps are zero

$$p_i^! = -\otimes[F] = 0 : K_i(\mathbb{Z}[\pi]) \longrightarrow K_i(\mathbb{Z}[\pi\times\rho])$$
$$(i = 0,1\ \ \rho = \pi_1(F))$$

so that $\widetilde{p}_i^! = 0$. Also, the algebraic L-theory transfer maps are given by the products with the round symmetric signatures

$$\begin{array}{l} p^! = -\otimes\sigma^*_{rh}(F) : L_m^p(\pi) \longrightarrow L_{m+n}^h(\pi\times\rho) \\ p^! = -\otimes\sigma^*_{rs}(F) : L_m^h(\pi) \longrightarrow L_{m+n}^s(\pi\times\rho) \ , \end{array}$$

and δ is given by product with the torsion $\tau(F) \in \{\pm\rho\} \subseteq K_1(\mathbb{Z}[\rho])$

$$\delta = -\otimes\tau(F) : \hat{H}^m(\mathbb{Z}_2;\widetilde{K}_0(\mathbb{Z}[\pi])) \longrightarrow \hat{H}^{m+n}(\mathbb{Z}_2;Wh(\pi\times\rho))$$

as in the case $F = S^1$ considered in [16].

§6. The L-groups of a polynomial extension

There are 4 ways of extending an involution $a \longmapsto \bar{a}$ on a ring A to an involution on the Laurent polynomial extension ring $A[z,z^{-1}]$, sending z to one of $z, z^{-1}, -z, -z^{-1}$. In each case it is possible to express $L_*(A[z,z^{-1}])$ (and indeed $L^*(A[z,z^{-1}])$) in terms of $L_*(A)$, and to relate such an expression to splitting theorems for manifolds – see Chapter 7 of Ranicki [24] for a general account of algebraic and geometric splitting theorems in L-theory. Only the case

$$A = \mathbb{Z}[\pi] \quad , \quad \bar{z} = z^{-1}$$

is considered here, for which $A[z,z^{-1}] = \mathbb{Z}[\pi][z,z^{-1}]$.

The geometric splittings of the L-groups $L_*(\pi \times \mathbb{Z})$ depend on the

realization theorem of $\left\{\begin{array}{l}\text{Wall [35]}\\ \text{Shaneson [29]}\\ \text{Pedersen and Ranicki [18]}\end{array}\right.$, by which every

element of $\left\{\begin{array}{l}L^s_n(\pi)\\ L^h_n(\pi)\\ L^p_n(\pi)\end{array}\right.$ $(n \geqslant 5$, π finitely presented) is the $\left\{\begin{array}{l}\text{simple}\\ \text{finite}\\ \text{projective}\end{array}\right.$

rel∂ surgery obstruction $\left\{\begin{array}{l}\sigma^s_*(f,b)\\ \sigma^h_*(f,b)\\ \sigma^p_*(f,b)\end{array}\right.$ of a normal map

$$(f,b) : (M,\partial M) \longrightarrow (X,\partial X)$$

from a compact n-dimensional manifold with boundary $(M,\partial M)$ to a

$\left\{\begin{array}{l}\text{simple}\\ \text{finite}\\ \text{finitely dominated}\end{array}\right.$ n-dimensional geometric Poincaré pair $(X,\partial X)$

equipped with a reference map $X \longrightarrow K(\pi,1)$, and such that the

restriction $\partial f = f| : \partial M \longrightarrow \partial X$ is a $\left\{\begin{array}{l}\text{simple}\\ -\\ -\end{array}\right.$ homotopy equivalence.

A morphism of groups

$$\phi : \pi \longrightarrow \Pi$$

induces functorially morphisms in the L-groups, given geometrically by

$$\phi_! : L_n^q(\pi) \longrightarrow L_n^q(\Pi) ;$$

$$\sigma_*^q((M,\partial M) \xrightarrow{(f,b)} (X,\partial X) \longrightarrow K(\pi,1))$$

$$\longmapsto \sigma_*^q((M,\partial M) \xrightarrow{(f,b)} (X,\partial X) \longrightarrow K(\pi,1) \xrightarrow{\phi} K(\Pi,1))$$

$$(q = s,h,p) ,$$

and algebraically by

$$\phi_! : L_n^q(\pi) \longrightarrow L_n^q(\Pi) ; \sigma_*^q(f,b) \longmapsto \mathbb{Z}[\Pi] \otimes_{\mathbb{Z}[\pi]} \sigma_*^q(f,b) .$$

In general $\phi_!$ will be written ϕ.

The geometric splitting of Shaneson [29]

$$L_n^s(\pi \times \mathbb{Z}) = L_n^s(\pi) \oplus L_{n-1}^h(\pi)$$

was obtained in the form of a split exact sequence

$$0 \longrightarrow L_n^s(\pi) \xrightarrow{\overline{\varepsilon}} L_n^s(\pi \times \mathbb{Z}) \xrightarrow{B} L_{n-1}^h(\pi) \longrightarrow 0$$

with $\overline{\varepsilon}$ the split injection of L-groups induced functorially from the split injection of groups $\overline{\varepsilon}:\pi \rightarrowtail \pi \times \mathbb{Z}$. The split surjection B was defined geometrically by

$$B : L_n^s(\pi \times \mathbb{Z}) \twoheadrightarrow L_{n-1}^h(\pi) ;$$

$$\sigma_*^s((M,\partial M) \xrightarrow{(f,b)} (X,\partial X) \times S^1 \longrightarrow K(\pi,1) \times S^1 = K(\pi \times \mathbb{Z},1))$$

$$\longmapsto \sigma_*^h((N,\partial N) \xrightarrow{(g,c)} (X,\partial X) \longrightarrow K(\pi,1))$$

using the splitting theorem of Farrell and Hsiang [5], [7] to represent every element of $L_n^s(\pi \times \mathbb{Z})$ as the rel∂ simple surgery obstruction $\sigma_*^s(f,b)$ of an n-dimensional normal map $(f,b):(M,\partial M) \longrightarrow (X,\partial X) \times S^1$ with $(X,\partial X)$ a finite (n-1)-dimensional geometric Poincaré pair, such that f is transverse regular at $(X,\partial X) \times \{pt.\} \subset (X,\partial X) \times S^1$ with the restriction defining an (n-1)-dimensional normal map

$$(g,c) = (f,b)| : (N,\partial N) = f^{-1}((X,\partial X) \times \{pt.\}) \longrightarrow (X,\partial X)$$

with $\partial f:\partial M \longrightarrow \partial X \times S^1$ a simple homotopy equivalence and $\partial g:\partial N \longrightarrow \partial X$ a homotopy equivalence. There was also defined in [29] a splitting map for B

$$\bar{B}' : L^h_{n-1}(\pi) \rightarrowtail L^s_n(\pi \times \mathbb{Z}) ;$$

$$\sigma^h_*((M,\partial M) \xrightarrow{(f,b)} (X,\partial X) \longrightarrow K(\pi,1))$$

$$\longmapsto \sigma^s_*((M,\partial M) \times S^1 \xrightarrow{(f,b)\times 1} (X,\partial X) \times S^1$$

$$\longrightarrow K(\pi,1) \times S^1 = K(\pi\times\mathbb{Z},1))$$

$$(= \sigma^h_*(f,b)\otimes\sigma^*_r(S^1) \text{ by Proposition 5.3 ii}))$$

Let $\varepsilon' : L^s_n(\pi\times\mathbb{Z}) \twoheadrightarrow L^s_n(\pi)$ be the geometric split surjection determined by $\bar{\varepsilon}, B, \bar{B}'$, so that there is defined a direct sum system

$$L^s_n(\pi) \underset{\varepsilon'}{\overset{\bar{\varepsilon}}{\rightleftarrows}} L^s_n(\pi \times \mathbb{Z}) \underset{\bar{B}'}{\overset{B}{\rightleftarrows}} L^h_{n-1}(\pi) \quad .$$

Although it was claimed in Ranicki [20] that ε' coincides with the split surjection induced functorially from the split surjection of groups $\varepsilon : \pi\times\mathbb{Z} \longrightarrow \pi$ (or equivalently $\mathbb{Z}[\pi][z,z^{-1}] \longrightarrow \mathbb{Z}[\pi]$; $z \longmapsto 1$) it does not do so in general. This may be seen by considering the composite

$$\varepsilon\bar{B}' : L^h_{n-1}(\pi) \xrightarrow{\bar{B}'} L^s_n(\pi\times\mathbb{Z}) \xrightarrow{\varepsilon} L^s_n(\pi) \ ,$$

which need not be zero. A generic element

$$\sigma^h_*((f,b) : (M,\partial M) \longrightarrow (X,\partial X)) \in L^h_{n-1}(\pi)$$

is sent by $\bar{B}'$ to

$$\bar{B}'(\sigma^h_*(f,b)) = \sigma^s_*((g,c) = (f,b) \times 1_{S^1} : (M,\partial M) \times S^1 \longrightarrow (X,\partial X) \times S^1)$$

$$\in L^h_n(\pi \times \mathbb{Z}) \ .$$

Now (g,c) is the boundary of the $(n+1)$-dimensional normal map

$$(f,b) \times 1_{(D^2,S^1)} : (M,\partial M) \times (D^2,S^1) \longrightarrow (X,\partial X) \times (D^2,S^1)$$

such that the target

$$(X,\partial X) \times (D^2,S^1) = (X \times D^2, X \times S^1 \cup_{\partial X \times S^1} \partial X \times D^2)$$

is a finite $(n+1)$-dimensional geometric Poincaré pair with simple boundary and

$$\tau((X,\partial X) \times (D^2,S^1)) = \tau(X,\partial X)\otimes\chi(D^2) + \chi(X)\otimes\tau(D^2,S^1)$$

$$= \tau(X,\partial X) \in Wh(\pi)$$

(by the relative version of Proposition 5.1). It follows that

$\varepsilon\bar{B}'\sigma_*^h(f,b) \in L_n^s(\pi)$ is the image of

$$\tau((X,\partial X)\times(D^2,S^1)) = \tau(X,\partial X)$$
$$\in \hat{H}^{n-1}(\mathbb{Z}_2;Wh(\pi)) = \hat{H}^{n+1}(\mathbb{Z}_2;Wh(\pi))$$

under the map $\hat{H}^{n+1}(\mathbb{Z}_2;Wh(\pi)) \longrightarrow L_n^s(\pi)$ in the Rothenberg exact sequence

$$\ldots \longrightarrow L_{n+1}^h(\pi) \longrightarrow \hat{H}^{n+1}(\mathbb{Z}_2;Wh(\pi)) \longrightarrow L_n^s(\pi) \longrightarrow L_n^h(\pi) \longrightarrow \ldots .$$

The discrepancy between ε and ε' will be expressed algebraically in Proposition 6.2 below; it is at most 2-torsion, and is 0 if $Wh(\pi) = 0$.

Novikov [17] initiated the development of analogues for algebraic L-theory of the techniques of Bass, Heller and Swan [2] and Bass [1] for the algebraic K-theory of polynomial extensions. In Ranicki [19], [20] the methods of [17] (which neglected 2-torsion) were refined to obtain for any group π algebraic isomorphisms

$$\begin{cases} B_L = \begin{pmatrix} \varepsilon \\ B \end{pmatrix} : L_n^s(\pi\times\mathbb{Z}) \longrightarrow L_n^s(\pi)\oplus L_{n-1}^h(\pi) \\ B_L = \begin{pmatrix} \varepsilon \\ B \end{pmatrix} : L_n^h(\pi\times\mathbb{Z}) \longrightarrow L_n^h(\pi)\oplus L_{n-1}^p(\pi) \end{cases}$$

with inverses

$$\begin{cases} B_L^{-1} = (\bar{\varepsilon}\ \bar{B}) : L_n^s(\pi)\oplus L_{n-1}^h(\pi) \longrightarrow L_n^s(\pi\times\mathbb{Z}) \\ B_L^{-1} = (\bar{\varepsilon}\ \bar{B}) : L_n^h(\pi)\oplus L_{n-1}^p(\pi) \longrightarrow L_n^h(\pi\times\mathbb{Z}) \end{cases} ,$$

by analogy with the isomorphism of [2]

$$B_K : Wh(\pi\times\mathbb{Z}) \longrightarrow Wh(\pi)\oplus\tilde{K}_0(\mathbb{Z}[\pi])\oplus\widetilde{Nil}(\mathbb{Z}[\pi])\oplus\widetilde{Nil}(\mathbb{Z}[\pi])$$

recalled in §3 above. The isomorphisms B_L define the <u>algebraically significant</u> splitting

$$\begin{cases} L_n^s(\pi\times\mathbb{Z}) = L_n^s(\pi)\oplus L_{n-1}^h(\pi) \\ L_n^h(\pi\times\mathbb{Z}) = L_n^h(\pi)\oplus L_{n-1}^p(\pi) \end{cases} .$$

As already indicated above this does not in general coincide with the geometric splitting of $L_n^s(\pi\times\mathbb{Z})$ due to Shaneson [29], although the split surjection $B: L_n^s(\pi\times\mathbb{Z}) \twoheadrightarrow L_{n-1}^h(\pi)$ of [29] agrees with the algebraic B of [20].

Pedersen and Ranicki [18,§4] claimed to be giving a geometric interpretation of the algebraically significant splitting $L_*^h(\pi\times\mathbb{Z}) = L_*^h(\pi)\oplus L_{*-1}^p(\pi)$. However, the composite

$$\varepsilon\overline{B}' : L_{n-1}^p(\pi) \xrightarrow{\overline{B}'} L_n^h(\pi\times\mathbb{Z}) \xrightarrow{\varepsilon} L_n^h(\pi)$$

of the geometric split injection

$$\overline{B}' : L_{n-1}^p(\pi) \rightarrowtail L_n^h(\pi\times\mathbb{Z}) ;$$

$$\sigma_*^p((f,b):(M,\partial M)\longrightarrow(X,\partial X))$$

$$\longmapsto \sigma_*^h((f,b)\times 1_{S^1} : (M,\partial M)\times S^1 \longrightarrow (X,\partial X)\times S^1)$$

$$(= \sigma_*^p(f,b)\otimes\sigma_r^*(S^1) \text{ by Proposition 5.3 ii))}$$

and the algebraic split surjection $\varepsilon : L_n^h(\pi\times\mathbb{Z}) \twoheadrightarrow L_n^h(\pi)$ need not be zero: there is defined a finitely dominated null-bordism with $\pi_1(X\times D^2) = \pi_1(X) = \pi$

$$(f,b)\times 1_{(D^2,S^1)} : (M,\partial M)\times(D^2,S^1) \longrightarrow (X,\partial X)\times(D^2,S^1)$$

of the relative (homotopy) finite surgery problem

$$(f,b)\times 1_{S^1} : (M,\partial M)\times S^1 \longrightarrow (X,\partial X)\times S^1 ,$$

with finiteness obstruction

$$[X\times D^2] = [X] \in \tilde{K}_0(\mathbb{Z}[\pi]) .$$

It follows that $\varepsilon\overline{B}'\sigma_*^p(f,b)\in L_n^h(\pi)$ is the image of $[X]\in\hat{H}^{n-1}(\mathbb{Z}_2;\tilde{K}_0(\mathbb{Z}[\pi])) = \hat{H}^{n+1}(\mathbb{Z}_2;\tilde{K}_0(\mathbb{Z}[\pi]))$ under the map $\hat{H}^{n+1}(\mathbb{Z}_2;\tilde{K}_0(\mathbb{Z}[\pi]))\longrightarrow L_n^h(\pi)$ in the generalized Rothenberg exact sequence

$$\ldots\longrightarrow L_{n+1}^p(\pi)\longrightarrow \hat{H}^{n+1}(\mathbb{Z}_2;\tilde{K}_0(\mathbb{Z}[\pi]))\longrightarrow L_n^h(\pi)\longrightarrow L_n^p(\pi)\longrightarrow\ldots .$$

Thus $\overline{B}'$ and ε do not in general belong to the same direct sum system. In fact ε belongs to the algebraically significant direct sum decomposition of $L_n^h(\pi\times\mathbb{Z})$ described above, while $\overline{B}'$ belongs to the geometrically defined direct sum decomposition

$$L_n^h(\pi) \underset{\varepsilon'}{\overset{\overline{\varepsilon}}{\rightleftarrows}} L_n^h(\pi\times\mathbb{Z}) \underset{\overline{B}'}{\overset{B}{\rightleftarrows}} L_{n-1}^p(\pi)$$

with B as defined in [18,§4] and ε' the split surjection determined by $\overline{\varepsilon},B,\overline{B}'$. It is the latter direct sum system which is meant when referring to "the geometric splitting $L_*^h(\pi\times\mathbb{Z}) = L_*^h(\pi)\oplus L_{*-1}^p(\pi)$ of [18]".

Define the <u>geometrically significant</u> splitting

$$\begin{cases} L_n^s(\pi\times\mathbb{Z}) = L_n^s(\pi)\oplus L_{n-1}^h(\pi) \\ L_n^h(\pi\times\mathbb{Z}) = L_n^h(\pi)\oplus L_{n-1}^p(\pi) \end{cases}$$

to be the one given by the algebraic isomorphism

$$\begin{cases} \beta_L' = \begin{pmatrix} \varepsilon' \\ B \end{pmatrix} : L_n^s(\pi\times\mathbb{Z}) \longrightarrow L_n^s(\pi)\oplus L_{n-1}^h(\pi) \\ \beta_L' = \begin{pmatrix} \varepsilon' \\ B \end{pmatrix} : L_n^h(\pi\times\mathbb{Z}) \longrightarrow L_n^h(\pi)\oplus L_{n-1}^p(\pi) \end{cases}$$

with inverse

$$\begin{cases} \beta_L'^{-1} = (\bar{\varepsilon}\ \bar{B}') : L_n^s(\pi)\oplus L_{n-1}^h(\pi) \longrightarrow L_n^s(\pi\times\mathbb{Z}) \\ \beta_L'^{-1} = (\bar{\varepsilon}\ \bar{B}') : L_n^h(\pi)\oplus L_{n-1}^p(\pi) \longrightarrow L_n^h(\pi\times\mathbb{Z}) \end{cases},$$

where

$$\begin{cases} \bar{B}' = -\otimes\sigma_r^*(S^1) : L_{n-1}^h(\pi) \rightarrowtail L_n^s(\pi\times\mathbb{Z}) \\ \bar{B}' = -\otimes\sigma_r^*(S^1) : L_{n-1}^p(\pi) \rightarrowtail L_n^h(\pi\times\mathbb{Z}) \end{cases}$$

and

$$\begin{cases} \varepsilon' = \varepsilon(1-\bar{B}'B) : L_n^s(\pi\times\mathbb{Z}) \twoheadrightarrow L_n^s(\pi) \\ \varepsilon' = \varepsilon(1-\bar{B}'B) : L_n^h(\pi\times\mathbb{Z}) \twoheadrightarrow L_n^h(\pi) \end{cases}.$$

<u>Proposition 6.1</u> The geometric splitting $\begin{cases} L_n^s(\pi\times\mathbb{Z}) = L_n^s(\pi)\oplus L_{n-1}^h(\pi) \\ L_n^h(\pi\times\mathbb{Z}) = L_n^h(\pi)\oplus L_{n-1}^p(\pi) \end{cases}$ of $\begin{cases} \text{Shaneson [29]} \\ \text{Pedersen and Ranicki [18]} \end{cases}$ is the geometrically significant splitting in algebra.

[]

The algebraically significant split injections $\begin{cases} \bar{B}: L_*^h(\pi) \rightarrowtail L_{*+1}^s(\pi\times\mathbb{Z}) \\ \bar{B}: L_*^p(\pi) \rightarrowtail L_{*+1}^h(\pi\times\mathbb{Z}) \end{cases}$ were defined in Ranicki [20] using the forms and formations of Ranicki [19]; for example

$$\bar{B} : L_{2i}^p(\pi) \rightarrowtail L_{2i+1}^h(\pi\times\mathbb{Z}) ;$$

$$(Q,\psi) \longmapsto (M\oplus M, \psi\oplus -\psi; \Delta, (1\oplus z)\Delta)\oplus(H_{(-)^i}(N);N,N)$$

sends a projective non-singular $(-)^i$-quadratic form over $\mathbb{Z}[\pi]$ (Q,ψ)

to a free non-singular $(-)^i$-quadratic formation over $\mathbb{Z}[\pi\times\mathbb{Z}] = \mathbb{Z}[\pi][z,z^{-1}]$ with $M = Q[z,z^{-1}]$ the induced f.g. projective $\mathbb{Z}[\pi\times\mathbb{Z}]$-module, $\Delta = \{(x,x) \in M\oplus M \mid x \in M\} \subset M\oplus M$ the diagonal lagrangian of $(M\oplus M, \psi\oplus-\psi)$, and $H_{(-)^i}(N) = (N\oplus N^*, \begin{pmatrix} 0 & 1 \\ 0 & 0 \end{pmatrix})$ the $(-)^i$-hyperbolic (alias hamiltonian) form on a f.g. projective $\mathbb{Z}[\pi\times\mathbb{Z}]$-module N such that $M\oplus N$ is a f.g. free $\mathbb{Z}[\pi\times\mathbb{Z}]$-module. The geometrically significant split injections $\begin{cases} \bar{B}' : L^h_*(\pi) \rightarrowtail L^s_{*+1}(\pi\times\mathbb{Z}) \\ \bar{B}' : L^p_*(\pi) \rightarrowtail L^h_{*+1}(\pi\times\mathbb{Z}) \end{cases}$ were defined in §10 of Ranicki [22] using algebraic Poincaré complexes. It is easy to translate from complexes to forms and formations (or the other way round); for example, in terms of forms and formations

$$\bar{B}' : L^p_{2i}(\pi) \rightarrowtail L^h_{2i+1}(\pi\times\mathbb{Z}) ;$$
$$(Q,\psi) \longmapsto (M\oplus M, \psi\oplus-\psi; \Delta, (1\oplus z)\Delta) \oplus (H_{(-)^i}(N); N, N^*) ,$$

making apparent the difference between $\bar{B}$ and $\bar{B}'$ in this case.

For any group π the exact sequence

$$0 \longrightarrow \hat{H}^0(\mathbb{Z}_2; K_0(\mathbb{Z})) \longrightarrow L^1_{rh}(\pi) \longrightarrow L^1_h(\pi) \longrightarrow 0$$

splits, with the injection

$$\hat{H}^0(\mathbb{Z}_2; K_0(\mathbb{Z})) = \mathbb{Z}_2 \rightarrowtail L^1_{rh}(\pi) \ ; \ 1 \longmapsto \mathbb{Z}[\pi]\otimes_{\mathbb{Z}} \varepsilon\sigma^*_r(S^1)$$

split by the rational semicharacteristic

$$L^1_r(\pi) \twoheadrightarrow \mathbb{Z}_2 \ ; \ (C,\phi) \longmapsto \chi_{1/2}(\mathbb{Z}\otimes_{\mathbb{Z}[\pi]} C; \mathbb{Q}) .$$

By the discussion at the end of Ranicki [22, §10]

$$L^1(\mathbb{Z}) = L^1(\{1\}) \oplus L^0(\{1\}) = \mathbb{Z}_2 \oplus \mathbb{Z} ,$$

with $(0,1) = \sigma^*(S^1) \in L^1(\mathbb{Z})$ the symmetric signature of S^1. Let $\sigma^*_q(S^1) \in L^1_r(\mathbb{Z})$ be the image of $\sigma^*(S^1) \in L^1(\mathbb{Z})$ under the splitting map $L^1(\mathbb{Z}) \rightarrowtail L^1_r(\mathbb{Z})$, so that $\sigma^*_q(S^1) = (1-\bar{\varepsilon}\varepsilon)\sigma^*_r(S^1)$ and $\varepsilon\sigma^*_q(S^1) = 0 \in L^1_r(\{1\})$. The algebraically significant injections are defined by

$$\begin{cases} \bar{B} = -\otimes\sigma^*_q(S^1) : L^h_n(\pi) \rightarrowtail L^s_{n+1}(\pi\times\mathbb{Z}) \\ \bar{B} = -\otimes\sigma^*_q(S^1) : L^p_n(\pi) \rightarrowtail L^h_{n+1}(\pi\times\mathbb{Z}) \end{cases} .$$

Now

$$\sigma^*_r(S^1) - \sigma^*_q(S^1) = \bar{\varepsilon}\varepsilon\sigma^*_r(S^1) \in L^1_r(\mathbb{Z}) ,$$

so that

$$\begin{cases} \bar{B}' - \bar{B} = -\otimes(\sigma^*_r(S^1) - \sigma^*_q(S^1)) = -\otimes\bar{\varepsilon}\varepsilon\sigma^*_r(S^1) : L^h_n(\pi) \longrightarrow L^s_{n+1}(\pi\times\mathbb{Z}) \\ \bar{B}' - \bar{B} = -\otimes(\sigma^*_r(S^1) - \sigma^*_q(S^1)) = -\otimes\bar{\varepsilon}\varepsilon\sigma^*_r(S^1) : L^p_n(\pi) \longrightarrow L^h_{n+1}(\pi\times\mathbb{Z}) \end{cases} .$$

By analogy with the map of algebraic K-groups defined in §3

$$\omega = -\otimes\tau(-1) \; : \; \tilde{K}_0(\mathbb{Z}[\pi]) \longrightarrow Wh(\pi)$$

define maps of algebraic L-groups

$$\begin{cases} \omega = -\otimes\varepsilon\sigma^*_r(S^1) \; : \; L^h_n(\pi) \longrightarrow L^s_{n+1}(\pi) \\ \omega = -\otimes\varepsilon\sigma^*_r(S^1) \; : \; L^p_n(\pi) \longrightarrow L^h_{n+1}(\pi) \end{cases} ,$$

where $\varepsilon\sigma^*_r(S^1) = (1,1) \in L^1_r(\{1\}) = \mathbb{Z}_2\oplus\mathbb{Z}_2$. As $\varepsilon\tau(S^1) = \tau(-1) \in K_1(\mathbb{Z}) = \mathbb{Z}_2$ the various maps ω together define a morphism of generalized Rothenberg exact sequences

$$\begin{array}{ccccccccc} \dots \longrightarrow & L^h_n(\pi) & \longrightarrow & L^p_n(\pi) & \longrightarrow & \hat{H}^n(\mathbb{Z}_2;\tilde{K}_0(\mathbb{Z}[\pi])) & \longrightarrow & L^h_{n-1}(\pi) & \longrightarrow \dots \\ & \downarrow{\scriptstyle\omega} & & \downarrow{\scriptstyle\omega} & & \downarrow{\scriptstyle\omega} & & \downarrow{\scriptstyle\omega} & \\ \dots \longrightarrow & L^s_{n+1}(\pi) & \longrightarrow & L^h_{n+1}(\pi) & \longrightarrow & \hat{H}^{n+1}(\mathbb{Z}_2;Wh(\pi)) & \longrightarrow & L^s_n(\pi) & \longrightarrow \dots \end{array} .$$

<u>Proposition 6.2</u> The algebraically and geometrically significant split injections of L-groups differ by

$$\begin{cases} \bar{B}' - \bar{B} = \bar{\varepsilon}\omega \; : \; L^h_n(\pi) \xrightarrow{\;\omega\;} L^s_{n+1}(\pi) \xrightarrow{\;\bar{\varepsilon}\;} L^s_{n+1}(\pi\times\mathbb{Z}) \\ \bar{B}' - \bar{B} = \bar{\varepsilon}\omega \; : \; L^p_n(\pi) \xrightarrow{\;\omega\;} L^h_{n+1}(\pi) \xrightarrow{\;\bar{\varepsilon}\;} L^h_{n+1}(\pi\times\mathbb{Z}) \end{cases} .$$

The split surjections differ by

$$\begin{cases} \varepsilon' - \varepsilon = \omega B \; : \; L^s_n(\pi\times\mathbb{Z}) \xrightarrow{\;B\;} L^h_{n-1}(\pi) \xrightarrow{\;\omega\;} L^s_n(\pi) \\ \varepsilon' - \varepsilon = \omega B \; : \; L^h_n(\pi\times\mathbb{Z}) \xrightarrow{\;B\;} L^p_{n-1}(\pi) \xrightarrow{\;\omega\;} L^h_n(\pi) \end{cases} .$$

The L-theory maps ω factor as

$$\begin{cases} \omega \; : \; L^h_n(\pi) \longrightarrow \hat{H}^n(\mathbb{Z}_2;Wh(\pi)) = \hat{H}^{n+2}(\mathbb{Z}_2;Wh(\pi)) \longrightarrow L^s_{n+1}(\pi) \\ \omega \; : \; L^p_n(\pi) \longrightarrow \hat{H}^n(\mathbb{Z}_2;\tilde{K}_0(\mathbb{Z}[\pi])) = \hat{H}^{n+2}(\mathbb{Z}_2;\tilde{K}_0(\mathbb{Z}[\pi])) \longrightarrow L^h_{n+1}(\pi) \end{cases} .$$

The K-theory map ω is the sum of the composites

$$\hat{H}^n(\mathbb{Z}_2;\tilde{K}_0(\mathbb{Z}[\pi])) \longrightarrow L^h_{n-1}(\pi) \longrightarrow \hat{H}^{n-1}(\mathbb{Z}_2;Wh(\pi)) = \hat{H}^{n+1}(\mathbb{Z}_2;Wh(\pi))$$

$$\hat{H}^n(\mathbb{Z}_2;\tilde{K}_0(\mathbb{Z}[\pi])) = H^{n+2}(\mathbb{Z}_2;\tilde{K}_0(\mathbb{Z}[\pi])) \longrightarrow L^h_{n+1}(\pi) \longrightarrow \hat{H}^{n+1}(\mathbb{Z}_2;Wh(\pi)) .$$

Proof: Let $\begin{cases} L_n^{h,s}(\pi) \\ L_n^{p,h}(\pi) \end{cases}$ $(n \geqslant 0)$ be the relative cobordism group of $\begin{cases} \text{(finite,simple)} \\ \text{(finitely dominated,finite)} \end{cases}$ n-dimensional quadratic Poincaré pairs over $\mathbb{Z}[\pi]$ $(f:C \longrightarrow D, (\delta\psi,\psi) \in Q_n(f))$, so that there is defined an exact sequence

$$\begin{cases} \ldots \longrightarrow L_n^s(\pi) \longrightarrow L_n^h(\pi) \longrightarrow L_n^{h,s}(\pi) \longrightarrow L_{n-1}^s(\pi) \longrightarrow \ldots \\ \ldots \longrightarrow L_n^h(\pi) \longrightarrow L_n^p(\pi) \longrightarrow L_n^{p,h}(\pi) \longrightarrow L_{n-1}^h(\pi) \longrightarrow \ldots \end{cases}$$

and there are defined isomorphisms

$$\begin{cases} L_n^{h,s}(\pi) \longrightarrow \hat{H}^n(\mathbb{Z}_2;Wh(\pi))\ ; \\ \qquad (f:C \longrightarrow D, (\delta\psi,\psi)) \longmapsto \tau((1+T)(\delta\psi,\psi)_0 : C(f)^{n-*} \longrightarrow D) \\ L_n^{p,h}(\pi) \longrightarrow \hat{H}^n(\mathbb{Z}_2;\tilde{K}_0(\mathbb{Z}[\pi]))\ ;\ (f:C \longrightarrow D, (\delta\psi,\psi)) \longmapsto [D]\ . \end{cases}$$

Product with the 2-dimensional symmetric Poincaré pair $\sigma^*(D^2,S^1)$ over $\mathbb{Z}$ defines isomorphisms of relative L-groups

$$\begin{cases} -\otimes\sigma^*(D^2,S^1)\ :\ L_n^{h,s}(\pi) \longrightarrow L_{n+2}^{h,s}(\pi) \\ -\otimes\sigma^*(D^2,S^1)\ :\ L_n^{p,h}(\pi) \longrightarrow L_{n+2}^{p,h}(\pi)\quad , \end{cases}$$

corresponding to the canonical 2-periodicity isomorphisms of the Tate $\mathbb{Z}_2$-cohomology groups

$$\begin{cases} \hat{H}^n(\mathbb{Z}_2;Wh(\pi)) \longrightarrow \hat{H}^{n+2}(\mathbb{Z}_2;Wh(\pi)) \\ \hat{H}^n(\mathbb{Z}_2;\tilde{K}_0(\mathbb{Z}[\pi])) \longrightarrow \hat{H}^{n+2}(\mathbb{Z}_2;\tilde{K}_0(\mathbb{Z}[\pi]))\ . \end{cases}$$

The boundary of $\sigma^*(D^2,S^1)$ is $\varepsilon\sigma_r^*(S^1)$.

[]

In particular, the algebraic and geometric splitting maps in L-theory differ in 2-torsion only, since $2\omega = 0$ (cf. Proposition 3.3).

The splitting maps in the algebraic and geometric splittings of $Wh(\pi\times\mathbb{Z})$ given in §3 and the duality involutions * are such that

$$\bar{\varepsilon}^* = *\bar{\varepsilon}\ :\ Wh(\pi) \longrightarrow Wh(\pi\times\mathbb{Z})$$
$$\varepsilon^* = *\varepsilon\ ,\ \varepsilon'^* = *\varepsilon'\ :\ Wh(\pi\times\mathbb{Z}) \longrightarrow Wh(\pi)$$
$$B^* = -*B\ :\ Wh(\pi\times\mathbb{Z}) \longrightarrow \tilde{K}_0(\mathbb{Z}[\pi])$$
$$\bar{B}^* = -*\bar{B}\ ,\ \bar{B}'^* = -*\bar{B}'\ :\ \tilde{K}_0(\mathbb{Z}[\pi]) \longrightarrow Wh(\pi\times\mathbb{Z})$$
$$\bar{\Delta}_\pm^* = *\bar{\Delta}_\mp\ :\ \widetilde{Nil}(\mathbb{Z}[\pi]) \longrightarrow Wh(\pi\times\mathbb{Z})$$
$$\Delta_\pm^* = *\Delta_\mp\ :\ Wh(\pi\times\mathbb{Z}) \longrightarrow \widetilde{Nil}(\mathbb{Z}[\pi])\ .$$

The involution $*: Wh(\pi\times\mathbb{Z}) \longrightarrow Wh(\pi\times\mathbb{Z})$ interchanges the two $\widetilde{Nil}$ summands, so that they do not appear in the Tate $\mathbb{Z}_2$-cohomology groups and there are defined two splittings

$$\hat{H}^n(\mathbb{Z}_2;Wh(\pi\times\mathbb{Z})) = \hat{H}^n(\mathbb{Z}_2;Wh(\pi))\oplus\hat{H}^{n-1}(\mathbb{Z}_2;\tilde{K}_0(\mathbb{Z}[\pi])) \quad ,$$

the <u>algebraically significant</u> direct sum decomposition

$$\hat{H}^n(\mathbb{Z}_2;Wh(\pi)) \underset{\varepsilon}{\overset{\bar{\varepsilon}}{\rightleftarrows}} \hat{H}^n(\mathbb{Z}_2;Wh(\pi\times\mathbb{Z})) \underset{\bar{B}}{\overset{B}{\rightleftarrows}} \hat{H}^{n-1}(\mathbb{Z}_2;\tilde{K}_0(\mathbb{Z}[\pi]))$$

and the <u>geometrically significant</u> direct sum decomposition

$$\hat{H}^n(\mathbb{Z}_2;Wh(\pi)) \underset{\varepsilon'}{\overset{\bar{\varepsilon}}{\rightleftarrows}} \hat{H}^n(\mathbb{Z}_2;Wh(\pi\times\mathbb{Z})) \underset{\bar{B}'}{\overset{B}{\rightleftarrows}} \hat{H}^{n-1}(\mathbb{Z}_2;\tilde{K}_0(\mathbb{Z}[\pi])) .$$

<u>Proposition 6.3</u> The Rothenberg exact sequence of a polynomial extension

$$\ldots \longrightarrow L_n^s(\pi\times\mathbb{Z}) \longrightarrow L_n^h(\pi\times\mathbb{Z}) \longrightarrow \hat{H}^n(\mathbb{Z}_2;Wh(\pi\times\mathbb{Z})) \longrightarrow L_{n-1}^s(\pi\times\mathbb{Z}) \longrightarrow \ldots$$

has two splittings as a direct sum of the exact sequences

$$\ldots \longrightarrow L_n^s(\pi) \longrightarrow L_n^h(\pi) \longrightarrow \hat{H}^n(\mathbb{Z}_2;Wh(\pi)) \longrightarrow L_{n-1}^s(\pi) \longrightarrow \ldots \ ,$$

$$\ldots \longrightarrow L_{n-1}^h(\pi) \longrightarrow L_{n-1}^p(\pi) \longrightarrow \hat{H}^{n-1}(\mathbb{Z}_2;\tilde{K}_0(\mathbb{Z}[\pi])) \longrightarrow L_{n-2}^h(\pi) \longrightarrow \ldots ,$$

an algebraically and a geometrically significant one.

[]

The split injection of exact sequences in the appendix of Munkholm and Ranicki [16] is the geometrically significant injection

$$\begin{array}{ccccccccc}
\ldots \longrightarrow & L_{n-1}^h(\pi) & \longrightarrow & L_{n-1}^p(\pi) & \longrightarrow & \hat{H}^{n-1}(\mathbb{Z}_2;\tilde{K}_0(\mathbb{Z}[\pi])) & \longrightarrow & L_{n-2}^h(\pi) & \longrightarrow \ldots \\
& \downarrow \bar{B}' & & \downarrow \bar{B}' & & \downarrow \bar{B}' & & \downarrow \bar{B}' & \\
\ldots \longrightarrow & L_n^s(\pi\times\mathbb{Z}) & \longrightarrow & L_n^h(\pi\times\mathbb{Z}) & \longrightarrow & \hat{H}^n(\mathbb{Z}_2;Wh(\pi\times\mathbb{Z})) & \longrightarrow & L_{n-1}^s(\pi\times\mathbb{Z}) & \longrightarrow \ldots .
\end{array}$$

As for algebraic K-theory (cf. the discussion just after Proposition 3.3) it is tempting to identify the geometrically significant split surjection $\begin{cases} \varepsilon' : L_n^s(\pi\times\mathbb{Z}) \longrightarrow L_n^s(\pi) \\ \varepsilon' : L_n^h(\pi\times\mathbb{Z}) \longrightarrow L_n^h(\pi) \end{cases}$ with the split surjection of L-groups induced functorially by the split surjection of rings with involution

$$\eta : \mathbb{Z}[\pi][z,z^{-1}] = \mathbb{Z}[\pi\times\mathbb{Z}] \longrightarrow \mathbb{Z}[\pi] \ ; \ \sum_{j=-\infty}^{\infty} a_j z^j \longmapsto \sum_{j=-\infty}^{\infty} a_j(-1)^j$$

and indeed

$$\begin{cases} \varepsilon'|(=1) = \eta| \ : \ \mathrm{im}(\bar{\varepsilon}:L_n^s(\pi) \rightarrowtail L_n^s(\pi\times\mathbb{Z})) \longrightarrow L_n^s(\pi) \\ \varepsilon'|(=1) = \eta| \ : \ \mathrm{im}(\bar{\varepsilon}:L_n^h(\pi) \rightarrowtail L_n^h(\pi\times\mathbb{Z})) \longrightarrow L_n^h(\pi) \ . \end{cases}$$

However, $\eta\sigma_r^*(S^1) = (1,0) \neq 0 \in L_r^1(\{1\}) = \mathbb{Z}_2 \oplus \mathbb{Z}_2$ (since the underlying $\mathbb{Z}$-module chain complex is $\mathbb{Z} \xrightarrow{2} \mathbb{Z}$) and in general

$$\begin{cases} \varepsilon'|(=0) \neq \eta| \ : \ \mathrm{im}(\bar{B}' = -\otimes\sigma_r^*(S^1):L_{n-1}^h(\pi) \rightarrowtail L_n^s(\pi\times\mathbb{Z})) \longrightarrow L_n^s(\pi) \\ \varepsilon'|(=0) \neq \eta| \ : \ \mathrm{im}(\bar{B}' = -\otimes\sigma_r^*(S^1):L_{n-1}^p(\pi) \rightarrowtail L_n^h(\pi\times\mathbb{Z})) \longrightarrow L_n^h(\pi) \end{cases}$$

so that

$$\begin{cases} \varepsilon' \neq \eta \ : \ L_n^s(\pi\times\mathbb{Z}) \longrightarrow L_n^s(\pi) \\ \varepsilon' \neq \eta \ : \ L_n^h(\pi\times\mathbb{Z}) \longrightarrow L_n^h(\pi) \ . \end{cases}$$

For $q = s,h,p$ the type q total surgery obstruction groups $\mathcal{S}_*^q(X)$ were defined in Ranicki [21] for any topological space X to fit into an exact sequence

$$\ldots \longrightarrow H_n(X;\underline{\mathbb{L}}_0) \xrightarrow{\sigma_*^q} L_n^q(\pi_1(X)) \longrightarrow \mathcal{S}_n^q(X) \longrightarrow H_{n-1}(X;\underline{\mathbb{L}}_0) \longrightarrow \ldots ,$$

with $\underline{\mathbb{L}}_0$ an algebraic 1-connective Ω-spectrum such that

$$\pi_*(\underline{\mathbb{L}}_0) = L_*(\{1\})$$

and σ_*^q an algebraic version of the Quinn assembly map. If X is a $\begin{cases} \text{simple} \\ \text{finite} \\ \text{finitely dominated} \end{cases}$ n-dimensional geometric Poincaré complex the total surgery obstruction $\begin{cases} s(X) \in \mathcal{S}_n^s(X) \\ s(X) \in \mathcal{S}_n^h(X) \\ s(X) \in \mathcal{S}_n^p(X) \end{cases}$ is defined, and is such that $s(X) = 0$ if (and for $n \geq 5$ only if) $\begin{cases} X \\ X \\ X \times S^1 \end{cases}$ is $\begin{cases} \text{simple} \\ - \\ - \end{cases}$ homotopy equivalent to a compact $\begin{cases} n- \\ n- \\ (n+1)- \end{cases}$ dimensional topological manifold. For a compact n-dimensional topological manifold M with $n \geq 5$ the exact sequence

$$\ldots \longrightarrow H_{n+1}(M;\underline{\mathbb{L}}_0) \xrightarrow{\sigma_*^q} L_{n+1}^q(\pi_1(M)) \longrightarrow \mathcal{S}_{n+1}^q(M) \longrightarrow H_n(M;\underline{\mathbb{L}}_0) \xrightarrow{\sigma_*^q} L_n^q(\pi_1(M))$$

is isomorphic to the type q Sullivan-Wall surgery exact sequence

$$\ldots \longrightarrow [M\times D^1, M\times S^0; G/TOP, *] \xrightarrow{\theta^q} L^q_{n+1}(\pi_1(M)) \longrightarrow \mathcal{S}^{q\,TOP}(M)$$
$$\longrightarrow [M, G/TOP] \xrightarrow{\theta^q} L^q_n(\pi_1(M))$$

with θ^q the type q surgery obstruction map and $\mathcal{S}^{q\,TOP}(M)$ the type q topological manifold structure set of M.

<u>Proposition 6.4</u> For any connected space X with $\pi_1(X) = \pi$ the commutative braid of algebraic surgery exact sequences of a polynomial extension

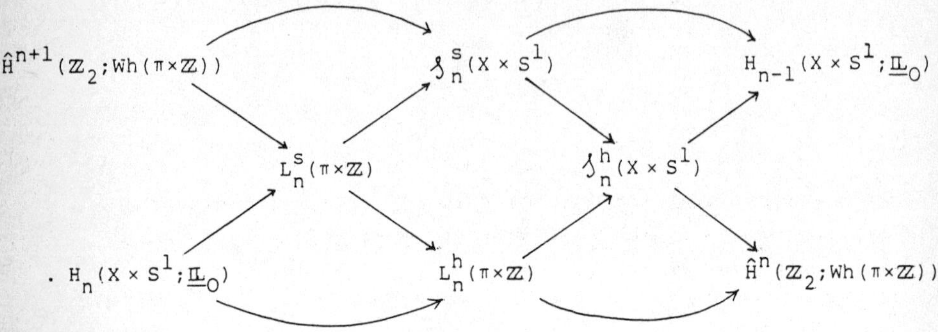

has a geometrically significant splitting as a direct sum of the braid

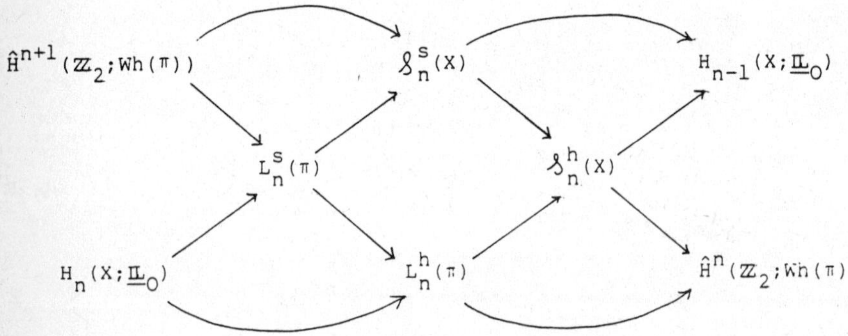

and the braid

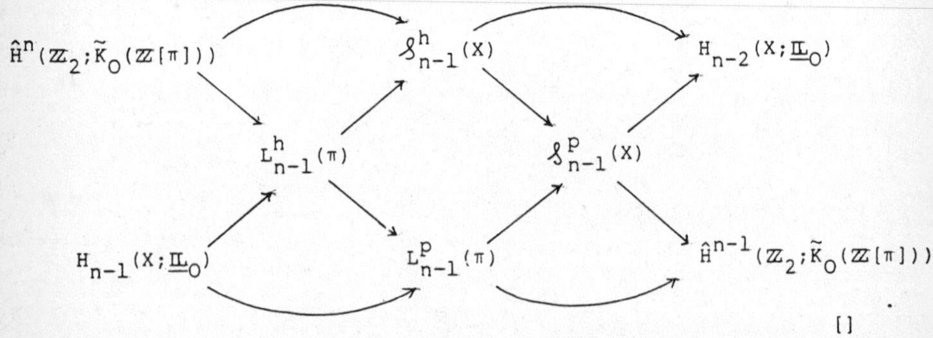

[]

It is appropriate to record here (in the terminology of this paper) a footnote from the preprint version of Cappell and Shaneson [3]: "it is not completely obvious that the maps given in Ranicki [20] give a splitting

$$L_n^s(\pi\times\mathbb{Z}) = L_n^s(\pi)\oplus L_{n-1}^h(\pi)$$

respected by the surgery map

$$\theta^s : [M\times S^1, G/TOP] = [M\times D^1, M\times S^0; G/TOP,*]\oplus[M,G/TOP] \longrightarrow L_{n+1}^s(\pi\times\mathbb{Z})$$

with M a compact n-dimensional topological manifold and $\pi = \pi_1(M)$."

Department of Mathematics,
Edinburgh University

REFERENCES

[1] H.Bass Algebraic K-theory Benjamin (1968)

[2] , A.Heller and R.G.Swan
The Whitehead group of a polynomial extension
Publ. Math. I.H.E.S. 22, 61 - 79 (1964)

[3] S.Cappell and J.Shaneson
Pseudo free actions I.,
Proceedings 1978 Arhus Algebraic Topology Conference,
Springer Lecture Notes 763, 395 - 447 (1979)

[4] F.T.Farrell
The obstruction to fibering a manifold over the circle
Indiana Univ. J. 21, 315 - 346 (1971)
Proc. I.C.M. Nice 1970, Vol.2, 69 - 72 (1971)

[5] and W.C.Hsiang
A geometric interpretation of the Künneth formula for algebraic K-theory
Bull. A.M.S. 74, 548 - 553 (1968)

[6] A formula for $K_1R_\alpha[T]$
Proc. Symp. Pure Maths. A.M.S. 17, 192 - 218 (1970)

[7] Manifolds with $\pi_1 = G\times_\alpha T$
Amer. J. Math. 95, 813 - 845 (1973)

[8] S.Ferry A simple-homotopy approach to the finiteness obstruction
Proc. 1981 Dubrovnik Shape Theory Conference
Springer Lecture Notes 870, 73 - 81 (1981)

[9] S.Gersten A product formula for Wall's obstruction
Am. J. Math. 88, 337 - 346 (1966)

[10] The torsion of a self equivalence
Topology 6, 411 - 414 (1967)

[11] I.Hambleton, A.Ranicki and L.Taylor
Round L-theory, to appear in J.Pure and Appl.Algebra

[12] K.Kwun and R.Szczarba
Product and sum theorems for Whitehead torsion
Ann. of Maths. 82, 183 - 190 (1965)

[13] W.Lück and A.Ranicki
Transfer maps and duality to appear

[14] M.Mather Counting homotopy types of manifolds
Topology 4, 93 - 94 (1965)

[15] H.J.Munkholm
Proper simple homotopy theory versus simple homotopy theory controlled over $\mathbb{R}^2$ to appear

[16] H.J.Munkholm and A.Ranicki
The projective class group transfer induced by an S^1-bundle
Proc. 1981 Ontario Topology Conference,
Canadian Math. Soc. Proc. 2, Vol.2, 461 - 484 (1982)

[17] S.Novikov The algebraic construction and properties of hermitian analogues of K-theory for rings with involution, from the point of view of the hamiltonian formalism. Some applications to differential topology and the theory of characteristic classes
Izv. Akad. Nauk SSSR, ser. mat. 34, 253-288, 478-500 (1970)

[18] E.Pedersen and A.Ranicki
Projective surgery theory Topology 19, 239 - 254 (1980)

[19] A.Ranicki Algebraic L-theory I. Foundations
Proc. Lond. Math. Soc. (3) 27, 101 - 125 (1973)

[20] II. Laurent extensions ibid., 126 - 158 (1973)

[21] The total surgery obstruction
Proc. 1978 Arhus Topology Conference, Springer Lecture Notes 763, 275 - 316 (1979)

[22] The algebraic theory of surgery I. Foundations
Proc. Lond. Math. Soc. (3) 40, 87 - 192 (1980)

[23] II. Applications to topology ibid., 193 - 287 (1980)

[24] Exact sequences in the algebraic theory of surgery
Mathematical Notes 26, Princeton (1981)

[25] The algebraic theory of torsion I. Foundations
Proc. 1983 Rutgers Topology Conference, Springer Lecture Notes 1126, 199 - 237 (1985)

[26] II. Products, to appear in J. of K-theory

[27] III. Lower K-theory preprint (1984)

[28] Splitting theorems in the algebraic theory of surgery
to appear

[29] J.Shaneson
Wall's surgery groups for $G \times \mathbb{Z}$
Ann. of Maths. 90, 296 - 334 (1969)

[30] L.Siebenmann
The obstruction to finding a boundary for an open manifold of dimension greater than five
Princeton Ph.D. thesis (1965)

[31] A torsion invariant for bands
Notices A.M.S. 68T-G7, 811 (1968)

[32] Infinite simple homotopy types
Indag. Math. 32, 479 - 495 (1970)

[33] A total Whitehead torsion obstruction to fibering over the circle Comm. Math. Helv. 45, 1 - 48 (1970)

[34] C.T.C.Wall
Finiteness conditions for CW complexes
I. Ann. of Maths. 81, 56 - 69 (1965)
II. Proc. Roy. Soc. A295, 129 - 139 (1966)

[35] Surgery on compact manifolds Academic Press (1970)

Coherence in Homotopy Group Actions

R. Schwänzl and R. M. Vogt

1. Introduction

In the effort to construct an action of a group G on a homotopy type one encounters the problem of having to realize a homotopy action of G on a space X by a genuine G-action on a space Y of the same homotopy type as X.

1.1 Definition: A *homotopy action* of a group G on a space X is a homomorphism $\alpha\colon G \longrightarrow \pi_0(\mathrm{Aut}X)$, where AutX is the space of self-homotopy equivalences of X. A *realization* of α is a G-space Y together with a homotopy equivalence $f\colon X \longrightarrow Y$ which is equivariant in the homotopy category Top_h. If Y is a free G-space, we call (Y,f) a *free realization*.

This problem has been solved by Cooke [C] for discrete groups:

1.2 Theorem: $\alpha\colon G \longrightarrow \pi_0(\mathrm{Aut}X)$ admits a realization iff there is a lift (up to homotopy),

$$\begin{array}{ccc} & & B(\mathrm{Aut}X) \\ & \nearrow & \downarrow \\ BG & \xrightarrow{\;B\alpha\;} & B(\pi_0\,\mathrm{Aut}X) \end{array}$$

where B denotes the classifying space functor.

A rational version has been studied by Oprea [O].
Zabrodsky took up this problem in [Z] with a different attitude. He investigated the relations induced by AutX on the space of homeomorphisms of a realization Y. He indicated an obstructions theory for realizing a homotopy G-map from a G-space to a homotopy G-space.

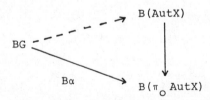

1.3 Definition: Let $\alpha\colon G \longrightarrow \pi_0(\mathrm{Aut}X)$ and $\beta\colon G \longrightarrow \pi_0(\mathrm{Aut}Y)$ be homotopy actions of G on X respecitively Y. A *homotopy G-map* from X to Y is a map $f\colon X \longrightarrow Y$ which is G-equivariant in the homotopy category. A *realization* of a homotopy G-map f: X $\longrightarrow$ Y is a homotopy commutative diagram

$$\begin{array}{ccc} X & \xrightarrow{f} & Y \\ {\scriptstyle h_X}\downarrow & & \downarrow{\scriptstyle h_Y} \\ X' & \xrightarrow{f'} & Y' \end{array}$$

where (X',h_X) and (Y',h_Y) are realizations of α and β and f' is a G-equivariant map.

A draw-back of Zabrodsky's theory is that he works in the category of based topological spaces, so that all group actions have to leave the base point fixed.
The aim of the present paper is to tackle these problems with the methods of homotopical coherence theory as developed in [B - V]. We interprete Cooke's obstructions as obstructions to higher coherence. Our proofs allow a generalization to topological groups. We deal with relative versions in the sense of [Z] and with relative versions with respect to subgroups.
Throughout this paper we work in the category *Top* of compactly generated spaces in the sense of [V1].

Organization of the paper: We introduce the notions of n-coherent homotopy G-actions and n-coherent homotopy G-maps (in Section 2) using the W-construction of [B - V] and state some of their fundamental properties, we formulate the main results (in Section 3) and discuss the universal property of the W-construction (in Section 4). In order to keep this paper fairly self-contained all constructions from coherence theory are executed and proofs of almost all statements are indicated so that a knowledge of the more complicated theory of [B - V] is not required. The proofs (in Sections 5) of our main results make use of homotopy-homomorphisms of monoids and functors of topologized categories which have close connections with Fuchs's theory of H_∞-maps and G_∞-maps [F1], [F2], [F3]. We discuss this relationship in a final Section 6.

Our interest in this subject was initiated by a problem posed to us by T. tom Dieck. We want to thank him for suggesting to apply coherence theory to this type of problems. Finally we want to draw the

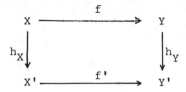

reader's attention to work of Dwyer and Kan [D - K]. We could equally
well have used their methods to obtain our results in the ∞-coherent
case, which after all is the most interesting one.

2. n-coherent homotopy actions and homotopy G-maps

Given a homotopy action $\alpha: G \longrightarrow \pi_o \, \text{Aut} X$ on X, we choose a representative $\overline{g} \in \alpha(g)$ and a path

$$w(g_1, g_2): \qquad \underset{\overline{g_1 \cdot g_2}}{\bullet} \rule{6cm}{0.4pt} \underset{\overline{g}_1 \circ \overline{g}_2}{\bullet}$$

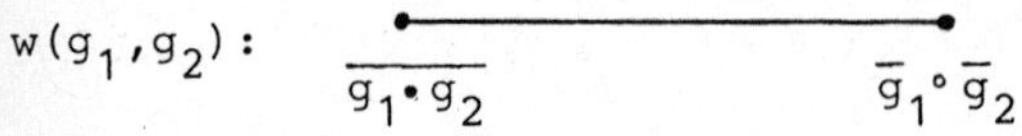

for each pair $(g_1, g_2) \in G \times G$. If $e \in G$ is the neutral element, we make
the spacial choices $\overline{e} = \text{id}_X$, and $w(e,g) = w(g,e) = $ trivial path on $\overline{g}$.
We call the resulting structure a *1-coherent homotopy G-action* on X.
Of course, a 1-coherent homotopy action is not uniquely determined
by a homotopy action.

Given three elements g_1, g_2, g_3 in G different from e, a 1-coherent
homotopy G-action on X gives rise to a loop $l(g_1, g_2, g_3)$ in Aut X

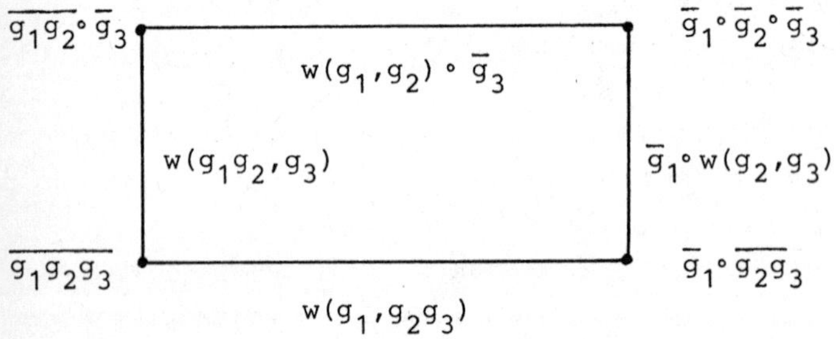

Sometimes it is possible to fill in all loops $l(g_1, g_2, g_3)$ by a disk
$d(g_1, g_2, g_3)$. We add these disks to the data and call the structure
thus obtained a *2-coherent homotopy action*. Playing this game with
more group elements we can define arbitrarily high coherence. We now
formalize this concept.

Let C be an arbitrary small category. Throughout this paper we assume
all small categories to have sets of objects but topologized morphism
spaces such that composition is continuous. We call C *well-pointed*
if ob C ⊂ mor C is a closed cofibration. Let *Cat* be the category of
such topologized categories. We construct a functor (see [B - V])

$$W\colon \mathcal{Cat} \longrightarrow \mathcal{Cat}$$

as follows: Let $C \in \mathcal{Cat}$; then ob WC = ob C, and

$$WC(A,B) = \coprod_{n\geq 0} C_{n+1}(A,B) \times I^n/\sim$$

where $C_{n+1}(A,B)$ is the space of all composable morphisms

$$A = A_0 \xrightarrow{f_0} A_1 \xrightarrow{f_1} A_2 \longrightarrow \cdots \xrightarrow{f_n} A_{n+1} = B$$

in C with the obvious subspace topology from $(\text{mor } C)^{n+1}$, and $I = [0,1]$. The relations are

(2.1) $(f_n,t_n,\ldots,f_1,t_1,f_0)$

(1)	$= (f_n,t_n,\ldots,f_i \circ f_{i-1}, t_{i-1},\ldots,f_1,t_1,f_0)$	if $t_i = 0$
(2)	$= (f_n,t_n,\ldots,f_1)$	if $f_0 = id$
(3)	$= (f_n,t_n,\ldots,f_{i+1},\max(t_{i+1},t_i),f_{i-1},\ldots,f_0)$	if $f_i = id$
(4)	$= (f_{n-1},t_{n-1},\ldots,f_1,t_1,f_0)$	if $f_n = id$

Composition in WC is given by
$(f_n,t_n,\ldots,f_0) \circ (g_k, u_k,\ldots,g_0) = (f_n,t_n,\ldots,f_0, 1, g_k,u_k,\ldots,g_0)$.

The n-skeleton subcategory W^nC of WC is the subcategory generated by all morphisms having a representative $(f_k, t_k,\ldots,f_0)$ with $k \leq n$.

2.2 Definition: An *n-coherent homotopy action* of a topological group G on a space X is a homomorphism $\alpha\colon W^nG \longrightarrow \text{Aut}X$ of topological monoids.

Explanation: A topological monoid can be considered as a topological category with one object and vice versa.

Since G is a group, an n-coherent homotopy G-action determines and is determined by a continuous functor $W^nG \longrightarrow \mathcal{Top}$ sending the unique object to X. We often call such a functor (and consequently $\alpha\colon W^nG \longrightarrow \text{Aut}X$) a W^nG-*structure* on X, or X a W^nG-*space*.

2.3 Notation: A C-*space* is a continuous functor $C \longrightarrow \mathcal{Top}$. A homomorphism of C-spaces is a natural transformation of such functors.

As indicated in the introduction we want to investigate maps which are homomorphisms up to homotopy, possibly with coherence conditions. To find the appropriate definition, observe that a natural transformation

$$\gamma\colon F_o \longrightarrow F_1\colon \mathcal{C} \longrightarrow \mathcal{Top}$$

of functors F_o, F_1 determines and is determined by a continuous functor $\mathcal{C} \times \mathcal{L}_1 \longrightarrow \mathcal{Top}$, where $\mathcal{L}_1$ is the category $0 \longrightarrow 1$. This leads to

2.4 <u>Definition:</u> Let $\alpha\colon W^nG \longrightarrow \mathcal{Top}$ and $\beta\colon W^nG \longrightarrow \mathcal{Top}$ be n-coherent homotopy actions of a topological group G on spaces X and Y. An *n-coherent homotopy G-map* from (X,α) to (Y,β) is a continuous functor $\gamma\colon W^n(G \times \mathcal{L}_1) \longrightarrow \mathcal{Top}$ with $\gamma|W^n(G\times 0) = \alpha$ and $\gamma|W^n(G\times 1) = \beta$. The map $\gamma((e \times (0 \to 1)))\colon X \longrightarrow Y$ is called the underlying map of γ.

We recall from [B-V; chapt.4] that homotopy classes (through functors) of ∞-coherent homotopy G-maps form a category, where G may be any well-pointed topological group. The same holds (by the same arguments) for discrete groups and n-coherent homotopy G-maps. Moreover we shall use [B-V; (4.20),(4.21)]:

2.5 <u>Proposition:</u> Let H be a subgroup of G such that $H \subset G$ is a closed cofibration. Let $\alpha'\colon WH \longrightarrow \mathcal{Top}$ and $\beta,\gamma\colon WG \longrightarrow \mathcal{Top}$ be ∞-coherent homotopy actions of H on X and of G on Y and Z. Suppose further we are given an ∞-coherent homotopy H-map $\rho'\colon (X,\alpha') \longrightarrow (Y,\beta|WH)$ and an ∞-coherent homotopy G-map $\lambda\colon (Y,\beta) \longrightarrow (Z,\gamma)$ whose underlying maps are homotopy equivalences. Then:

(1) α' extends to a WG-structure α and ρ' to an ∞-coherent homotopy G-map $\rho\colon (X,\alpha) \longrightarrow (Y,\beta)$

(2) Any homotopy inverse $\kappa'\colon W(H \times \mathcal{L}_1) \longrightarrow \mathcal{Top}$ of $\lambda|(WH \times \mathcal{L}_1)$ extends to a homotopy inverse $\kappa\colon W(G \times \mathcal{L}_1) \longrightarrow \mathcal{Top}$ of λ.

2.6 <u>Remark:</u> Of course, Definition 2.4 still makes sense if G is replaced by an arbitrary topological category $\mathcal{C}$, and Proposition 2.5 holds with G replaced by an arbitrary well-pointed category $\mathcal{C}$ and H replaced by a subcategory $\mathcal{D}$ of $\mathcal{C}$ such that $\operatorname{mor}\mathcal{D} \subset \operatorname{mor}\mathcal{C}$ is a closed cofibration.

3. Main results

Throughout this section let G be a discrete group unless stated otherwise.

We first interprete the obstructions to a lift of $B\alpha$ in (1.2) as obstructions to higher coherence

3.1 Theorem: A homotopy action $\alpha\colon G \longrightarrow \pi_0(\mathrm{Aut}X)$ of G on X is induced by an n-coherent homotopy action iff there is a lift up to homotopy

$$\begin{array}{ccc} B^{n+1}G & \overset{\beta}{\dashrightarrow} & B(\mathrm{Aut}X) \\ \cap & \simeq & \downarrow \\ BG & \xrightarrow{\;B\alpha\;} & B(\pi_0\,\mathrm{Aut}X) \end{array}$$

where $B^{n+1}G$ is the $(n+1)$-skeleton of BG.

An ∞-coherent homotopy action can always be realized (see (3.2)) so that (3.1) and (3.2) imply Cooke's result (1.2):

3.2 Theorem: Given an ∞-coherent homotopy action $\beta\colon WG \longrightarrow \mathrm{Aut}X$, there is a free G-space Y_β and an ∞-coherent homotopy G-map $i_\beta\colon X \longrightarrow Y_\beta$ with the following properties

(1) i_β embeds X as a strong deformation retract

(2) any ∞-coherent homotopy G-map $\rho\colon (X,\beta) \longrightarrow (Z,\gamma)$ into a genuine G-space Z factors uniquely as through i_β and a genuine G-equivariant map $Y_\beta \longrightarrow Z$.

If one starts with a G-space X and drags it through the machines of (3.1) and (3.2) Cooke already showed that one ends up with X made free. We prove a corresponding result in our set-up by giving a complete classification of all free realizations of a given homotopy action: Let $\alpha\colon G \longrightarrow \pi_0(\mathrm{Aut}X)$ be a homotopy action of G on X. We call two realizations (Y,f) and (Z,g) of α *equivalent* iff there is a G-homotopy equivalence $h\colon Y \longrightarrow Z$, i.e. a homotopy equivalence in the category of G-spaces and equivariant maps, such that $h\circ f \simeq g$.

3.3 <u>Theorem:</u> There is a bijective correspondence between the equivalence classes of free realizations (Y,f) of a homotopy G-action α on X and the homotopy classes of lifts

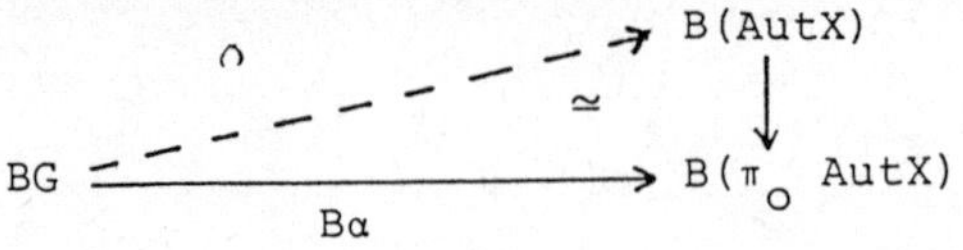

As a generalisation we now consider the case that a homotopy G-action extends a given genuine H-action of a subgroup H of G.

3.4. <u>Theorem:</u> Let $H \subset G$ be a subgroup of G. Let $\alpha\colon G \longrightarrow \pi_o(\mathrm{Aut}X)$ be a homotopy action of G on X such that $\alpha|H$ is induced by a genuine H-structure $\beta\colon H \longrightarrow \mathrm{Aut}X$. Then α is induced by an n-coherent homotopy action $\gamma\colon W^nG \longrightarrow \mathrm{Aut}X$ extending β iff there is a filler up to homotopy

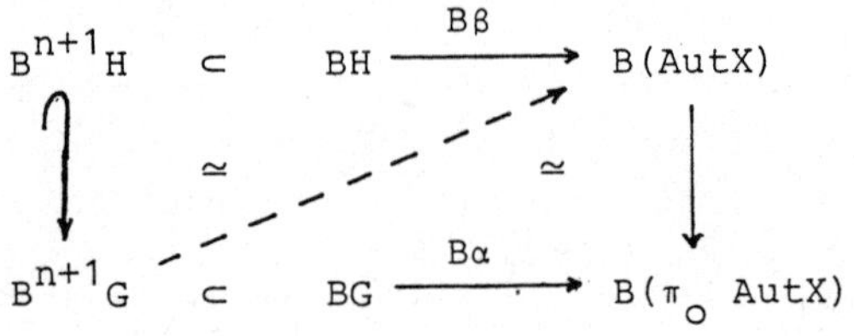

Of course, (3.2) has its analogue in the relative case.

3.5 <u>Theorem:</u> Let $\gamma\colon WG \longrightarrow \mathrm{Aut}X$ be an ∞-coherent homotopy action of G on X extending a strict H-action β. Then there is a free G-space Y and an H-equivariant map $f\colon Y \longrightarrow X$ which is an ordinary homotopy equivalence and whose H-structure extends to an ∞-coherent homotopy G-map.

<u>Remark:</u> We have defined realizations as maps $X \longrightarrow Y$ into a G-space. Since f in (3.5) is H-equivariant and Y is free we cannot expect to obtain such a map from X to Y unless X is H-free. In this case, we indeed may choose f as H-map from X to Y by (2.5) and (4.5) below.

3.6 <u>Corollary:</u> Let H be a p-Sylow subgroup of a finite group G, and let X be a p-local space of the homotopy type of a CW-complex with an H-action compatible with a homotopy G-action α on X. If $\mathrm{Aut}_1X \subset \mathrm{Aut}\,X$ denotes the component of id_X and if $H^*(BH;\{\pi_{*-2}\mathrm{Aut}_1X\})$ coincides with its G-invariant part [Br; p.84] then there is a free

G-space Y and a H-equivariant "realization" $f:Y \to X$ of α (i.e. f is H-equivariant and a homotopy equivalence).

We now turn to relative versions in the sense of [Z]. We need some preparations to state our results:
A topological space X with a right M_o-action and a left M_1-action of topological monoids M_o and M_1 gives rise to a category $\mathcal{C}(M_1,X,M_o)$ with two objects 0,1 and morphism spaces $mor(i,i) = M_i$, $mor(0,1) = X$ and $mor(1,0) = \emptyset$. Composition is defined by monoid multiplication and the actions. Conversely, any such category $\mathcal{C}$ makes $\mathcal{C}(0,1)$ into a right $\mathcal{C}(0,0)$ - and left $\mathcal{C}(1,1)$-space.
For a topological group G let $\hat{G}$ denote the space G with its left and right G-action from multiplication. Since $\mathcal{C}(G,\hat{G},G) = G \times \mathcal{L}_1$, an n-coherent homotopy G-map is a functor

$$\overline{\alpha}: W^n\mathcal{C}(G,\hat{G},G) \longrightarrow \mathcal{Top}.$$

Since G is a group such functors $\overline{\alpha}$ with $\overline{\alpha}(0) = X$ and $\overline{\alpha}(1) = Y$ are in 1-1 correspondence with functors

$$\hat{\alpha}: W^n\mathcal{C}(G,\hat{G},G) \longrightarrow \mathcal{C}(\mathrm{Aut}Y,\ F(X,Y),\ \mathrm{Aut}X)$$

where $F(X,Y)$ is the space of maps from X to Y.
In particular, a homotopy G-map $(X,\alpha_o) \longrightarrow (Y,\alpha_1)$ of spaces with homotopy G-actions is a functor

$$\hat{\alpha}: \mathcal{C}(G,\hat{G},G) \longrightarrow \mathcal{C}(\pi_o(\mathrm{Aut}Y),\ \pi_o F(X,Y),\ \pi_o(\mathrm{Aut}X))$$

extending $\alpha_o: G \longrightarrow \pi_o(\mathrm{Aut}X)$ and $\alpha_1: G \longrightarrow \pi_o(\mathrm{Aut}Y)$.
This functor defines a map $\alpha: \hat{G} \longrightarrow \pi_o F(X,Y)$ of the left $G \times G^{op}$-space $\hat{G}$ to the left $\pi_o(\mathrm{Aut}Y) \times \pi_o(\mathrm{Aut}X^{op})$-space $\pi_o F(X,Y)$ which is equivariant with respect to the homomorphism $\alpha_1 \times \alpha_o{}^{op}$. The pair $(\alpha_1 \times \alpha_o{}^{op},\alpha)$ and the obvious projections induce maps of 2-sided bar constructions [M; section 7]

(3.7)
$$\begin{array}{ccc}
BG^{op} & \xrightarrow{\quad B\alpha_o{}^{op}\quad} & B(\pi_o \mathrm{Aut}X^{op}) \\
\uparrow p_o & \curvearrowright & \uparrow q_o \\
B(*,G \times G^{op},\hat{G}) & \xrightarrow{\quad B\alpha \quad} & B(*,\pi_o(\mathrm{Aut}Y) \times \pi_o(\mathrm{Aut}X^{op}),\pi_o F(X,Y)) \\
\downarrow p_1 & \curvearrowright & \downarrow q_1 \\
BG & \xrightarrow{\quad B\alpha_1 \quad} & B(\pi_o \mathrm{Aut}Y)
\end{array}$$

(Recall $BG = B(*,G,*)$).

3.8 <u>Theorem:</u> Let G be discrete. A homotopy G-map

$$\hat{\alpha}: C(G,\hat{G},G) \longrightarrow C(\pi_o(\mathrm{Aut}Y),\pi_o F(X,Y),\pi_o(\mathrm{Aut}X))$$

from a homotopy G-space (X,α_o) to a homotopy G-space (Y,α_1) is induced by an n-coherent homotopy G-map

$$\gamma: W^n C(G,\hat{G},G) \longrightarrow C(\mathrm{Aut}Y,\ F(X,Y),\ \mathrm{Aut}X)$$

iff there is a lift (up to homotopy)

$$\begin{array}{ccc}
B^{n+1}G^{op} & \xrightarrow{\ f_{n+1}\ } & B(\mathrm{Aut}X^{op}) \\
\uparrow{\scriptstyle p_o} & \simeq & \uparrow{\scriptstyle q_o} \\
B^n(*,G\times G^{op},\hat{G}) & \xrightarrow{\ h_n\ } & B(*,\mathrm{Aut}Y\times \mathrm{Aut}X^{op},\ F(X,Y)) \\
\downarrow{\scriptstyle p_1} & \simeq & \downarrow{\scriptstyle q_1} \\
B^{n+1}G & \xrightarrow{\ g_{n+1}\ } & B(\mathrm{Aut}Y)
\end{array}$$

of (3.7) on the indicated skeletons.
Moreover, if f_{n+1} and g_{n+1} are obtained from WG-stuctures on X and Y according to (3.1), γ can be chosen to be compatible with these structures.

The analogue of (3.2) is

3.9 <u>Theorem:</u> Given an ∞-coherent homotopy G-map

$$\gamma: WC(G,\hat{G},G) \longrightarrow C(\mathrm{Aut}Y,\ F(X,Y),\ \mathrm{Aut}X)$$

there exists a homotopy commutative diagram

$$\begin{array}{ccc}
X & \xrightarrow{\ f\ } & Y \\
{\scriptstyle i_X}\downarrow & \simeq & \downarrow{\scriptstyle i_Y} \\
X' & \xrightarrow{\ f'\ } & Y'
\end{array}$$

where f is the underlying map of γ (i.e. $f = \gamma((e,o \to 1))$),

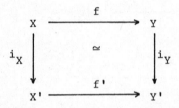

i_X and i_Y are the underlying maps of ∞-coherent homotopy G-maps which embed X and Y as strong deformation retracts into free G-spaces X' and Y', and f' is a strict G-map. Moreover, if X and Y are G-spaces there are homotopy inverses of i_X and i_Y which are G-equivariant.

This answers the realization problem for homotopy G-maps.

3.10 Extensions of our results:

(1) The proofs will show that in the most interesting case of infinite coherence our results hold for any well-pointed topological group G of the homotopy type of a CW-complex. If G is not well-pointed we have to substitute WG by WG', where G' is the monoid obtained from G by attaching a whisker.

3.11 Theorem: If $n = \infty$ all our results hold for a (well-pointed) topological group G of the homotopy type of a CW-complex. In the cases (3.4) and (3.5) well-pointed subgroups H of G of the homotopy type of a CW-complex are admitted if $H \subset G$ are closed cofibrations.

(2) In the case of finite coherence, an analysis of our proofs gives results similar to (3.1), (3.4), and (3.8) for finite-dimensional CW-groups but with dimension shifts. The details are left to the reader.

(3) It is not difficult to state and prove classification results of the type of (3.3) in the relative cases.

(4) In (3.5) one often wants the stronger result that we have an H-equivariant realization in the strong sense, i.e. a realization in the category of H-spaces. We prove this in the case that X is H-free. If this does not hold one has to take care of the fixed point structure of X which makes the analysis more complicated. We shall deal with this problem in a subsequent paper [S-V].

4. Basic properties of the W-construction

The correspondence $(f_n, t_n, \ldots, f_o) \longrightarrow f_n \circ f_{n-1} \circ \ldots \circ f_o$ defines a natural transformation $\varepsilon\colon W \longrightarrow \mathrm{Id}$. Pulling back a G-structure via ε to a WG-structure we can make the notion of an n-coherent homotopy G-map into a genuine G-space (see (3.2)) formally precise. The same holds for n-coherent homotopy G-actions extending genuine H-actions in (3.4).

4.1 <u>Proposition:</u> (1) $\varepsilon\colon WC \longrightarrow C$ is a homotopy equivalence (on morphism spaces).

(2) If C is well-pointed, $\varepsilon_n = \varepsilon | W^nC\colon W^nC \longrightarrow C$ is n-connected.

<u>Proof:</u> ε has a natural, non-functorial section $\eta\colon C \longrightarrow WC$ sending f to (f), and

$$h_t(f_n,t_n,\ldots,f_1,t_1,f_o) = (f_n,t\cdot t_n,\ldots,f_1,t\cdot t_1,f_o)$$

is a fibrewise deformation of $\eta\circ\varepsilon$ to the identity. This proves (1). $W^{r+1}C$ is obtained from W^rC by attaching (r+1)-cubes $C_{r+2}(A,B)\times I^{r+1}$ along $DC_{r+2}(A,B)\times I^{r+1}\cup C_{r+2}(A,B)\times\partial I^{r+1}$ and products of those cubes as upper faces of some higher dimensional cubes. Here $DC_{r+2}(A,B)$ is the space of all strings $(f_{r+2},\ldots,f_o)$ containing an identity. Since $DC_{r+2}(A,B)\subset C_{r+2}(A,B)$ is a closed cofibration, the homotopy excision theorem implies that $W^rC \longrightarrow W^{r+1}C$ is r-connected. Hence the inclusion $W^nC \longrightarrow WC$ is n-connected. So (2) follows from (1).

Proposition 4.1 can be interpreted as follows: The relations in C hold in WC up to a contractible choice of homotopies. An inspection of the relations (2.1) shows that WC is obtained from the free category W^oC on the graph defined by C by putting back the relations up to compatible homotopies. The next result will show that WC is universal with respect to the properties in (4.1).

Let $V\subset WC$ be a subcategory, and $V_{n+1}(A,B)\subset C_{n+1}(A,B)\times I^n$ the subspace of all elements respresenting morphisms in V. We call V an *admissible subcategory* of WC provided each morphism in V that decomposes in WC also decomposes in V, and

$$V_{n+1}(A,B)\cup C_{n+1}(A,B)\times\partial I^n\cup DC_{n+1}(A,B)\times I^n\subset C_{n+1}(A,B)\times I^n$$

is a closed cofibration for all n,A,B. Note that the empty subcategory is admissible if C is well-pointed.

4.2 <u>Proposition:</u> Consider the diagram of categories and functors

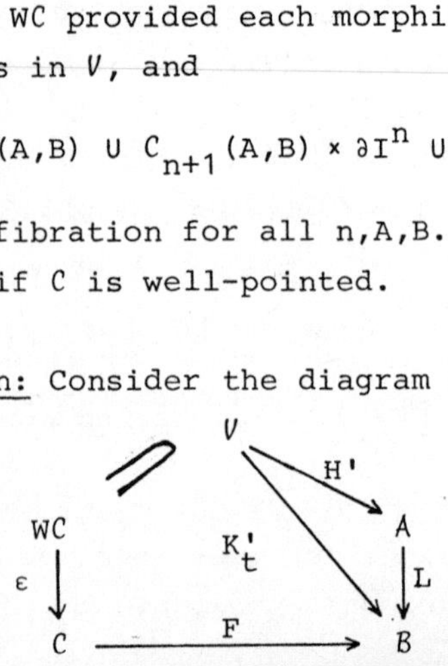

$t\in I$

Assume (i) V is admissible

(ii) L is a homotopy equivalence

(iii) K_t' is a homotopy through functors from $F\circ(\varepsilon|V)$ to $L\circ H'$.

Then there exist extensions $H: WC \longrightarrow A$ and $K_t: WC \longrightarrow B$ of H' and K_t' such that $K_t: F\circ\varepsilon \simeq L\circ H$. Moreover, any two such extensions are homotopic rel$V$.

The extensions are constructed by induction over the n-skeletons W^nC. For details see [B-V, p.84 ff]. This proves the universality of ε.

4.3 Proposition: Consider the diagram of categories and functors

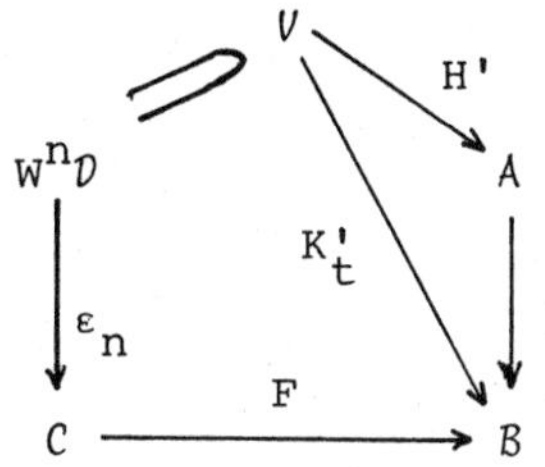

$t \in I$

where $W^n D \subset WC$ is the subcategory generated by V and W^nC, and $\varepsilon_n = \varepsilon|W^n D$.

Assume (O) C has discrete morphism spaces

(i) V is an admissible subcategory of WC

(ii) L is n-connected

(iii) K_t' is a homotopy through functors from $F\circ(\varepsilon_n|V)$ to $L\circ H'$.

Then there are extensions $H: W^n D \longrightarrow A$ and $K_t: W^n D \longrightarrow B$ of H' and K_t' such that $K_t: F\circ\varepsilon_n \simeq L\circ H$. Moreover, the restrictions of any two extensions to $W^{n-1}D$ are homotopic rel V.

Note that the morphism spaces of $W^n D$ are CW-complexes so that (4.3) is an immediate consequence of classical homotopy theory.

Another important result is the homotopy extension property of the W-construction. We use the terminology of (4.3).

4.4 Proposition: Let V be an admissible subcategory of WC and let n be a natural number or ∞. Suppose we are given a functor $F_o: W^n D \longrightarrow E$ and a homotopy through functors $H_t: V \longrightarrow E$ such that $H_o = F_o|V$. Then there exists an extension F_t of F_o and H_t.

This follows directly from the definition of an admissible subcategory.

We now turn to the problem of "realizing" a WC-space by a C-space: Let $Y: C \longrightarrow Top$ be a C-space. From (2.6) we deduce that any collection of homotopy equivalences $f_A: X_A \longrightarrow Y(A)$, $A \in \text{ob } C$, can be extended to a homotopy C-map $X \longrightarrow Y$. In particular, the correspondence $A \longrightarrow X_A$ extends to a WC-structure X. For the proof of (3.2) we need the converse of this fact.

4.5 <u>Proposition:</u> There is a functor M from the category of WC-spaces and homomorphisms to the category of C-spaces and homomorphisms together with an ∞-coherent homotopy C-map $i_X: X \longrightarrow MX$ with the following properties

(i) $i_X(A): X(A) \longrightarrow MX(A)$ embeds X(A) as a strong deformation retract into $MX(A), A \in \text{ob } C$

(ii) Any ∞-coherent homotopy C-map $\alpha: X \longrightarrow Y$ from a WC-space X to a C-space Y factors uniquely as $\alpha = h \circ i_X$, where $h: MX \longrightarrow Y$ is a homomorphism of C-spaces.

<u>Proof:</u> Define

$$MX(B) = \coprod_A W(C \times L_1)((A,0),(B,1)) \times X(A)/\sim$$

with the relation

$$(a \circ b \circ c, x) \sim (\varepsilon(a) \circ b,\ X(c)(x))$$

if $a \in W(C \times 1)$ and $c \in W(C \times 0)$. The ∞-coherent homotopy C-map i_X is given by the adjunctions of the projections

$$W(C \times L_1)((A,0),(B,1)) \times X(A) \longrightarrow MX(B).$$

Its underlying map is

$$X(A) \longrightarrow MX(A) \qquad x \longrightarrow ((id_A, 0 \rightarrow 1; x).$$

The C-structure on MX is the obvious left action of C on MX, and the universal property of i_X follows from the construction.

It remains to show that X is a strong deformation retract of MX. For this we filter MX(A) by skeletons F_n. For convenience we use the symbol

$$(f_n,t_n,\dots,f_{i+1},t_{i+1},\vec{f_i},t_i,\dots,f_o;x)$$

for the representative

$$((f_n,id_1),t_n,\dots,(f_{i+1},id_1),t_{i+1},(f_i,0\to 1),t_i,\dots,(f_o,id_o);x)$$

of an element of MX. Let $K \subset MX(A)$ denote the space of all those elements which have a representative of the form $(\vec{f_k},t_k,\dots,f_o;x)$. In a first step we deform MX(A) into K. Since $F_{n-1} \subset F_n$ is a closed cofibration, it suffices to construct deformations of $F_n \cup K$ into $F_{n-1} \cup K$. Observe that $(f_n,t_n,\dots,\vec{f_i},t_i,\dots,f_o;x)$ represents an element in $F_{n-1} \cup K$ iff $i = n$, or some $f_j = id$, $j \neq i$, or $(t_n,\dots,t_1) \in 0 \times I^{n-1} \cup I \times \partial I^{n-1}$. Since the latter space is a deformation retract of I^n the required deformation of $F_n \cup K$ to $F_{n-1} \cup K$ exists. The deformation h_t of K into X(A) is defined by

$$h_t(\vec{f_k},t_k,\dots,f_o;x) = (\vec{id_A},t,f_k,t_k,\dots,f_o;x).$$

5. Proofs

Part of the proof of (3.1) in the case of $n=\infty$ consists of constructing a homomorphism $WG \longrightarrow AutX$ from a map $BG \longrightarrow B(AutX)$, i.e. we have to pass from the classifying space of a monoid back to the monoid itself. One way of doing this is to compare the fibers of the "universal G-fibration" $p_G\colon EG \longrightarrow BG$, where $EG = B(*,G,G)$ is a free contractible right G-space, and the path space fibration $\pi\colon P(BG;*,BG) \longrightarrow BG$. Here $P(X;A,B)$ denotes the space of Moore paths in X, starting in A and ending in B. Its elements are pairs $(\omega,r) \in F(\mathbb{R}_+,X) \times \mathbb{R}_+$ such that $\omega(o) \in A$, $\omega(r) \in B$, and $\omega(t) = \omega(r)$ for $t \geq r$.

The inclusion $G \subset EG$ of the simplicial 0-skeleton is an equivariant map of right G-spaces. Using the G-structure, we define a monoid structure on $P(EG;e,G)$ by setting $(\rho,s)+(\nu,r) = (\omega,r+s)$ with

(5.1)
$$\omega(t) = \begin{cases} \nu(t) & 0 \leq t \leq r \\ \rho(t-r)\cdot\nu(r) & r \leq t \leq r+s \end{cases}$$

The endpoint projection $\pi\colon P(EG;e,G) \to G$ is a homomorphism. $P(EG;e,G)$ is the homotopy fiber of $G \subset EG$. Since EG is contractible, π is a homotopy equivalence. Hence, from (4.2) we obtain

5.2 Proposition: If G is a well-pointed topological monoid, there is a homotopy commutative diagram of homomorphisms

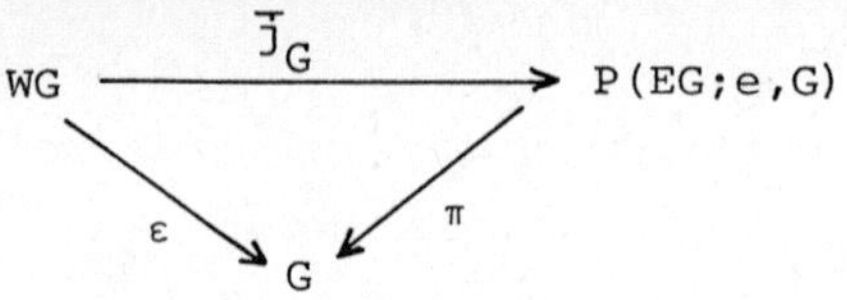

Moreover, $\bar{j}_G$ is natural up to homotopy with respect to homomorphisms $G \longrightarrow H$.

Convention: Homotopies of homomorphisms or functors are always homotopies through homomorphisms or functors.

The last statement of (5.2) is a consequence of the uniqueness part of (4.2) applied to the following diagram of homomorphisms

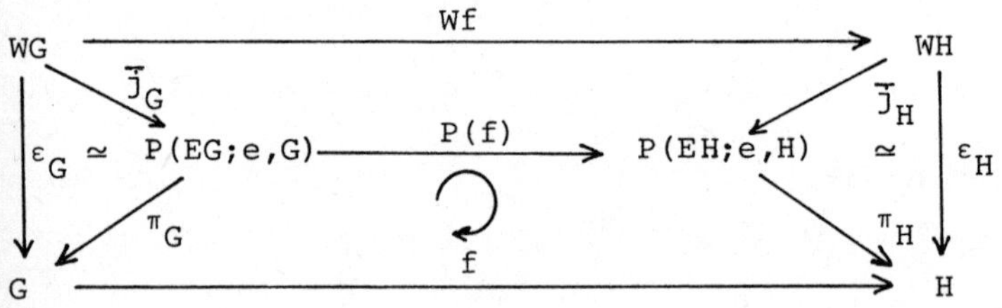

Since $\pi_H \circ P(f) \circ \bar{j}_G \simeq \pi_H \circ \bar{j}_H \circ Wf$, both homomorphisms $P(f) \circ \bar{j}_G$ and $\bar{j}_H \circ Wf$ lift $f \circ \varepsilon_G$ (up to homotopy) and hence are homotopic.

(5.2) together with the following well-known fact establishes the comparison of fibers mentioned above.

5.3 Proposition: If G is a grouplike well-pointed topological monoid, the homomorphism $P(p_G)$: $P(EG;,e,G) \longrightarrow \Omega BG := P(BG;*,*)$ is a homotopy equivalence (as a map).

Remark: We call a monoid G *grouplike* if its multiplication admits a homotopy inverse. If G is of the homotopy type of a CW-complex this is equivalent to the usual definition that $\pi_o G$ be a group [tD-K-P; (12.7)]

Hence, for well-pointed grouplike monoids G we have a homomorphism

(5.4) $$j_G\colon WG \longrightarrow \Omega BG$$

which is a homotopy equivalence (as a map) and natural in G up to homotopy.
Applying (4.2) twice we obtain homomorphisms l_G und k_G which are homotopy equivalences (as maps)

(5.5)

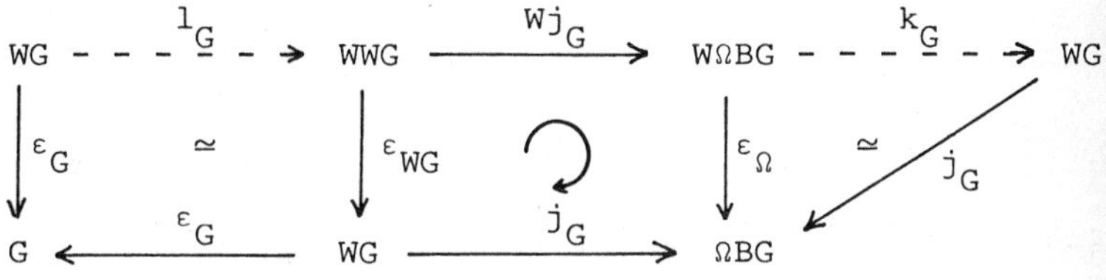

The uniqueness part of (4.2) implies that

(5.6) $$\varepsilon_{WG}\circ l_G \simeq \mathrm{id} \qquad\qquad k_G\circ Wj_G\circ l_G \simeq \mathrm{id}$$

Moreover, k_G and l_G are natural up to homotopy in G. For l_G this is clear from (4.2). For k_G it follows from the diagram

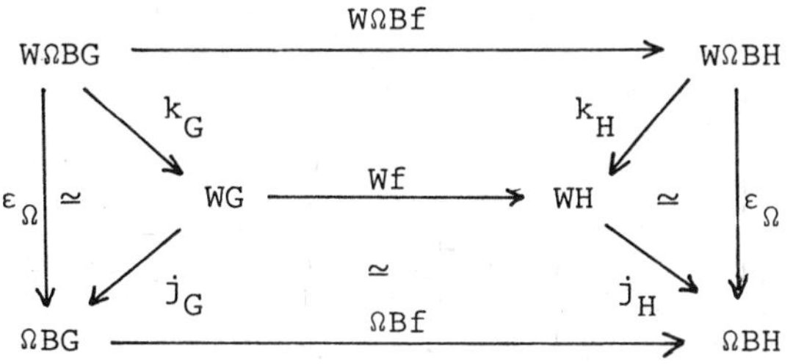

All these results hold for well-behaved monoids. But if X is too big, AutX could be nasty. In this case we substitute it by the CW-monoid R(AutX) where R is the topological realization of the simplicial complex functor. The back adjunction R(AutX) ⟶ AutX is a homomorphism and a weak equivalence. Since in all our statements (including (3.11)) BG is of the homotopy type of a CW-complex, each map BG ⟶ B(AutX) factors uniquely up to homotopy through BR(AutX). Moreover, each homomorphism WG ⟶ AutX factors uniquely up to homotopy through R(AutX). This follows from the fact that (4.2) also holds if L is a weak equivalence and mor $\mathcal{C}$ is of the homotopy type of a CW-complex.
So from now on we assume that AutX is a CW-monoid.

5.7 Proofs of (3.1) and (3.4):

(3.1) follows from (3.4). We prove (3.4). Suppose $\alpha: G \longrightarrow \pi_0(\mathrm{Aut}X)$ is induced by an n-coherent homotopy action $\gamma: W^nG \longrightarrow \mathrm{Aut}X$, compatible with $\beta \circ \varepsilon|WH$. Let $W^n\mathcal{D} \subset WG$ be the subcategory generated by W^nG and WH. Since WG is obtained from $W^n\mathcal{D}$ by attaching cubes of dimensions greater than n, the functor ε defines a commutative square

$$\begin{array}{ccc} WH & \subset & W^n\mathcal{D} \\ \downarrow{\scriptstyle \varepsilon_H} & & \downarrow{\scriptstyle \varepsilon'} \\ H & \subset & G \end{array}$$

with ε_H an equivalence and ε' n-connected (see 4.1). We obtain a map of pairs

$$(B\varepsilon', B\varepsilon_H): (BW^n\mathcal{D}, BWH) \longrightarrow (BG, BH)$$

with $B\varepsilon_H$ a homotopy equivalence and $B\varepsilon'$ (n+1)-connected. Hence the inclusion $(B^{n+1}G \cup BH, BH) \subset (BG, BH)$ factors up to homotopy

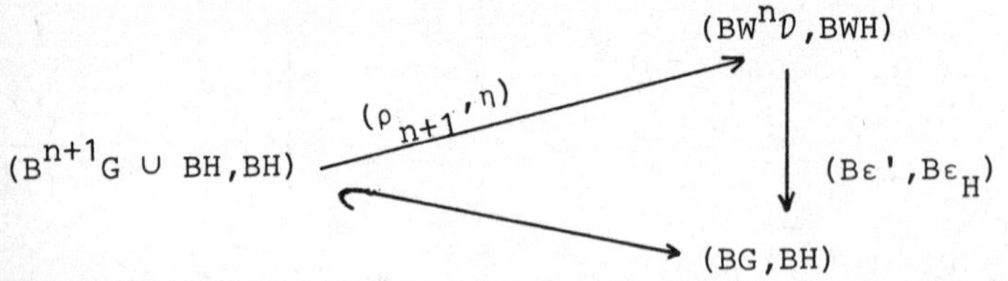

where η is any chosen homotopy inverse of $B\varepsilon_H$. The composite $B(\gamma \cup \beta \circ \varepsilon) \circ \rho_{n+1}$ is a required filler.

Conversely, suppose we are given a filler f. Since $B^{n+1}H \subset B^{n+1}G$ is a cofibration we may assume that f and $B\beta$ together define a map $B^{n+1}G \cup BH \longrightarrow B(\mathrm{Aut}X)$, which we also denote by f. Consider the diagram

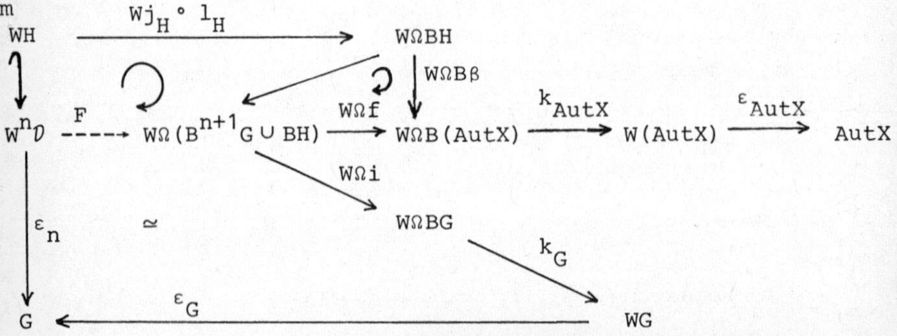

where i: $B^{n+1}G \cup BH \subset BG$ is the inclusion. Since it is (n+1)-connected, $W\Omega i$ and hence the composite $\varepsilon_G \circ k_G \circ W\Omega i$ is n-connected. By (4.3), F exists. (5.6) and the naturality of k_G provide homotopies

$$\varepsilon_{AutX} \circ k_{AutX} \circ W\Omega B\beta \circ Wj_H \circ l_H \simeq \varepsilon_{AutX} \circ W\beta \circ k_H \circ Wj_H \circ l_H \simeq \beta \circ \varepsilon_H.$$

By (4.4), we can extend this homotopy to a homotopy of homomorphisms from $\varepsilon_{AutX} \circ k_{AutX} \circ W\Omega f \circ F$ to a functor $\gamma: W^nD \longrightarrow AutX$ with $\gamma | WH = \beta \circ \varepsilon_H$.

5.8 <u>Proof of (3.2) and (3.5):</u> (3.2) is a special case of (4.5). For (3.5) we apply (4.5) to obtain a free H-space M_HX and a free G-space M_GX together with an ∞-coherent homotopy H-map i_H: $X \longrightarrow M_HX$ and an ∞-coherent homotopy G-map i_G: $X \longrightarrow M_GX$. Since $W(H \times L_1) \subset W(G \times L_1)$ we have a cofibration

$$j: M_HX \longrightarrow M_GX$$

which is H-equivariant. Since $j \circ i_H = i_G$ as maps of spaces, j is a homotopy equivalence and hence an H-equivariant homotopy equivalence, because both spaces are H-free. By (4.5.2), the retraction r: $M_HX \longrightarrow X$ can be chosen to be H-equivariant. If j^{-1} denotes an H-equivariant homotopy inverse of j, the composite

$$r \circ j^{-1}: M_GX \longrightarrow M_HX \longrightarrow X$$

is the required H-equivariant map. By (2.5.2) its H-structure can be extended to an ∞-coherent homotopy G-map, because it is homotopy inverse to i_G.

5.9 <u>Proof of (3.3):</u> Let $l(\gamma)$: BG $\longrightarrow$ B(AutX) be the lift obtained from γ: WG $\longrightarrow$ AutX, and let a(f): WG $\longrightarrow$ AutX be the functor induced by f: BG $\longrightarrow$ B(AutX). By construction

$$l(\gamma) = B\gamma \circ \lambda_G$$

$$a(f) = \varepsilon_{AutX} \circ k_{AutX} \circ W\Omega f \circ Wj_G \circ l_G$$

where λ_G is a chosen homotopy inverse of $B\varepsilon_G$. The diagram

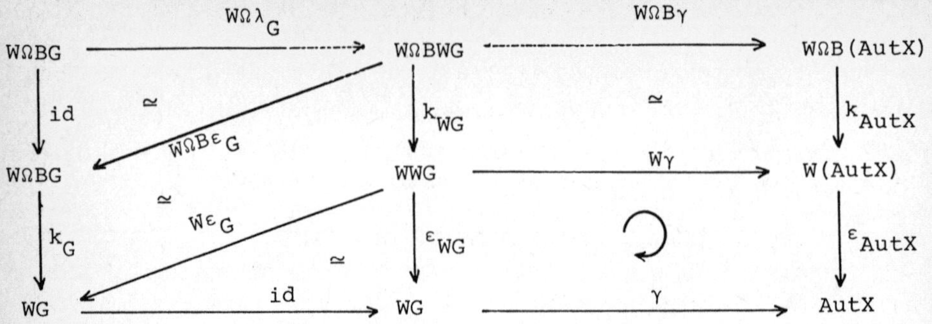

implies that $a(l(\gamma)) \simeq \gamma$.

By definition, $l \circ a: [BG,B(AutX)] \longrightarrow [BG,B(AutX)]$ is the composite

$$[BG,B(AutX)] \xrightarrow{BW\Omega} [BW\Omega BG, BW\Omega B(AutX)] \xrightarrow[\cong]{w} [BWG,B(AutX)] \xleftarrow[\cong]{B\varepsilon_G^*} [BG,B(AutX)]$$

where $[\ ,]$ denotes homotopy classes,

and $w[h] = [B(\varepsilon_{AutX} \circ k_{AutX}) \circ h \circ B(Wj_G \circ 1_G)]$.

Consider

$$
\begin{array}{ccccc}
BW\Omega BG & \xrightarrow{B\varepsilon_\Omega} & B\Omega BG & \xrightarrow{u_{BG}} & BG \\
\downarrow{\scriptstyle BW\Omega f} \quad \text{I} & & \downarrow{\scriptstyle B\Omega f} \quad \text{II} & & \downarrow{\scriptstyle f} \\
BW\Omega B(AutX) & \xrightarrow{B\varepsilon_\Omega} & B\Omega B(AutX) & \xrightarrow{u_{B(AutX)}} & B(AutX)
\end{array}
$$

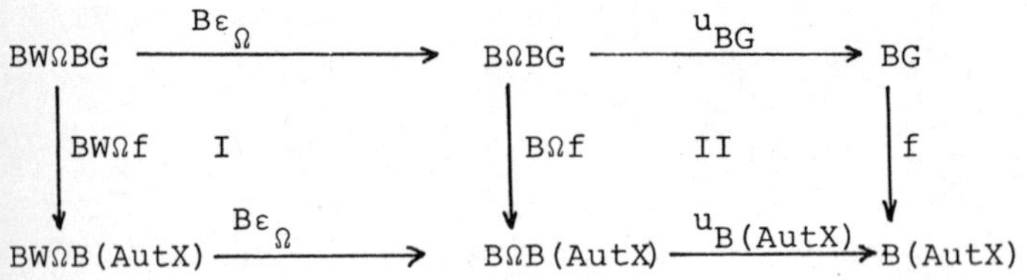

Clearly (I) commutes. By $[M;(14.3)]$ there is a homotopy equivalence $u_X: B\Omega X \longrightarrow X$, X a connected space of the homotopy type of a CW-complex, which is natural up to homotopy. Hence $BW\Omega$ is bijective. This proves that l is inverse to a.

We have shown

5.10 <u>Proposition:</u> $a: [BG,B(AutX)] \longrightarrow [WG,AutX]$ is bijective with inverse l.

Any WG-structure $\gamma: WG \longrightarrow AutX$ inducing $\alpha: G \longrightarrow \pi_0(AutX)$ determines a free realization $i_\gamma: X \longrightarrow M_\gamma X$ by (4.5). Conversely, a free realization $f: X \longrightarrow Y$ of α by (2.5) gives rise to a WG-structure $\rho(f)$ on X, which induces α. We shall show below that i and ρ induce maps

$$i\colon [WG,\mathrm{Aut}X]_\alpha \rightleftarrows \mathrm{Real}(\alpha)\colon \rho$$

where Real(α) is the set of equivalence classes of free realizations of α and $[WG,\mathrm{Aut}X]_\alpha$ the set of homotopy classes of WG-structures inducing α.
If $\overline{\gamma} = \rho(i_\gamma)$, we are given ∞-coherent homotopy G-maps

$$(X,\gamma) \longrightarrow M_\gamma X \longleftarrow (X,\overline{\gamma})$$

having the same homotopy equivalence as underlying map. By (2.5), we obtain a composite ∞-coherent homotopy G-map $(X,\gamma) \longrightarrow (X,\overline{\gamma})$ whose underlying map may be chosen to be the identity (composites are defined up to homotopy). Conversely, given a free realization $f\colon X \longrightarrow Y$, we by (2.5) and (4.5.2) have a commutative diagram of ∞-coherent homotopy G-maps

$$\begin{array}{ccc} & (X,\rho(f)) & \\ {\scriptstyle i_\rho}\swarrow & & \searrow{\scriptstyle f} \\ M_{\rho(f)}X & \overset{h}{\dashrightarrow} & Y \end{array}$$

with h strictly G-equivariant. Since $M_{\rho(f)}X$ and Y are free, h is a G-equivariant homotopy equivalence. Hence $i \circ \rho = \mathrm{id}$. So (3.3) is proved once we have shown.

5.11 Lemma: Two WG-structures α and β on X are homotopic iff there is an ∞-coherent homotopy G-map $(X,\alpha) \longrightarrow (X,\beta)$ with id_X as underlying map.

Proof: Suppose $\alpha \simeq \beta$. In (4.4) let $C = G \times L_1$ and V be the subcategory of WC generated by $W(G \times 0)$, $W(G \times 1)$ and the morphism $((e,0\to 1))$. We extend the identity homotopy G-map $(X,\alpha) \longrightarrow (X,\alpha)$ and the homotopy on V given by the constant homotopy on $W(G \times 0)$ and $((e,0\to 1))$, and by $\alpha \simeq \beta$ on $W(G \times 1)$, to obtain an ∞-coherent homotopy G-map $(X,\alpha) \longrightarrow (X,\beta)$ over id_X.
Conversely, suppose id_X has the structure of an ∞-coherent homotopy G-map $\gamma\colon W(G \times L_1) \longrightarrow \mathit{Top}$ from (X,α) to (X,β). For the rest of the proof we have to recall the basic idea of the proof of (2.5). Let Is be the category

$$0 \underset{j}{\overset{i}{\rightleftarrows}} 1$$

A homotopy inverse of γ is constructed by first extending the underlying map to a functor $\mu\colon W\mathcal{I}\mathfrak{s} \longrightarrow \mathcal{T}op$ and then extending μ and γ to $\nu\colon W(G \times \mathcal{I}\mathfrak{s}) \longrightarrow \mathcal{T}op$. The inclusion of L_1 into $\mathcal{I}\mathfrak{s}$ as j defines the homotopy inverse. In our case we may choose μ to be the constant functor. Let C' be the full subcategory of $W(G \times \mathcal{I}\mathfrak{s})$ consisting of the object 0, and let C be obtained from C' by adding the relation

$$(f_n, t_n, \ldots, f_o) = f_n \circ f_{n-1} \circ \ldots \circ f_n$$

if each of the f_k is of the form (e,i) or (e,j). By our choice of μ, the functor ν induces a functor $\lambda\colon C \longrightarrow \mathcal{T}op$. The functors $F_o, F_1\colon WG \longrightarrow C$

$$F_o(g_n, t_n, \ldots, g_o) = ((g_n, id_o), t_n, \ldots, (g_o, id_o))$$

$$F_1(g_n, t_n, \ldots, g_o) = ((e,j), 1, (g_n, id_1), t_n, \ldots, (g_o, id_1), 1, (e,i))$$

both make

$$\begin{array}{ccc} WG & \dashrightarrow & C \\ & \searrow^{\varepsilon} \quad {}^{\varepsilon'}\swarrow & \\ & G & \end{array}$$

commute and hence are homotopic (4.2). By construction, $\nu \circ F_o = \alpha$ and $\nu \circ F_1 = \beta$.

5.12 <u>Proof of (3.6):</u> By (3.4) and (3.5) we have to find a filler for

$$\begin{array}{ccc} BH & \xrightarrow{B\beta} & B(\mathrm{Aut}X) \\ \cap & & \downarrow \\ BG & \xrightarrow{B\alpha} & B(\pi_o \mathrm{Aut}X) \end{array}$$

The obstructions for its existence lie in $H^n(BG, BH; \{\pi_{n-2}(\mathrm{Aut}_1 X)\})$, $n \geq 3$, where $\mathrm{Aut}_1 X$ is the component of the identity in AutX. By [C;Cor.2.2], $\pi_{n-2}(\mathrm{Aut}_1 X)$ is p-local. Hence the transfer ensures the vanishing of $H^n(BG, BH; \{\pi_{n-2}(\mathrm{Aut}_1 X)\})$ for $n \geq 3$.

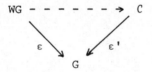

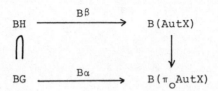

The idea of the proof of (3.8) is the same as the one of (3.1): We have to pass from BG back to G (or rather WG) and from $B(*,G\times H^{op},X)$ to X, where X is a left G-right H-space. For BG this has been done in the proof of (3.1), for $B(*,G\times H^{op},X)$ we compare the fiber sequence

$$X \longrightarrow B(*,G\times H^{op},X) \longrightarrow BG\times BH^{op}$$

with the fiber sequence

$$\mathrm{Fib} \longrightarrow B(*,G\times H^{op},X) \longrightarrow BG\times BH^{op}$$

where Fib is the homotopy fiber. We represent Fib by a space having a natural action of the Moore-loops $\Omega BG\times \Omega BH^{op}$, and, similar to the proof of (3.1), we construct a functor

$$WC(G,X,H) \longrightarrow C(\Omega BG,\mathrm{Fib},(\Omega BH^{op})^{op})$$

extending j_G and $(j_{H^{op}})^{op}$.
We proceed as far as possible in analogy to (3.1): Let G and H be well-pointed monoids. Let $EX = B(*,G\times H^{op},X)$, $E_GX = B(*,G,X)$, and $E^HX = B(*,H^{op},X)$. The injections $G\to G\times H^{op}$ and $H^{op}\to G\times H^{op}$ make E_GX and E^HX subspaces of EX with intersection X, the simplicial 0-skeleton of all three spaces. We have pairings

$$EG\times E^HX \cong B(*,G\times H^{op},G\times X) \longrightarrow B(*,G\times H^{op},X) = EX$$
$$EH^{op}\times E_GX \cong B(*,G\times H^{op},H^{op}\times X) \longrightarrow B(*,G\times H^{op},X) = EX$$

which commute on $E_GX\cap E^HX = X$ and extend the pairing on the 0-skeletons given by the $G\times H^{op}$-action on X.
Let $Sq\subset F(\mathbb{R}_+\times\mathbb{R}_+,EX)\times\mathbb{R}_+\times\mathbb{R}_+$ be the subspace of all "Moore-squares" (w,r,s) in EX such that

$$w(t,u) = \begin{cases} w(r,u)\in E_GX & \text{for } t\ge r \text{ and all } u\\ w(t,s)\in E^HX & \text{for } s\ge u \text{ and all } t\ .\end{cases}$$

Consequently, $w(r,s)\in X$. We define a left $P(EG;e,G)\times P(EH^{op};e,H^{op})$ - action on Sq by

$$((\omega,l),(\nu,k)) * (w,r,s) = (v,r+k,s+l)$$

where

$$v(t,u) = \begin{cases} w(t,u) & 0 \le t \le r, \quad 0 \le u \le s \\ \nu(t-r) \cdot w(r,u) & r \le t \le r+k, \ 0 \le u \le s \\ \omega(u-s) \cdot w(t,s) & 0 \le t \le r, \quad s \le u \le s+1 \\ \omega(u-s) \cdot \nu(t-r) \cdot w(r,s) & r \le t \le r+k, \ s \le u \le s+1 \end{cases}$$

where $\cdot$ denotes the pairings

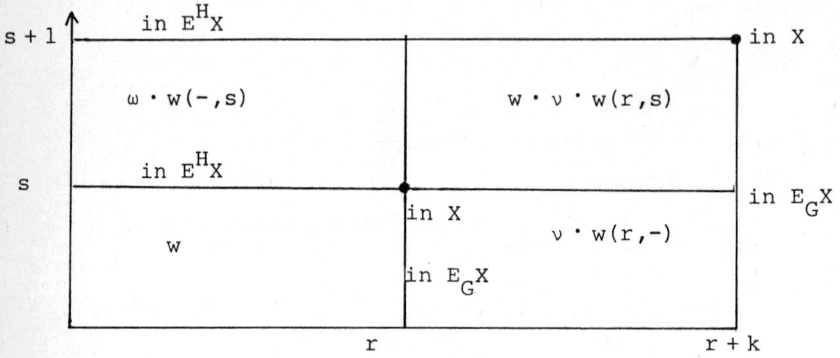

The endpoint projection π: Sq $\longrightarrow$ X, $(w,r,s) \longrightarrow w(r,s)$ together with the endpoint projections $P(EG;e,G) \longrightarrow G$ and $P(EH^{op};e,H^{op}) \longrightarrow H^{op}$ define a functor

$$\pi\colon C(P(EG;e,G), Sq, (P(EH^{op};e,H^{op})^{op}) \longrightarrow C(G,X,H)$$

which is a homotopy equivalence (on morphism spaces). Hence we obtain the analogue of (5.2).

5.13 <u>Lemma:</u> If G and H are well-pointed topological spaces and X is a left G-right H-space, there is a diagram of categories and functors

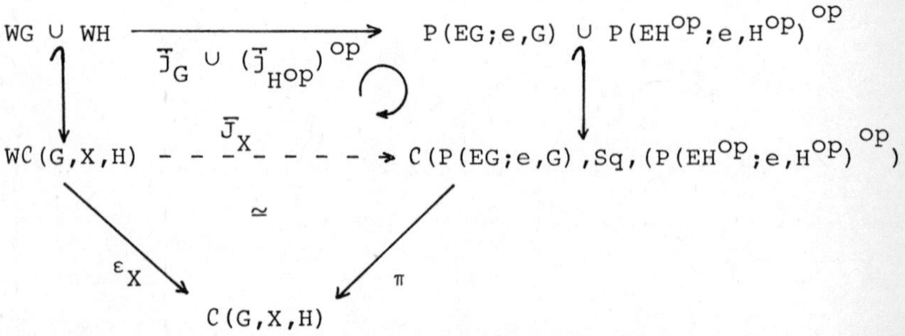

where $\bar{J}$ is the functor of (5.2).

Let $p_G\colon EX \longrightarrow BG$ and $p^H\colon EX \longrightarrow BH^{op}$. As model for the h-fiber of $(p_G,p^H)\colon EX \longrightarrow BG \times BH^{op}$ we take the space

$$Fib(p_G,p^H) = \{(\omega,\nu,z) \in P(BG;BG,*) \times P(BH^{op};BH^{op},*) \times EX; \omega(0) = p_G(z), \nu(0) = p^H(z)\}$$

There is an obvious left action of $\Omega BG \times \Omega BH^{op}$ on $Fib(p_G,p^H)$. The map $Sq \longrightarrow Fib(p_G,p^H)$ sending (w,r,s) to the triple $(p_G \circ w(0,-), p^H \circ w(-,0), w(0,0))$ together with the maps of (5.3) define a functor

$$Sq(p_G,p^H)\colon C(P(EG;e,G), Sq, (P(H^{op};e,H^{op}))^{op}) \longrightarrow C(\Omega BG, Fib(p_G,p^H), (\Omega BH^{op})^{op})$$

5.14 <u>Lemma:</u> If G and H are group-like, $Sq(p_G,p^H)$ is a homotopy equivalence.

This follows immediately from (5.3) and [P; Thm.], where $Fib(p_G,p^H)$ is proved to be homotopy equivalent to X.
Hence if G and H are grouplike and well-pointed we obtain functors

$$J_X\colon WC(G,X,H) \longrightarrow C(\Omega BG, Fib(p_G,p^H), (\Omega BH^{op})^{op})$$

$$K_X\colon WC(\Omega BG, Fib(p_G,p^H), (\Omega BH^{op})^{op}) \longrightarrow WC(G,X,H)$$

$$L_X\colon WC(G,X,H) \longrightarrow WWC(G,X,H)$$

like in (5.4) and (5.5). They are homotopy equivalences, natural up to homotopy in (G,X,H), and satisfy (5.6).

5.15 <u>Proof of (3.8):</u> Suppose
$\hat{\alpha}\colon C(G,\hat{G},G) \longrightarrow C(\pi_0(AutY), \pi_0 F(X,Y), \pi_0(AutX))$ is induced by an n-coherent homotopy G-map
$\hat{\gamma}\colon W^n C(G,\hat{G},G) \longrightarrow C(AutY, F(X,Y), AutX)$. Then $\hat{\gamma}$ defines a map (compare (3.7))

$$(5.16)\qquad \begin{array}{ccc} BW^nG^{op} & \xrightarrow{\;B\gamma_0\;} & B(AutX^{op}) \\ \uparrow p_0 & \circlearrowright & \uparrow q_0 \\ B(*,W^nG \times W^nG^{op}, W^n\hat{G}) & \xrightarrow{\;B\gamma\;} & B(*,AutY \times AutX^{op}, F(X,Y)) \\ \downarrow p_1 & \circlearrowright & \downarrow q_1 \\ BW^nG & \xrightarrow{\;B\gamma_1\;} & B(AutY) \end{array}$$

which sits over (3.7). By [P; Thm.] the rows in the following diagram are h-fibration sequences

$$\begin{array}{ccccc}
W^n\hat{G} & \text{———} & B(*,W^nG \times W^nG^{op},W^n\hat{G}) & \text{———} & B(W^nG \times W^nG^{op}) \\
\Big\downarrow \varepsilon_n & & \Big\downarrow B(\varepsilon_{\hat{G}})_n & & \Big\downarrow B\varepsilon_n \\
\hat{G} & \longrightarrow & B(*,G \times G^{op},\hat{G}) & \longrightarrow & B(G \times G^{op})
\end{array}$$

Hence $B(\varepsilon_{\hat{G}})_n$ is n-connected because ε_n is n-connected and $B\varepsilon_n$ is (n+1)-connected. So there exist maps

$$k_n\colon B^n(*,G \times G^{op},\hat{G}) \longrightarrow B(*,W^nG \times W^nG^{op},W^n\hat{G})$$
$$r_{n+1}\colon B^{n+1}G \longrightarrow BW^nG$$

such that $B(\varepsilon_{\hat{G}})_n \circ k_n \simeq j_n$ and $B\varepsilon_n \circ r_{n+1} \simeq i_{n+1}$, where $i_{n+1}\colon B^{n+1}G \subset BG$ and $j_n\colon B^n(*,G \times G^{op},\hat{G}) \subset B(*,G \times G^{op},\hat{G})$. The diagram

$$\begin{array}{ccc}
B^{n+1}G^{op} & \xrightarrow{\;r^{op}_{n+1}\;} & BW^nG^{op} \\
\Big\uparrow p_o & & \Big\uparrow p_o \\
B^n(*,G \times G^{op},\hat{G}) & \xrightarrow{\;k_n\;} & B(*,W^nG \times W^nG^{op},W^n\hat{G}) \\
\Big\downarrow p_1 & & \Big\downarrow p_1 \\
B^{n+1}G & \xrightarrow{\;r_{n+1}\;} & BW^nG
\end{array}$$

commutes up to homotopy because

$$(B\varepsilon_n)_*\colon [B^n(*,G \times G^{op},\hat{G}),\ BW^nG] \longrightarrow [B^n(*,G \times G^{op},\hat{G}),BG]$$

is bijective. Together with (5.16) it provides the required lift. Conversely, suppose we are given lifts $f_{n+1}\colon B^{n+1}G^{op} \longrightarrow B(\mathrm{Aut}X^{op})$, $g_{n+1}\colon B^{n+1}G \longrightarrow B(\mathrm{Aut}Y)$, and $h_n\colon B^n(*,G \times G^{op},\hat{G}) \longrightarrow B(*,\mathrm{Aut}Y \times \mathrm{Aut}X^{op},F(X,Y))$ as in (3.8). The inclusions of skeletons and the triple (g_{n+1},h_n,f_{n+1}) give rise to maps of h-fibration sequences

$$\begin{array}{ccccc}
\mathrm{Fib}(p_G,p^G) & \longrightarrow & B(*,G \times G^{op},\hat{G}) & \longrightarrow & BG \times BG^{op} \\
\Big\uparrow \mathrm{Fib}(j) & & \Big\uparrow j_n & & \Big\uparrow i_{n+1} \times i^{op}_{n+1} \\
\mathrm{Fib}^n(p_G,p^G) & \longrightarrow & B^n(*,\ G \times G^{op},\hat{G}) & \longrightarrow & B^{n+1}G \times B^{n+1}G^{op} \\
\Big\downarrow \mathrm{Fib}(h) & & \Big\downarrow h_n & & \Big\downarrow g_{n+1} \times f_{n+1} \\
\mathrm{Fib}(p_Y,p^X) & \longrightarrow & B(*,\mathrm{Aut}Y \times \mathrm{Aut}X^{op},F(X,Y)) & \longrightarrow & B(\mathrm{Aut}Y) \times B(\mathrm{Aut}X^{op})
\end{array}$$

We take the model described above as h-fiber. This diagram in turn defines functors

$$\mathcal{C}(\Omega B^{n+1}G, Fib^n(p_G, p^G), (\Omega B^{n+1}G^{op})^{op}) \begin{array}{l} \xrightarrow{S} \mathcal{C}(\Omega BG, Fib(p_G, p^G), (\Omega BG^{op})^{op}) \\ \xrightarrow{T} \mathcal{C}(\Omega B(AutY), Fib(p_Y, p^X), (\Omega B(AutX^{op})^{op}) \end{array}$$

Since i_{n+1} is (n+1)-connected and j_n is n-connected, Fib(j) is n-connected. Hence the functor S is n-connected. The rest of the proof now is exactly the same as in (5.7).

(5.17) The proof of (3.9) is just another application of (4.5).

6. Final remarks

The methods of § 5 are related to the theories of H_∞-maps of topological monoids and G_∞-maps of G-spaces in the sense of Fuchs [F2]. An analysis of the definitions (1.3) and (1.4) of [F2] shows that an H_∞-map from a monoid G to a monoid H can be interpreted as a homomorphism FG $\longrightarrow$ H and a G_∞-map from a G-space X to an H-space Y as a "functor" $F\mathcal{C}(G,X,\{e\}) \longrightarrow \mathcal{C}(H,Y,\{e\})$, where $F\mathcal{C}$ is the semicategory (i.e. category without identities) obtained from a category (or semicategory) $\mathcal{C}$ in the same way as $W\mathcal{C}$ but with relations (2.1.2),-,(2.1.4) dropped. For our purposes we need the stronger structure $W\mathcal{C}$.

If $\mathcal{C}$ is a well-pointed category, with our methods it is easy to show that a "functor" $F\mathcal{C} \longrightarrow \mathcal{D}$ into a category $\mathcal{D}$ is homotopic to a "functor" which factors through the projection "functor" $F\mathcal{C} \longrightarrow W\mathcal{C}$. Hence our results in § 5 give quick proofs of many results of [F1], [F2], [F3] and make explicit constructions unnecessary. In particular, the preparations for the proof of (3.8) in § 5, applied to the case H = {e} where Moore squares may be replaced by the more familiar Moore paths, can be used to correct a flaw in [F3; Section 5].

References

[B-V] J.M. Boardman and R.M.Vogt, Homotopy invariant algebraic structures on topological spaces, Springer Lecture Notes in Math. 347 (1973)

[C] G. Cooke, Replacing homotopy actions by topological actions, Trans. Amer. Math. Soc. 237 (1978), 391-406

[tD-K-P] T. tom Dieck, K.H. Kamps, and D. Puppe, Homotopietheorie, Springer Lecture Notes in Math. 157, (1970)

[D-K] W. Dwyer and D. Kan, Equivariant homotopy classification, J. Pure and Applied Algebra 35 (1985), 269-285

[F1] M. Fuchs, Verallgemeinerte Homotopie-Homomorphismen und klassifizierende Räume, Math. Ann. 161 (1965), 197-230

[F2] ———, Homotopy equivalences in equivariant topology, Proc. Amer. Math. Soc. 58 (1976), 347-352

[F3] ———, Equivariant maps up to homotopy and Borel spaces, Publ. Math. Universitat Autònoma de Barcelona 28 (1984), 79-102

[M] J.P. May, Classifying spaces and fibrations, Memoirs A.M.S. 155 (1975)

[O] J.F. Oprea, Lifting homotopy actions in rational homotopy theory, J. Pure and Applied Algebra 32 (1984), 177-190

[P] V. Puppe, A remark on homotopy fibrations, Manuscripta Math. 12 (1974), 113-120

[S-V] R. Schwänzl and R.M. Vogt, Relative realizations of homotopy actions, in preparation

[V1] R.M. Vogt, Convenient categories of topological spaces for homotopy theory, Arch. der Math. 22 (1971), 545-555

[V2] ———, Homotopy limits and colimits, Math. Z. 134 (1973), 11-52

[Z] A. Zabrodsky, On George Cooke's theory of homotopy and topological actions, Canadian Math. Soc. Conf. Proc., Vol. 2, Part 2 (1982), 313-317

EXISTENCE OF COMPACT FLAT RIEMANNIAN MANIFOLDS WITH THE FIRST BETTI NUMBER EQUAL TO ZERO

Andrzej Szczepański
Gdańsk, Poland

0. Let M^n be a compact flat Riemannian manifold of dimension n. From Bieberbach's Theorems (see [3,8]) we know that its fundamental group $\pi_1(M) = \Gamma$ has the following properties:

1) Γ is a torsion free, discrete and cocompact subgroup of $E(n)$, the group of isometries of R^n.

   In particular, Γ acts freely and properly discontinuously as a group of Euclidean motions

2) There exists a short exact sequence

$$0 \to Z^n \to \Gamma \to G \to 1 \qquad (*)$$

   where Z^n is a maximal abelian subgroup in Γ and G is finite.

The sequence (*) defines by conjugation a faithful representation $p : G \to GL(n,Z)$ and is classified by an element $\alpha \in H^2_p(G,Z^n)$.

<u>Lemma 0.1</u> [2]. Let Z^n be a G-module. The extension of G by Z^n corresponding to $\alpha \in H^2(G,Z^n)$ is torsion free if and only if $\mathrm{res}^G_H \alpha \neq 0$, where H runs over representatives of conjugacy classes of subgroups of prime order ∎

We have the following construction due to E. Calabi

<u>Theorem 0.2</u> [1,8]. If M is an n-dimensional flat manifold with $b_1(M) = q > 0$ then there exist an (n-q)-dimensional flat manifold N and a finite abelian group F of affine automorphisms of N of rank $\leq q$ so that

$$M = N \times T^q/F ,$$

where T^q is a flat q-torus on which F acts by isometries ∎

This construction suggests a programme for an inductive classification of flat manifolds with positive first Betti number. Those with $b_1 = 0$ must necessarily be handled separately.

<u>Remark 0.3.</u> It can be proved [5] that $b_1(M) = 0$ if and only if $\dim_Q[Q^n]^G = 0$, where G acts on Z^n by conjugation in the short exact sequence

$$0 \to Z^n \to \pi_1(M) \to G \to 1$$

and

$$Q^n = Z^n \otimes_Z Q .$$

Definition 0.4 [5]. Let H be finite group. We say H is primitive if H is the holonomy group of a flat manifold M with $b_1(M) = 0$. Recently H. Hiller and C.H. Sah [5] have determined the primitive groups.

Theorem 0.5. A finite group H is primitive if and only if no cyclic Sylow p-subgroup of H has a normal complement ∎

In this note we shall consider properties of the short exact sequence (*) for $G = Z_n$ (cyclic), $G = D_n$ (dihedral), $G = Q(2^n)$ (generalized quaternion 2-group).

1. Let $g(G)$ denote the smallest degree of a faithful integral representation of G. It is easy to see that such a "minimal" integral representation has no fixed points. Therefore we can ask the following question:

Question 1.1. Suppose $0 \to Z^n \to \Gamma \to G \to 1$ is a short exact sequence, such that the integral representation induced by conjugation $G \to GL(n,Z)$ is "minimal" and faithful.

Can Γ be a fundamental group of a flat manifold?

Conjecture 1.2. Suppose $0 \to Z^n \to \Gamma \to G \to 1$ is a short exact sequence and the integral representation induced by conjugation is irreducible and faithful. Then Γ is not a fundamental group of a flat manifold.

For generalized quaternion 2-groups $Q(2^n)$ the conjecture and our question coincide.

Now we formulate our main result.

Theorem 1.3. If $G = Z_n$, $G = D_n$, $G = Q(2^n)$ then the answer to the question 1.1 is negative.

Proof.

1. Let $G = Z_n$ be a cyclic group. The number $g(Z_n)$ is equal [4]:

a) $g(Z_{p^k}) = p^k - p^{k-1}$ for any k, where p-prime number

b) if m and n are relatively prime then $g(Z_{m,n}) = g(Z_m) + g(Z_n)$, unless $m = 2$ and n is odd, in which case $g(Z_{2n}) = g(Z_n)$.

From this and from the fact that the representation of Z_n of degree $g(Z_n)$ has no fixed points we have that $H^2(Z_n, Z^{g(Z_n)}) = 0$. Now the theorem follows from lemma 0.1.

2. Let $G = D_n = \langle x,y | x^n = 1, yxy^{-1} = x^{-1}, y^2 = 1\rangle$. We shall sketch the proof that

$$g(D_n) = g(Z_n) . \tag{**}$$

It is well known [7] that $g(D_p) = g(Z_p) = p-1$. For $n = p^k$ $(k > 1)$ the result (**) follows from the inclusion $D_{p^k} \supset D_{p^{k-1}}$ and theorem about the dimension of the induced representation. Finally for an arbitrary n the equality (**) follows from the first part of the proof /for cyclic groups/ and the definition of the Dihedral group. Now we may consider a homomorphism:

$$\mathrm{res}^{D_n}_{Z_n} : H^2(D_n, Z^{g(D_n)}) \to H^2(Z_n, Z^{g(D_n)}) = 0$$

of abelian groups where the second one is equal to zero by (**).

The theorem follows from lemma 0.1.

3. Let $G = Q(2^n)$, a generalized quaternion 2-group. It is well known that $g(Q(2^n)) = 2^n$. From the preprint of [6] it can be proved that minimal dimension of a flat manifolds with $b_1 = 0$ and $Q(2^n)$ as holonomy group is equal to 2^n+3 . It completes the proof of the theorem

■

REFERENCES:

[1] CALABI, E.: Closed locally euclidean four dimensional manifolds, Bull. Amer. Math. Soc. 63, 135 (1957)

[2] CHARLAP, L.S.: Compact flat Riemannian manifolds I. Ann. Math. 81, 15-30 (1965)

[3] FARKAS, D.R.: Crystallographic groups and their mathematics. Rocky mountain J. Math. 1i. 4.511-551 (1981)

[4] HILLER, H.: Minimal dimension of flat manifolds with abelian holonomy - preprint

[5] HILLER, H., SAH, C.H.: Holonomy of flat manifolds with $b_1 = 0$, to appear in the Quaterly J. Math.

[6] HILLER, H., MARCINIAK, Z., SAH, C.H., SZCZEPANSKI, A.: Holonomy of flat manifolds with $b_1 = 0$,II - preprint

[7] PU, L.: Integral representations of non-abelian groups of order pq , Mich. Math. J. 12, 231-246 (1965)

[8] WOLF, J.A.: Spaces of constant curvature, Boston, Perish 1974

WHICH GROUPS HAVE STRANGE TORSION?

Steven H. Weintraub
Department of Mathematics
Louisiana State University
Baton Rouge, Louisiana 70803-4918
U.S.A.

The purpose of this note is to ask what we think is a natural question, and to provide some examples which suggest that it should have an interesting answer.

1. STRANGE TORSION

DEFINITION 1. A group G has strange p-torsion if

a) $H^*(G;\mathbb{Z})$ has p-torsion, but

b) G does not have an element of order p.

It has strange torsion if it has strange p-torsion for some p. (We take coefficients in $\mathbb{Z}$ as a trivial $\mathbb{Z}G$-module.)

There are admittedly some reasonably natural groups which have strange torsion:

EXAMPLE -4. Let B_k be Artin's braid group on k strands (in $\mathbb{R}^2$) and let B_∞ be the direct limit $B_\infty = \varinjlim B_k$. Then B_∞ is torsion-free but for every prime p, $H^i(B_\infty;\mathbb{Z})$ has p-torsion for arbitrarily large i. This is a result of F. Cohen [CLM, III. Appendix].

EXAMPLE -3. If G is a one-relator group, then $H^i(G;\mathbb{Z})$ may have strange torsion for $i = 2$ (but not for $i \neq 2$). This follows from Lyndon's computation [Ly].

EXAMPLE -2. (A special case of example -3.) G = the fundamental group of a non-orientable surface of genus $g > 1$, or of the mapping torus of $f: S^1 \to S^1$ by $z \to z^n$, $n \neq 0,1,2$.

EXAMPLE -1. Many Bieberbach groups, e.g. the following group considered by A. Szczepanski: $1 \to \mathbb{Z}^3 \to G \to \mathbb{Z}/2 + \mathbb{Z}/2 \to 1$ where the two generators a and b of $\mathbb{Z}_2 + \mathbb{Z}_2$ act on $\mathbb{Z}^3$ by $a(x,y,z) = (x,-y,-z)$, $b(x,y,z) = (-x,y,-z)$.

On the other hand, here are some examples of groups which do not have strange torsion:

EXAMPLE 0. All finite groups. (The existence of the transfer implies that the cohomology of a finite group is annihilated by multiplication by the order of the group.)

EXAMPLE 1. Any subgroup of $SL_2(\mathbb{Z})$ or $PSL_2(\mathbb{Z})$. This follows from the following well-known theorem (There are some point-set theoretical conditions here, which we suppress.):

THEOREM 1. Let a group G act on a contractible space X with the isotropy group G_x of x finite for every $x \in X$. Then if p is prime to $|G_x|$ for all x, $H^*(G;\mathbb{Z}_p)$ is isomorphic to $H^*(X/G;\mathbb{Z}_p)$.

Proof: Let EG be a contractible space on which G acts freely. Then G acts freely on $X \times EG$ by the diagonal action, so $H^*(G;\mathbb{Z}_p) = H^*((X \times EG)/G;\mathbb{Z}_p)$. Let $f: X \to Y = X/G$ and $\pi: X \times_G EG = (X \times EG)/G \to Y$ be the projections. Then $H^*(\pi^{-1}(y);\mathbb{Z}_p) = H^*(BG_x;\mathbb{Z}_p) = H^*(pt;\mathbb{Z}_p)$ (by example 0) for all y, where $y = f(x)$, so by the Vietoris-Begle mapping theorem, $H^*(G;\mathbb{Z}_p) = H^*(X/G;\mathbb{Z}_p)$.

COROLLARY 2. If $H^*(X/G;\mathbb{Z})$ has no p-torsion, G has no strange p-torsion. (In particular, this holds for G acting on $\mathbb{R}^2$ in an orientation-preserving way.)

The following examples require a lot more work:

EXAMPLE 2. $G = Sp_4(\mathbb{Z})$ and $G = \Gamma(2)$, the principal congruence subgroup of level 2, as well as $PSp_4(\mathbb{Z})$ and $P\Gamma(2)$. This is proven in [LW].

EXAMPLE 3. $G = SL_3(\mathbb{Z})$. This follows from Soule's computation [So].

The interesting thing about example n, $n > 0$, is that the non-existence of strange torsion is proven geometrically, by studying a G-action on a contractible space X satisfying the hypothesis of Theorem 1. Thus we ask the question:

QUESTION 1. Which groups have strange torsion?

2. VERY STRANGE TORSION

S. Jackowski has suggested that it might be better to ask about very strange torsion.

DEFINITION 2. A group G has very strange p-torsion if

a) $H^i(G:\mathbb{Z})$ has p-torsion for i arbitrarily large, but

b) G does not have an element of order p.

It has very strange torsion if it has very strange p-torsion for some p.

In this connection we have the following well-known result.

THEOREM 2. Let G be a group with $vcd(G) < \infty$. Then G has no very strange torsion.

Proof. Recall the following from [S]: A group G' has finite cohomological dimension $n = cd(G') < \infty$ if for every module M, $H^i(G';M) = 0$ for $i > n$. A group G has virtually finite cohomological dimension, $vcd(G) < \infty$, if G has a subgroup G' of finite index with $cd(G') < \infty$. In this case we set $vcd(G) = n = cd(G')$, and $vcd(G)$ is well defined (i.e. independent of the choice of G').

If $n = vcd(G) < \infty$, we have Farrell cohomology $\hat{H}^i(G:\mathbb{Z})$ defined, with the property that $\hat{H}^i(G:\mathbb{Z}) = H^i(G:\mathbb{Z})$ for $i > n$. Furthermore, by [B, p. 280, ex. 2] $\hat{H}^i(G:\mathbb{Z})$ has p-torsion only for primes for which G has an element of order p, so G has no strange torsion above dimension n.

There are many important classes of groups G for which $vcd(G) < \infty$. A host of examples are given in [B, Sec. VIII.9]. In particular, all arithmetic groups G satisfy $vcd(G) < \infty$.

Example n has finite vcd for $n \geq -2$. Example -3 has strange torsion but not very strange torsion, while example -4 has very strange torsion. Thus we conclude with the question:

QUESTION 2. Which groups have very strange torsion?

References

[B] Brown, K. Cohomology of Groups. Springer, Berlin, 1982.

[CLM] Cohen, F. R., Lada, T. J., and May, J. P. The homology of iterated loop spaces, Lecture notes in math. no. 533, Springer, Berlin, 1976.

[LW] Lee, R., and Weintraub, S. H. Cohomology of $Sp_4(\mathbb{Z})$ and related groups and spaces, Topology 24(1985), 391-410.

[Ly] Lyndon, R. C. Cohomology theory of groups with a single defining relation, Ann. Math. 52(1950), 650-665.

[Q] Quillen, D. The spectrum of an equivariant cohomology ring, Ann. of Math. 94(1971), 549-602.

[S] Serre, J. -P. Cohomologie des groupes discrets, in Prospects in Mathematics, Ann. of Math. Studies vol. 70, Princeton Univ. Press, Princeton NJ, 1971, 77-169.

[So] Soulé, C. Cohomology of $SL_3(\mathbb{Z})$, Topology 17(1978), 1-22.

Vol. 1062: J. Jost, Harmonic Maps Between Surfaces. X, 133 pages. 1984.

Vol. 1063: Orienting Polymers. Proceedings, 1983. Edited by Ericksen. VII, 166 pages. 1984.

Vol. 1064: Probability Measures on Groups VII. Proceedings, 1983. Edited by H. Heyer. X, 588 pages. 1984.

Vol. 1065: A. Cuyt, Padé Approximants for Operators: Theory and Applications. IX, 138 pages. 1984.

Vol. 1066: Numerical Analysis. Proceedings, 1983. Edited by D.F. Griffiths. XI, 275 pages. 1984.

Vol. 1067: Yasuo Okuyama, Absolute Summability of Fourier Series and Orthogonal Series. VI, 118 pages. 1984.

Vol. 1068: Number Theory, Noordwijkerhout 1983. Proceedings. Edited by H. Jager. V, 296 pages. 1984.

Vol. 1069: M. Kreck, Bordism of Diffeomorphisms and Related Topics. III, 144 pages. 1984.

Vol. 1070: Interpolation Spaces and Allied Topics in Analysis. Proceedings, 1983. Edited by M. Cwikel and J. Peetre. III, 239 pages. 1984.

Vol. 1071: Padé Approximation and its Applications, Bad Honnef 1983. Prodeedings. Edited by H. Werner and H.J. Bünger. VI, 264 pages. 1984.

Vol. 1072: F. Rothe, Global Solutions of Reaction-Diffusion Systems. V, 216 pages. 1984.

Vol. 1073: Graph Theory, Singapore 1983. Proceedings. Edited by K.M. Koh and H.P. Yap. XIII, 335 pages. 1984.

Vol. 1074: E.W. Stredulinsky, Weighted Inequalities and Degenerate Elliptic Partial Differential Equations. III, 143 pages. 1984.

Vol. 1075: H. Majima, Asymptotic Analysis for Integrable Connections with Irregular Singular Points. IX, 159 pages. 1984.

Vol. 1076: Infinite-Dimensional Systems. Proceedings, 1983. Edited by F. Kappel and W. Schappacher. VII, 278 pages. 1984.

Vol. 1077: Lie Group Representations III. Proceedings, 1982–1983. Edited by R. Herb, R. Johnson, R. Lipsman, J. Rosenberg. XI, 454 pages. 1984.

Vol. 1078: A.J.E.M. Janssen, P. van der Steen, Integration Theory. V, 224 pages. 1984.

Vol. 1079: W. Ruppert. Compact Semitopological Semigroups: An Intrinsic Theory. V, 260 pages. 1984

Vol. 1080: Probability Theory on Vector Spaces III. Proceedings, 1983. Edited by D. Szynal and A. Weron. V, 373 pages. 1984.

Vol. 1081: D. Benson, Modular Representation Theory: New Trends and Methods. XI, 231 pages. 1984.

Vol. 1082: C.-G. Schmidt, Arithmetik Abelscher Varietäten mit komplexer Multiplikation. X, 96 Seiten. 1984.

Vol. 1083: D. Bump, Automorphic Forms on GL (3,IR). XI, 184 pages. 1984.

Vol. 1084: D. Kletzing, Structure and Representations of Q-Groups. VI, 290 pages. 1984.

Vol. 1085: G.K. Immink, Asymptotics of Analytic Difference Equations. V, 134 pages. 1984.

Vol. 1086: Sensitivity of Functionals with Applications to Engineering Sciences. Proceedings, 1983. Edited by V. Komkov. V, 130 pages. 1984

Vol. 1087: W. Narkiewicz, Uniform Distribution of Sequences of Integers in Residue Classes. VIII, 125 pages. 1984.

Vol. 1088: A.V. Kakosyan, L.B. Klebanov, J.A. Melamed, Characterization of Distributions by the Method of Intensively Monotone Operators. X, 175 pages. 1984.

Vol. 1089: Measure Theory, Oberwolfach 1983. Proceedings. Edited by D. Kölzow and D. Maharam-Stone. XIII, 327 pages. 1984.

Vol. 1090: Differential Geometry of Submanifolds. Proceedings, 1984. Edited by K. Kenmotsu. VI, 132 pages. 1984.

Vol. 1091: Multifunctions and Integrands. Proceedings, 1983. Edited by G. Salinetti. V, 234 pages. 1984.

Vol. 1092: Complete Intersections. Seminar, 1983. Edited by S. Greco and R. Strano. VII, 299 pages. 1984.

Vol. 1093: A. Prestel, Lectures on Formally Real Fields. XI, 125 pages. 1984.

Vol. 1094: Analyse Complexe. Proceedings, 1983. Edité par E. Amar, R. Gay et Nguyen Thanh Van. IX, 184 pages. 1984.

Vol. 1095: Stochastic Analysis and Applications. Proceedings, 1983. Edited by A. Truman and D. Williams. V, 199 pages. 1984.

Vol. 1096: Théorie du Potentiel. Proceedings, 1983. Edité par G. Mokobodzki et D. Pinchon. IX, 601 pages. 1984.

Vol. 1097: R.M. Dudley, H. Kunita, F. Ledrappier, École d'Éte de Probabilités de Saint-Flour XII – 1982. Edité par P.L. Hennequin. X, 396 pages. 1984.

Vol. 1098: Groups – Korea 1983. Proceedings. Edited by A.C. Kim and B.H. Neumann. VII, 183 pages. 1984.

Vol. 1099: C.M. Ringel, Tame Algebras and Integral Quadratic Forms. XIII, 376 pages. 1984.

Vol. 1100: V. Ivrii, Precise Spectral Asymptotics for Elliptic Operators Acting in Fiberings over Manifolds with Boundary. V, 237 pages. 1984.

Vol. 1101: V. Cossart, J. Giraud, U. Orbanz, Resolution of Surface Singularities. Seminar. VII, 132 pages. 1984.

Vol. 1102: A. Verona, Stratified Mappings – Structure and Triangulability. IX, 160 pages. 1984.

Vol. 1103: Models and Sets. Proceedings, Logic Colloquium, 1983, Part I. Edited by G.H. Müller and M.M. Richter. VIII, 484 pages. 1984.

Vol. 1104: Computation and Proof Theory. Proceedings, Logic Colloquium, 1983, Part II. Edited by M.M. Richter, E. Börger, W. Oberschelp, B. Schinzel and W. Thomas. VIII, 475 pages. 1984.

Vol. 1105: Rational Approximation and Interpolation. Proceedings, 1983. Edited by P.R. Graves-Morris, E.B. Saff and R.S. Varga. XII, 528 pages. 1984.

Vol. 1106: C.T. Chong, Techniques of Admissible Recursion Theory. IX, 214 pages. 1984.

Vol. 1107: Nonlinear Analysis and Optimization. Proceedings, 1982. Edited by C. Vinti. V, 224 pages. 1984.

Vol. 1108: Global Analysis – Studies and Applications I. Edited by Yu.G. Borisovich and Yu.E. Gliklikh. V, 301 pages. 1984.

Vol. 1109: Stochastic Aspects of Classical and Quantum Systems. Proceedings, 1983. Edited by S. Albeverio, P. Combe and M. Sirugue-Collin. IX, 227 pages. 1985.

Vol. 1110: R. Jajte, Strong Limit Theorems in Non-Commutative Probability. VI, 152 pages. 1985.

Vol. 1111: Arbeitstagung Bonn 1984. Proceedings. Edited by F. Hirzebruch, J. Schwermer and S. Suter. V, 481 pages. 1985.

Vol. 1112: Products of Conjugacy Classes in Groups. Edited by Z. Arad and M. Herzog. V, 244 pages. 1985.

Vol. 1113: P. Antosik, C. Swartz, Matrix Methods in Analysis. IV, 114 pages. 1985.

Vol. 1114: Zahlentheoretische Analysis. Seminar. Herausgegeben von E. Hlawka. V, 157 Seiten. 1985.

Vol. 1115: J. Moulin Ollagnier, Ergodic Theory and Statistical Mechanics. VI, 147 pages. 1985.

Vol. 1116: S. Stolz, Hochzusammenhängende Mannigfaltigkeiten und ihre Ränder. XXIII, 134 Seiten. 1985.

Vol. 1117: D.J. Aldous, J.A. Ibragimov, J. Jacod, Ecole d'Été de Probabilités de Saint-Flour XIII – 1983. Édité par P.L. Hennequin. IX, 409 pages. 1985.

Vol. 1118: Grossissements de filtrations: exemples et applications. Seminaire, 1982/83. Edité par Th. Jeulin et M. Yor. V, 315 pages. 1985.

Vol. 1119: Recent Mathematical Methods in Dynamic Programming. Proceedings, 1984. Edited by I. Capuzzo Dolcetta, W.H. Fleming and T. Zolezzi. VI, 202 pages. 1985.

Vol. 1120: K. Jarosz, Perturbations of Banach Algebras. V, 118 pages. 1985.

Vol. 1121: Singularities and Constructive Methods for Their Treatment. Proceedings, 1983. Edited by P. Grisvard, W. Wendland and J.R. Whiteman. IX, 346 pages. 1985.

Vol. 1122: Number Theory. Proceedings, 1984. Edited by K. Alladi. VII, 217 pages. 1985.

Vol. 1123: Séminaire de Probabilités XIX 1983/84. Proceedings. Edité par J. Azéma et M. Yor. IV, 504 pages. 1985.

Vol. 1124: Algebraic Geometry, Sitges (Barcelona) 1983. Proceedings. Edited by E. Casas-Alvero, G.E. Welters and S. Xambó-Descamps. XI, 416 pages. 1985.

Vol. 1125: Dynamical Systems and Bifurcations. Proceedings, 1984. Edited by B.L.J. Braaksma, H.W. Broer and F. Takens. V, 129 pages. 1985.

Vol. 1126: Algebraic and Geometric Topology. Proceedings, 1983. Edited by A. Ranicki, N. Levitt and F. Quinn. V, 523 pages. 1985.

Vol. 1127: Numerical Methods in Fluid Dynamics. Seminar. Edited by F. Brezzi, VII, 333 pages. 1985.

Vol. 1128: J. Elschner, Singular Ordinary Differential Operators and Pseudodifferential Equations. 200 pages. 1985.

Vol. 1129: Numerical Analysis, Lancaster 1984. Proceedings. Edited by P.R. Turner. XIV, 179 pages. 1985.

Vol. 1130: Methods in Mathematical Logic. Proceedings, 1983. Edited by C.A. Di Prisco. VII, 407 pages. 1985.

Vol. 1131: K. Sundaresan, S. Swaminathan, Geometry and Nonlinear Analysis in Banach Spaces. III, 116 pages. 1985.

Vol. 1132: Operator Algebras and their Connections with Topology and Ergodic Theory. Proceedings, 1983. Edited by H. Araki, C.C. Moore, Ş. Strătilă and C. Voiculescu. VI, 594 pages. 1985.

Vol. 1133: K.C. Kiwiel, Methods of Descent for Nondifferentiable Optimization. VI, 362 pages. 1985.

Vol. 1134: G.P. Galdi, S. Rionero, Weighted Energy Methods in Fluid Dynamics and Elasticity. VII, 126 pages. 1985.

Vol. 1135: Number Theory, New York 1983–84. Seminar. Edited by D.V. Chudnovsky, G.V. Chudnovsky, H. Cohn and M.B. Nathanson. V, 283 pages. 1985.

Vol. 1136: Quantum Probability and Applications II. Proceedings, 1984. Edited by L. Accardi and W. von Waldenfels. VI, 534 pages. 1985.

Vol. 1137: Xiao G., Surfaces fibrées en courbes de genre deux. IX, 103 pages. 1985.

Vol. 1138: A. Ocneanu, Actions of Discrete Amenable Groups on von Neumann Algebras. V, 115 pages. 1985.

Vol. 1139: Differential Geometric Methods in Mathematical Physics. Proceedings, 1983. Edited by H. D. Doebner and J. D. Hennig. VI, 337 pages. 1985.

Vol. 1140: S. Donkin, Rational Representations of Algebraic Groups. VII, 254 pages. 1985.

Vol. 1141: Recursion Theory Week. Proceedings, 1984. Edited by H.-D. Ebbinghaus, G.H. Müller and G.E. Sacks. IX, 418 pages. 1985.

Vol. 1142: Orders and their Applications. Proceedings, 1984. Edited by I. Reiner and K. W. Roggenkamp. X, 306 pages. 1985.

Vol. 1143: A. Krieg, Modular Forms on Half-Spaces of Quaternions. XIII, 203 pages. 1985.

Vol. 1144: Knot Theory and Manifolds. Proceedings, 1983. Edited by D. Rolfsen. V, 163 pages. 1985.

Vol. 1145: G. Winkler, Choquet Order and Simplices. VI, 143 pages 1985.

Vol. 1146: Séminaire d'Algèbre Paul Dubreil et Marie-Paule Mallia Proceedings, 1983–1984. Edité par M.-P. Malliavin. IV, 420 pa 1985.

Vol. 1147: M. Wschebor, Surfaces Aléatoires. VII, 111 pages. 1985.

Vol. 1148: Mark A. Kon, Probability Distributions in Quantum Statistica Mechanics. V, 121 pages. 1985.

Vol. 1149: Universal Algebra and Lattice Theory. Proceedings, 1984 Edited by S. D. Comer. VI, 282 pages. 1985.

Vol. 1150: B. Kawohl, Rearrangements and Convexity of Level Sets i PDE. V, 136 pages. 1985.

Vol 1151: Ordinary and Partial Differential Equations. Proceeding 1984. Edited by B.D. Sleeman and R.J. Jarvis. XIV, 357 pages. 1985

Vol. 1152: H. Widom, Asymptotic Expansions for Pseudodifferentia Operators on Bounded Domains. V, 150 pages. 1985.

Vol. 1153: Probability in Banach Spaces V. Proceedings, 1984. Editec by A. Beck, R. Dudley, M. Hahn, J. Kuelbs and M. Marcus. VI, 45 pages. 1985.

Vol. 1154: D.S. Naidu, A.K. Rao, Singular Pertubation Analysis o Discrete Control Systems. IX, 195 pages. 1985.

Vol. 1155: Stability Problems for Stochastic Models. Proceedings 1984. Edited by V.V. Kalashnikov and V.M. Zolotarev. VI, 44 pages. 1985.

Vol. 1156: Global Differential Geometry and Global Analysis 1984. Proceedings, 1984. Edited by D. Ferus, R.B. Gardner, S. Helgason and U. Simon. V, 339 pages. 1985.

Vol. 1157: H. Levine, Classifying Immersions into $\mathbb{R}^4$ over Stable Maps of 3-Manifolds into $\mathbb{R}^2$. V, 163 pages. 1985.

Vol. 1158: Stochastic Processes – Mathematics and Physics. Proceedings, 1984. Edited by S. Albeverio, Ph. Blanchard and L. Streit. VI, 230 pages. 1986.

Vol. 1159: Schrödinger Operators, Como 1984. Seminar. Edited by S. Graffi. VIII, 272 pages. 1986.

Vol. 1160: J.-C. van der Meer, The Hamiltonian Hopf Bifurcation. VI, 115 pages. 1985.

Vol. 1161: Harmonic Mappings and Minimal Immersions, Montecatini 1984. Seminar. Edited by E. Giusti. VII, 285 pages. 1985.

Vol. 1162: S.J.L. van Eijndhoven, J. de Graaf, Trajectory Spaces, Generalized Functions and Unbounded Operators. IV, 272 pages. 1985.

Vol. 1163: Iteration Theory and its Functional Equations. Proceedings, 1984. Edited by R. Liedl, L. Reich and Gy. Targonski. VIII, 231 pages. 1985.

Vol. 1164: M. Meschiari, J.H. Rawnsley, S. Salamon, Geometry Seminar "Luigi Bianchi" II – 1984. Edited by E. Vesentini. VI, 224 pages. 1985.

Vol. 1165: Seminar on Deformations. Proceedings, 1982/84. Edited by J. Ławrynowicz. IX, 331 pages. 1985.

Vol. 1166: Banach Spaces. Proceedings, 1984. Edited by N. Kalton and E. Saab. VI, 199 pages. 1985.

Vol. 1167: Geometry and Topology. Proceedings, 1985. Edited by J. Alexander and J. Harer. VI, 292 pages. 1985.

Vol. 1168: S.S. Agaian, Hadamard Matrices and their Applications. III, 227 pages. 1985.

Vol. 1169: W.A. Light, E.W. Cheney, Approximation Theory in Tensor Product Spaces. VII, 157 pages. 1985.

Vol. 1170: B.S. Thomson, Real Functions. VII, 229 pages. 1985.

Vol. 1171: Polynômes Orthogonaux et Applications. Proceedings, 1984. Edité par C. Brezinski, A. Draux, A.P. Magnus, P. Maroni et A. Ronveaux. XXXVII, 584 pages. 1985.

Vol. 1172: Algebraic Topology, Göttingen 1984. Proceedings. Edited by L. Smith. VI, 209 pages. 1985.